MW00667841

**Statistical Procedures
for Agricultural Research**

STATISTICAL PROCEDURES FOR AGRICULTURAL RESEARCH

Second Edition

KWANCHAI A. GOMEZ
Head, Department of Statistics
The International Rice Research Institute
Los Baños, Laguna, Philippines

ARTURO A. GOMEZ
Professor of Agronomy
University of the Philippines at Los Baños
College, Laguna, Philippines

AN INTERNATIONAL RICE RESEARCH INSTITUTE BOOK

A Wiley-Interscience Publication
JOHN WILEY & SONS,
New York · Chichester · Brisbane · Toronto · Singapore

First edition published in the Philippines in 1976 by the
International Rice Research Institute.

Library of Congress Cataloging in Publication Data:

Gomez, Kwanchai A.
   Statistical procedures for agricultural research.

   "An International Rice Research Institute book."
   "A Wiley-Interscience publication."
   Previously published as: Statistical procedures for
agricultural research with emphasis on rice/K. A. Gomez,
A. A. Gomez.
   Includes index.
   1. Agriculture—Research—Statistical methods.
2. Rice—Research—Statistical methods. 3. Field
experiments—Statistical methods. I. Gomez, Arturo A.
II. Gomez, Kwanchai A. Statistical procedures for agri-
cultural research with emphasis on rice. III. Title.
S540.S7G65 1983   630'.72   83-14556

ISBN 0-471-87092-7

Printed in the United States of America

10 9 8 7 6 5 4 3 2 1

To our son, Victor

Preface

There is universal acceptance of statistics as an essential tool for all types of research. That acceptance and ever-proliferating areas of research specialization have led to corresponding increases in the number and diversity of available statistical procedures. In agricultural research, for example, there are different statistical techniques for crop and animal research, for laboratory and field experiments, for genetic and physiological research, and so on. Although this diversity indicates the availability of appropriate statistical techniques for most research problems, it also indicates the difficulty of matching the best technique to a specific experiment. Obviously, this difficulty increases as more procedures develop.

Choosing the correct statistical procedure for a given experiment must be based on expertise in statistics and in the subject matter of the experiment. Thorough knowledge of only one of the two is not enough. Such a choice, therefore, should be made by:

- A subject matter specialist with some training in experimental statistics
- A statistician with some background and experience in the subject matter of the experiment
- The joint effort and cooperation of a statistician and a subject matter specialist

For most agricultural research institutions in the developing countries, the presence of trained statisticians is a luxury. Of the already small number of such statisticians, only a small fraction have the interest and experience in agricultural research necessary for effective consultation. Thus, we feel the best alternative is to give agricultural researchers a statistical background so that they can correctly choose the statistical technique most appropriate for their experiment. The major objective of this book is to provide the developing-country researcher that background.

For research institutions in the developed countries, the shortage of trained statisticians may not be as acute as in the developing countries. Nevertheless, the subject matter specialist must be able to communicate with the consulting statistician. Thus, for the developed-country researcher, this volume should help forge a closer researcher–statistician relationship.

We have tried to create a book that any subject matter specialist can use. First, we chose only the simpler and more commonly used statistical procedures in agricultural research, with special emphasis on field experiments with crops. In fact, our examples are mostly concerned with rice, the most important crop in Asia and the crop most familiar to us. Our examples, however, have applicability to a wide range of annual crops. In addition, we have used a minimum of mathematical and statistical theories and jargon and a maximum of actual examples.

This is a second edition of an International Rice Research Institute publication with a similar title and we made extensive revisions to all but three of the original chapters. We added four new chapters. The primary emphases of the working chapters are as follows:

Chapters 2 to 4 cover the most commonly used experimental designs for single-factor, two-factor, and three-or-more-factor experiments. For each design, the corresponding randomization and analysis of variance procedures are described in detail.

Chapter 5 gives the procedures for comparing specific treatment means: LSD and DMRT for pair comparison, and single and multiple $d.f.$ contrast methods for group comparison.

Chapters 6 to 8 detail the modifications of the procedures described in Chapters 2 to 4 necessary to handle the following special cases:

- Experiments with more than one observation per experimental unit
- Experiments with missing values or in which data violate one or more assumptions of the analysis of variance
- Experiments that are repeated over time or site

Chapters 9 to 11 give the three most commonly used statistical techniques for data analysis in agricultural research besides the analysis of variance. These techniques are regression and correlation, covariance, and chi-square. We also include a detailed discussion of the common misuses of the regression and correlation analysis.

Chapters 12 to 14 cover the most important problems commonly encountered in conducting field experiments and the corresponding techniques for coping with them. The problems are:

- Soil heterogeneity
- Competition effects
- Mechanical errors

Chapter 15 describes the principles and procedures for developing an appropriate sampling plan for a replicated field experiment.

Chapter 16 gives the problems and procedures for research in farmers' fields. In the developing countries where farm yields are much lower than experiment-station yields, the appropriate environment for comparing new and existing technologies is the actual farmers' fields and not the favorable environment of the experiment stations. This poses a major challenge to existing statistical procedures and substantial adjustments are required.

Chapter 17 covers the serious pitfalls and provides guidelines for the presentation of research results. Most of these guidelines were generated from actual experience.

We are grateful to the International Rice Research Institute (IRRI) and the University of the Philippines at Los Baños (UPLB) for granting us the study leaves needed to work on this edition; and the Food Research Institute, Stanford University, and the College of Natural Resources, University of California at Berkeley, for being our hosts during our leaves.

Most of the examples were obtained from scientists at IRRI. We are grateful to them for the use of their data.

We thank the research staff of IRRI's Department of Statistics for their valuable assistance in searching and processing the suitable examples; and the secretarial staff for their excellent typing and patience in proofreading the manuscript. We are grateful to Walter G. Rockwood who suggested modifications to make this book more readable.

We appreciate permission from the Literary Executor of the late Sir Ronald A. Fisher, F.R.S., Dr. Frank Yates, F.R.S., and Longman Group Ltd., London to reprint Table III, "Distribution of *t* Probability," from their book *Statistical Tables for Biological, Agricultural and Medical Research* (6th edition, 1974).

KWANCHAI A. GOMEZ
ARTURO A. GOMEZ

Los Baños, Philippines
September 1983

Contents

Statistical Procedures
for Agricultural Research

CHAPTER 1

Elements of Experimentation

In the early 1950s, a Filipino journalist, disappointed with the chronic shortage of rice in his country, decided to test the yield potential of existing rice cultivars and the opportunity for substantially increasing low yields in farmers' fields. He planted a single rice seed—from an ordinary farm—on a well-prepared plot and carefully nurtured the developing seedling to maturity. At harvest, he counted more than 1000 seeds produced by the single plant. The journalist concluded that Filipino farmers who normally use 50 kg of grains to plant a hectare, could harvest 50 tons (0.05×1000) from a hectare of land instead of the disappointingly low national average of 1.2 t/ha.

As in the case of the Filipino journalist, agricultural research seeks answers to key questions in agricultural production whose resolution could lead to significant changes and improvements in existing agricultural practices. Unlike the journalist's experiment, however, scientific research must be designed precisely and rigorously to answer these key questions.

In agricultural research, the key questions to be answered are generally expressed as a statement of hypothesis that has to be verified or disproved through experimentation. These hypotheses are usually suggested by past experiences, observations, and, at times, by theoretical considerations. For example, in the case of the Filipino journalist, visits to selected farms may have impressed him as he saw the high yield of some selected rice plants and visualized the potential for duplicating that high yield uniformly on a farm and even over many farms. He therefore hypothesized that rice yields in farmers' fields were way below their potential and that, with better husbandry, rice yields could be substantially increased.

Another example is a Filipino maize breeder who is apprehensive about the low rate of adoption of new high-yielding hybrids by farmers in the Province of Mindanao, a major maize-growing area in the Philippines. He visits the maize-growing areas in Mindanao and observes that the hybrids are more vigorous and more productive than the native varieties in disease-free areas. However, in many fields infested with downy mildew, a destructive and prevalent maize disease in the area, the hybrids are substantially more severely diseased than the native varieties. The breeder suspects, and therefore hypothe-

1

sizes, that the new hybrids are not widely grown in Mindanao primarily because they are more susceptible to downy mildew than the native varieties.

Theoretical considerations may play a major role in arriving at a hypothesis. For example, it can be shown theoretically that a rice crop removes more nitrogen from the soil than is naturally replenished during one growing season. One may, therefore, hypothesize that in order to maintain a high productivity level on any rice farm, supplementary nitrogen must be added to every crop.

Once a hypothesis is framed, the next step is to design a procedure for its verification. This is the experimental procedure, which usually consists of four phases:

1. Selecting the appropriate materials to test
2. Specifying the characters to measure
3. Selecting the procedure to measure those characters
4. Specifying the procedure to determine whether the measurements made support the hypothesis

In general, the first two phases are fairly easy for a subject matter specialist to specify. In our example of the maize breeder, the test materials would probably be the native and the newly developed varieties. The characters to be measured would probably be disease infection and grain yield. For the example on maintaining productivity of rice farms, the test variety would probably be one of the recommended rice varieties and the fertilizer levels to be tested would cover the suspected range of nitrogen needed. The characters to be measured would include grain yield and other related agronomic characters.

On the other hand, the procedures regarding how the measurements are to be made and how these measurements can be used to prove or disprove a hypothesis depend heavily on techniques developed by statisticians. These two tasks constitute much of what is generally termed the design of an experiment, which has three essential components:

1. Estimate of error
2. Control of error
3. Proper interpretation of results

1.1 ESTIMATE OF ERROR

Consider a plant breeder who wishes to compare the yield of a new rice variety A to that of a standard variety B of known and tested properties. He lays out two plots of equal size, side by side, and sows one to variety A and the other to variety B. Grain yield for each plot is then measured and the variety with higher yield is judged as better. Despite the simplicity and commonsense appeal of the procedure just outlined, it has one important flaw. It presumes that any difference between the yields of the two plots is caused by the varieties and nothing else. This certainly is not true. Even if the same variety were planted on both plots, the yield would differ. Other factors, such as soil fertility, moisture, and damage by insects, diseases, and birds also affect rice yields.

Because these other factors affect yields, a satisfactory evaluation of the two varieties must involve a procedure that can separate varietal difference from other sources of variation. That is, the plant breeder must be able to design an experiment that allows him to decide whether the difference observed is caused by varietal difference or by other factors.

The logic behind the decision is simple. Two rice varieties planted in two adjacent plots will be considered different in their yielding ability only if the observed yield difference is larger than that expected if both plots were planted to the same variety. Hence, the researcher needs to know not only the yield difference between plots planted to different varieties, but also the yield difference between plots planted to the same variety.

The difference among experimental plots treated alike is called *experimental error*. This error is the primary basis for deciding whether an observed difference is real or just due to chance. Clearly, every experiment must be designed to have a measure of the experimental error.

1.1.1 Replication

In the same way that at least two plots of the same variety are needed to determine the difference among plots treated alike, experimental error can be measured only if there are at least two plots planted to the same variety (or receiving the same treatment). Thus, to obtain a measure of experimental error, replication is needed.

1.1.2 Randomization

There is more involved in getting a measure of experimental error than simply planting several plots to the same variety. For example, suppose, in comparing two rice varieties, the plant breeder plants varieties A and B each in four plots as shown in Figure 1.1. If the area has a unidirectional fertility gradient so that there is a gradual reduction of productivity from left to right, variety B would then be handicapped because it is always on the right side of variety A and always in a relatively less fertile area. Thus, the comparison between the yield performances of variety A and variety B would be biased in favor of A. A part of the yield difference between the two varieties would be due to the difference in the fertility levels and not to the varietal difference.

To avoid such bias, varieties must be assigned to experimental plots so that a particular variety is not consistently favored or handicapped. This can be achieved by randomly assigning varieties to the experimental plots. Random-

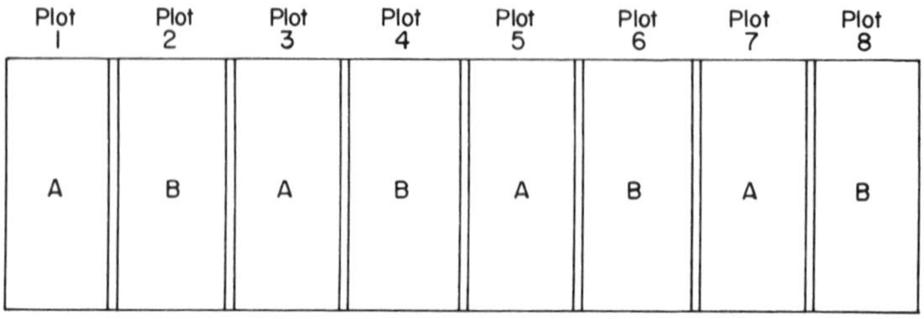

Figure 1.1 A systematic arrangement of plots planted to two rice varieties *A* and *B*. This scheme does not provide a valid estimate of experimental error.

ization ensures that each variety will have an equal chance of being assigned to any experimental plot and, consequently, of being grown in any particular environment existing in the experimental site.

1.2   CONTROL OF ERROR

Because the ability to detect existing differences among treatments increases as the size of the experimental error decreases, a good experiment incorporates all possible means of minimizing the experimental error. Three commonly used techniques for controlling experimental error in agricultural research are:

1.   Blocking
2.   Proper plot technique
3.   Data analysis

1.2.1   Blocking

By putting experimental units that are as similar as possible together in the same group (generally referred to as a block) and by assigning all treatments into each block separately and independently, variation among blocks can be measured and removed from experimental error. In field experiments where substantial variation within an experimental field can be expected, significant reduction in experimental error is usually achieved with the use of proper blocking. We emphasize the importance of blocking in the control of error in Chapters 2–4, with blocking as an important component in almost all experimental designs discussed.

1.2.2   Proper Plot Technique

For almost all types of experiment, it is absolutely essential that all other factors aside from those considered as treatments be maintained uniformly for all experimental units. For example, in variety trials where the treatments consist solely of the test varieties, it is required that all other factors such as soil nutrients, solar energy, plant population, pest incidence, and an almost infinite number of other environmental factors are maintained uniformly for all plots in the experiment. Clearly, the requirement is almost impossible to satisfy. Nevertheless, it is essential that the most important ones be watched closely to ensure that variability among experimental plots is minimized. This is the primary concern of a good plot technique.

For field experiments with crops, the important sources of variability among plots treated alike are soil heterogeneity, competition effects, and mechanical errors. The techniques appropriate for coping with each of these important sources of variation are discussed in Chapters 12–14.

1.2.3 Data Analysis

In cases where blocking alone may not be able to achieve adequate control of experimental error, proper choice of data analysis can help greatly. Covariance analysis is most commonly used for this purpose. By measuring one or more *covariates*—the characters whose functional relationships to the character of primary interest are known—the analysis of covariance can reduce the variability among experimental units by adjusting their values to a common value of the covariates. For example, in an animal feeding trial, the initial weight of the animals usually differs. Using this initial weight as the covariate, final weight after the animals are subjected to various feeds (i.e., treatments) can be adjusted to the values that would have been attained had all experimental animals started with the same body weight. Or, in a rice field experiment where rats damaged some of the test plots, covariance analysis with rat damage as the covariate can adjust plot yields to the levels that they should have been with no rat damage in any plot.

1.3 PROPER INTERPRETATION OF RESULTS

An important feature of the design of experiments is its ability to uniformly maintain all environmental factors that are not a part of the treatments being evaluated. This uniformity is both an advantage and a weakness of a controlled experiment. Although maintaining uniformity is vital to the measurement and reduction of experimental error, which are so essential in hypothesis testing, this same feature greatly limits the applicability and generalization of the experimental results, a limitation that must always be considered in the interpretation of results.

Consider the plant breeder's experiment comparing varieties A and B (Section 1.1). It is obvious that the choice of management practices (such as fertilization and weed control) or of the site and crop season in which the trial is conducted (such as in a rainy or dry environment) will greatly affect the relative performance of the two varieties. In rice and maize, for example, it has been shown that the newly developed, improved varieties are greatly superior to the native varieties when both are grown in a good environment and with good management; but the improved varieties are no better, or even poorer, when both are grown by the traditional farmer's practices.

Clearly the result of an experiment is, strictly speaking, applicable only to conditions that are the same as, or similar to, that under which the experiment was conducted. This limitation is especially troublesome because most agricultural research is done on experiment stations where average productivity is higher than that for ordinary farms. In addition, the environment surrounding a single experiment can hardly represent the variation over space and time that is so typical of commercial farms. Consequently, field experiments with crops are usually conducted for several crop seasons and years, in research stations and on farmers' fields, to insure that the results will apply over a wide range of environments. This is our primary concern in Chapters 8 and 16.

CHAPTER 2
Single-Factor Experiments

Experiments in which only a single factor varies while all others are kept constant are called single-factor experiments. In such experiments, the treatments consist solely of the different levels of the single variable factor. All other factors are applied uniformly to all plots at a single prescribed level. For example, most crop variety trials are single-factor experiments in which the single variable factor is variety and the factor levels (i.e., treatments) are the different varieties. Only the variety planted differs from one experimental plot to another and all management factors, such as fertilizer, insect control, and water management, are applied uniformly to all plots. Other examples of single-factor experiment are:

- Fertilizer trials where several rates of a single fertilizer element are tested.
- Insecticide trials where several insecticides are tested.
- Plant-population trials where several plant densities are tested.

There are two groups of experimental design that are applicable to a single-factor experiment. One group is the family of *complete block designs*, which is suited for experiments with a small number of treatments and is characterized by blocks, each of which contains at least one complete set of treatments. The other group is the family of *incomplete block designs*, which is suited for experiments with a large number of treatments and is characterized by blocks, each of which contains only a fraction of the treatments to be tested.

We describe three complete block designs (completely randomized, randomized complete block, and latin square designs) and two incomplete block designs (lattice and group balanced block designs). For each design, we illustrate the procedures for randomization, plot layout, and analysis of variance with actual experiments.

7

2.1 COMPLETELY RANDOMIZED DESIGN

A completely randomized design (CRD) is one where the treatments are assigned completely at random so that each experimental unit has the same chance of receiving any one treatment. For the CRD, any difference among experimental units receiving the same treatment is considered as experimental error. Hence, the CRD is only appropriate for experiments with homogeneous experimental units, such as laboratory experiments, where environmental effects are relatively easy to control. For field experiments, where there is generally large variation among experimental plots, in such environmental factors as soil, the CRD is rarely used.

2.1.1 Randomization and Layout

The step-by-step procedures for randomization and layout of a CRD are given here for a field experiment with four treatments A, B, C, and D, each replicated five times.

☐ STEP 1.   Determine the total number of experimental plots (n) as the product of the number of treatments (t) and the number of replications (r); that is, $n = (r)(t)$. For our example, $n = (5)(4) = 20$.

☐ STEP 2.   Assign a plot number to each experimental plot in any convenient manner; for example, consecutively from 1 to n. For our example, the plot numbers $1,\dots,20$ are assigned to the 20 experimental plots as shown in Figure 2.1.

☐ STEP 3.   Assign the treatments to the experimental plots by any of the following randomization schemes:

    A.   By table of random numbers. The steps involved are:

        STEP A_1.   Locate a starting point in a table of random numbers (Appendix A) by closing your eyes and pointing a finger to any position

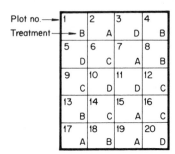

Figure 2.1 A sample layout of a completely randomized design with four treatments (A, B, C, and D) each replicated five times.

in a page. For our example, the starting point is at the intersection of the sixth row and the twelfth (single) column, as shown here.

Appendix A.   Table of Random Numbers

14620	95430	12951	81953	17629
09724	85125	48477	42783	70473
56919	17803	95781	85069	61594
97310	78209	51263	52396	82681
07585	28040	26939	64531	70570
		↘		
25950	85189	69374	37904	06759
82937	16405	81497	20863	94072
60819	27364	59081	72635	49180
59041	38475	03615	84093	49731
74208	69516	79530	47649	53046
39412	03642	87497	29735	14308
48480	50075	11804	24956	72182
95318	28749	49512	35408	21814
72094	16385	90185	72635	86259
63158	49753	84279	56496	30618
19082	73645	09182	73649	56823
15232	84146	87729	65584	83641
94252	77489	62434	20965	20247
72020	18895	84948	53072	74573
48392	06359	47040	05695	79799
37950	77387	35495	48192	84518
09394	59842	39573	51630	78548
34800	28055	91570	99154	39603
36435	75946	85712	06293	85621
28187	31824	52265	80494	66428

STEP A_2.   Using the starting point obtained in step A_1, read downward vertically to obtain $n = 20$ distinct three-digit random numbers. Three-digit numbers are preferred because they are less likely to include ties than one- or two-digit numbers. For our example, starting at the intersection of the sixth row and the twelfth column, the 20 distinct three-digit random numbers are as shown here together with their corresponding sequence of appearance.

Random Number	Sequence	Random Number	Sequence
937	1	918	11
149	2	772	12
908	3	243	13
361	4	494	14
953	5	704	15
749	6	549	16
180	7	957	17
951	8	157	18
018	9	571	19
427	10	226	20

STEP A_3.   Rank the n random numbers obtained in step A_2 in ascending or descending order. For our example, the 20 random numbers are ranked from the smallest to the largest, as shown in the following:

Random Number	Sequence	Rank	Random Number	Sequence	Rank
937	1	17	918	11	16
149	2	2	772	12	14
908	3	15	243	13	6
361	4	7	494	14	9
953	5	19	704	15	12
749	6	13	549	16	10
180	7	4	957	17	20
951	8	18	157	18	3
018	9	1	571	19	11
427	10	8	226	20	5

STEP A_4.   Divide the n ranks derived in step A_3 into t groups, each consisting of r numbers, according to the sequence in which the random numbers appeared. For our example, the 20 ranks are divided into four groups, each consisting of five numbers, as follows:

Group Number	Ranks in the Group				
1	17,	2,	15,	7,	19
2	13,	4,	18,	1,	8
3	16,	14,	6,	9,	12
4	10,	20,	3,	11,	5

STEP A_5. Assign the t treatments to the n experimental plots, by using the group number of step A_4 as the treatment number and the corresponding ranks in each group as the plot number in which the corresponding treatment is to be assigned. For our example, the first group is assigned to treatment A and plots numbered 17, 2, 15, 7, and 19 are assigned to receive this treatment; the second group is assigned to treatment B with plots numbered 13, 4, 18, 1, and 8; the third group is assigned to treatment C with plots numbered 16, 14, 6, 9, and 12; and the fourth group to treatment D with plots numbered 10, 20, 3, 11, and 5. The final layout of the experiment is shown in Figure 2.1.

B.   By drawing cards. The steps involved are:

STEP B_1.   From a deck of ordinary playing cards, draw n cards, one at a time, mixing the remaining cards after every draw. This procedure cannot be used when the total number of experimental units exceeds 52 because there are only 52 cards in a pack.

For our example, the 20 selected cards and the corresponding sequence in which each card was drawn may be shown below:

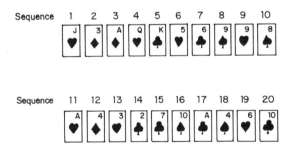

STEP B_2.   Rank the 20 cards drawn in step B_1 according to the suit rank (♣ ♦ ♥ ♠) and number of the card (2 is lowest, A is highest).

For our example, the 20 cards are ranked from the smallest to the largest as follows:

Sequence	1	2	3	4	5	6	7	8	9	10
Rank	14	7	9	15	5	11	2	19	13	18

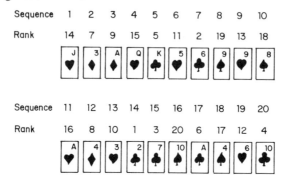

Sequence	11	12	13	14	15	16	17	18	19	20
Rank	16	8	10	1	3	20	6	17	12	4

STEP B_3. Assign the t treatments to the n plots by using the rank obtained in step B_2 as the plot number. Follow the procedure in steps A_4 and A_5. For our example, the four treatments are assigned to the 20 experimental plots as follows:

Treatment			Plot Assignment		
A	14,	7,	9,	15,	5
B	11,	2,	19,	13,	18
C	16,	8,	10,	1,	3
D	20,	6,	17,	12,	4

C. By drawing lots. The steps involved are:

STEP C_1. Prepare n identical pieces of paper and divide them into t groups, each group with r pieces of paper. Label each piece of paper of the same group with the same letter (or number) corresponding to a treatment. Uniformly fold each of the n labeled pieces of paper, mix them thoroughly, and place them in a container. For our example, there should be 20 pieces of paper, five each with treatments *A*, *B*, *C*, and *D* appearing on them.

STEP C_2. Draw one piece of paper at a time, without replacement and with constant shaking of the container after each draw to mix its content. For our example, the label and the corresponding sequence in which each piece of paper is drawn may be as follows:

Treatment label:	*D*	*B*	*A*	*B*	*C*	*A*	*D*	*C*	*B*	*D*
Sequence:	1	2	3	4	5	6	7	8	9	10
Treatment label:	*D*	*A*	*A*	*B*	*B*	*C*	*D*	*C*	*C*	*A*
Sequence:	11	12	13	14	15	16	17	18	19	20

STEP C_3. Assign the treatments to plots based on the corresponding treatment label and sequence, drawn in step C_2. For our example, treatment *A* would be assigned to plots numbered 3, 6, 12, 13, and 20;

treatment B to plots numbered 2, 4, 9, 14, and 15; treatment C to plots numbered 5, 8, 16, 18, and 19; and treatment D to plots numbered 1, 7, 10, 11, and 17.

2.1.2 Analysis of Variance

There are two sources of variation among the n observations obtained from a CRD trial. One is the treatment variation, the other is experimental error. The relative size of the two is used to indicate whether the observed difference among treatments is real or is due to chance. The treatment difference is said to be real if treatment variation is sufficiently larger than experimental error.

A major advantage of the CRD is the simplicity in the computation of its analysis of variance, especially when the number of replications is not uniform for all treatments. For most other designs, the analysis of variance becomes complicated when the loss of data in some plots results in unequal replications among treatments tested (see Chapter 7, Section 7.1).

2.1.2.1 Equal Replication. The steps involved in the analysis of variance for data from a CRD experiment with an equal number of replications are given below. We use data from an experiment on chemical control of brown planthoppers and stem borers in rice (Table 2.1).

☐ STEP 1. Group the data by treatments and calculate the treatment totals (T) and grand total (G). For our example, the results are shown in Table 2.1.

☐ STEP 2. Construct an outline of the analysis of variance as follows:

Source of Variation	Degree of Freedom	Sum of Squares	Mean Square	Computed F	Tabular F 5%	1%
Treatment						
Experimental error						
Total						

☐ STEP 3. Using t to represent the number of treatments and r, the number of replications, determine the degree of freedom $(d.f.)$ for each source of variation as follows:

$$\text{Total } d.f. = (r)(t) - 1 = (4)(7) - 1 = 27$$
$$\text{Treatment } d.f. = t - 1 = 7 - 1 = 6$$
$$\text{Error } d.f. = t(r - 1) = 7(4 - 1) = 21$$

The error $d.f.$ can also be obtained through subtraction as:

$$\text{Error } d.f. = \text{Total } d.f. - \text{Treatment } d.f. = 27 - 6 = 21$$

Table 2.1   Grain Yield of Rice Resulting from Use of Different Foliar and Granular Insecticides for the Control of Brown Planthoppers and Stem Borers, from a CRD Experiment with 4 (*r*) Replications and 7 (*t*) Treatments

Treatment	Grain Yield, kg/ha				Treatment Total (*T*)	Treatment Mean
Dol-Mix (1 kg)	2,537	2,069	2,104	1,797	8,507	2,127
Dol-Mix (2 kg)	3,366	2,591	2,211	2,544	10,712	2,678
DDT + γ-BHC	2,536	2,459	2,827	2,385	10,207	2,552
Azodrin	2,387	2,453	1,556	2,116	8,512	2,128
Dimecron-Boom	1,997	1,679	1,649	1,859	7,184	1,796
Dimecron-Knap	1,796	1,704	1,904	1,320	6,724	1,681
Control	1,401	1,516	1,270	1,077	5,264	1,316
Grand total (*G*)					57,110	
Grand mean						2,040

□ STEP 4.   Using X_i to represent the measurement of the ith plot, T_i as the total of the ith treatment, and n as the total number of experimental plots [i.e., $n = (r)(t)$], calculate the correction factor and the various sums of squares (SS) as:

$$\text{Correction factor } (C.F.) = \frac{G^2}{n}$$

$$\text{Total } SS = \sum_{i=1}^{n} X_i^2 - C.F.$$

$$\text{Treatment } SS = \frac{\sum_{i=1}^{t} T_i^2}{r}$$

$$\text{Error } SS = \text{Total } SS - \text{Treatment } SS$$

Throughout this book, we use the symbol Σ to represent "the sum of." For example, the expression $G = X_1 + X_2 + \cdots + X_n$ can be written as $G = \sum_{i=1}^{n} X_i$ or simply $G = \Sigma X$. For our example, using the T values and the G value from Table 2.1, the sums of squares are computed as:

$$C.F. = \frac{(57,110)^2}{(4)(7)} = 116,484,004$$

$$\text{Total } SS = \left[(2,537)^2 + (2,069)^2 + \cdots + (1,270)^2 + (1,077)^2 \right]$$
$$- 116,484,004$$
$$= 7,577,412$$

$$\text{Treatment } SS = \frac{(8,507)^2 + (10,712)^2 + \cdots + (5,264)^2}{4} - 116,484,004$$

$$= 5,587,174$$

$$\text{Error } SS = 7,577,412 - 5,587,174 = 1,990,238$$

☐ STEP 5. Calculate the mean square (MS) for each source of variation by dividing each SS by its corresponding $d.f.$:

$$\text{Treatment } MS = \frac{\text{Treatment } SS}{t - 1}$$

$$= \frac{5,587,174}{6} = 931,196$$

$$\text{Error } MS = \frac{\text{Error } SS}{t(r - 1)}$$

$$= \frac{1,990,238}{(7)(3)} = 94,773$$

☐ STEP 6. Calculate the F value for testing significance of the treatment difference as:

$$F = \frac{\text{Treatment } MS}{\text{Error } MS}$$

$$= \frac{931,196}{94,773} = 9.83$$

Note here that the F value should be computed only when the error $d.f.$ is large enough for a reliable estimate of the error variance. As a general guideline, the F value should be computed only when the error $d.f.$ is six or more.

☐ STEP 7. Obtain the tabular F values from Appendix E, with f_1 = treatment $d.f. = (t - 1)$ and f_2 = error $d.f. = t(r - 1)$. For our example, the tabular F values with $f_1 = 6$ and $f_2 = 21$ degrees of freedom are 2.57 for the 5% level of significance and 3.81 for the 1% level.

☐ STEP 8. Enter all the values computed in steps 3 to 7 in the outline of the analysis of variance constructed in step 2. For our example, the result is shown in Table 2.2.

☐ STEP 9. Compare the computed F value of step 6 with the tabular F values of step 7, and decide on the significance of the difference among treatments using the following rules:

1. If the computed F value is larger than the tabular F value at the 1% level of significance, the treatment difference is said to be *highly signifi-*

Table 2.2   Analysis of Variance (CRD with Equal Replication) of Rice Yield Data in Table 2.1[a]

Source of Variation	Degree of Freedom	Sum of Squares	Mean Square	Computed F^b	Tabular F 5%	1%
Treatment	6	5,587,174	931,196	9.83**	2.57	3.81
Experimental error	21	1,990,238	94,773			
Total	27	7,577,412				

[a]cv = 15.1%.
[b]** = significant at 1% level.

cant. Such a result is generally indicated by placing two asterisks on the computed F value in the analysis of variance.

2.   If the computed F value is larger than the tabular F value at the 5% level of significance but smaller than or equal to the tabular F value at the 1% level of significance, the treatment difference is said to be *significant*. Such a result is indicated by placing one asterisk on the computed F value in the analysis of variance.

3.   If the computed F value is smaller than or equal to the tabular F value at the 5% level of significance, the treatment difference is said to be *nonsignificant*. Such a result is indicated by placing *ns* on the computed F value in the analysis of variance.

Note that a nonsignificant F test in the analysis of variance indicates the failure of the experiment to detect any difference among treatments. It does not, in any way, prove that all treatments are the same, because the failure to detect treatment difference, based on the nonsignificant F test, could be the result of either a very small or nil treatment difference or a very large experimental error, or both. Thus, whenever the F test is nonsignificant, the researcher should examine the size of the experimental error and the numerical difference among treatment means. If both values are large, the trial may be repeated and efforts made to reduce the experimental error so that the difference among treatments, if any, can be detected. On the other hand, if both values are small, the difference among treatments is probably too small to be of any economic value and, thus, no additional trials are needed.

For our example, the computed F value of 9.83 is larger than the tabular F value at the 1% level of significance of 3.81. Hence, the treatment difference is said to be highly significant. In other words, chances are less than 1 in 100 that all the observed differences among the seven treatment means could be due to chance. It should be noted that such a significant F test verifies the existence of some differences among the treatments tested but does not specify the particular pair (or pairs) of treatments that differ significantly. To obtain this information, procedures for comparing treatment means, discussed in Chapter 3, are needed.

☐ STEP 10.   Compute the grand mean and the coefficient of variation cv as follows:

$$\text{Grand mean} = \frac{G}{n}$$

$$cv = \frac{\sqrt{\text{Error } MS}}{\text{Grand mean}} \times 100$$

For our example,

$$\text{Grand mean} = \frac{57,110}{28} = 2,040$$

$$cv = \frac{\sqrt{94,773}}{2,040} \times 100 = 15.1\%$$

The *cv* indicates the degree of precision with which the treatments are compared and is a good index of the reliability of the experiment. It expresses the experimental error as percentage of the mean; thus, the higher the *cv* value, the lower is the reliability of the experiment. The *cv* value is generally placed below the analysis of variance table, as shown in Table 2.2.

The *cv* varies greatly with the type of experiment, the crop grown, and the character measured. An experienced researcher, however, can make a reasonably good judgement on the acceptability of a particular *cv* value for a given type of experiment. Our experience with field experiments in transplanted rice, for example, indicates that, for data on rice yield, the acceptable range of *cv* is 6 to 8% for variety trials, 10 to 12% for fertilizer trials, and 13 to 15% for insecticide and herbicide trials. The *cv* for other plant characters usually differs from that of yield. For example, in a field experiment where the *cv* for rice yield is about 10%, that for tiller number would be about 20% and that for plant height, about 3%.

2.1.2.2 Unequal Replication.

2.1.2.2 Unequal Replication.   Because the computational procedure for the CRD is not overly complicated when the number of replications differs among treatments, the CRD is commonly used for studies where the experimental material makes it difficult to use an equal number of replications for all treatments. Some examples of these cases are:

· Animal feeding experiments where the number of animals for each breed is not the same.

- Experiments for comparing body length of different species of insect caught in an insect trap.
- Experiments that are originally set up with an equal number of replications but some experimental units are likely to be lost or destroyed during experimentation.

The steps involved in the analysis of variance for data from a CRD experiment with an unequal number of replications are given below. We use data from an experiment on performance of postemergence herbicides in dryland rice (Table 2.3).

☐ STEP 1.   Follow steps 1 and 2 of Section 2.1.2.1.

☐ STEP 2.   Using t to represent the number of treatments and n for the total number of observations, determine the degree of freedom for each source of variation, as follows:

$$\text{Total } d.f. = n - 1$$

$$= 40 - 1 = 39$$

$$\text{Treatment } d.f. = t - 1$$

$$= 11 - 1 = 10$$

$$\text{Error } d.f. = \text{Total } d.f. - \text{Treatment } d.f.$$

$$= 39 - 10 = 29$$

☐ STEP 3.   With the treatment totals (T) and the grand total (G) of Table 2.3, compute the correction factor and the various sums of squares, as follows:

$$C.F. = \frac{G^2}{n}$$

$$= \frac{(103,301)^2}{40} = 266,777,415$$

$$\text{Total } SS = \sum_{i=1}^{n} X_i^2 - C.F.$$

$$= \left[(3,187)^2 + (4,610)^2 + \cdots + (1,030)^2 \right] - 266,777,415$$

$$= 20,209,724$$

Table 2.3 Grain Yield of Rice Grown in a Dryland Field with Different Types, Rates, and Times of Application of Postemergence Herbicides, from a CRD Experiment with Unequal Number of Replications

Treatment Type	Rate,[a] kg a.i./ha	Time of application,[b] DAS	Grain Yield, kg/ha				Treatment Total (T)	Treatment Mean
Propanil/Bromoxynil	2.0/0.25	21	3,187	4,610	3,562	3,217	14,576	3,644
Propanil/2,4-D-B	3.0/1.00	28	3,390	2,775	2,875		9,040	3,013
Propanil/Bromoxynil	2.0/0.25	14	2,797	3,001	2,505	3,490	11,793	2,948
Propanil/Ioxynil	2.0/0.50	14	2,832	3,103	3,448	2,255	11,638	2,910
Propanil/CHCH	3.0/1.50	21	2,233	2,743	2,727		7,703	2,568
Phenyedipham	1.5	14	2,952	2,272	2,470		7,594	2,565
Propanil/Bromoxynil	2.0/0.25	28	2,858	2,895	2,458	1,723	9,934	2,484
Propanil/2,4-D-IPE	3.0/1.00	28	2,308	2,335	1,975		6,518	2,205
Propanil/Ioxynil	2.0/0.50	28	2,013	1,788	2,248	2,115	8,164	2,041
Handweeded twice	—	15 and 35	3,202	3,060	2,240	2,690	11,192	2,733
Control	—	—	1,192	1,652	1,075	1,030	4,349	1,037
Grand total (G)							105,301	
Grand mean								2,533

[a] a.i. = active ingredient.
[b] DAS = days after seeding.

19

Table 2.4   Analysis of Variance (CRD with Unequal Replication) of Grain Yield Data in Table 2.3[a]

Source of Variation	Degree of Freedom	Sum of Squares	Mean Square	Computed F[b]	Tabular F 5%	1%
Treatment	10	15,090,304	1,509,030	8.55**	2.18	3.00
Experimental error	29	5,119,420	176,532			
Total	39	20,209,724				

[a]$cv = 16.3\%$.
[b]** = significant at 1% level.

$$\text{Treatment } SS = \sum_{i=1}^{t} \frac{T_i^2}{r_i} - C.F.$$

$$= \left[\frac{(14,576)^2}{4} + \frac{(9,040)^2}{3} + \cdots + \frac{(4,949)^2}{4} \right] - 266,777,415$$

$$= 15,090,304$$

$$\text{Error } SS = \text{Total } SS - \text{Treatment } SS$$

$$= 20,209,724 - 15,090,304 = 5,119,420$$

☐ STEP 4.   Follow steps 5 to 10 of Section 2.1.2.1. The completed analysis of variance for our example is given in Table 2.4. The result of the F test indicates a highly significant difference among treatment means.

2.2   RANDOMIZED COMPLETE BLOCK DESIGN

The randomized complete block (RCB) design is one of the most widely used experimental designs in agricultural research. The design is especially suited for field experiments where the number of treatments is not large and the experimental area has a predictable productivity gradient. The primary distinguishing feature of the RCB design is the presence of blocks of equal size, each of which contains all the treatments.

2.2.1   Blocking Technique

The primary purpose of blocking is to reduce experimental error by eliminating the contribution of known sources of variation among experimental units. This is done by grouping the experimental units into blocks such that variability within each block is minimized and variability among blocks is maximized. Because only the variation within a block becomes part of the experimental error, blocking is most effective when the experimental area has a predictable pattern of variability. With a predictable pattern, plot shape and block orientation can be chosen so that much of the variation is accounted for by the difference among blocks, and experimental plots within the same block are kept as uniform as possible.

There are two important decisions that have to be made in arriving at an appropriate and effective blocking technique. These are:

- The selection of the source of variability to be used as the basis for blocking.
- The selection of the block shape and orientation.

An ideal source of variation to use as the basis for blocking is one that is large and highly predictable. Examples are:

- Soil heterogeneity, in a fertilizer or variety trial where yield data is the primary character of interest.
- Direction of insect migration, in an insecticide trial where insect infestation is the primary character of interest.
- Slope of the field, in a study of plant reaction to water stress.

After identifying the specific source of variability to be used as the basis for blocking, the size and shape of the blocks must be selected to maximize variability among blocks. The guidelines for this decision are:

1. When the gradient is unidirectional (i.e., there is only one gradient), use long and narrow blocks. Furthermore, orient these blocks so their length is perpendicular to the direction of the gradient.
2. When the fertility gradient occurs in two directions with one gradient much stronger than the other, ignore the weaker gradient and follow the preceding guideline for the case of the unidirectional gradient.
3. When the fertility gradient occurs in two directions with both gradients equally strong and perpendicular to each other, choose one of these alternatives:
   - Use blocks that are as square as possible.
   - Use long and narrow blocks with their length perpendicular to the direction of one gradient (see guideline 1) and use the covariance technique (see Chapter 10, Section 10.1.1) to take care of the other gradient.
   - Use the latin square design (see Section 2.3) with two-way blockings, one for each gradient.
4. When the pattern of variability is not predictable, blocks should be as square as possible.

Whenever blocking is used, the identity of the blocks and the purpose for their use must be consistent throughout the experiment. That is, whenever a source of variation exists that is beyond the control of the researcher, he should assure that such variation occurs among blocks rather than within blocks. For example, if certain operations such as application of insecticides or data collection cannot be completed for the whole experiment in one day, the task should be completed for all plots of the same block in the same day. In this way, variation among days (which may be enhanced by weather factors) becomes a part of block variation and is, thus, excluded from the experimental error. If more than one observer is to make measurements in the trial, the same observer should be assigned to make measurements for all plots of the same block (see also Chapter 14, Section 14.8). In this way, the variation among observers, if any, would constitute a part of block variation instead of the experimental error.

2.2.2 Randomization and Layout

The randomization process for a RCB design is applied separately and independently to each of the blocks. We use a field experiment with six treatments *A, B, C, D, E, F* and four replications to illustrate the procedure.

□ STEP 1.   Divide the experimental area into *r* equal blocks, where *r* is the number of replications, following the blocking technique described in Sec-

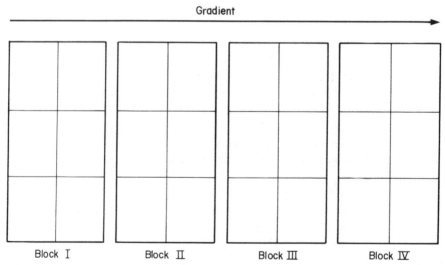

Figure 2.2 Division of an experimental area into four blocks, each consisting of six plots, for a randomized complete block design with six treatments and four replications. Blocking is done such that blocks are rectangular and perpendicular to the direction of the unidirectional gradient (indicated by the arrow).

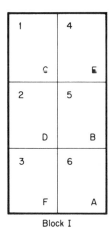

Block I

Figure 2.3 Plot numbering and random assignment of six treatments (*A, B, C, D, E,* and *F*) to the six plots in the first block of the field layout of Fig. 2.2.

tion 2.2.1. For our example, the experimental area is divided into four blocks as shown in Figure 2.2. Assuming that there is a unidirectional fertility gradient along the length of the experimental field, block shape is made rectangular and perpendicular to the direction of the gradient.

☐ STEP 2. Subdivide the first block into *t* experimental plots, where *t* is the number of treatments. Number the *t* plots consecutively from 1 to *t*, and assign *t* treatments at random to the *t* plots following any of the randomization schemes for the CRD described in Section 2.1.1. For our example, block I is subdivided into six equal-sized plots, which are numbered consecutively from top to bottom and from left to right (Figure 2.3); and, the six treatments are assigned at random to the six plots using the table of random numbers (see Section 2.1.1, step 3A) as follows:

- Select six three-digit random numbers. We start at the intersection of the sixteenth row and twelfth column of Appendix A and read downward vertically, to get the following:

Random Number	Sequence
918	1
772	2
243	3
494	4
704	5
549	6

- Rank the random numbers from the smallest to the largest, as follows:

Random Number	Sequence	Rank
918	1	6
772	2	5
243	3	1
494	4	2
704	5	4
549	6	3

- Assign the six treatments to the six plots by using the sequence in which the random numbers occurred as the treatment number and the corresponding rank as the plot number to which the particular treatment is to be assigned. Thus, treatment A is assigned to plot 6, treatment B to plot 5, treatment C to plot 1, treatment D to plot 2, treatment E to plot 4, and treatment F to plot 3. The layout of the first block is shown in Figure 2.3.

□ STEP 3.   Repeat step 2 completely for each of the remaining blocks. For our example, the final layout is shown in Figure 2.4.

It is worthwhile, at this point, to emphasize the major difference between a CRD and a RCB design. Randomization in the CRD is done without any restriction, but for the RCB design, all treatments must appear in each block. This difference can be illustrated by comparing the RCB design layout of Figure 2.4 with a hypothetical layout of the same trial based on a CRD, as

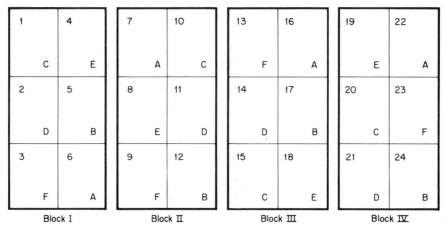

Figure 2.4   A sample layout of a randomized complete block design with six treatments (A, B, C, D, E, and F) and four replications.

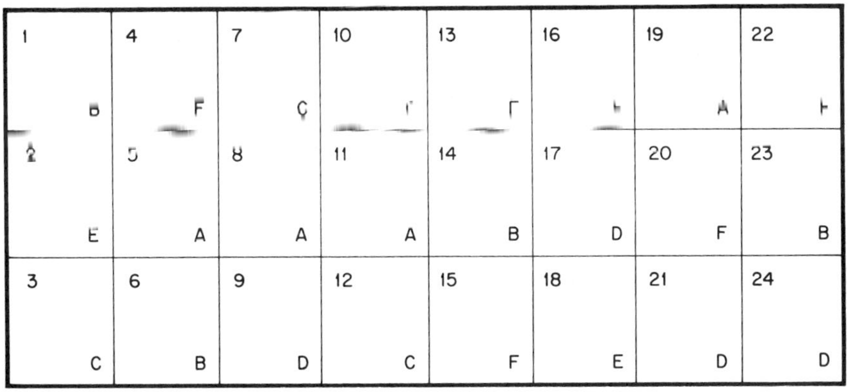

1	4	7	10	13	16	19	22
B	F	C	r	r	b	A	F
2	5	8	11	14	17	20	23
E	A	A	A	B	D	F	B
3	6	9	12	15	18	21	24
C	B	D	C	F	E	D	D

Figure 2.5 A hypothetical layout of a completely randomized design with six treatments (A, B, C, D, E, and F) and four replications.

shown in Figure 2.5. Note that each treatment in a CRD layout can appear anywhere among the 24 plots in the field. For example, in the CRD layout, treatment A appears in three adjacent plots (plots 5, 8, and 11). This is not possible in a RCB layout.

2.2.3 Analysis of Variance

There are three sources of variability in a RCB design: treatment, replication (or block), and experimental error. Note that this is one more than that for a CRD, because of the addition of replication, which corresponds to the variability among blocks.

To illustrate the steps involved in the analysis of variance for data from a RCB design we use data from an experiment that compared six rates of seeding of a rice variety IR8 (Table 2.5).

☐ STEP 1. Group the data by treatments and replications and calculate treatment totals (T), replication totals (R), and grand total (G), as shown in Table 2.5.

☐ STEP 2. Outline the analysis of variance as follows:

Source of Variation	Degree of Freedom	Sum of Squares	Mean Square	Computed F	Tabular F 5%	Tabular F 1%
Replication						
Treatment						
Error						
Total						

Table 2.5   Grain Yield of Rice Variety IR8 with Six Different Rates of Seeding, from a RCB Experiment with Four Replications

Treatment, kg seed/ha	Grain Yield, kg/ha				Treatment Total (T)	Treatment Mean
	Rep. I	Rep. II	Rep. III	Rep. IV		
25	5,113	5,398	5,307	4,678	20,496	5,124
50	5,346	5,952	4,719	4,264	20,281	5,070
75	5,272	5,713	5,483	4,749	21,217	5,304
100	5,164	4,831	4,986	4,410	19,391	4,848
125	4,804	4,848	4,432	4,748	18,832	4,708
150	5,254	4,542	4,919	4,098	18,813	4,703
Rep. total (R)	30,953	31,284	29,846	26,947		
Grand total (G)					119,030	
Grand mean						4,960

□ STEP 3.   Using r to represent the number of replications and t, the number of treatments, determine the degree of freedom for each source of variation as:

Total $d.f. = rt - 1 = 24 - 1 = 23$
Replication $d.f. = r - 1 = 4 - 1 = 3$
Treatment $d.f. = t - 1 = 6 - 1 = 5$
Error $d.f. = (r - 1)(t - 1) = (3)(5) = 15$

Note that as in the CRD, the error $d.f.$ can also be computed by subtraction, as follows:

Error $d.f.$ = Total $d.f.$ − Replication $d.f.$ − Treatment $d.f.$

$$= 23 - 3 - 5 = 15$$

□ STEP 4.   Compute the correction factor and the various sums of squares (SS) as follows:

$$C.F. = \frac{G^2}{rt}$$

$$= \frac{(119,030)^2}{(4)(6)} = 590,339,204$$

$$\text{Total } SS = \sum_{i=1}^{t} \sum_{j=1}^{r} X_{ij}^2 - C.F.$$

$$= \left[(5,113)^2 + (5,398)^2 + \cdots + (4,098)^2 \right] - 590,339,204$$

$$= 4,801,068$$

$$\text{Replication } SS = \frac{\sum_{j=1}^{r} R_j^2}{t} - C.F.$$

$$= \frac{(30,953)^2 + (31,284)^2 + (29,846)^2 + (26,947)^2}{6}$$

$$- 590,339,204$$

$$= 1,944,361$$

$$\text{Treatment } SS = \frac{\sum_{i=1}^{t} T_i^2}{r} - C.F.$$

$$= \frac{(20,496)^2 + \cdots + (18,813)^2}{4} - 590,339,204$$

$$= 1,198,331$$

$$\text{Error } SS = \text{Total } SS - \text{Replication } SS - \text{Treatment } SS$$

$$= 4,801,068 - 1,944,361 - 1,198,331 = 1,658,376$$

☐ STEP 5.   Compute the mean square for each source of variation by dividing each sum of squares by its corresponding degree of freedom as:

$$\text{Replication } MS = \frac{\text{Replication } SS}{r - 1}$$

$$= \frac{1,944,361}{3} = 648,120$$

$$\text{Treatment } MS = \frac{\text{Treatment } SS}{t - 1}$$

$$= \frac{1,198,331}{5} = 239,666$$

$$\text{Error } MS = \frac{\text{Error } SS}{(r - 1)(t - 1)}$$

$$= \frac{1,658,376}{15} = 110,558$$

☐ STEP 6.   Compute the F value for testing the treatment difference as:

$$F = \frac{\text{Treatment } MS}{\text{Error } MS}$$

$$= \frac{239,666}{110,558} = 2.17$$

☐ STEP 7.   Compare the computed F value with the tabular F values (from Appendix E) with f_1 = treatment $d.f.$ and f_2 = error $d.f.$ and make conclusions following the guidelines given in step 9 of Section 2.1.2.1.

For our example, the tabular F values with $f_1 = 5$ and $f_2 = 15$ degrees of freedom are 2.90 at the 5% level of significance and 4.56 at the 1% level. Because the computed F value of 2.17 is smaller than the tabular F value at the 5% level of significance, we conclude that the experiment failed to show any significant difference among the six treatments.

☐ STEP 8.   Compute the coefficient of variation as:

$$cv = \frac{\sqrt{\text{Error } MS}}{\text{Grand mean}} \times 100$$

$$= \frac{\sqrt{110,558}}{4,960} \times 100 = 6.7\%$$

☐ STEP 9.   Enter all values computed in steps 3 to 8 in the analysis of variance outline of step 2. The final result is shown in Table 2.6.

Table 2.6   Analysis of Variance (RCB) of Grain Yield Data in Table 2.5[a]

Source of Variation	Degree of Freedom	Sum of Squares	Mean Square	Computed F[b]	Tabular F 5%	Tabular F 1%
Replication	3	1,944,361	648,120			
Treatment	5	1,198,331	239,666	2.17[ns]	2.90	4.56
Error	15	1,658,376	110,558			
Total	23	4,801,068				

[a] cv = 6.7%.
[b] ns = not significant.

2.2.4 Block Efficiency

Blocking maximizes the difference among blocks, leaving the difference among plots of the same block as small as possible. Thus, the result of every RCB experiment should be examined to see how this objective has been achieved. The procedure for doing this is presented with the same data we used in Section 2.2.3 (Table 2.3).

☐ STEP 1. Determine the level of significance of the replication variation by computing the F value for replication as:

$$F(\text{replication}) = \frac{\text{Replication } MS}{\text{Error } MS}$$

and test its significance by comparing it to the tabular F values with $f_1 = (r - 1)$ and $f_2 = (r - 1)(t - 1)$ degrees of freedom. Blocking is considered effective in reducing the experimental error if $F(\text{replication})$ is significant (i.e., when the computed F value is greater than the tabular F value).

For our example, the computed F value for testing block difference is computed as:

$$F(\text{replication}) = \frac{648,120}{110,558} = 5.86$$

and the tabular F values with $f_1 = 3$ and $f_2 = 15$ degrees of freedom are 3.29 at the 5% level of significance and 5.42 at the 1% level. Because the computed F value is larger than the tabular F value at the 1% level of significance, the difference among blocks is highly significant.

☐ STEP 2. Determine the magnitude of the reduction in experimental error due to blocking by computing the relative efficiency ($R.E.$) parameter as:

$$R.E. = \frac{(r - 1)E_b + r(t - 1)E_e}{(rt - 1)E_e}$$

where E_b is the replication mean square and E_e is the error mean square in the RCB analysis of variance.

If the error $d.f.$ is less than 20, the $R.E.$ value should be multiplied by the adjustment factor k defined as:

$$k = \frac{[(r - 1)(t - 1) + 1][t(r - 1) + 3]}{[(r - 1)(t - 1) + 3][t(r - 1) + 1]}$$

Note that in the equation for $R.E.$, E_e in the denominator is the error for the RCB design, and the numerator is the comparable error had the CRD been used. Because the difference in the magnitude of experimental error between a CRD and a RCB design is essentially due to blocking, the value of the relative efficiency is indicative of the gain in precision due to blocking.

For our example, the $R.E.$ value is computed as:

$$R.E. = \frac{(3)(648{,}120) + 4(5)(110{,}558)}{(24 - 1)(110{,}558)} = 1.63$$

Because the error $d.f.$ is only 15, the adjustment factor is computed as:

$$k = \frac{[(3)(5) + 1][6(3) + 3]}{[(3)(5) + 3][6(3) + 1]} = 0.982$$

and the adjusted $R.E.$ value is computed as:

$$\text{Adjusted } R.E. = (k)(R.E.)$$

$$= (0.982)(1.63)$$

$$= 1.60$$

The results indicate that the use of the RCB design instead of a CRD design increased experimental precision by 60%.

2.3   LATIN SQUARE DESIGN

The major feature of the latin square (LS) design is its capacity to simultaneously handle two known sources of variation among experimental units. It treats the sources as two independent blocking criteria, instead of only one as in the RCB design. The two-directional blocking in a LS design, commonly referred to as row-blocking and column-blocking, is accomplished by ensuring that every treatment occurs only once in each row-block and once in each column-block. This procedure makes it possible to estimate variation among row-blocks as well as among column-blocks and to remove them from experimental error.

Some examples of cases where the LS design can be appropriately used are:

· Field trials in which the experimental area has two fertility gradients running perpendicular to each other, or has a unidirectional fertility gradient but also has residual effects from previous trials (see also Chapter 10, Section 10.1.1.2).
· Insecticide field trials where the insect migration has a predictable direction that is perpendicular to the dominant fertility gradient of the experimental field.
· Greenhouse trials in which the experimental pots are arranged in straight line perpendicular to the glass or screen walls, such that the difference

among rows of pots and the distance from the glass wall (or screen wall) are expected to be the two major sources of variability among the experimental pots.

Laboratory trials with replication over time, such that the difference among experimental units conducted at the same time and among those conducted over time constitute the two known sources of variability.

The presence of row-blocking and column-blocking in a LS design, while useful in taking care of two independent sources of variation, also becomes a major restriction in the use of the design. This is so because the requirement that all treatments appear in each row-block and in each column-block can be satisfied only if the number of replications is equal to the number of treatments. As a result, when the number of treatments is large the design becomes impractical because of the large number of replications required. On the other hand, when the number of treatments is small the degree of freedom associated with the experimental error becomes too small for the error to be reliably estimated.

Thus, in practice, the LS design is applicable only for experiments in which the number of treatments is not less than four and not more than eight. Because of such limitation, the LS design has not been widely used in agricultural experiments despite its great potential for controlling experimental error.

2.3.1 Randomization and Layout

The process of randomization and layout for a LS design is shown below for an experiment with five treatments A, B, C, D, and E.

□ STEP 1. Select a sample LS plan with five treatments from Appendix K. For our example, the 5×5 latin square plan from Appendix K is:

A	B	C	D	E
B	A	E	C	D
C	D	A	E	B
D	E	B	A	C
E	C	D	B	A

□ STEP 2. Randomize the row arrangement of the plan selected in step 1, following one of the randomization schemes described in Section 2.1.1. For this experiment, the table-of-random-numbers method of Section 2.1.1 is applied.

- Select five three-digit random numbers from Appendix A; for example, 628, 846, 475, 902, and 452.

- Rank the selected random numbers from lowest to highest:

Random Number	Sequence	Rank
628	1	3
846	2	4
475	3	2
902	4	5
452	5	1

- Use the rank to represent the existing row number of the selected plan and the sequence to represent the row number of the new plan. For our example, the third row of the selected plan (rank = 3) becomes the first row (sequence = 1) of the new plan; the fourth row of the selected plan becomes the second row of the new plan; and so on. The new plan, after the row randomization is:

C	D	A	E	B
D	E	B	A	C
B	A	E	C	D
E	C	D	B	A
A	B	C	D	E

☐ STEP 3.   Randomize the column arrangement, using the same procedure used for row arrangement in step 2. For our example, the five random numbers selected and their ranks are:

Random Number	Sequence	Rank
792	1	4
032	2	1
947	3	5
293	4	3
196	5	2

The rank will now be used to represent the column number of the plan obtained in step 2 (i.e., with rearranged rows) and the sequence will be used to represent the column number of the final plan.

For our example, the fourth column of the plan obtained in step 2 becomes the first column of the final plan, the first column of the plan of step 2 becomes the second column of the final plan, and so on. The final plan, which becomes the layout of the experiment is:

Row Number	Column Number				
	1	2	3	4	5
1	E	C	B	A	D
2	A	D	C	B	E
3	C	B	D	E	A
4	B	E	A	D	C
5	D	A	E	C	B

2.3.2 Analysis of Variance

There are four sources of variation in a LS design, two more than that for the CRD and one more than that for the RCB design. The sources of variation are row, column, treatment, and experimental error.

To illustrate the computation procedure for the analysis of variance of a LS design, we use data on grain yield of three promising maize hybrids (A, B, and D) and of a check (C) from an advanced yield trial with a 4×4 latin square design (Table 2.7).

The step-by-step procedures in the construction of the analysis of variance are:

□ STEP 1. Arrange the raw data according to their row and column designations, with the corresponding treatment clearly specified for each observation, as shown in Table 2.7.

□ STEP 2. Compute row totals (R), column totals (C), and the grand total (G) as shown in Table 2.7. Compute treatment totals (T) and treatment

Table 2.7 Grain Yield of Three Promising Maize Hybrids (A, B, and D) and a Check Variety (C) from an Experiment with Latin Square Design

Row Number	Grain Yield, t/ha				Row Total (R)
	Col. 1	Col. 2	Col. 3	Col. 4	
1	1.640(B)	1.210(D)	1.425(C)	1.345(A)	5.620
2	1.475(C)	1.185(A)	1.400(D)	1.290(B)	5.350
3	1.670(A)	0.710(C)	1.665(B)	1.180(D)	5.225
4	1.565(D)	1.290(B)	1.655(A)	0.660(C)	5.170
Column total (C)	6.350	4.395	6.145	4.475	
Grand total (G)					21.365

means as follows:

Treatment	Total	Mean
A	5.855	1.464
B	5.885	1.471
C	4.270	1.068
D	5.355	1.339

☐ STEP 3. Outline the analysis of variance as follows:

Source of Variation	Degree of Freedom	Sum of Squares	Mean Square	Computed F	Tabular F 5%	1%
Row						
Column						
Treatment						
Error						
Total						

☐ STEP 4. Using t to represent the number of treatments, determine the degree of freedom for each source of variation as:

Total $d.f. = t^2 - 1 = 16 - 1 = 15$

Row $d.f. = $ Column $d.f. = $ Treatment $d.f. = t - 1 = 4 - 1 = 3$

Error $d.f. = (t - 1)(t - 2) = (4 - 1)(4 - 2) = 6$

The error $d.f$ can also be obtained by subtraction as:

Error $d.f. = $ Total $d.f. - $ Row $d.f - $ Column $d.f. - $ Treatment $d.f.$

$= 15 - 3 - 3 - 3 = 6$

☐ STEP 5. Compute the correction factor and the various sums of squares as:

$$C.F. = \frac{G^2}{t^2}$$

$$= \frac{(21.365)^2}{16} = 28.528952$$

Total $SS = \Sigma X^2 - C.F.$

$$= \left[(1.640)^2 + (1.210)^2 + \cdots + (0.660)^2 \right] \quad 18.528952$$

$$= 1.413923$$

Row $SS = \dfrac{\Sigma R^2}{i} - C.F.$

$$= \dfrac{(5.620)^2 + (5.350)^2 + (5.225)^2 + (5.170)^2}{4}$$

$$- 28.528952$$

$$= 0.030154$$

Column $SS = \dfrac{\Sigma C^2}{t} - C.F.$

$$= \dfrac{(6.350)^2 + (4.395)^2 + (6.145)^2 + (4.475)^2}{4}$$

$$- 28.528952$$

$$= 0.827342$$

Treatment $SS = \dfrac{\Sigma T^2}{t} - C.F.$

$$= \dfrac{(5.855)^2 + (5.885)^2 + (4.270)^2 + (5.355)^2}{4}$$

$$- 28.528952$$

$$= 0.426842$$

Error SS = Total SS − Row SS − Column SS − Treatment SS

$$= 1.413923 - 0.030154 - 0.827342 - 0.426842$$

$$= 0.129585$$

☐ STEP 6. Compute the mean square for each source of variation by dividing the sum of squares by its corresponding degree of freedom:

$$\text{Row } MS = \dfrac{\text{Row } SS}{t - 1}$$

$$= \dfrac{0.030154}{3} = 0.010051$$

$$\text{Column } MS = \frac{\text{Column } SS}{t - 1}$$

$$= \frac{0.827342}{3} = 0.275781$$

$$\text{Treatment } MS = \frac{\text{Treatment } SS}{t - 1}$$

$$= \frac{0.426842}{3} = 0.142281$$

$$\text{Error } MS = \frac{\text{Error } SS}{(t - 1)(t - 2)}$$

$$= \frac{0.129585}{(3)(2)} = 0.021598$$

□ STEP 7. Compute the F value for testing the treatment effect as:

$$F = \frac{\text{Treatment } MS}{\text{Error } MS}$$

$$= \frac{0.142281}{0.021598} = 6.59$$

□ STEP 8. Compare the computed F value with the tabular F value, from Appendix E, with f_1 = treatment $d.f. = t - 1$ and f_2 = error $d.f. = (t - 1)(t - 2)$ and make conclusions following the guidelines in step 9 of Section 2.1.2.1.

For our example, the tabular F values, from Appendix E, with $f_1 = 3$ and $f_2 = 6$ degrees of freedom, are 4.76 at the 5% level of significance and 9.78 at the 1% level. Because the computed F value is higher than the tabular F value at the 5% level of significance but lower than the tabular F value at the 1% level, the treatment difference is significant at the 5% level of significance.

□ STEP 9. Compute the coefficient of variation as:

$$cv = \frac{\sqrt{\text{Error } MS}}{\text{Grand mean}} \times 100$$

$$= \frac{\sqrt{0.021598}}{1.335} \times 100 = 11.0\%$$

□ STEP 10. Enter all values computed in steps 4 to 9 in the analysis of variance outline of step 3, as shown in Table 2.8.

Note that although the F test in the analysis of variance indicates significant differences among the mean yields of the four maize varieties

Table 2.8 Analysis of Variance (LS Design) of Grain Yield Data in Table 2.7[a]

Source of Variation	Degree of Freedom	Sum of Squares	Mean Square	Computed F^b	Tabular F 5%	1%
Row	3	0.030154	0.010051			
Column	3	0.827342	0.275781			
Treatment	3	0.426842	0.142281	6.59*	4.76	9.78
Error	6	0.129585	0.021598			
Total	15	1.413923				

[a]$cv = 11.0\%$.
[b]* = significant at 5% level.

tested, it does not identify the specific pairs or groups of varieties that differed. For example, the *F* test is not able to answer the question of whether every one of the three hybrids gave significantly higher yield than that of the check variety or whether there is any significant difference among the three hybrids. To answer these questions, the procedures for mean comparisons discussed in Chapter 5 should be used.

2.3.3 Efficiencies of Row- and Column-Blockings

As in the RCB design, where the efficiency of one-way blocking indicates the gain in precision relative to the CRD (see Section 2.2.4), the efficiencies of both row- and column-blockings in a LS design indicate the gain in precision relative to either the CRD or the RCB design. The procedures are:

☐ STEP 1. Test the level of significance of the differences among row- and column-blocks:

A. Compute the *F* values for testing the row difference and column difference as:

$$F(\text{row}) = \frac{\text{Row } MS}{\text{Error } MS}$$

$$= \frac{0.010051}{0.021598} < 1$$

$$F(\text{column}) = \frac{\text{Column } MS}{\text{Error } MS}$$

$$= \frac{0.275781}{0.021598} = 12.77$$

B. Compare each of the computed *F* values that is larger than 1 with the tabular *F* values (from Appendix E) with $f_1 = t - 1$ and $f_2 = (t - 1)(t - 2)$ degrees of freedom. For our example, the computed

F(row) value is smaller than 1 and, hence, is not significant. For the computed *F*(column) value, the corresponding tabular *F* values with $f_1 = 3$ and $f_2 = 6$ degrees of freedom are 4.76 at the 5% level of significance and 9.78 at the 1% level. Because the computed *F*(column) value is greater than both tabular *F* values, the difference among column-blocks is significant at the 1% level. These results indicate the success of column-blocking, but not that of row-blocking, in reducing experimental error.

□ STEP 2.   Compute the relative efficiency parameter of the LS design relative to the CRD or RCB design:

- The relative efficiency of a LS design as compared to a CRD:

$$R.E.(\text{CRD}) = \frac{E_r + E_c + (t - 1)E_e}{(t + 1)E_e}$$

where E_r is the row mean square, E_c is the column mean square, and E_e is the error mean square in the LS analysis of variance; and *t* is the number of treatments.

For our example, the R.E. is computed as:

$$R.E.(\text{CRD}) = \frac{0.010051 + 0.275781 + (4 - 1)(0.021598)}{(4 + 1)(0.021598)}$$

$$= 3.25$$

This indicates that the use of a LS design in the present example is estimated to increase the experimental precision by 225%. This result implies that, if the CRD had been used, an estimated 2.25 times more replications would have been required to detect the treatment difference of the same magnitude as that detected with the LS design.

- The relative efficiency of a LS design as compared to a RCB design can be computed in two ways—when rows are considered as blocks, and when columns are considered as blocks, of the RCB design. These two relative efficiencies are computed as:

$$R.E.(\text{RCB, row}) = \frac{E_r + (t - 1)E_e}{(t)(E_e)}$$

$$R.E.(\text{RCB, column}) = \frac{E_c + (t - 1)E_e}{(t)(E_e)}$$

where E_r, E_c, E_e, and *t* are as defined in the preceding formula.

When the error $d.f.$ in the LS analysis of variance is less than 20, the $R.E.$ value should be multiplied by the adjustment factor k defined as

$$k = \frac{[(t-1)(t-2)+1][(t-1)^2+3]}{[(t-1)(t-2)+3][(t-1)^2+1]}$$

For our example, the values of the relative efficiency of the LS design compared to a RCB design with rows as blocks and with columns as blocks are computed as:

$$R.E.(\text{RCB, row}) = \frac{0.010051 + (4-1)(0.021598)}{4(0.021598)}$$

$$= 0.87$$

$$R.E.(\text{RCB, column}) = \frac{0.275781 + (4-1)(0.021598)}{4(0.021598)}$$

$$= 3.94$$

Because the error $d.f.$ of the LS design is only 6, the adjustment factor k is computed as:

$$k = \frac{[(4-1)(4-2)+1][(4-1)^2+3]}{[(4-1)(4-2)+3][(4-1)^2+1]} = 0.93$$

And, the adjusted R.E. values are computed as:

$$R.E.(\text{RCB, row}) = (0.87)(0.93) = 0.81$$

$$R.E.(\text{RCB, column}) = (3.94)(0.93) = 3.66$$

The results indicate that the additional column-blocking, made possible by the use of a LS design, is estimated to have increased the experimental precision over that of the RCB design with rows as blocks by 266%; whereas the additional row-blocking in the LS design did not increase precision over the RCB design with columns as blocks. Hence, for this trial, a RCB design with columns as blocks would have been as efficient as a LS design.

2.4 LATTICE DESIGN

Theoretically, the complete block designs, such as the RCB and the LS designs discussed in Sections 2.2 and 2.3, are applicable to experiments with any number of treatments. However, these complete block designs become less efficient as the number of treatments increases, primarily because block size increases proportionally with the number of treatments, and the homogeneity of experimental plots within a large block is difficult to maintain. That is, the experimental error of a complete block design is generally expected to increase with the number of treatments.

An alternative set of designs for single-factor experiments having a large number of treatments is the *incomplete block designs*, one of which is the lattice design. As the name implies, each block in an incomplete block design does not contain all treatments and a reasonably small block size can be maintained even if the number of treatments is large. With smaller blocks, the homogeneity of experimental units in the same block is easier to maintain and a higher degree of precision can generally be expected.

The improved precision with the use of an incomplete block design is achieved with some costs. The major ones are:

- Inflexible number of treatments or replications or both
- Unequal degrees of precision in the comparison of treatment means
- Complex data analysis

Although there is no concrete rule as to how large the number of treatments should be before the use of an incomplete block design should be considered, the following guidelines may be helpful:

Variability in the Experimental Material.   The advantage of an incomplete block design over the complete block design is enhanced by an increased variability in the experimental material. In general, whenever block size in a RCB design is too large to maintain a reasonable level of uniformity among experimental units within the same block, the use of an incomplete block design should be seriously considered. For example, in irrigated rice paddies where the experimental plots are expected to be relatively homogeneous, a RCB design would probably be adequate for a variety trial with as many as, say, 25 varieties. On the other hand, with the same experiment on a dryland field, where the experimental plots are expected to be less homogeneous, a lattice design may be more efficient.

Computing Facilities and Services.   Data analysis for an incomplete block design is more complex than that for a complete block design. Thus, in situations where adequate computing facilities and services are not easily available, incomplete block designs may have to be considered only as the last measure.

In general, an incomplete block design, with its reduced block size, is expected to give a higher degree of precision than a complete block design. Thus, the use of an incomplete block design should generally be preferred so long as the resources required for its use (e.g., more replications, inflexible number of treatments, and more complex analysis) can be satisfied.

The lattice design is the incomplete block design most commonly used in agricultural research. There is sufficient flexibility in the design to make its application simpler than most other incomplete block designs. This section is devoted primarily to two of the most commonly used lattice designs, the balanced lattice and the partially balanced lattice designs. Both require that the number of treatments must be a perfect square.

2.4.1 Balanced Lattice

The balanced lattice design is characterized by the following basic features:

1. The number of treatments (t) must be a perfect square (i.e., $t = k^2$, such as 25, 36, 49, 64, 81, 100, etc.). Although this requirement may seem stringent at first, it is usually easy to satisfy in practice. As the number of treatments becomes large, adding a few more or eliminating some less important treatments is usually easy to accomplish. For example, if a plant breeder wishes to test the performance of 80 varieties in a balanced lattice design, all he needs to do is add one more variety for a perfect square. Or if he has 82 or 83 varieties to start he can easily eliminate one or two.

2. The block size (k) is equal to the square root of the number of treatments (i.e., $k = t^{1/2}$).

3. The number of replications (r) is one more than the block size [i.e., $r = (k + 1)$]. That is, the number of replications required is 6 for 25 treatments, 7 for 36 treatments, 8 for 49 treatments, and so on.

2.4.1.1 Randomization and Layout. We illustrate the randomization and layout of a balanced lattice design with a field experiment involving nine treatments. There are four replications, each consisting of three incomplete blocks with each block containing three experimental plots. The steps to follow are:

☐ STEP 1. Divide the experimental area into $r = (k + 1)$ replications, each containing $t = k^2$ experimental plots. For our example, the experimental area is divided into $r = 4$ replications, each containing $t = 9$ experimental plots, as shown in Figure 2.6.

☐ STEP 2. Divide each replication into k incomplete blocks, each containing k experimental plots. In choosing the shape and size of the incomplete block, follow the blocking technique discussed in Section 2.2.1 to achieve maximum homogeneity among plots in the same incomplete block. For our example, each replication is divided into $k = 3$ incomplete blocks, each containing $k = 3$ experimental plots (Figure 2.6).

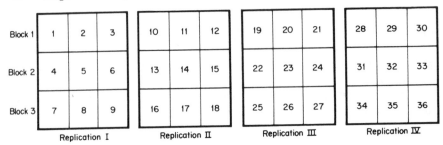

Figure 2.6 Division of the experimental area, consisting of 36 plots (1,2,...,36) into four replications, each containing three incomplete blocks of three plots each, as the first step in laying out a 3 × 3 balanced lattice design.

□ STEP 3.   Select from Appendix L a basic balanced lattice plan corresponding to the number of treatments to be tested. For our example, the basic plan for the 3 × 3 balanced lattice design is shown in Table 2.9.

□ STEP 4.   Randomize the replication arrangement of the selected basic plan, following an appropriate randomization scheme of Section 2.1.1. For our example, the table-of-random-numbers method is applied:

· Select four three-digit random numbers from Appendix A; for example, 372, 217, 963, and 404.

· Rank them from lowest to highest as:

Random Number	Sequence	Rank
372	1	2
217	2	1
963	3	4
404	4	3

· Use the sequence to represent the existing replication number of the basic plan and the rank to represent the replication number of the new

Table 2.9   Basic Plan of a 3 × 3 Balanced Lattice Design Involving Nine Treatments (1, 2,..., 9) in Blocks of Three Units and Four Replications

Incomplete Block Number	Treatment Number			
	Rep. I	Rep. II	Rep. III	Rep. IV
1	1 2 3	1 4 7	1 5 9	1 6 8
2	4 5 6	2 5 8	2 6 7	2 4 9
3	7 8 9	3 6 9	3 4 8	3 5 7

plan. Thus, the first replication of the basic plan (sequence = 1) becomes the second replication of the new plan (rank = 2), the second replication of the basic plan becomes the first replication of the new plan, and so on. The outcome of the new plan at this step is:

Incomplete Block Number	Treatment Number			
	Rep. I	Rep. II	Rep. III	Rep. IV
1	1 4 7	1 2 3	1 6 8	1 5 9
2	2 5 8	4 5 6	2 4 9	2 6 7
3	3 6 9	7 8 9	3 5 7	3 4 8

☐ STEP 5. Randomize the incomplete blocks within each replication following an appropriate randomization scheme of Section 2.1.1. For our example, the same randomization scheme used in step 4 is used to randomly reassign three incomplete blocks in each of the four replications. After four independent randomization processes, the reassigned incomplete blocks may be shown as:

Incomplete Block Number in Basic Plan	Reassigned Incomplete Block Number in New Plan			
	Rep. I	Rep. II	Rep. III	Rep. IV
1	3	2	3	1
2	2	1	1	3
3	1	3	2	2

As shown, for replication I, block 1 of the basic plan becomes block 3 of the new plan, block 2 retains the same position, and block 3 of the basic plan becomes block 1 of the new plan. For replication II, block 1 of the basic plan becomes block 2 of the new plan, block 2 of the basic plan becomes block 1 of the new plan, and so on. The outcome of the new plan at this step is:

Incomplete Block Number	Treatment Number			
	Rep. I	Rep. II	Rep. III	Rep. IV
1	3 6 9	4 5 6	3 5 7	1 5 9
2	2 5 8	1 2 3	1 6 8	3 4 8
3	1 4 7	7 8 9	2 4 9	2 6 7

☐ STEP 6. Randomize the treatment arrangement within each incomplete block. For our example, randomly reassign the three treatments in each of the 12 incomplete blocks, following the same randomization scheme used in steps 4 and 5. After 12 independent randomization processes, the reassigned treatment sequences may be shown as:

Treatment Sequence in Basic Plan	Reassigned Treatment Sequence in New Plan					
	Rep. I			Rep. II		
	Block 1	Block 2	Block 3	Block 1	Block 2	Block 3
1	2	3	2	2	3	3
2	3	2	3	1	2	2
3	1	1	1	3	1	1

Treatment Sequence in Basic Plan	Reassigned Treatment Sequence in New Plan					
	Rep. III			Rep. IV		
	Block 1	Block 2	Block 3	Block 1	Block 2	Block 3
1	3	3	1	1	3	2
2	2	1	2	3	1	3
3	1	2	3	2	2	1

In this case, for incomplete block 1 of replication I, treatment sequence 1 of the basic plan (treatment 3) becomes treatment sequence 2 of the new plan, treatment sequence 2 of the basic plan (treatment 6) becomes treatment sequence 3 of the new plan, and treatment sequence 3 of the basic plan (treatment 9) becomes treatment sequence 1 of the new plan, and so on. The outcome of the new plan at this step is:

Incomplete Block Number	Treatment Number			
	Rep. I	Rep. II	Rep. III	Rep. IV
1	9 3 6	5 4 6	7 5 3	1 9 5
2	8 5 2	3 2 1	6 8 1	4 8 3
3	7 1 4	9 8 7	2 4 9	7 2 6

☐ STEP 7. Apply the final outcome of the randomization process of step 6 to the field layout of Figure 2.6 resulting in the final layout of Figure 2.7. Note that an important feature in the layout of a balanced lattice design is that every pair of treatments occurs together only once in the same block. For example, treatment 1 appears only once with treatments 4 and 7 in block 3

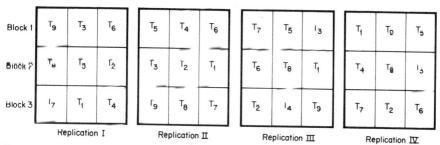

Block 1	T_9	T_3	T_6		T_5	T_4	T_6		T_7	T_5	I_3		T_1	T_0	T_5
Block 2	T_4	T_5	I_2		I_3	T_2	T_1		T_6	T_8	T_1		T_4	T_8	I_3
Block 3	I_7	T_1	T_4		I_9	T_8	T_7		T_2	I_4	T_9		T_7	T_2	T_6

Replication I Replication II Replication III Replication IV

Figure 2.7 A sample layout of a 3×3 balanced lattice design, involving nine treatments $(T_1, T_2, \ldots, T_9)$.

of replication I; with treatments 2 and 3 in block 2 of replication II; with treatments 6 and 8 in block 2 of replication III; and with treatments 5 and 9 in block 1 of replication IV. As a consequence of this feature, the degree of precision for comparing each pair of treatments in a balanced lattice design is the same for all pairs.

2.4.1.2 Analysis of Variance.   There are four sources of variation that can be accounted for in a balanced lattice design: replication, treatment, incomplete block, and experimental error. Relative to the RCB design, the incomplete block is an additional source of variation and reflects the differences among incomplete blocks of the same replication.

The computational procedure for the analysis of variance of a balanced lattice design is illustrated using data on tiller count from a field experiment involving 16 rice fertilizer treatments. The experiment followed a 4×4 balanced lattice design with five replications. The data are shown in Table 2.10, with the blocks and treatments rearranged according to the basic plan for the 4×4 balanced lattice design of Appendix L. Such a rearrangement is not necessary for the computation of the analysis of variance but we do it here to facilitate the understanding of the analytical procedure to be presented. The steps involved are:

☐ STEP 1.   Calculate the block totals (B) and replication totals (R), as shown in Table 2.10.

☐ STEP 2.   Calculate the treatment totals (T) and the grand total (G), as shown in column 2 of Table 2.11.

☐ STEP 3.   For each treatment, calculate the B_t value as the sum of block totals over all blocks in which the particular treatment appears. For example, treatment 5 in our example was tested in blocks 2, 5, 10, 15, and 20 (Table 2.10). Thus, B_t for treatment 5 is computed as the sum of the block totals of blocks 2, 5, 10, 15, and 20, or $B_5 = 616 + 639 + 654 + 675 + 827 = 3,411$.

Table 2.10   Tiller Number per Square Meter from 16 Fertilizer Treatments Tested in a 4 × 4 Balanced Lattice Design[a]

Block Number	Tiller, no./m²				Block Total (B)	Block Number	Tiller, no./m²				Block Total (B)
	Rep. I						*Rep. II*				
	(1)	(2)	(3)	(4)			(1)	(5)	(9)	(13)	
1	147	152	167	150	616	5	140	165	182	152	639
	(5)	(6)	(7)	(8)			(10)	(2)	(14)	(6)	
2	127	155	162	172	616	6	97	155	192	142	586
	(9)	(10)	(11)	(12)			(7)	(15)	(3)	(11)	
3	147	100	192	177	616	7	155	182	192	192	721
	(13)	(14)	(15)	(16)			(16)	(8)	(12)	(4)	
4	155	195	192	205	747	8	182	207	232	162	783
Rep. total R_1					2595	Rep. total R_2					2729
	Rep. III						*Rep. IV*				
	(1)	(6)	(11)	(16)			(1)	(14)	(7)	(12)	
9	155	162	177	152	646	13	220	202	175	205	802
	(5)	(2)	(15)	(12)			(13)	(2)	(11)	(8)	
10	182	130	177	165	654	14	205	152	180	187	724
	(9)	(14)	(3)	(8)			(5)	(10)	(3)	(16)	
11	137	185	152	152	626	15	165	150	200	160	675
	(13)	(10)	(7)	(4)			(9)	(6)	(15)	(4)	
12	185	122	182	192	681	16	155	177	185	172	689
Rep. total R_3					2607	Rep. total R_4					2890
	Rep. V										
	(1)	(10)	(15)	(8)							
17	147	112	177	147	583						
	(9)	(2)	(7)	(16)							
18	180	205	190	167	742						
	(13)	(6)	(3)	(12)							
19	172	212	197	192	773						
	(5)	(14)	(11)	(4)							
20	177	220	205	225	827						
Rep. total R_5					2925						

[a] The values enclosed in parentheses correspond to the treatment numbers.

The B_t values for all 16 treatments are shown in column 3 of Table 2.11. Note that the sum of B_t values over all treatments must equal $(k)(G)$, where k is the block size.

□ STEP 4.   For each treatment, calculate:

$$W = kT - (k + 1)B_t + G$$

Table 2.11   Computations of the Adjusted and Unadjusted Treatment Totals for the 4 × 4 Balanced Lattice Data in Table 2.10

Treatment Number	Treatment Total (T)	Block Total (B_t)	$W = 4T - 5B_t + G$	$T' = T + \mu W$	$M' = \dfrac{T'}{5}$
1	809	3,286	552	829	166
2	794	3,322	312	805	161
3	908	3,411	323	920	184
4	901	3,596	−630	878	176
5	816	3,411	−45	814	163
6	848	3,310	588	869	174
7	864	3,562	−608	842	168
8	865	3,332	546	885	177
9	801	3,312	390	815	163
10	581	3,141	365	594	119
11	946	3,534	−140	941	188
12	971	3,628	−510	953	191
13	869	3,564	−598	848	170
14	994	3,588	−218	986	197
15	913	3,394	428	928	186
16	866	3,593	−755	839	168
Sum	13,746 (G)	54,984	0	—	—

For our example, the W value for treatment 5 is computed as:

$$W_5 = 4(816) - (5)(3{,}411) + 13{,}746 = -45$$

The W values for all 16 treatments are presented in column 4 of Table 2.11. Note that the sum of W values over all treatments must be zero.

□ STEP 5.   Construct an outline of the analysis of variance, specifying the sources of variation and their corresponding degrees of freedom as:

Source of Variation	Degree of Freedom	Sum of Squares	Mean Square
Replication	$k = 4$		
Treatment(unadj.)	$k^2 - 1 = 15$		
Block(adj.)	$k^2 - 1 = 15$		
Intrablock error	$(k - 1)(k^2 - 1) = 45$		
Treatment(adj.)	$[(k^2 - 1) = 15]$		
Effective error	$[(k - 1)(k^2 - 1) = 45]$		
Total	$k^2(k + 1) - 1 = 79$		

☐ STEP 6. Compute the total SS, the replication SS, and the treatment (unadjusted) SS as:

$$C.F. = \frac{G^2}{(k^2)(k+1)}$$

$$= \frac{(13,746)^2}{(16)(5)} = 2,361,906$$

Total $SS = \sum X^2 - C.F.$

$$= \left[(147)^2 + (152)^2 + \cdots + (225)^2\right] - 2,361,906$$

$$= 58,856$$

Replication $SS = \dfrac{\sum R^2}{k^2} - C.F.$

$$= \frac{(2,595)^2 + (2,729)^2 + \cdots + (2,925)^2}{16} - 2,361,906$$

$$= 5,946$$

Treatment(unadj.) $SS = \dfrac{\sum T^2}{(k+1)} - C.F.$

$$= \frac{(809)^2 + (794)^2 + \cdots + (866)^2}{5} - 2,361,906$$

$$= 26,995$$

☐ STEP 7. Compute the block(adjusted) SS (i.e., the sum of squares for block within replication adjusted for treatment effects) as:

$$Block(adj.)\ SS = \frac{\sum\limits_{i=1}^{t} W_i^2}{(k^3)(k+1)}$$

$$= \frac{(552)^2 + (312)^2 + \cdots + (-755)^2}{(64)(5)}$$

$$= 11,382$$

☐ STEP 8. Compute the intrablock error SS as:

Intrablock error SS Total SS − Replication SS

$$- \text{Treatment(unadj.)} SS - \text{Block(adj.)} SS$$

$$= 58,856 \quad 5,946 - 26,995 - 11,382$$

$$= 14,533$$

☐ STEP 9. Compute the block(adj.) mean square and the intrablock error mean square as:

$$\text{Block(adj.)} MS = \frac{\text{Block(adj.)} SS}{k^2 - 1}$$

$$= \frac{11,382}{15} = 759$$

$$\text{Intrablock error } MS = \frac{\text{Intrablock error } SS}{(k - 1)(k^2 - 1)}$$

$$= \frac{14,533}{(3)(15)} = 323$$

☐ STEP 10. For each treatment, calculate the adjusted treatment total T' as:

$$T' = T + \mu W$$

where

$$\mu = \frac{\text{Block(adj.)} MS - \text{Intrablock error } MS}{k^2 [\text{Block(adj.)} MS]}$$

Note that if the intrablock error MS is greater than the block(adj.) MS, μ is taken to be zero and no adjustment for treatment nor any further adjustment is necessary. The F test for significance of treatment effect is then made in the usual manner as the ratio of the treatment(unadj.) MS and intrablock error MS, and steps 10 to 14 and step 17 can be ignored.

For our example, the intrablock error MS is smaller than the block(adj.) MS. Hence, the adjustment factor μ is computed as:

$$\mu = \frac{759 - 323}{16(759)} = 0.0359$$

The T' value for treatment 5, for example, is computed as $T_5' = 816 + (0.0359)(-45) = 814$. The results of T' values for all 16 treatments are shown in column 5 of Table 2.11.

☐ STEP 11.   For each treatment, calculate the adjusted treatment mean M' as:

$$M' = \frac{T'}{k+1}$$

For our example, the M' value for treatment 5 is computed as $M_5' = 814/5 = 163$. The results of M' values for all 16 treatments are presented in the last column of Table 2.11.

☐ STEP 12.   Compute the adjusted treatment mean square as:

$$\text{Treatment(adj.) } MS = \left[\frac{1}{(k+1)(k^2-1)}\right]\left[\sum T'^2 - \frac{G^2}{k^2}\right]$$

$$= \left[\frac{1}{(5)(15)}\right]\left\{\left[(829)^2 + (805)^2 + \cdots + (839)^2\right]\right.$$

$$\left. - \frac{(13,746)^2}{16}\right\}$$

$$= 1,602$$

☐ STEP 13.   Compute the effective error MS are:

$$\text{Effective error } MS = (\text{Intrablock error } MS)(1 + k\mu)$$

$$= 323[1 + 4(0.0359)]$$

$$= 369$$

Compute the corresponding cv value as:

$$cv = \frac{\sqrt{\text{Effective error } MS}}{\text{Grand mean}} \times 100$$

$$= \frac{\sqrt{369}}{172} \times 100 = 11.2\%$$

☐ STEP 14.   Compute the F value for testing the treatment difference as:

$$F = \frac{\text{Treatment(adj.) } MS}{\text{Effective error } MS}$$

$$= \frac{1,602}{369} = 4.34$$

☐ STEP 15.   Compare the computed F value to the tabular F values, from Appendix E, with $f_1 = (k^2 - 1) = 15$ and $f_2 = (k - 1)(k^2 - 1) = 45$ degrees of freedom. Because the computed F value is larger than the tabular F value at the 1% level of significance, the treatment difference is judged to be highly significant.

☐ STEP 16.   Enter all values computed in steps 6 to 9 and 12 to 15 in the analysis of variance outline of step 5. The final result is shown in Table 2.12.

☐ STEP 17.   Estimate the gain in precision of a balanced lattice design relative to the RCB design as:

$$R.E. = \frac{100\left[\text{Block(adj.) } SS + \text{Intrablock error } SS\right]}{k(k^2 - 1)(\text{Effective error } MS)}$$

$$= \frac{100(11,382 + 14,533)}{(4)(16 - 1)(369)} = 117\%$$

Table 2.12   Analysis of Variance (a 4 × 4 Balanced Lattice Design) of Tiller Number Data in Table 2.10[a]

Source of Variation	Degree of Freedom	Sum of Squares	Mean Square	Computed F[b]	Tabular F 5%	1%
Replication	4	5,946				
Treatment(unadj.)	15	26,995				
Block(adj.)	15	11,382	759			
Intrablock error	45	14,533	323			
Treatment(adj.)	(15)	—	1,602	4.34**	1.90	2.47
Effective error	(45)	—	369			
Total	79	58,856				

[a]$cv = 11.2\%$.
[b]** = significant at 1% level.

That is, the use of the 4 × 4 balanced lattice design is estimated to have increased the experimental precision by 17% over that which would have been obtained with a RCB design.

2.4.2 Partially Balanced Lattice

The partially balanced lattice design is similar to the balanced lattice design but allows for a more flexible choice of the number of replications. While the partially balanced lattice design requires that the number of treatments must be a perfect square and that the block size is equal to the square root of this treatment number, the number of replications is not prescribed as a function of the number of treatments. In fact, any number of replications can be used in a partially balanced lattice design.

With two replications, the partially balanced lattice design is referred to as a *simple lattice*; with three replications, a *triple lattice*; with four replications, a *quadruple lattice*; and so on. However, such flexibility in the choice of the number of replications results in a loss of symmetry in the arrangement of treatments over blocks (i.e., some treatment pairs never appear together in the same incomplete block). Consequently, the treatment pairs that are tested in the same incomplete block are compared with a level of precision that is higher than for those that are not tested in the same incomplete block. Because there is more than one level of precision for comparing treatment means, data analysis becomes more complicated.

2.4.2.1 *Randomization and Layout.* The procedures for randomization and layout of a partially balanced lattice design are similar to those for a balanced lattice design described in Section 2.4.1.1, except for the modification in the number of replications. For example, with a 3 × 3 simple lattice (i.e., a partially balanced lattice with two replications) the same procedures we described in Section 2.4.1.1 can be followed using only the first two replications. With a triple lattice (i.e., a partially balanced lattice with three replications) the first three replications of the basic plan of the corresponding balanced lattice design would be used.

When the number of replications (r) of a partially balanced lattice design exceeds three and is an even number, the basic plan can be obtained:

- as the first r replications of the basic plan of the balanced lattice design having the same number of treatments, or
- as the first r/p replications of the basic plan of the balanced lattice design having the same number of treatments, repeated p times (with rerandomization each time).

For example, for a 5 × 5 quadruple lattice design (i.e., a partially balanced lattice design with four replications) the basic plan can be obtained either as

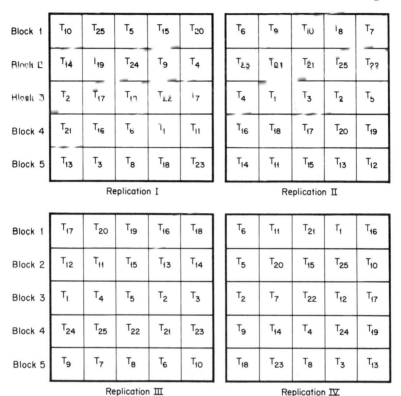

Figure 2.8 A sample layout of a 5 × 5 quadruple lattice design *with two repetitions* (replications I and IV; and replications II and III), involving 25 treatments ($T_1, T_2, \ldots, T_{25}$).

the first four replications of the 5 × 5 balanced lattice design or as the 5 × 5 simple lattice design repeated twice (i.e., $p = 2$).

In general, the procedure of using the basic plan without repetition is slightly preferred because it comes closer to the symmetry achieved in a balanced lattice design. For a partially balanced lattice design with p repetitions, the process of randomization will be done p times, separately and independently. For example, for the 5 × 5 quadruple lattice design with $p = 2$, the process of randomization is applied twice—as if there were two 5 × 5 simple lattice designs.

Two sample field layouts of a 5 × 5 quadruple lattice design, one with repetition and another without repetition, are shown in Figures 2.8 and 2.9.

2.4.2.2 Analysis of Variance.   The procedure for the analysis of variance of a partially balanced lattice design is discussed separately for a case with repetition and one without repetition. A 9 × 9 triple lattice design is used to illustrate the case without repetition; a 5 × 5 quadruple lattice is used to illustrate the case with repetition.

Block 1	T_{14}	T_{12}	T_{11}	T_{13}	T_{15}
Block 2	T_4	T_5	T_3	T_1	T_2
Block 3	T_{20}	T_{17}	T_{18}	T_{16}	T_{19}
Block 4	T_{22}	T_{24}	T_{25}	T_{21}	T_{23}
Block 5	T_8	T_7	T_{10}	T_9	T_6

Replication I

T_{17}	T_{23}	T_{10}	T_4	T_{11}
T_{15}	T_{22}	T_3	T_{16}	T_9
T_{21}	T_8	T_2	T_{14}	T_{20}
T_{24}	T_{12}	T_6	T_5	T_{18}
T_7	T_{13}	T_1	T_{25}	T_{19}

Replication II

Block 1	T_1	T_{12}	T_{20}	T_9	T_{23}
Block 2	T_8	T_5	T_{19}	T_{11}	T_{22}
Block 3	T_{16}	T_{24}	T_{10}	T_{13}	T_2
Block 4	T_7	T_4	T_{15}	T_{21}	T_{18}
Block 5	T_6	T_{14}	T_3	T_{17}	T_{25}

Replication III

T_6	T_{11}	T_1	T_{16}	T_{21}
T_{17}	T_2	T_{12}	T_{22}	T_7
T_9	T_4	T_{14}	T_{24}	T_{19}
T_{23}	T_8	T_3	T_{18}	T_{13}
T_{15}	T_{20}	T_{10}	T_{25}	T_5

Replication IV

Figure 2.9 A sample layout of a 5×5 quadruple lattice design *without repetition*, involving 25 treatments $(T_1, T_2, \ldots, T_{25})$.

2.4.2.2.1 Design without Repetition. To illustrate the analysis of variance of a partially balanced lattice design without repetition, we use a 9×9 triple lattice design that evaluates the performance of 81 rice varieties. The yield data, rearranged according to the basic plan of Appendix L, are given in Table 2.13. The steps in the analysis of variance procedure are:

□ STEP 1.   Calculate the block totals (B) and the replication totals (R) as shown in Table 2.13. Then, compute the grand total:

$$G = R_1 + R_2 + R_3$$

$$= 323.25 + 300.62 + 301.18 = 925.05$$

□ STEP 2.   Calculate the treatment totals (T) as shown in Table 2.14.

Table 2.13   Grain Yield Data from a Trial of 81 Upland Rice Varieties Conducted in a 9 × 9 Triple Lattice Design[a]

Block Number				Grain Yield, t/ha						Block Total (B)
					Rep. I					
	(1)	(2)	(3)	(4)	(5)	(6)	(7)	(8)	(9)	
1	2.70	1.60	4.45	2.91	2.78	3.32	1.70	4.72	4.79	28.97
	(10)	(11)	(12)	(13)	(14)	(15)	(16)	(17)	(18)	
2	4.20	5.22	3.96	1.51	3.48	4.69	1.57	2.61	3.16	30.40
	(19)	(20)	(21)	(22)	(23)	(24)	(25)	(26)	(27)	
3	4.63	3.33	6.31	6.08	1.86	4.10	5.72	5.87	4.20	42.10
	(28)	(29)	(30)	(31)	(32)	(33)	(34)	(35)	(36)	
4	3.74	3.05	5.16	4.76	3.75	3.66	4.52	4.64	5.36	38.64
	(37)	(38)	(39)	(40)	(41)	(42)	(43)	(44)	(45)	
5	4.76	4.43	5.36	4.73	5.30	3.93	3.37	3.74	4.06	39.68
	(46)	(47)	(48)	(49)	(50)	(51)	(52)	(53)	(54)	
6	3.45	2.56	2.39	2.30	3.54	3.66	1.20	3.34	4.04	26.48
	(55)	(56)	(57)	(58)	(59)	(60)	(61)	(62)	(63)	
7	3.99	4.48	2.69	3.95	2.59	3.99	4.37	4.24	3.70	34.00
	(64)	(65)	(66)	(67)	(68)	(69)	(70)	(71)	(72)	
8	5.29	3.58	2.14	5.54	5.14	5.73	3.38	3.63	5.08	39.51
	(73)	(74)	(75)	(76)	(77)	(78)	(79)	(80)	(81)	
9	3.76	6.45	3.96	3.64	4.42	6.57	6.39	3.39	4.89	43.47
Rep. total R_1										323.25
					Rep. II					
	(1)	(10)	(19)	(28)	(37)	(46)	(55)	(64)	(73)	
1	3.06	2.08	2.95	3.75	4.08	3.88	2.14	3.68	2.85	28.47
	(2)	(11)	(20)	(29)	(38)	(47)	(56)	(65)	(74)	
2	1.61	5.30	2.75	4.06	3.89	2.60	4.19	3.14	4.82	32.36
	(3)	(12)	(21)	(30)	(39)	(48)	(57)	(66)	(75)	
3	4.19	3.33	4.67	4.99	4.58	3.17	2.69	2.57	3.82	34.01
	(4)	(13)	(22)	(31)	(40)	(49)	(58)	(67)	(76)	
4	2.99	2.50	4.87	3.71	4.85	2.87	3.79	5.28	3.32	34.18
	(5)	(14)	(23)	(32)	(41)	(50)	(59)	(68)	(77)	
5	3.81	3.48	1.87	4.34	4.36	3.24	3.62	4.49	3.62	32.83
	(6)	(15)	(24)	(33)	(42)	(51)	(60)	(69)	(78)	
6	3.34	3.30	3.68	3.84	4.25	3.90	3.64	5.09	6.10	37.14
	(7)	(16)	(25)	(34)	(43)	(52)	(61)	(70)	(79)	
7	2.98	2.69	5.55	3.52	4.03	1.20	4.36	3.18	6.77	34.28
	(8)	(17)	(26)	(35)	(44)	(53)	(62)	(71)	(80)	
8	4.20	2.69	5.14	4.32	3.47	3.41	3.74	3.67	2.27	32.91
	(9)	(18)	(27)	(36)	(45)	(54)	(63)	(72)	(81)	
9	4.75	2.59	3.94	4.51	3.10	3.59	2.70	4.40	4.86	34.44
Rep. total R_2										300.62

Table 2.13   (*Continued*)

Block Number	Grain Yield, t/ha									Block Total (B)
					Rep. III					
	(1)	(12)	(20)	(34)	(45)	(53)	(58)	(70)	(77)	
1	3.52	2.18	3.50	3.30	3.88	2.45	3.75	4.45	4.14	31.17
	(2)	(10)	(21)	(35)	(43)	(54)	(59)	(67)	(78)	
2	.79	3.58	4.83	3.63	3.02	4.20	3.59	5.06	6.51	35.21
	(3)	(11)	(19)	(36)	(44)	(52)	(60)	(68)	(76)	
3	4.69	5.33	4.43	5.31	4.13	1.98	4.66	4.50	4.50	39.53
	(4)	(15)	(23)	(28)	(39)	(47)	(61)	(72)	(80)	
4	3.06	4.30	2.02	3.57	5.80	2.58	4.27	4.84	2.74	33.18
	(5)	(13)	(24)	(29)	(37)	(48)	(62)	(70)	(81)	
5	3.79	.88	3.40	4.92	2.12	1.89	3.73	3.51	3.50	27.74
	(6)	(14)	(22)	(30)	(38)	(46)	(63)	(71)	(79)	
6	3.34	3.94	5.72	5.34	4.47	4.18	2.70	3.96	3.48	37.13
	(7)	(18)	(26)	(31)	(42)	(50)	(55)	(66)	(74)	
7	2.35	2.87	5.50	2.72	4.20	2.87	2.99	1.62	5.33	30.45
	(8)	(16)	(27)	(32)	(40)	(51)	(56)	(64)	(75)	
8	4.51	1.26	4.20	3.19	4.76	3.35	3.61	4.52	3.38	32.78
	(9)	(17)	(25)	(33)	(41)	(49)	(57)	(65)	(73)	
9	4.21	3.17	5.03	3.34	5.31	3.05	3.19	2.63	4.06	33.99
Rep. total R_3										301.18

[a] The values enclosed in parentheses correspond to the treatment numbers.

□ STEP 3.   Construct an outline of the analysis of variance of a 9×9 triple lattice design as:

Source of Variation	Degree of Freedom	Sum of Squares	Mean Square
Replication	$r - 1 = 2$		
Block(adj.)	$r(k - 1) = 24$		
Treatment(unadj.)	$k^2 - 1 = 80$		
Intrablock error	$(k - 1)(rk - k - 1) = 136$		
Treatment(adj.)	$[(k^2 - 1) = (80)]$		
Total	$(r)(k^2) - 1 = 242$		

Here, r is the number of replications and k is the block size.

Table 2.14 Treatment Totals Computed from Data in Table 2.13

Treatment No.	Total (T)	Treatment No.	Total (T)	Treatment No.	Total (T)	Treatment No.	Total (T)	Treatment No.	Total (T)	Treatment No.	Total (T)	Treatment No.	Total (T)	Treatment No.	Total (T)	Treatment No.	Total (T)
1	9.28	2	4.00	3	13.33	4	8.96	5	10.38	6	10.00	7	7.03	8	13.43	9	13.75
10	9.86	11	15.85	12	9.47	13	4.89	14	10.90	15	12.29	16	5.52	17	8.47	18	8.62
19	12.01	20	9.58	21	15.81	22	16.67	23	5.75	24	11.18	25	16.30	26	16.5_	27	12.34
28	11.06	29	12.03	30	15.49	31	11.19	32	11.28	33	10.84	34	11.34	35	12.59	36	15.18
37	10.96	38	12.79	39	15.74	40	14.34	41	14.97	42	12.38	43	10.42	44	11.34	45	11.04
46	11.51	47	7.74	48	7.45	49	8.22	50	9.65	51	10.91	52	4.38	53	9.20	54	11.83
55	9.12	56	12.28	57	8.57	58	11.49	59	9.80	60	12.29	61	13.00	62	11.71	63	9.10
64	13.49	65	9.35	66	6.33	67	15.88	68	14.13	69	15.27	70	10.07	71	11.26	72	14.32
73	10.67	74	16.60	75	11.16	76	11.46	77	12.18	78	19.18	79	16.64	80	8.40	81	13.25

□ STEP 4.   Compute the total SS, replication SS, and treatment (unadj.) SS in the standard manner:

$$C.F. = \frac{G^2}{(r)(k^2)}$$

$$= \frac{(925.05)^2}{(3)(81)} = 3,521.4712$$

Total $SS = \sum X^2 - C.F.$

$$= \left[(2.70)^2 + (1.60)^2 + \cdots + (4.06)^2\right] - 3,521.4712$$

$$= 308.9883$$

Replication $SS = \dfrac{\sum R^2}{k^2} - C.F.$

$$= \frac{(323.25)^2 + (300.62)^2 + (301.18)^2}{81} - 3,521.4712$$

$$= 4.1132$$

Treatment(unadj.) $SS = \dfrac{\sum T^2}{r} - C.F.$

$$= \frac{(9.28)^2 + (4.00)^2 + \cdots + (13.25)^2}{3} - 3,521.4712$$

$$= 256.7386$$

□ STEP 5.   For each block, calculate:

$$C_b = M - rB$$

where M is the sum of treatment totals for all treatments appearing in that particular block and B is the block total. For example, block 2 of replication II contained treatments 2, 11, 20, 29, 38, 47, 56, 65, and 74 (Table 2.13). Hence, the M value for block 2 of replication II is:

$$M = T_2 + T_{11} + T_{20} + T_{29} + T_{38} + T_{47} + T_{56} + T_{65} + T_{74}$$

$$= 4.00 + 15.85 + \cdots + 16.60 = 100.22$$

and the corresponding C_b value is:

$$C_b = 100.22 - 3(32.36) = 3.14$$

The C_b values for the 27 blocks are presented in Table 2.15.

Table 2.15 The C_b Values Computed from a 9×9 Triple Lattice Design Data in Tables 2.13 and 2.14

Rep. I		Rep. II		Rep. III	
Block Number	C_b	Block Number	C_b	Block Number	C_b
1	3.25	1	12.55	1	5.34
2	− 5.33	2	3.14	2	3.74
3	− 10.15	3	1.32	3	− 8.62
4	− 4.92	4	0.56	4	− 2.28
5	− 5.06	5	0.55	5	8.70
6	1.45	6	2.92	6	2.97
7	− 4.64	7	− 8.14	7	6.08
8	− 8.43	8	4.18	8	6.41
9	− 10.87	9	6.11	9	− 0.83
Total	− 44.70	Total	23.19	Total	21.51

□ STEP 6.   For each replication, calculate the sum of C_b values over all blocks (i.e., R_c):

For replication I,

$$R_c(\text{I}) = 3.25 - 5.33 + \cdots - 10.87 = -44.70$$

For replication II,

$$R_c(\text{II}) = 12.55 + 3.14 + \cdots + 6.11 = 23.19$$

For replication III,

$$R_c(\text{III}) = 5.34 + 3.74 + \cdots - 0.83 = 21.51$$

Note that the R_c values should add to zero (i.e., $-44.70 + 23.19 + 21.51 = 0$).

□ STEP 7.   Calculate the block(adj.) SS as:

$$\text{Block(adj.) } SS = \frac{\sum C_b^2}{(k)(r)(r-1)} - \frac{\sum R_c^2}{(k^2)(r)(r-1)}$$

$$= \frac{(3.25)^2 + (-5.33)^2 + \cdots + (-0.83)^2}{(9)(3)(3-1)}$$

$$- \frac{(-44.70)^2 + (23.19)^2 + (21.51)^2}{(81)(3)(3-1)}$$

$$= 12.1492$$

☐ STEP 8.   Calculate the intrablock error SS as:

$$\text{Intrablock error } SS = \text{Total } SS - \text{Replication } SS - \text{Treatment(unadj.) } SS$$

$$- \text{Block(adj.) } SS$$

$$= 308.9883 - 4.1132 - 256.7386 - 12.1492$$

$$= 35.9873$$

☐ STEP 9.   Calculate the intrablock error mean square and block(adj.) mean square as:

$$\text{Intrablock error } MS = \frac{\text{Intrablock error } SS}{(k-1)(rk-k-1)}$$

$$= \frac{35.9873}{(9-1)[(3)(9)-9-1]} = 0.2646$$

$$\text{Block(adj.) } MS = \frac{\text{Block(adj.) } SS}{r(k-1)}$$

$$= \frac{12.1492}{3(9-1)} = 0.5062$$

☐ STEP 10.   Calculate the adjustment factor μ. For a triple lattice, the formula is

$$\mu = \frac{\dfrac{1}{MSE} - \dfrac{2}{3MSB - MSE}}{k\left(\dfrac{2}{MSE} + \dfrac{2}{3MSB - MSE}\right)}$$

where MSE is the intrablock error mean square and MSB is the block(adj.) mean square.

Note that if MSB is less than MSE, μ is taken to be zero and no further adjustment is made. The F test for significance of treatment effect is made in the usual manner as the ratio of treatment(unadj.) MS and intrablock error MS, and steps 10 to 14 and step 17 can be ignored.

For our example, the *MSB* value of 0.5062 is larger than the *MSE* value of 0.2646; and, thus, the adjustment factor is computed as:

$$\mu = \frac{\left[\dfrac{1}{0.2646} - \dfrac{2}{3(0.5062)} - 0.2646\right]}{9\left[\dfrac{2}{0.2646} + \dfrac{2}{3(0.5062) - 0.2646}\right]}$$

$$= 0.0265$$

☐ STEP 11. For each treatment, calculate the adjusted treatment total T' as:

$$T' = T + \mu \sum C_b$$

where the summation runs over all blocks in which the particular treatment appears. For example, the adjusted treatment total for treatment number 2 is computed as:

$$T'_2 = 4.00 + 0.0265(3.25 + 3.14 + 3.74) = 4.27$$

Note that for mean comparisons (see Chapter 5) the adjusted treatment means are used. They are computed simply by dividing these individual adjusted treatment totals by the number of replications.

☐ STEP 12. Compute the adjusted treatment *SS*:

Treatment(adj.) SS = Treatment(unadj.) SS − A

$$A = \left[\frac{1}{MSE} - \frac{2}{(3MSB - MSE)}\right]$$

$$\times \left[(MSE)B_u - (k-1)(MSE)(rMSB - MSE)\right]$$

$$B_u = \frac{\sum B^2}{k} - \frac{\sum R^2}{k^2}$$

For our example,

$$B_u = \frac{(28.97)^2 + (30.40)^2 + \cdots + (33.99)^2}{9}$$

$$- \frac{(323.25)^2 + (300.62)^2 + (301.18)^2}{81}$$

$$= 49.4653$$

$$A = \left[\frac{1}{0.2646} - \frac{2}{3(0.5062) - 0.2646} \right]$$

$$\times \{ 0.2646(49.4653)$$

$$- 8(0.2646)[3(0.5062) - 0.2646] \}$$

$$= 22.7921$$

$$\text{Treatment(adj.) } SS = 256.7386 - 22.7921$$

$$= 233.9465$$

☐ STEP 13. Compute the treatment(adj.) mean square as:

$$\text{Treatment(adj.) } MS = \frac{\text{Treatment(adj.) } SS}{k^2 - 1}$$

$$= \frac{233.9465}{80} = 2.9243$$

☐ STEP 14. Compute the F test for testing the significance of treatment difference as:

$$F = \frac{\text{Treatment(adj.) } MS}{\text{Intrablock error } MS}$$

$$= \frac{2.9243}{0.2646} = 11.05$$

Compute the corresponding cv value as:

$$cv = \frac{\sqrt{\text{Intrablock } MS}}{\text{Grand mean}} \times 100$$

$$= \frac{\sqrt{0.2646}}{3.81} \times 100 = 13.5\%$$

☐ STEP 15. Compare the computed F value to the tabular F values of Appendix E, with $f_1 = (k^2 - 1) = 80$ and $f_2 = (k - 1)(rk - k - 1) = 136$ degrees of freedom. Because the computed F value is greater than the tabular F value at the 1% level of significance, the F test indicates a highly significant treatment difference.

☐ STEP 16. Enter all values computed in steps 4 to 9 and 12 to 15 in the analysis of variance outline of step 3. The final result is shown in Table 2.16.

Table 2.16 Analysis of Variance (a 9 × 9 Triple Lattice Design) of Data in Table 2.13[a]

Source of Variation	Degree of Freedom	Sum of Squares	Mean Square	Computed F^b	Tabular F 5%	Tabular F 1%
Replication	2	4.1132				
Block(adj.)	24	12.1492	0.5062			
Treatment(unadj.)	80	256.7386				
Intrablock error	136	35.9873	0.2646			
Treatment(adj.)	(80)	233.9465	2.9243	11.05**	1.38	1.57
Total	242	308.9883				

[a]$cv = 13.5\%$.
[b]** = significant at 1% level.

□ STEP 17. Estimate the gain in precision of a partially balanced lattice design relative to the RCB design, as follows:

A. Compute the effective error mean square. For a partially balanced lattice design, there are two error terms involved: one for comparisons between treatments appearing in the same block [i.e., Error $MS(1)$] and another for comparisons between treatments not appearing in the same block [i.e., Error $MS(2)$]. For a triple lattice, the formulas are:

$$\text{Error } MS(1) = \frac{MSE}{k} \left[\frac{\dfrac{6}{MSE}}{\dfrac{2}{MSE} + \dfrac{2}{3MSB - MSE}} + (k - 2) \right]$$

$$\text{Error } MS(2) = \frac{MSE}{k} \left[\frac{\dfrac{9}{MSE}}{\dfrac{2}{MSE} + \dfrac{2}{3MSB - MSE}} + (k - 3) \right]$$

With a large experiment, these two values may not differ much. And, for simplicity, the *average error MS* may be computed and used for comparing any pair of means (i.e., without the need to distinguish whether or not the pair of treatments appeared together in the same block or not). For a triple lattice, the formula is:

$$\text{Av. error } MS = \frac{MSE}{(k + 1)} \left[\frac{\dfrac{9}{MSE}}{\dfrac{2}{MSE} + \dfrac{2}{3MSB - MSE}} + (k - 2) \right]$$

For our example, the value of the two error mean squares are computed as:

$$\text{Error } MS(1) = \frac{(0.2646)}{9} \left[\frac{\dfrac{6}{0.2646}}{\dfrac{2}{0.2646} + \dfrac{2}{3(0.5062) - 0.2646}} + 7 \right]$$

$$= 0.2786$$

$$\text{Error } MS(2) = \frac{(0.2646)}{9} \left[\frac{\dfrac{9}{0.2646}}{\dfrac{2}{0.2646} + \dfrac{2}{3(0.5062) - 0.2646}} + 6 \right]$$

$$= 0.2856$$

As expected, the two values of error MS do not differ much and, hence, the average error MS can be used. It is computed as:

$$\text{Av. error } MS = \frac{0.2646}{10} \left[\frac{\dfrac{9}{0.2646}}{\dfrac{2}{0.2646} + \dfrac{2}{3(0.5062) - 0.2646}} + 7 \right]$$

$$= 0.2835$$

B. Compute the efficiency of the partially balanced lattice design relative to a comparable RCB design as:

$$R.E. = \left[\frac{\text{Block(adj.) } SS + \text{Intrablock error } SS}{r(k - 1) + (k - 1)(rk - k - 1)} \right] \left[\frac{100}{\text{Error } MS} \right]$$

For our example, the three values of the relative efficiency corresponding to Error $MS(1)$, Error $MS(2)$, and Av. error MS are computed as:

$$R.E.(1) = \left(\frac{12.1492 + 35.9873}{24 + 136} \right) \left(\frac{100}{0.2786} \right) = 108.0\%$$

$$R.E.(2) = \left(\frac{12.1492 + 35.9873}{24 + 136} \right) \left(\frac{100}{0.2856} \right) = 105.3\%$$

$$R.E.(\text{av.}) = \left(\frac{12.1492 + 35.9873}{24 + 136} \right) \left(\frac{100}{0.2835} \right) = 106.1\%$$

2.4.2.2.2 Design with Repetition. For the analysis of variance of a partially balanced lattice design with repetition, we use a 5×5 quadruple lattice whose basic plan is obtained by repeating a simple lattice design (i.e., base design) twice. Data on grain yield for the 25 rice varieties used as treatments (rearranged according to the basic plan) are shown in Table 2.17. Note that replications I and II are from the first two replications of the basic plan of the 5×5 balanced lattice design (Appendix L) and replications III and IV are repetition of replications I and II.

The steps involved in the analysis of variance are:

☐ STEP 1. Calculate the block totals (B) and replication totals (R) as shown in Table 2.17. Then, compute the grand total (G) as $G = \Sigma R = 147,059 + 152,078 + 151,484 + 155,805 = 606,426$.

☐ STEP 2. Calculate the treatment totals (T) as shown in Table 2.18.

☐ STEP 3. Construct an outline of the analysis of variance of a partially balanced lattice design with repetition as:

Source of Variation	Degree of Freedom	Sum of Squares	Mean Square
Replication	$(n)(p) - 1 = 3$		
Block(adj.)	$(n)(p)(k - 1) = 16$		
Component(a)	$[n(p - 1)(k - 1) = 8]$		
Component(b)	$[n(k - 1) = 8]$		
Treatment(unadj.)	$(k^2 - 1) = 24$		
Intrablock error	$(k - 1)(npk - k - 1) = 56$		
Treatment(adj.)	$[k^2 - 1 = 24]$		
Total	$(n)(p)(k^2) - 1 = 99$		

Here, n is the number of replications in the base design and p is the number of repetitions (i.e., the number of times the base design is repeated). As before, k is the block size. In our example, the base design is simple lattice so that $n = 2$ and, because this base design is used twice, $p = 2$.

☐ STEP 4. Compute the total SS, replication SS, and treatment(unadj.) SS, in the standard manner as:

$$C.F. = \frac{G^2}{(n)(p)(k^2)}$$

$$= \frac{(606,426)^2}{(2)(2)(25)} = 3,677,524,934$$

Table 2.17 Grain Yield Data from a Rice Variety Trial Conducted in a 5 × 5 Quadruple Lattice Design with Repetition

Block Number	Rep. I			Rep. II			Rep. III			Rep. IV		
	Treatment Number	Yield, kg/ha	Block Total (B)	Treatment Number	Yield, kg/ha	Block Total (B)	Treatment Number	Yield, kg/ha	Block Total (B)	Treatment Number	Yield, kg/ha	Block Total (B)
1	1	4,723		1	6,262		1	5,975		1	5,228	
	2	4,977		6	5,690		2	5,915		6	5,302	
	3	6,247		11	6,498		3	6,914		11	5,190	
	4	5,325		16	8,011		4	6,389		16	7,127	
	5	7,139	28,411	21	5,887	32,348	5	7,542	32,735	21	5,323	28,170
2	6	5,444		2	5,038		6	4,750		2	5,681	
	7	5,567		7	4,615		7	5,983		7	6,146	
	8	5,809		12	5,520		8	5,339		12	6,032	
	9	5,086		17	6,063		9	4,615		17	7,066	
	10	6,849	28,755	22	6,486	27,722	10	5,336	26,023	22	6,680	31,605
3	11	5,237		3	6,057		11	5,073		3	6,750	
	12	5,174		8	6,397		12	6,110		8	6,567	
	13	5,395		13	5,214		13	6,001		13	5,786	
	14	5,112		18	7,093		14	5,486		18	7,159	
	15	5,637	26,555	23	7,002	31,763	15	6,415	29,085	23	7,268	33,530

4	16	5,793		4	5,291		16	6,064		4	6,02⊑
	17	6,008		9	4,864		17	6,405		9	5,13≡
	18	6,864		14	5,453		18	6,856		14	6,413
	19	5,026		19	4,917		19	4,654		19	5,75⊐
	20	6,348		24	6,318		20	5,986		24	6,85⊑
		30,039			26,843			29,965			⊐0,185
5	21	5,321		5	7,685		21	5,750		5	7,1⊑3
	22	6,870		10	5,985		22	6,539		10	5,62≡
	23	7,512		15	6,107		23	7,576		15	6,31⊐
	24	6,648		20	6,710		24	7,372		20	6,52⊐
	25	6,948		25	6,915		25	6,439		25	6,6⊏⊏
		33,299			33,402			33,676			⊏⊑,315
Rep. total (R)		147,059			152,078			151,484			⊏5,805

Table 2.18   Treatment Totals Computed from Data in Table 2.17

Treatment		Treatment		Treatment		Treatment		Treatment	
Number	Total (T)	Number	Total (T)	Number	Total (T)	Number	Total (T)	Number	Total (T)
1	22,188	2	21,611	3	25,968	4	23,025	5	29,539
6	21,186	7	22,311	8	24,112	9	19,701	10	23,796
11	21,998	12	22,836	13	22,396	14	22,464	15	24,469
16	26,995	17	25,542	18	27,972	19	20,357	20	25,573
21	22,281	22	26,575	23	29,358	24	27,194	25	26,979

Total $SS = \sum X^2 - C.F.$

$$= \left[(4{,}723)^2 + (4{,}977)^2 + \cdots + (6{,}677)^2 \right] - 3{,}677{,}524{,}934$$

$$= 63{,}513{,}102$$

Replication $SS = \dfrac{\sum R^2}{k^2} - C.F.$

$$= \frac{(147{,}059)^2 + (152{,}078)^2 + (151{,}484)^2 + (155{,}805)^2}{25}$$

$$- 3{,}677{,}524{,}934$$

$$= 1{,}541{,}779$$

Treatment(unadj.) $SS = \dfrac{\sum T^2}{(n)(p)} - C.F.$

$$= \frac{(22{,}188)^2 + (21{,}611)^2 + \cdots + (26{,}979)^2}{(2)(2)}$$

$$- 3{,}677{,}524{,}934$$

$$= 45{,}726{,}281$$

□ STEP 5.   For each block in each repetition, compute the S value as the sum of block totals over all replications in that repetition and, for each S value, compute the corresponding C value, as:

$$C = \sum T - nS$$

where n is as defined in step 3, T is the treatment total, and the summation is made only over the treatments appearing in the block corresponding to the particular S value involved.

For our example, there are two repetitions, each consisting of two replications—replication I and replication III in repetition 1 and replication II and replication IV in repetition 2. Hence, the first S value, corresponding to block 1 from replications I and III, is computed as $S = 28,411 + 32,735 = 61,146$. Because the five treatments in block 1 of repetition 1 are treatments 1, 2, 3, 4, and 5 (Table 2.17), the first C value, corresponding to the first S value, is computed as:

$$C = (22,188 + 21,611 + 25,968 + 23,025 + 29,539) - 2(61,146)$$

$$= 122,331 - 122,292 = 39$$

The computations of all S values and C values are illustrated and shown in Table 2.19. Compute the total C values over all blocks in a repetition (i.e., $R_j, j = 1, \ldots, p$). For our example, the two R_j values are 9,340 for repetition 1 and $-9,340$ for repetition 2. The sum of all R_j values must be zero.

□ STEP 6. Let B denote the block total; D, the sum of S values for each repetition; and A, the sum of block totals for each replication. Compute the

Table 2.19 Computation of the S Values and the Corresponding C Values, Based on Block Totals (Table 2.17) Rearranged in Pairs of Blocks Containing the Same Set of Treatments

Block Number	Block Total		S	C
	1st Replication	2nd Replication		
Repetition 1				
1	28,411	32,735	61,146	39
2	28,755	26,023	54,778	1,550
3	26,555	29,085	55,640	2,883
4	30,039	29,965	60,004	6,431
5	33,299	33,676	66,975	−1,563
Total	147,059	151,484	298,543	9,340
Repetition 2				
1	32,348	28,170	60,518	−6,388
2	27,722	31,605	59,327	221
3	31,763	33,530	65,293	−780
4	26,843	30,185	57,028	−1,315
5	33,402	32,315	65,717	−1,078
Total	152,078	155,805	307,883	−9,340

two components, (a) and (b), of the block(adj.) SS as:

A.   Component(a) $SS = X - Y - Z$

  where

$$X = \frac{\sum B^2}{k} - \frac{\sum\limits_{j=1}^{p} D_j^2}{pk^2}$$

$$Y = \frac{\sum S^2}{pk} - \frac{\sum\limits_{j=1}^{p} D_j^2}{pk^2}$$

$$Z = \frac{\sum A^2}{k^2} - \frac{\sum\limits_{j=1}^{p} D_j^2}{pk^2}$$

For our example, the parameters and the component(a) SS are computed as:

$$X = \frac{(28{,}411)^2 + (28{,}755)^2 + \cdots + (32{,}315)^2}{5}$$

$$- \frac{(298{,}543)^2 + (307{,}883)^2}{(2)(25)}$$

$$= 3{,}701{,}833{,}230 - 3{,}678{,}397{,}290$$

$$= 23{,}435{,}940$$

$$Y = \frac{(61{,}146)^2 + (54{,}778)^2 + \cdots + (65{,}717)^2}{(2)(5)} - 3{,}678{,}397{,}290$$

$$= 15{,}364{,}391$$

$$Z = \frac{(147{,}059)^2 + (152{,}078)^2 + (151{,}484)^2 + (155{,}805)^2}{25}$$

$$- 3{,}678{,}397{,}290$$

$$= 669{,}423$$

Component(a) $SS = 23{,}435{,}940 - 15{,}364{,}391 - 669{,}423$

$$= 7{,}402{,}126$$

B.   Component(b) $SS = \dfrac{\sum C^2}{(k)(n)(p)(n-1)} - \dfrac{\sum R_j^2}{(k^2)(n)(n)(n-1)}$

$$= \frac{(29)^2 + (1,550)^2 + \cdots + (-1,078)^2}{(5)(2)(2)(1)}$$

$$- \frac{(9,340)^2 + (-9,340)^2}{(25)(2)(2)(1)}$$

$$= 3,198,865$$

☐ STEP 7.   Compute the block(adj.) SS as the sum of component(a) SS and component(b) SS computed in step 6:

$$\text{Block(adj.) } SS = \text{Component}(a) \text{ } SS + \text{Component}(b) \text{ } SS$$

$$= 7,402,126 + 3,198,865$$

$$= 10,600,991$$

☐ STEP 8.   Compute the intrablock error SS as:

$$\text{Intrablock error } SS = \text{Total } SS - \text{Replication } SS - \text{Treatment(unadj.) } SS$$

$$- \text{Block(adj.) } SS$$

$$= 63,513,102 - 1,541,779 - 45,726,281 - 10,600,991$$

$$= 5,644,051$$

☐ STEP 9.   Compute the block(adj.) mean square and the intrablock error mean square as:

$$MSB = \frac{\text{Block(adj.) } SS}{np(k-1)}$$

$$= \frac{10,600,991}{(2)(2)(4)} = 662,562$$

$$MSE = \frac{\text{Intrablock error } SS}{(k-1)(npk-k-1)}$$

$$= \frac{5,644,051}{(4)(20-5-1)} = 100,787$$

□ STEP 10.   Compute the adjustment factor μ as:

$$\mu = \frac{p(MSB - MSE)}{k[p(n-1)MSB + (p-1)MSE]}$$

$$= \frac{2(662{,}562 - 100{,}787)}{5[2(662{,}562) + (100{,}787)]}$$

$$= 0.15759$$

□ STEP 11.   For each treatment, compute the adjusted treatment total as:

$$T' = T + \mu\sum C$$

where the summation runs over all blocks in which the particular treatment appears. For example, using the data of Tables 2.18 and 2.19, the adjusted treatment total for treatment number 1, which appeared in block 1 of both repetitions, is computed as:

$$T' = 22{,}188 + 0.15759(39 - 6{,}388) = 21{,}187$$

The results of all T' values are shown in Table 2.20.

□ STEP 12.   Compute the adjusted treatment sum of squares as:

Treatment(adj.) SS = Treatment(unadj.) $SS - A$

$$A = k(n-1)\mu\left[\frac{(n)(Y)}{(n-1)(1+k\mu)} - \text{Component}(b)\,SS\right]$$

where Y is as defined by formula in step 6.

Table 2.20   Adjusted Treatment Totals Computed from Data in Tables 2.18 and 2.19

Treatment		Treatment		Treatment		Treatment		Treatment	
Number	Total (T')	Number	Total (T')	Number	Total (T')	Number	Total (T')	Number	Total (T')
1	21,187	2	21,652	3	25,851	4	22,824	5	29,375
6	20,424	7	22,590	8	24,233	9	19,738	10	23,870
11	21,446	12	23,325	13	22,727	14	22,711	15	24,754
16	27,002	17	26,590	18	28,863	19	21,163	20	26,417
21	21,028	22	26,364	23	28,989	24	26,740	25	26,563

For our example, we have

$$A = 5(1)(0.15739)\left\{ \frac{?(15,364,391)}{[1 + 5(0.15719)]} - 3,198,865 \right\}$$

$$= 11,021,636$$

$$\text{Treatment(adj.) } SS = 45,726,281 - 11,021,636$$

$$= 34,704,645$$

☐ STEP 13. Compute the adjusted treatment mean square as:

$$\text{Treatment(adj.) } MS = \frac{\text{Treatment(adj.) } SS}{k^2 - 1}$$

$$= \frac{34,704,645}{25 - 1}$$

$$= 1,446,027$$

☐ STEP 14. Compute the *F* value as:

$$F = \frac{\text{Treatment(adj.) } MS}{\text{Intrablock error } MS}$$

$$= \frac{1,446,027}{100,787} = 14.35$$

Compute the corresponding *cv* value as:

$$cv = \frac{\sqrt{\text{Intrablock error } MS}}{\text{Grand mean}} \times 100$$

$$= \frac{\sqrt{100,787}}{6,064} \times 100 = 5.2\%$$

☐ STEP 15. Compare the computed *F* value with the tabular *F* value, from Appendix E, with $f_1 = (k^2 - 1) = 24$ and $f_2 = (k - 1)(npk - k - 1) = 56$ degrees of freedom, at a desired level of significance. Because the computed *F* value is greater than the corresponding tabular *F* value at the 1% level of significance, a highly significant difference among treatments is indicated.

☐ STEP 16. Enter all values computed in steps 4 to 9 and 12 to 14 in the analysis of variance outline of step 3. The final result is shown in Table 2.21.

Table 2.21   Analysis of Variance (a 5 × 5 Quadruple Lattice Design) of Data in Table 2.17[a]

Source of Variation	Degree of Freedom	Sum of Squares	Mean Square	Computed F^b	Tabular F 5%	1%
Replication	3	1,541,779				
Block(adj.)	16	10,600,991	662,562			
Component(*a*)	(8)	7,402,126				
Component(*b*)	(8)	3,198,865				
Treatment(unadj.)	24	45,726,281				
Intrablock error	56	5,644,051	100,787			
Treatment(adj.)	(24)	34,704,645	1,446,027	14.35**	1.72	2.14
Total	99	63,513,102				

[a]$cv = 5.2\%$.
[b]** = significant at 1% level.

☐ STEP 17.   Compute the values of the two effective error mean square as:

A.   For comparing treatments appearing in the same block:

$$\text{Error } MS(1) = MSE\left[1 + (n - 1)\mu\right]$$

$$= 100,787[1 + (2 - 1)(0.15759)]$$

$$= 116,670$$

B.   For comparing treatments not appearing in the same block:

$$\text{Error } MS(2) = MSE(1 + n\mu)$$

$$= 100,787[1 + 2(0.15759)]$$

$$= 132,553$$

Note that when the average effective error *MS* is to be used (see step 17 of Section 2.4.2.2.1), compute it as:

$$\text{Av. error } MS = MSE\left[1 + \frac{(n)(k)(\mu)}{k + 1}\right]$$

$$= 100,787\left[1 + \frac{2(5)(0.15759)}{6}\right]$$

$$= 127,259$$

☐ STEP 18. Compute the efficiency relative to the RCB design as:

$$R.E. = \left| \frac{\text{Block(adj.) } SS + \text{Intrablock error } SS}{(n)(p)(k-1) + (k-1)(np/k \quad \lambda \quad 1)} \right| \left[\frac{100}{\text{Error } MS} \right]$$

where Error *MS* refers to the appropriate effective error *MS*.

For our example, the three values of the relative efficiency corresponding to Error *MS*(1), Error *MS*(2), and Av. error *MS* are computed as:

$$R.E.\,(1) = \left[\frac{10,600,991 + 5,644,051}{72} \right] \left[\frac{100}{116,670} \right]$$

$$= 193.4\%$$

$$R.E.\,(2) = \left[\frac{10,600,991 + 5,644,051}{72} \right] \left[\frac{100}{132,553} \right]$$

$$= 170.2\%$$

$$R.E.\,(\text{av.}) = \left(\frac{10,600,991 + 5,644,051}{72} \right) \left(\frac{100}{127,259} \right)$$

$$= 177.3\%$$

2.5 GROUP BALANCED BLOCK DESIGN

The primary feature of the group balanced block design is the grouping of treatments into homogeneous blocks based on selected characteristics of the treatments. Whereas the lattice design achieves homogeneity within blocks by grouping *experimental plots* based on some known patterns of heterogeneity in the experimental area, the group balanced block design achieves the same objective by grouping *treatments* based on some known characteristics of the treatments.

In a group balanced block design, treatments belonging to the same group are always tested in the same block, but those belonging to different groups are never tested together in the same block. Hence, the precision with which the different treatments are compared is not the same for all comparisons. Treatments belonging to the same group are compared with a higher degree of precision than those belonging to different groups.

The group balanced block design is commonly used in variety trials where varieties with similar morphological characters are put together in the same group. Two of the most commonly used criteria for grouping of varieties are:

- Plant height, in order to avoid the expected large competition effects (see Chapter 13, Section 13.1.2) when plants with widely different heights are grown in adjacent plots.

- Growth duration, in order to minimize competition effects and to facilitate harvest operations.

Another type of trials using the group balanced block design is that involving chemical insect control in which treatments may be subdivided into similar spray operations to facilitate the field application of chemicals.

We outline procedures for randomization, layout, and analysis of variance for a group balanced block design, using a trial involving 45 rice varieties with three replications. Based on growth duration, varieties are divided into group *A* for varieties with less than 105 days in growth duration, group *B* for 105 to 115 days, and group *C* for longer than 115 days. Each group consists of 15 varieties.

2.5.1 Randomization and Layout

The steps involved in the randomization and layout are:

☐ STEP 1.   Based on the prescribed grouping criterion, group the treatments into *s* groups, each consisting of *t/s* treatments, where *t* is the total number of treatments. For our example, the varieties are grouped into three groups, *A*, *B*, and *C* each consisting of 15 varieties, according to their expected growth duration.

☐ STEP 2.   Divide the experimental area into *r* replications, each consisting of *t* experimental plots. For our example, the experimental area is divided into three replications, each consisting of (3)(15) = 45 experimental plots.

☐ STEP 3.   Divide each replication into *s* blocks, each consisting of *t/s* experimental plots. For our example, each of the three replications is divided into three blocks, each consisting of 45/3 = 15 experimental plots.

☐ STEP 4.   Using one of the randomization schemes described in Section 2.1.1, assign the *s* groups at random to the *s* blocks of the first replication. Then, independently repeat the process for the remaining replications.

For our example, the varietal groups *A*, *B*, and *C* are assigned at random to the three blocks of replication I, then replication II, and finally replication III. The result is shown in Figure 2.10.

☐ STEP 5.   To each of the three blocks per replication, assign at random the *t/s* treatments belonging to the group that was assigned in step 4 to the particular block. For our example, starting with the first block of replication I, randomly assign the 15 varieties of group *A* to the 15 plots in the block. Repeat this process for the remaining eight blocks, independently of each other. The final result is shown in Figure 2.11.

2.5.2 Analysis of Variance

The steps in the analysis of variance of a group balanced block design are shown below using the data in Table 2.22 and the layout in Figure 2.11.

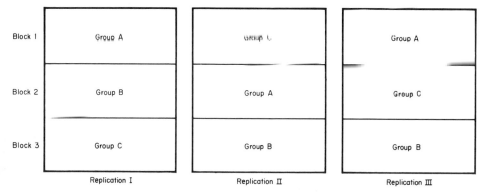

Figure 2.10 Random assignment of three groups of varieties (*A*, *B*, and *C*) into three blocks in each of the three replications, representing the first step in the randomization process of a group balanced block design.

7	8	1	4	14
12	3	13	10	9
6	11	2	5	15
21	20	27	18	22
16	25	17	24	30
26	19	28	29	23
41	31	32	44	43
36	37	40	35	38
33	39	42	45	34

Replication I

38	39	43	44	33
31	42	36	32	45
40	34	35	41	37
1	11	9	4	6
12	10	7	15	13
3	2	8	5	14
17	27	22	25	28
26	18	21	19	20
16	23	24	30	29

Replication II

2	1	13	5	15
11	4	3	14	7
9	12	10	6	8
31	33	35	38	44
36	34	37	42	41
39	43	32	45	40
16	25	23	20	17
22	19	30	28	21
18	24	26	29	27

Replication III

Block 1 / Block 2 / Block 3

Figure 2.11 A sample layout of a group balanced block design involving 45 varieties, divided into three groups, each consisting of 15 varieties, tested in three replications.

Table 2.22   Grain Yield Data of 45 Rice Varieties Tested in a Group Balanced Block Design, with 15 Varieties per Group[a]

Variety Number	Grain Yield, t/ha			Variety Total (*T*)
	Rep. I	Rep. II	Rep. III	
1	4.252	3.548	3.114	10.914
2	3.463	2.720	2.789	8.972
3	3.228	2.797	2.860	8.885
4	4.153	3.672	3.738	11.563
5	3.672	2.781	2.788	9.241
6	3.337	2.803	2.936	9.076
7	3.498	3.725	2.627	9.850
8	3.222	3.142	2.922	9.286

77

Table 2.22 (*Continued*)

Variety Number	Grain Yield, t/ha			Variety Total (T)
	Rep. I	Rep. II	Rep. III	
9	3.161	3.108	2.779	9.048
10	3.781	3.906	3.295	10.982
11	3.763	3.709	3.612	11.084
12	3.177	3.742	2.933	9.852
13	3.000	2.843	2.776	8.619
14	4.040	3.251	3.220	10.511
15	3.790	3.027	3.125	9.942
16	3.955	3.030	3.000	9.985
17	3.843	3.207	3.285	10.335
18	3.558	3.271	3.154	9.983
19	3.488	3.278	2.784	9.550
20	2.957	3.284	2.816	9.057
21	3.237	2.835	3.018	9.090
22	3.617	2.985	2.958	9.560
23	4.193	3.639	3.428	11.260
24	3.611	3.023	2.805	9.439
25	3.328	2.955	3.031	9.314
26	4.082	3.089	2.987	10.158
27	4.063	3.367	3.931	11.361
28	3.597	3.211	3.238	10.046
29	3.268	3.913	3.057	10.238
30	4.030	3.223	3.867	11.120
31	3.943	3.133	3.357	10.433
32	2.799	3.184	2.746	8.729
33	3.479	3.377	4.036	10.892
34	3.498	2.912	3.479	9.889
35	3.431	2.879	3.505	9.815
36	4.140	4.107	3.563	11.810
37	4.051	4.206	3.563	11.820
38	3.647	2.863	2.848	9.358
39	4.262	3.197	3.680	11.139
40	4.256	3.091	3.751	11.098
41	4.501	3.770	3.825	12.096
42	4.334	3.666	4.222	12.222
43	4.416	3.824	3.096	11.336
44	3.578	3.252	4.091	10.921
45	4.270	3.896	4.312	12.478
Rep. total (R)	166.969	148.441	146.947	
Grand total (G)				462.357

[a]Group *A* consists of varieties 1–15, group *B* consists of varieties 16–30, and group *C* consists of varieties 31–45.

☐ STEP 1. Outline the analysis of variance of a group balanced block design with t treatments, s groups, and r replications as:

Source of Variation	Degree of Freedom	Sum of Squares	Mean Square
Replication	$r - 1$		
Treatment group	$s - 1$		
Error(a)	$(r - 1)(s - 1)$		
Treatments within group 1	$\dfrac{t}{s} - 1$		
Treatments within group 2	$\dfrac{t}{s} - 1$		
⋮	⋮		
Treatments within group s	$\dfrac{t}{s} - 1$		
Error(b)	$s(r - 1)\left(\dfrac{t}{s} - 1\right)$		
Total	$(r)(t) - 1$		

☐ STEP 2. Compute the treatment totals (T), replication totals (R), and the grand total (G), as shown in Table 2.22.

☐ STEP 3. Construct the replication × group two-way table of totals (RS) and compute the group totals (S), as shown in Table 2.23. Then compute the correction factor, total SS, replication SS, group SS, and error(a) SS as:

$$C.F. = \frac{G^2}{rt}$$

$$= \frac{(462.357)^2}{(3)(45)} = 1{,}583.511077$$

$$\text{Total } SS = \sum X^2 - C.F.$$

$$= \left[(4.252)^2 + \cdots + (4.312)^2\right] - 1{,}583.511077$$

$$= 29.353898$$

$$\text{Replication } SS = \frac{\sum R^2}{t} - C.F.$$

$$= \frac{(166.969)^2 + (148.441)^2 + (146.947)^2}{45}$$

$$- 1{,}583.511077$$

$$= 5.528884$$

Table 2.23   The Replication × Group Table of Yield Totals Computed from Data in Table 2.22

Group	Yield Total (RS) Rep. I	Rep. II	Rep. III	Group total (S)
A	53.537	48.774	45.514	147.825
B	54.827	48.310	47.359	150.496
C	58.605	51.357	54.074	164.036

$$\text{Group } SS = \frac{\sum S^2}{rt/s} - C.F.$$

$$= \frac{\left[(147.825)^2 + (150.496)^2 + (164.036)^2\right]}{(3)(45)/(3)}$$

$$-1,583.511077$$

$$= 3.357499$$

$$\text{Error}(a)\ SS = \frac{\sum (RS)^2}{t/s} - C.F. - \text{Replication } SS - \text{Group } SS$$

$$= \frac{\left[(53.537)^2 + \cdots + (54.074)^2\right]}{(45)/(3)} - 1,583.511077$$

$$-5.528884 - 3.357499$$

$$= 0.632773$$

☐ STEP 4.   Compute the sum of squares among treatments within the ith group as:

$$\text{Treatments within group } i\ SS = \frac{\displaystyle\sum_{j=1}^{t/s} T_{ij}^2}{r} - \frac{S_i^2}{rt/s}$$

where T_{ij} is the total of the jth treatment in the ith group and S_i is the total of the ith group.

For our example, the sum of squares among varieties within each of the three groups is computed as:

$$\text{Varieties within group } A\ SS = \frac{\sum T_A^2}{3} - \frac{S_A^2}{(3)(45)/(3)}$$

$$= \frac{(10.914)^2 + \cdots + (9.942)^2}{3} - \frac{(147.825)^2}{45}$$

$$= 4.154795$$

Varieties within group B SS $= \dfrac{\sum T_B^2}{3} - \dfrac{S_B^2}{(3)(45)/(3)}$

$$= \frac{(9.985)^2 + \quad + (11.120)^2}{3} - \frac{(150.196)^2}{45}$$

$$= 2.591050$$

Varieties within group C SS $= \dfrac{\sum T_C^2}{3} - \dfrac{S_C^2}{(3)(45)/3}$

$$= \frac{(10.433)^2 + \cdots + (12.478)^2}{3} - \frac{(164.036)^2}{45}$$

$$= 5.706299$$

Here, T_A, T_B, and T_C refer to the treatment totals, and S_A, S_B, and S_C refer to the group totals, of group A, group B, and group C, respectively.

☐ STEP 5. Compute the error(b) SS as:

Error(b) SS $=$ Total SS $-$ (the sum of all other SS)

$$= 29.353898 - (5.528884 + 3.357499 + 0.632773$$

$$+ 4.154795 + 2.591050 + 5.706299)$$

$$= 7.382598$$

☐ STEP 6. Compute the mean square for each source of variation by dividing the SS by its $d.f.$ as:

$$\text{Replication } MS = \frac{\text{Replication } SS}{r - 1}$$

$$= \frac{5.528884}{2} = 2.764442$$

$$\text{Group } MS = \frac{\text{Group } SS}{s - 1}$$

$$= \frac{3.357499}{2} = 1.678750$$

$$\text{Error}(a) \, MS = \frac{\text{Error}(a) \, SS}{(r - 1)(s - 1)}$$

$$= \frac{0.632773}{(2)(2)} = 0.158193$$

$$\text{Varieties within group } A \; MS = \frac{\text{Varieties within group } A \; SS}{(t/s) - 1}$$

$$= \frac{4.154795}{14} = 0.296771$$

$$\text{Varieties within group } B \; MS = \frac{\text{Varieties within group } B \; SS}{(t/s) - 1}$$

$$= \frac{2.591050}{14} = 0.185075$$

$$\text{Varieties within group } C \; MS = \frac{\text{Varieties within group } C \; SS}{(t/s) - 1}$$

$$= \frac{5.706299}{14} = 0.407593$$

$$\text{Error}(b) \; MS = \frac{\text{Error}(b) \; SS}{s(r - 1)[(t/s) - 1]}$$

$$= \frac{7.382598}{84} = 0.087888$$

□ STEP 7.   Compute the following F values:

$$F(\text{group}) = \frac{\text{Group } MS}{\text{Error}(a) \; MS}$$

$$= \frac{1.678750}{0.158193} = 10 \; 61*$$

$$F(\text{varieties within group } A) = \frac{\text{Varieties within group } A \; MS}{\text{Error}(b) \; MS}$$

$$= \frac{0.296771}{0.087888} = 3.38$$

$$F(\text{varieties within group } B) = \frac{\text{Varieties within group } B \; MS}{\text{Error}(b) \; MS}$$

$$= \frac{0.185075}{0.087888} = 2.11$$

$$F(\text{varieties within group } C) = \frac{\text{Varieties within group } C \; MS}{\text{Error}(b) \; MS}$$

$$= \frac{0.407593}{0.087888} = 4.64$$

*Although the error (a) d.f. of 4 is not adequate for valid test of significance (see Section 2.1.2.1, step 6), for illustration purposes, such a deficiency has been ignored.

□ STEP 8. For each computed F value, obtain its corresponding tabular F value, from Appendix E, at the prescribed level of significance, with $f_1 = d.f.$ of the numerator *MS* and $f_2 = d.f.$ of the denominator *MS*.

For our example, the tabular F values corresponding to the computed $F(\text{group})$ value, with $f_1 = 2$ and $f_2 = 4$ degrees of freedom, are 6.94 at the 5% level of significance and 18.00 at the 1% level; those corresponding to each of the three computed $F(\text{varieties within group})$ values, with $f_1 = 14$ and $f_2 = 84$ degrees of freedom, are 1.81 at the 5% level of significance and 2.31 at the 1% level.

□ STEP 9. Compute the two coefficients of variation corresponding to the two values of the error mean square as:

$$cv(a) = \frac{\sqrt{\text{Error}(a)\ MS}}{\text{Grand mean}} \times 100$$

$$= \frac{\sqrt{0.158193}}{3.425} \times 100 = 11.6\%$$

$$cv(b) = \frac{\sqrt{\text{Error}(b)\ MS}}{\text{Grand mean}} \times 100$$

$$= \frac{\sqrt{0.087888}}{3.425} \times 100 = 8.7\%$$

□ STEP 10. Enter all values obtained in steps 3 to 9 in the analysis of variance outline of step 1, as shown in Table 2.24. Results indicate a significant difference among the means of the three groups of varieties and significant differences among the varieties in each of the three groups.

Table 2.24 Analysis of Variance (Group Balanced Block Design) for Data in Table 2.22[a]

Source of Variation	Degree of Freedom	Sum of Squares	Mean Square	Computed F[b]	Tabular F 5%	Tabular F 1%
Replication	2	5.528884	2.764442			
Varietal group	2	3.357499	1.678750	10.61*	6.94	18.00
Error(a)	4	0.632773	0.158193			
Varieties within group A	14	4.154795	0.296771	3.38**	1.81	2.31
Varieties within group B	14	2.591050	0.185075	2.11*	1.81	2.31
Varieties within group C	14	5.706299	0.407593	4.64**	1.81	2.31
Error(b)	84	7.382598	0.087888			
Total	134	29.353898				

[a] $cv\ (a) = 11.6\%$, $cv\ (b) = 8.7\%$.
[b] ** = significant at 1% level, * = significant at 5% level.

CHAPTER 3

Two-Factor Experiments

Biological organisms are simultaneously exposed to many growth factors during their lifetime. Because an organism's response to any single factor may vary with the level of the other factors, single-factor experiments are often criticized for their narrowness. Indeed, the result of a single-factor experiment is, strictly speaking, applicable only to the particular level in which the other factors were maintained in the trial.

Thus, when response to the factor of interest is expected to differ under different levels of the other factors, avoid single-factor experiments and consider instead the use of a factorial experiment designed to handle simultaneously two or more variable factors.

3.1 INTERACTION BETWEEN TWO FACTORS

Two factors are said to interact if the effect of one factor changes as the level of the other factor changes. We shall define and describe the measurement of the interaction effect based on an experiment with two factors A and B, each with two levels (a_0 and a_1 for factor A and b_0 and b_1 for factor B). The four treatment combinations are denoted by $a_0 b_0$, $a_1 b_0$, $a_0 b_1$, and $a_1 b_1$. In addition, we define and describe the measurement of the *simple effect* and the *main effect* of each of the two factors A and B because these effects are closely related to, and are in fact an immediate step toward the computation of, the interaction effect.

To illustrate the computation of these three types of effects, consider the two sets of data presented in Table 3.1 for two varieties X and Y and two nitrogen rates N_0 and N_1; one set with no interaction and another with interaction.

☐ STEP 1. Compute the simple effect of factor A as the difference between its two levels at a given level of factor B. That is:

- The simple effect of A at $b_0 = a_1 b_0 - a_0 b_0$
- The simple effect of A at $b_1 = a_1 b_1 - a_0 b_1$

Table 3.1 Two Hypothetical Sets of 2×2 Factorial Data: One with, and Another without, Interaction between Two Factors (Variety and Nitrogen Rate)

	Rice Yield, t/ha		
Variety	0 kg N/ha (N_0)	60 kg N/ha (N_1)	Av.
	No interaction		
X	1.0	3.0	2.0
Y	2.0	4.0	3.0
Av.	1.5	3.5	
	Interaction present		
X	1.0	1.0	1.0
Y	2.0	4.0	3.0
Av.	1.5	2.5	

In the same manner, compute the simple effect of factor B at each of the two levels of factor A as:

- The simple effect of B at $a_0 = a_0b_1 - a_0b_0$
- The simple effect of B at $a_1 = a_1b_1 - a_1b_0$

For our example (Table 3.1), the computations based on data of the set with interaction are:

Simple effect of variety at $N_0 = 2.0 - 1.0 = 1.0$ t/ha

Simple effect of variety at $N_1 = 4.0 - 1.0 = 3.0$ t/ha

Simple effect of nitrogen of $X = 1.0 - 1.0 = 0.0$ t/ha

Simple effect of nitrogen of $Y = 4.0 - 2.0 = 2.0$ t/ha

And the computations based on data of the set without interaction are:

Simple effect of variety at $N_0 = 2.0 - 1.0 = 1.0$ t/ha

Simple effect of variety at $N_1 = 4.0 - 3.0 = 1.0$ t/ha

Simple effect of nitrogen of $X = 3.0 - 1.0 = 2.0$ t/ha

Simple effect of nitrogen of $Y = 4.0 - 2.0 = 2.0$ t/ha

☐ STEP 2.   Compute the main effect of factor A as the average of the simple effects of factor A over all levels of factor B as:

The main effect of $A = (1/2)$ (simple effect of A at b_0

$+$ simple effect of A at b_1)

$$= (1/2)[(a_1b_0 - a_0b_0) + (a_1b_1 - a_0b_1)]$$

In the same manner, compute the main effect of factor B as:

The main effect of $B = (1/2)$(simple effect of B at a_0

$+$ simple effect of B at a_1)

$$= (1/2)[(a_0b_1 - a_0b_0) + (a_1b_1 - a_1b_0)]$$

For our example, the computations based on data of the set with interaction are:

Main effect of variety $= (1/2)(1.0 + 3.0) = 2.0 \text{ t/ha}$

Main effect of nitrogen $= (1/2)(0.0 + 2.0) = 1.0 \text{ t/ha}$

And the computations based on data without interaction are:

Main effect of variety $= (1/2)(1.0 + 1.0) = 1.0 \text{ t/ha}$

Main effect of nitrogen $= (1/2)(2.0 + 2.0) = 2.0 \text{ t/ha}$

☐ STEP 3.   Compute the interaction effect between factor A and factor B as a function of the difference between the simple effects of A at the two levels of B or the difference between the simple effects of B at the two levels of A:

$A \times B = (1/2)$(simple effect of A at $b_1 -$ simple effect of A at b_0)

$$= (1/2)[(a_1b_1 - a_0b_1) - (a_1b_0 - a_0b_0)]$$

or,

$A \times B = (1/2)$(simple effect of B at $a_1 -$ simple effect of B at a_0)

$$= (1/2)[(a_1b_1 - a_1b_0) - (a_0b_1 - a_0b_0)]$$

For our example, the computations of the variety × nitrogen interaction effect based on data of the set with interaction are:

$$V \times N = (1/2)(\text{simple effect of variety at } N_1$$

$$- \text{simple effect of variety at } N_0)$$

$$= (1/2)(3.0 - 1.0) = 1.0 \text{ t/ha}$$

or,

$$V \times N = (1/2)(\text{simple effect of nitrogen of } Y$$

$$- \text{simple effect of nitrogen of } X)$$

$$= (1/2)(2.0 - 0.0) = 1.0 \text{ t/ha}$$

And the computations of the variety × nitrogen interaction effect based on data of the set without interaction are:

$$V \times N = 1/2(1.0 - 1.0) = 0.0 \text{ t/ha}$$

or,

$$V \times N = 1/2(2.0 - 2.0) = 0.0 \text{ t/ha}$$

A graphical representation of the nitrogen response of the two varieties is shown in Figure 3.1a for the no-interaction data and in Figure 3.1c for the with-interaction data having an interaction effect of 1.0 t/ha. Cases with lower and higher interaction effects than 1.0 t/ha are illustrated in Figures 3.1b and 3.1d. Figure 3.1b shows the nitrogen response to be positive for both varieties but with higher response for variety Y (2.0 t/ha) than for variety X (1.0 t/ha), giving an interaction effect of 0.5 t/ha. Figure 3.1d shows a large positive nitrogen response for X (2.0 t/ha) and an equally large but negative response for variety Y, giving an interaction effect of 2.0 t/ha.

From the foregoing numerical computation and graphical representations of the interaction effects, three points should be noted:

1. An interaction effect between two factors can be measured only if the two factors are tested together in the same experiment (i.e., in a factorial experiment).

2. When interaction is absent (as in Figure 3.1a) the simple effect of a factor is the same for all levels of the other factors and equals the main effect. For our example, the simple effects of variety at N_0 and N_1 are both 1.0 t/ha, which is the same as its main effect. That is, when interaction is absent, the results from separate single-factor experiments (i.e., one for each factor) are

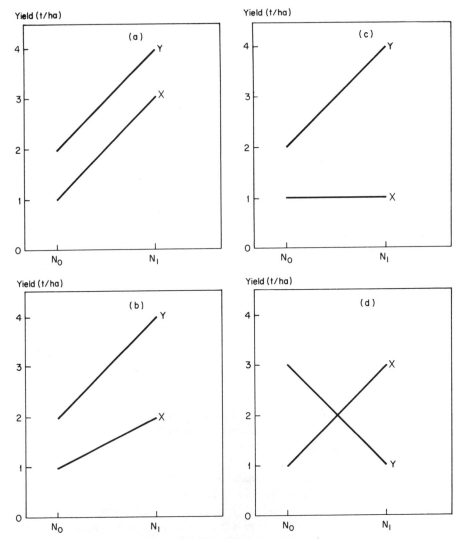

Figure 3.1 Graphical representation of the different magnitudes of interaction between varieties (X and Y) and nitrogen rates (N_0 and N_1) with (a) showing no interaction, (b) and (c) showing intermediate interactions, and (d) showing high interaction.

equivalent to those from a factorial experiment with all factors tested together. In our example, the varietal effect would have been estimated at 1.0 t/ha regardless of whether:

- The two varieties are tested under N_0 in a single-factor experiment,
- The two varieties are tested under N_1 in a single-factor experiment, or
- The two varieties are tested in combination with the two nitrogen rates in a two-factor experiment.

3. When interaction is present (as in Figures 3.1b, 3.1c, and 3.1d) the simple effect of a factor changes as the level of the other factor changes. Consequently, the main effect is different from the simple effects. For example, in Figure 3.1c, the simple effects of nitrogen are 0.0 t/ha for variety X and 2.0 t/ha for variety Y, and its main effect is thus 1.0 t/ha. In other words, although there was a large response to nitrogen application in variety Y, there was none in variety X. Or, in Figure 3.1d, variety Y outyielded variety X by 2.0 t/ha under N_0 but gave a 2.0 t/ha lower yield under N_1. If the mean yields of the two varieties were calculated over the two nitrogen rates, the two variety means would be the same (i.e., 2.5 t/ha). Thus, if we look at the difference between these two variety means (i.e., main effect of variety), we would have concluded that there was no varietal difference. It is therefore clear that when an interaction effect between two factors is present:

- The simple effects and not the main effects should be examined.
- The result from a single-factor experiment is applicable only to the particular level in which the other factors were maintained in the experiment and there can be no generalization of the result to cover any other levels.

3.2   FACTORIAL EXPERIMENT

An experiment in which the treatments consist of all possible combinations of the selected levels in two or more factors is referred to as a factorial experiment.* For example, an experiment involving two factors, each at two levels, such as two varieties and two nitrogen rates, is referred to as a 2×2 or a 2^2 factorial experiment. Its treatments consist of the following four possible combinations of the two levels in each of the two factors.

Treatment Number	Treatment Combination	
	Variety	N rate, kg/ha
1	X	0
2	X	60
3	Y	0
4	Y	60

If the 2^2 factorial experiment is expanded to include a third factor, say weed control at two levels, the experiment becomes a $2 \times 2 \times 2$ or a 2^3 factorial

---

*The term *complete factorial experiment* is sometimes used when the treatments include all combinations of the selected levels of the variable factors. In contrast, the term incomplete factorial experiment is used when only a fraction of all the combinations is tested. Throughout this book, however, we refer to complete factorial experiments as factorial experiments and use the term incomplete factorial, otherwise.

experiment, with the following eight treatment combinations:

Treatment Number	Treatment Combination		
	Variety	N rate, kg/ha	Weed Control
1	X	0	With
2	X	0	Without
3	X	60	With
4	X	60	Without
5	Y	0	With
6	Y	0	Without
7	Y	60	With
8	Y	60	Without

Note that the term *factorial* describes a specific way in which the treatments are formed and does not, in any way, refer to the experimental design used. For example, if the foregoing 2^3 factorial experiment is in a randomized complete block design, then the correct description of the experiment would be 2^3 *factorial experiment in a randomized complete block design.*

The total number of treatments in a factorial experiment is the product of the levels in each factor; in the 2^2 factorial example, the number of treatments is $2 \times 2 = 4$, in the 2^3 factorial the number of treatments is $2 \times 2 \times 2 = 8$. The number of treatments increases rapidly with an increase in the number of factors or an increase in the levels in each factor. For a factorial experiment involving five varieties, four nitrogen rates, and three weed-control methods, the total number of treatments would be $5 \times 4 \times 3 = 60$.

Thus, avoid indiscriminate use of factorial experiments because of their large size, complexity, and cost. Furthermore, it is not wise to commit oneself to a large experiment at the beginning of the investigation when several small preliminary experiments may offer promising results. For example, a plant breeder has collected 30 new rice varieties from a neighboring country and wants to assess their reaction to the local environment. Because the environment is expected to vary in terms of soil fertility, moisture levels, and so on, the ideal experiment would be one that tests the 30 varieties in a factorial experiment involving such other variable factors as fertilizer, moisture level, and population density. Such an experiment, however, becomes extremely large as variable factors other than varieties are added. Even if only one factor, say nitrogen fertilizer with three levels, were included the number of treatments would increase from 30 to 90.

Such a large experiment would mean difficulties in financing, in obtaining an adequate experimental area, in controlling soil heterogeneity, and so on.

Thus, the more practical approach would be to test the 30 varieties first in a single-factor experiment, and then use the results to select a few varieties for further studies in more detail. For example, the initial single-factor experiment may show that only five varieties are outstanding enough to warrant further testing. These five varieties could then be put into a factorial experiment with three levels of nitrogen, resulting in an experiment with 15 treatments rather than the 90 treatments needed with a factorial experiment with 30 varieties. Thus, although a factorial experiment provides valuable information on interaction, and is without question more informative than a single-factor experiment, practical consideration may limit its use.

For most factorial experiments, the number of treatments is usually too large for an efficient use of a complete block design. Furthermore, incomplete block designs such as the lattice designs (Chapter 2, Section 2.4) are not appropriate for factorial experiments. There are, however, special types of design, developed specifically for factorial experiments, that are comparable to the incomplete block designs for single-factor experiments. Such designs, which are suitable for two-factor experiments and are commonly used in agricultural research, are discussed here.

3.3 COMPLETE BLOCK DESIGN

Any of the complete block designs discussed in Chapter 2 for single-factor experiments is applicable to a factorial experiment. The procedures for randomization and layout of the individual designs are directly applicable by simply ignoring the factor composition of the factorial treatments and considering all the treatments as if they were unrelated. For the analysis of variance, the computations discussed for individual designs are also directly applicable. However, additional computational steps are required to partition the treatment sum of squares into factorial components corresponding to the main effects of individual factors and to their interactions. The procedure for such partitioning is the same for all complete block designs and is, therefore, illustrated for only one case, namely, that of a randomized complete block (RCB) design.

We illustrate the step-by-step procedures for the analysis of variance of a two-factor experiment in a RCB design with an experiment involving five rates of nitrogen fertilizer, three rice varieties, and four replications. The list of the 15 factorial treatment combinations is shown in Table 3.2, the experimental layout in Figure 3.2, and the data in Table 3.3.

☐ STEP 1. Denote the number of replications by r, the level of factor A (i.e., variety) by a, and the level of factor B (i.e., nitrogen) by b. Construct the

Table 3.2 The 3 × 5 Factorial Treatment Combinations of Three Rice Varieties and Five Nitrogen Levels

Nitrogen Level, kg/ha	Factorial Treatment Combination		
	6966 (V_1)	P1215936 (V_2)	Milfor 6(2) (V_3)
$0(N_0)$	N_0V_1	N_0V_2	N_0V_3
$40(N_1)$	N_1V_1	N_1V_2	N_1V_3
$70(N_2)$	N_2V_1	N_2V_2	N_2V_3
$100(N_3)$	N_3V_1	N_3V_2	N_3V_3
$130(N_4)$	N_4V_1	N_4V_2	N_4V_3

Table 3.3 Grain Yield of Three Rice Varieties Tested with Five Levels of Nitrogen in a RCB Design[a]

Nitrogen Level, kg/ha	Grain Yield, t/ha				Treatment Total (T)
	Rep. I	Rep. II	Rep. III	Rep. IV	
			V_1		
N_0	3.852	2.606	3.144	2.894	12.496
N_1	4.788	4.936	4.562	4.608	18.894
N_2	4.576	4.454	4.884	3.924	17.838
N_3	6.034	5.276	5.906	5.652	22.868
N_4	5.874	5.916	5.984	5.518	23.292
			V_2		
N_0	2.846	3.794	4.108	3.444	14.192
N_1	4.956	5.128	4.150	4.990	19.224
N_2	5.928	5.698	5.810	4.308	21.744
N_3	5.664	5.362	6.458	5.474	22.958
N_4	5.458	5.546	5.786	5.932	22.722
			V_3		
N_0	4.192	3.754	3.738	3.428	15.112
N_1	5.250	4.582	4.896	4.286	19.014
N_2	5.822	4.848	5.678	4.932	21.280
N_3	5.888	5.524	6.042	4.756	22.210
N_4	5.864	6.264	6.056	5.362	23.546
Rep. total (R)	76.992	73.688	77.202	69.508	
Grand total (G)					297.390

[a]For description of treatments, see Table 3.2.

Rep. I

V_3N_2	V_2N_1	V_1N_4	V_1N_1	V_2N_3
V_3N_0	V_1N_3	V_3N_4	V_1N_2	V_3N_1
V_2N_4	V_3N_1	V_2N_0	V_1N_0	V_2N_2

Rep. II

V_2N_3	V_3N_3	V_1N_1	V_2N_0	V_2N_1
V_1N_3	V_3N_2	V_1N_2	V_1N_4	V_2N_4
V_1N_0	V_3N_4	V_2N_2	V_3N_1	V_3N_0

Rep. III

V_1N_1	V_3N_0	V_1N_0	V_3N_1	V_1N_4
V_2N_2	V_1N_2	V_1N_3	V_2N_4	V_3N_4
V_2N_0	V_3N_2	V_2N_1	V_2N_3	V_3N_3

Rep. IV

V_1N_2	V_2N_2	V_2N_4	V_1N_0	V_2N_0
V_1N_3	V_3N_1	V_1N_4	V_1N_1	V_2N_3
V_3N_0	V_2N_1	V_3N_2	V_3N_3	V_3N_4

Figure 3.2 A sample layout of a 3×5 factorial experiment involving three varieties (V_1, V_2, and V_3) and five nitrogen rates (N_0, N_1, N_2, N_3, and N_4) in a randomized complete block design with four replications.

outline of the analysis of variance as:

Source of Variation	Degree of Freedom	Sum of Squares	Mean Square	Computed F	Tabular F 5%	1%
Replication	$r - 1 = 3$					
Treatment	$ab - 1 = 14$					
Variety (A)	$a - 1 = (2)$					
Nitrogen (B)	$b - 1 = (4)$					
$A \times B$	$(a - 1)(b - 1) = (8)$					
Error	$(r - 1)(ab - 1) = 42$					
Total	$rab - 1 = 59$					

□ STEP 2. Compute treatment totals (T), replication totals (R), and the grand total (G), as shown in Table 3.3; and compute the total SS, replication SS, treatment SS, and error SS, following the procedure described in

Chapter 2, Section 2.2.3:

$$C.F. = \frac{G^2}{rab}$$

$$= \frac{(297.390)^2}{(4)(3)(5)} = 1{,}474.014$$

Total $SS = \Sigma X^2 - C.F.$

$$= \left[(3.852)^2 + (2.606)^2 + \cdots + (5.362)^2\right] - 1{,}474.014$$

$$= 53.530$$

Replication $SS = \dfrac{\Sigma R^2}{ab} - C.F.$

$$= \frac{(76.992)^2 + \cdots + (69.508)^2}{(3)(5)} - 1{,}474.014$$

$$= 2.599$$

Treatment $SS = \dfrac{\Sigma T^2}{r} - C.F.$

$$= \frac{(12.496)^2 + \cdots + (23.546)^2}{4} - 1{,}474.014$$

$$= 44.578$$

Error SS = Total SS − Replication SS − Treatment SS

$$= 53.530 - 2.599 - 44.578$$

$$= 6.353$$

The preliminary analysis of variance, with the various SS just computed, is as shown in Table 3.4.

☐ STEP 3.   Construct the factor A × factor B two-way table of totals, with factor A totals and factor B totals computed. For our example, the variety × nitrogen table of totals (AB) with variety totals (A) and nitrogen totals (B) computed is shown in Table 3.5.

Table 3.4 Preliminary Analysis of Variance for Data in Table 3.3

Source of Variation	Degree of Freedom	Sum of Squares	Mean Square	Computed F^a	Tabular F 5%	1%
Replication	3	2.599	0.866	5.74**	2.83	4.29
Treatment	14	44.578	3.184	21.09**	1.94	2.54
Error	42	6.353	0.151			
Total	59	53.530				

[a]** = significant at 1% level.

□ STEP 4. Compute the three factorial components of the treatment sum of squares as:

$$A\ SS = \frac{\sum A^2}{rb} - C.F.$$

$$= \frac{(95.388)^2 + (100.840)^2 + (101.162)^2}{(4)(5)} - 1{,}474.014$$

$$= 1.052$$

$$B\ SS = \frac{\sum B^2}{ra} - C.F.$$

$$= \frac{(41.800)^2 + \cdots + (69.560)^2}{(4)(3)} - 1{,}474.014$$

$$= 41.234$$

Table 3.5 The Variety × Nitrogen Table of Totals from Data in Table 3.3

Nitrogen	Yield Total (AB) V_1	V_2	V_3	Nitrogen Total (B)
N_0	12.496	14.192	15.112	41.800
N_1	18.894	19.224	19.014	57.132
N_2	17.838	21.744	21.280	60.862
N_3	22.868	22.958	22.210	68.036
N_4	23.292	22.722	23.546	69.560
Variety total (A)	95.388	100.840	101.162	297.390

$$A \times B \ SS = \text{Treatment } SS - A \ SS - B \ SS$$
$$= 44.578 - 1.052 - 41.234$$
$$= 2.292$$

□ STEP 5.   Compute the mean square for each source of variation by dividing the SS by its $d.f.$:

$$A \ MS = \frac{A \ SS}{a - 1}$$

$$= \frac{1.052}{2} = 0.526$$

$$B \ MS = \frac{B \ SS}{b - 1}$$

$$= \frac{41.234}{4} = 10.308$$

$$A \times B \ MS = \frac{A \times B \ SS}{(a - 1)(b - 1)}$$

$$= \frac{2.292}{(2)(4)} = 0.286$$

$$\text{Error } MS = \frac{\text{Error } SS}{(r - 1)(ab - 1)}$$

$$= \frac{6.353}{(3)[(3)(5) - 1]} = 0.151$$

□ STEP 6.   Compute the F value for each of the three factorial components as:

$$F(A) = \frac{A \ MS}{\text{Error } MS}$$

$$= \frac{0.526}{0.151} = 3.48$$

$$F(B) = \frac{B \ MS}{\text{Error } MS}$$

$$= \frac{10.308}{0.151} = 68.26$$

$$F(A \times B) = \frac{A \times B \ MS}{\text{Error } MS}$$

$$= \frac{0.286}{0.151} = 1.89$$

Table 3.6 Analysis of Variance of Data in Table 3.3 from a 3 × 5 Factorial Experiment in RCB Design[a]

Source of Variation	Degree of Freedom	Sum of Squares	Mean Square	Computed F[b]	Tabular F 5%	Tabular F 1%
Replication	3	2.599	0.866	5.74**	2.83	4.29
Treatment	14	44.578	3.184	21.09**	1.94	2.54
Variety(A)	(2)	1.052	0.526	3.48*	3.22	5.15
Nitrogen(B)	(4)	41.234	10.308	68.26**	2.59	3.80
$A \times B$	(8)	2.292	0.286	1.89[ns]	2.17	2.96
Error	42	6.353	0.151			
Total	59	53.530				

[a] $cv = 7.8\%$.
[b] ** = significant at 1% level, * = significant at 5% level, [ns] = not significant.

☐ STEP 7. Compare each of the computed F values with the tabular F value, from Appendix E, with $f_1 = d.f.$ of the numerator MS and $f_2 = d.f.$ of the denominator MS, at a prescribed level of significance. For example, the computed $F(A)$ value is compared with the tabular F values (with $f_1 = 2$ and $f_2 = 42$ degrees of freedom) of 3.22 at the 5% level of significance and 5.15 at the 1% level. The result indicates that the main effect of factor A (variety) is significant at the 5% level of significance.

☐ STEP 8. Compute the coefficient of variation as:

$$cv = \frac{\sqrt{\text{Error } MS}}{\text{Grand mean}} \times 100$$

$$= \frac{\sqrt{0.151}}{4.956} \times 100 = 7.8\%$$

☐ STEP 9. Enter all values obtained in steps 4 to 8 in the preliminary analysis of variance of step 2, as shown in Table 3.6. The results show a nonsignificant interaction between variety and nitrogen, indicating that the varietal difference was not significantly affected by the nitrogen level applied and that the nitrogen effect did not differ significantly with the varieties tested. Main effects both of variety and of nitrogen were significant.

3.4 SPLIT-PLOT DESIGN

The split-plot design is specifically suited for a two-factor experiment that has more treatments than can be accommodated by a complete block design. In a split-plot design, one of the factors is assigned to the *main plot*. The assigned factor is called the *main-plot factor*. The main plot is divided into *subplots* to which the second factor, called the *subplot factor*, is assigned. Thus, each main plot becomes a block for the subplot treatments (i.e., the levels of the subplot factor).

With a split-plot design, the precision for the measurement of the effects of the main-plot factor is sacrificed to improve that of the subplot factor. Measurement of the main effect of the subplot factor and its interaction with the main-plot factor is more precise than that obtainable with a randomized complete block design. On the other hand, the measurement of the effects of the main-plot treatments (i.e., the levels of the main-plot factor) is less precise than that obtainable with a randomized complete block design.

Because, with the split-plot design, plot size and precision of measurement of the effects are not the same for both factors, the assignment of a particular factor to either the main plot or the subplot is extremely important. To make such a choice, the following guidelines are suggested:

1. Degree of Precision. For a greater degree of precision for factor *B* than for factor *A*, assign factor *B* to the subplot and factor *A* to the main plot. For example, a plant breeder who plans to evaluate 10 promising rice varieties with three levels of fertilization in a 10 × 3 factorial experiment would probably wish to have greater precision for varietal comparison than for fertilizer response. Thus, he would designate variety as the subplot factor and fertilizer as the main-plot factor.

On the other hand, an agronomist who wishes to study fertilizer responses of the 10 promising varieties developed by the plant breeder would probably want greater precision for fertilizer response than for varietal effect and would assign variety to main plot and fertilizer to subplot.

2. Relative Size of the Main Effects. If the main effect of one factor (factor *B*) is expected to be much larger and easier to detect than that of the other factor (factor *A*), factor *B* can be assigned to the main plot and factor *A* to the subplot. This increases the chance of detecting the difference among levels of factor *A* which has a smaller effect. For example, in a fertilizer × variety experiment, the researcher may assign variety to the subplot and fertilizer to the main plot because he expects the fertilizer effect to be much larger than the varietal effect.

3. Management Practices. The cultural practices required by a factor may dictate the use of large plots. For practical expediency, such a factor may be assigned to the main plot. For example, in an experiment to evaluate water management and variety, it may be desirable to assign water management to the main plot to minimize water movement between adjacent plots, facilitate the simulation of the water level required, and reduce border effects. Or, in an experiment to evaluate the performance of several rice varieties with different fertilizer rates, the researcher may assign the main plot to fertilizer to minimize the need to separate plots receiving different fertilizer levels.

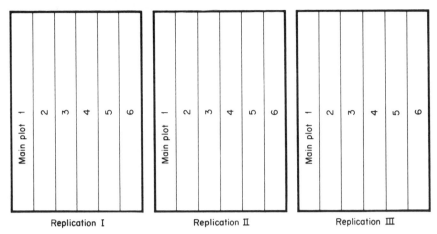

Figure 3.3 Division of the experimental area into three blocks (replications) each consisting of six main plots, as the first step in laying out of a split-plot experiment involving three replications and six main-plot treatments.

In a split-plot design, both the procedure for randomization and that for analysis of variance are accomplished in two stages—one on the main-plot level and another on the subplot level. At each level, the procedures of the randomized complete block design*, as described in Chapter 2, are applicable.

3.4.1 Randomization and Layout

There are two separate randomization processes in a split-plot design—one for the main plot and another for the subplot. In each replication, main-plot treatments are first randomly assigned to the main plots followed by a random assignment of the subplot treatments within each main plot. Each is done by any of the randomization schemes of Chapter 2, Section 2.1.1.

The steps in the randomization and layout of a split-plot design are shown, using a as the number of main-plot treatments, b as the number of subplot treatments, and r as the number of replications. For illustration, a two-factor experiment involving six levels of nitrogen (main-plot treatments) and four rice varieties (subplot treatments) in three replications is used.

☐ STEP 1.   Divide the experimental area into $r = 3$ blocks, each of which is further divided into $a = 6$ main plots, as shown in Figure 3.3.

---

*The assignment of the main-plot factor can, in fact, follow any of the complete block designs, namely, completely randomized design, randomized complete block, and latin square; but we consider only the randomized complete block because it is the most appropriate and the most commonly used for agricultural experiments.

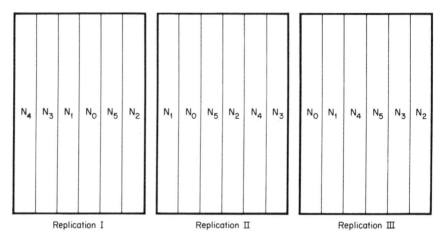

Figure 3.4   Random assignment of six nitrogen levels (N_0, N_1, N_2, N_3, N_4, and N_5) to the six main plots in each of the three replications of Figure 3.3.

☐ STEP 2.   Following the RCB randomization procedure with $a = 6$ treatments and $r = 3$ replications (Chapter 2, Section 2.2.2) randomly assign the 6 nitrogen treatments to the 6 main plots in each of the 3 blocks. The result may be as shown in Figure 3.4.

☐ STEP 3.   Divide each of the $(r)(a) = 18$ main plots into $b = 4$ subplots and, following the RCB randomization procedure for $b = 4$ treatments and

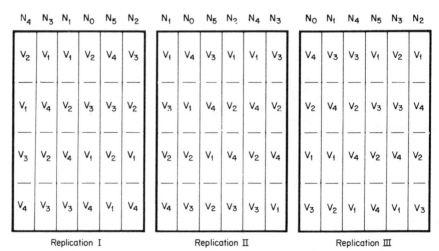

Figure 3.5   A sample layout of a split-plot design involving four rice varieties (V_1, V_2, V_3, and V_4) as subplot treatments and six nitrogen levels (N_0, N_1, N_2, N_3, N_4, and N_5) as main-plot treatments, in three replications.

$(r)(a) = 18$ replications, randomly assign the 4 varieties to the 4 subplots in each of the 18 main plots. The result may be as shown in Figure 3.5.

Note that field layout of a split plot design as illustrated by Figure 3.5 has the following important features:

1. The size of the main plot is b times the size of the subplot. In our example with 4 varieties $(b = 4)$ the size of the main plot is 4 times the subplot size.

2. Each main-plot treatment is tested r times whereas each subplot treatment is tested $(a)(r)$ times. Thus, the number of times a subplot treatment is tested will always be larger than that for the main plot and is the primary reason for more precision for the subplot treatments relative to the main-plot treatments. In our example, each of the 6 levels of nitrogen was tested 3 times but each of the 4 varieties was tested 18 times.

3.4.2 Analysis of Variance

The analysis of variance of a split-plot design is divided into the *main-plot analysis* and the *subplot analysis*. We show the computations involved in the analysis with data from the two-factor experiment (six levels of nitrogen and four rice varieties) shown in Figure 3.5. Grain yield data are shown in Table 3.7.

Let A denote the main-plot factor and B, the subplot factor. Compute analysis of variance:

☐ STEP 1. Construct an outline of the analysis of variance for a split-plot design as:

Source of Variation	Degree of Freedom	Sum of Squares	Mean Square	Computed F	Tabular F 5%	1%
Replication	$r - 1 = 2$					
Main-plot factor (A)	$a - 1 = 5$					
Error(a)	$(r - 1)(a - 1) = 10$					
Subplot factor (B)	$b - 1 = 3$					
$A \times B$	$(a - 1)(b - 1) = 15$					
Error(b)	$a(r - 1)(b - 1) = 36$					
Total	$rab - 1 = 71$					

☐ STEP 2. Construct two tables of totals:

  A. The replication × factor A two-way table of totals, with the replication totals, factor A totals, and grand total computed. For our example, the

Table 3.7 Grain Yield Data of Four Rice Varieties Grown with Six Levels of Nitrogen in a Split-Plot Design with Three Replications

	Grain Yield, kg/ha		
Variety	Rep. I	Rep. II	Rep. III
$N_0(0 \ kg \ N/ha)$			
V_1(IR8)	4,430	4,478	3,850
V_2(IR5)	3,944	5,314	3,660
V_3(C4-63)	3,464	2,944	3,142
V_4(Peta)	4,126	4,482	4,836
$N_1(60 \ kg \ N/ha)$			
V_1	5,418	5,166	6,432
V_2	6,502	5,858	5,586
V_3	4,768	6,004	5,556
V_4	5,192	4,604	4,652
$N_2(90 \ kg \ N/ha)$			
V_1	6,076	6,420	6,704
V_2	6,008	6,127	6,642
V_3	6,244	5,724	6,014
V_4	4,546	5,744	4,146
$N_3(120 \ kg \ N/ha)$			
V_1	6,462	7,056	6,680
V_2	7,139	6,982	6,564
V_3	5,792	5,880	6,370
V_4	2,774	5,036	3,638
$N_4(150 \ kg \ N/ha)$			
V_1	7,290	7,848	7,552
V_2	7,682	6,594	6,576
V_3	7,080	6,662	6,320
V_4	1,414	1,960	2,766
$N_5(180 \ kg \ N/ha)$			
V_1	8,452	8,832	8,818
V_2	6,228	7,387	6,006
V_3	5,594	7,122	5,480
V_4	2,248	1,380	2,014

Table 3.8   The Replication × Nitrogen Table of Yield Totals Computed from Data in Table 3.7

Nitrogen	Yield Total (RA) Rep. I	Rep. II	Rep. III	Nitrogen Total (A)
N_0	15,964	17,218	15,488	48,670
N_1	21,880	21,632	22,226	65,738
N_2	22,874	24,015	23,506	70,395
N_3	22,167	24,954	23,252	70,373
N_4	23,466	23,064	23,214	69,744
N_5	22,522	24,721	22,318	69,561
Rep. total (R)	128,873	135,604	130,004	
Grand total (G)				394,481

replication × nitrogen table of totals (RA), with the replication totals (R), nitrogen totals (A), and the grand total (G) computed, is shown in Table 3.8.

B.   The factor A × factor B two-way table of totals, with factor B totals computed. For our example, the nitrogen × variety table of totals (AB), with the variety totals (B) computed, is shown in Table 3.9.

☐ STEP 3.   Compute the correction factor and sums of squares for the main-plot analysis as:

$$C.F. = \frac{G^2}{rab}$$

$$= \frac{(394,481)^2}{(3)(6)(4)} = 2,161,323,047$$

Table 3.9   The Nitrogen × Variety Table of Yield Totals Computed from Data in Table 3.7

Nitrogen	Yield Total (AB) V_1	V_2	V_3	V_4
N_0	12,758	12,918	9,550	13,444
N_1	17,016	17,946	16,328	14,448
N_2	19,200	18,777	17,982	14,436
N_3	20,198	20,685	18,042	11,448
N_4	22,690	20,852	20,062	6,140
N_5	26,102	19,621	18,196	5,642
Variety total (B)	117,964	110,799	100,160	65,558

Total $SS = \Sigma X^2 - C.F.$

$$= \left[(4,430)^2 + \cdots + (2,014)^2\right] - 2,161,323,047$$

$$= 204,747,916$$

Replication $SS = \dfrac{\Sigma R^2}{ab} - C.F.$

$$= \frac{(128,873)^2 + (135,604)^2 + (130,004)^2}{(6)(4)} - 2,161,323,047$$

$$= 1,082,577$$

A (nitrogen) $SS = \dfrac{\Sigma A^2}{rb} - C.F.$

$$= \frac{(48,670)^2 + \cdots + (69,561)^2}{(3)(4)} - 2,161,323,047$$

$$= 30,429,200$$

Error(a) $SS = \dfrac{\Sigma(RA)^2}{b} - C.F.-$ Replication $SS - A$ SS

$$= \frac{(15,964)^2 + \cdots + (22,318)^2}{(4)} - 2,161,323,047$$

$$-1,082,577 - 30,429,200$$

$$= 1,419,678$$

☐ STEP 4.   Compute the sums of squares for the subplot analysis as:

B (variety) $SS = \dfrac{\Sigma B^2}{ra} - C.F.$

$$= \frac{(117,964)^2 + \cdots + (65,558)^2}{(3)(6)} - 2,161,323,047$$

$$= 89,888,101$$

$A \times B$ (nitrogen $\times$ variety) $SS - \dfrac{\Sigma(AB)^2}{r} - C.F. - B\ SS - A\ SS$

$$= \frac{(12{,}758)^2 + \cdots + (5{,}642)^2}{3}$$

$$- 2{,}161{,}323{,}047$$

$$- 89{,}888{,}101 - 30{,}429{,}200$$

$$= 69{,}343{,}487$$

Error(b) SS = Total SS − (sum of all other SS)

$$= 204{,}747{,}916 - (1{,}082{,}577 + 30{,}429{,}200 + 1{,}419{,}678$$

$$+ 89{,}888{,}101 + 69{,}343{,}487)$$

$$= 12{,}584{,}873$$

☐ STEP 5.   For each source of variation, compute the mean square by dividing the SS by its corresponding $d.f.$:

$$\text{Replication } MS = \frac{\text{Replication } SS}{r - 1}$$

$$= \frac{1{,}082{,}577}{2} = 541{,}228$$

$$A\ MS = \frac{A\ SS}{a - 1}$$

$$= \frac{30{,}429{,}200}{5} = 6{,}085{,}840$$

$$\text{Error}(a)\ MS = \frac{\text{Error}(a)\ SS}{(r - 1)(a - 1)}$$

$$= \frac{1{,}419{,}678}{10} = 141{,}968$$

$$B\ MS = \frac{B\ SS}{b - 1}$$

$$= \frac{89{,}888{,}101}{3} = 29{,}962{,}700$$

$$A \times B \; MS = \frac{A \times B \; SS}{(a-1)(b-1)}$$

$$= \frac{69{,}343{,}487}{15} = 4{,}622{,}899$$

$$\text{Error}(b) \; MS = \frac{\text{Error}(b) \; SS}{a(r-1)(b-1)}$$

$$= \frac{12{,}584{,}873}{36} = 349{,}580$$

☐ STEP 6.   Compute the F value for each effect that needs to be tested, by dividing each mean square by its corresponding error term:

$$F(A) = \frac{A \; MS}{\text{Error}(a) \; MS}$$

$$= \frac{6{,}085{,}840}{141{,}968} = 42.87$$

$$F(B) = \frac{B \; MS}{\text{Error}(b) \; MS}$$

$$= \frac{29{,}962{,}700}{349{,}580} = 85.71$$

$$F(A \times B) = \frac{A \times B \; MS}{\text{Error}(b) \; MS}$$

$$= \frac{4{,}622{,}899}{349{,}580} = 13.22$$

☐ STEP 7.   For each effect whose computed F value is not less than 1, obtain the corresponding tabular F value, from Appendix E, with $f_1 = d.f.$ of the numerator MS and $f_2 = d.f.$ of the denominator MS, at the prescribed level of significance. For example, the tabular F values for $F(A \times B)$ are 1.96 at the 5% level of significance and 2.58 at the 1% level.

☐ STEP 8.   Compute the two coefficients of variation, one corresponding to the main-plot analysis and another corresponding to the subplot analysis:

$$cv(a) = \frac{\sqrt{\text{Error}(a) \; MS}}{\text{Grand mean}} \times 100$$

$$= \frac{\sqrt{141{,}968}}{5{,}479} \times 100 = 6.9\%$$

$$cv(b) = \frac{\sqrt{\text{Error}(b)\ MS}}{\text{Grand mean}} \times 100$$

$$= \frac{\sqrt{349{,}580}}{5{,}479} \times 100 = 10.8\%$$

The value of $cv(a)$ indicates the degree of precision attached to the main-plot factor. The value of $cv(b)$ indicates the precision of the subplot factor and its interaction with the main-plot factor. The value of $cv(b)$ is expected to be smaller than that of $cv(a)$ because, as indicated earlier, the factor assigned to the main plot is expected to be measured with less precision than that assigned to the subplot. This trend does not always hold, however, as shown by this example in which the value of $cv(b)$ is larger than that of $cv(a)$. The cause for such an unexpected outcome is beyond the scope of this book. If such results occur *frequently*, a competent statistician should be consulted.

☐ STEP 9. Enter all values obtained from steps 3 to 8 in the analysis of variance outline of step 1, as shown in Table 3.10; and compare each of the computed F values with its corresponding tabular F values and indicate its significance by the appropriate asterisk notation (see Chapter 2, Section 2.1.2).

For our example, all the three effects (the two main effects and the interaction effect) are highly significant. With a significant interaction, caution must be exercised when interpreting the results (see Section 3.1). For proper comparisons between treatment means when the interaction effect is present, see Chapter 5, Section 5.2.4.

Table 3.10 Analysis of Variance of Data in Table 3.7 from a 4 × 6 Factorial Experiment in a Split-Plot Design[a]

Source of Variation	Degree of Freedom	Sum of Squares	Mean Square	Computed F[b]	Tabular F 5%	1%
Replication	2	1,082,577	541,228			
Nitrogen (A)	5	30,429,200	6,085,840	42.87**	3.33	5.64
Error(a)	10	1,419,678	141,968			
Variety (B)	3	89,888,101	29,962,700	85.71**	2.86	4.38
$A \times B$	15	69,343,487	4,622,899	13.22**	1.96	2.58
Error(b)	36	12,584,873	349,580			
Total	71	204,747,916				

[a]$cv(a) = 6.9\%$, $cv(b) = 10.8\%$.
[b]** = significant at 1% level.

3.5 STRIP-PLOT DESIGN

The strip-plot design is specifically suited for a two-factor experiment in which the desired precision for measuring the interaction effect between the two factors is higher than that for measuring the main effect of either one of the two factors. This is accomplished with the use of three plot sizes:

1. *Vertical-strip plot* for the first factor—the *vertical factor*
2. *Horizontal-strip plot* for the second factor—the *horizontal factor*
3. *Intersection plot* for the interaction between the two factors

The vertical-strip plot and the horizontal-strip plot are always perpendicular to each other. However, there is no relationship between their sizes, unlike the case of main plot and subplot of the split-plot design. The intersection plot is, of course, the smallest. Thus, in a strip-plot design, the degrees of precision associated with the main effects of both factors are sacrificed in order to improve the precision of the interaction effect.

3.5.1 Randomization and Layout

The procedure for randomization and layout of a strip-plot design consists of two independent randomization processes—one for the horizontal factor and another for the vertical factor. The order in which these two processes are performed is immaterial.

Let A represent the horizontal factor and B the vertical factor, and a and b represent their levels. As in all previous cases, r represents the number of replications. We illustrate the steps involved with a two-factor experiment involving six rice varieties (horizontal treatments) and three nitrogen rates (vertical treatments) tested in a strip-plot design with three replications.

☐ STEP 1. Assign horizontal plots by dividing the experimental area into $r = 3$ blocks and dividing each of those into $a = 6$ horizontal strips. Follow the randomization procedure for a randomized complete block design with $a = 6$ treatments and $r = 3$ replications (see Chapter 2, Section 2.2.2) and randomly assign the six varieties to the six horizontal strips in each of the three blocks, separately and independently. The result is shown in Figure 3.6.

☐ STEP 2. Assign vertical plots by dividing each block into $b = 3$ vertical strips. Follow the randomization procedure for a randomized complete block with $b = 3$ treatments and $r = 3$ replications (see Chapter 2, Section 2.2.2) and randomly assign the three nitrogen rates to the three vertical strips in each of the three blocks, separately and independently. The final layout is shown in Figure 3.7.

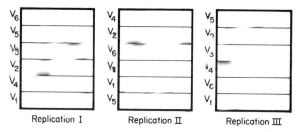

Figure 3.6 Random assignment of six varieties (V_1, V_2, V_3, V_4, V_5, and V_6) to the horizontal strips in a strip-plot design with three replications.

3.5.2 Analysis of Variance

The analysis of variance of a strip-plot design is divided into three parts: the *horizontal-factor analysis*, the *vertical-factor analysis*, and the *interaction analysis*. We show the computational procedure with data from a two-factor experiment involving six rice varieties (horizontal factor) and three nitrogen levels (vertical factor) tested in three replications. The field layout is shown in Figure 3.7; the data is in Table 3.11.

☐ STEP 1. Construct an outline of the analysis of variance for a strip-plot design as:

Source of Variation	Degree of Freedom	Sum of Squares	Mean Square	Computed F	Tabular F 5%	Tabular F 1%
Replication	$r - 1 = 2$					
Horizontal factor (A)	$a - 1 = 5$					
Error(a)	$(r - 1)(a - 1) = 10$					
Vertical factor (B)	$b - 1 = 2$					
Error(b)	$(r - 1)(b - 1) = 4$					
$A \times B$	$(a - 1)(b - 1) = 10$					
Error(c)	$(r - 1)(a - 1)(b - 1) = 20$					
Total	$rab - 1 = 53$					

☐ STEP 2. Construct three tables of totals:

1. The replication × horizontal-factor table of totals with replication totals, horizontal-factor totals, and grand total computed. For our example, the replication × variety table of totals (RA) with replication totals (R), variety totals (A), and the grand total (G) computed is shown in Table 3.12.

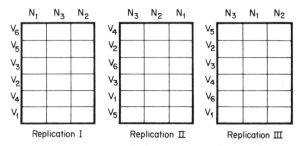

Figure 3.7 A sample layout of a strip-plot design with six varieties (V_1, V_2, V_3, V_4, V_5, and V_6) as horizontal treatments and three nitrogen rates (N_1, N_2, and N_3) as vertical treatments, in three replications.

Table 3.11 Data on Grain Yield of Six Varieties of Rice, Broadcast Seeded and Grown with Three Nitrogen Rates in a Strip-plot Design with Three Replications

Nitrogen Rate, kg/ha	Grain Yield, kg/ha		
	Rep. I	Rep. II	Rep. III
IR8(V_1)			
0 (N_1)	2,373	3,958	4,384
60 (N_2)	4,076	6,431	4,889
120 (N_3)	7,254	6,808	8,582
IR127-80(V_2)			
0	4,007	5,795	5,001
60	5,630	7,334	7,177
120	7,053	8,284	6,297
IR305-4-12(V_3)			
0	2,620	4,508	5,621
60	4,676	6,672	7,019
120	7,666	7,328	8,611
IR400-2-5(V_4)			
0	2,726	5,630	3,821
60	4,838	7,007	4,816
120	6,881	7,735	6,667
IR665-58(V_5)			
0	4,447	3,276	4,582
60	5,549	5,340	6,011
120	6,880	5,080	6,076
Peta (V_6)			
0	2,572	3,724	3,326
60	3,896	2,822	4,425
120	1,556	2,706	3,214

2. The replication × vertical-factor table of totals with the vertical-factor totals computed. For our example, the replication × nitrogen table of totals (*RB*) with nitrogen totals (*B*) computed is shown in Table 3.13.

3. The horizontal factor × vertical-factor table of totals. For our example, the variety × nitrogen table of totals (*AB*) is shown in Table 3.14.

□ STEP 3.   Compute the correction factor and the total sum of squares as:

$$C.F. = \frac{G^2}{rab}$$

$$= \frac{(285,657)^2}{(3)(6)(3)} = 1,511,109,660$$

Total $SS = \Sigma X^2 - C.F.$

$$= \left[(2,373)^2 + \cdots + (3,214)^2 \right] - 1,511,109,660$$

$$= 167,005,649$$

□ STEP 4.   Compute the sums of squares for the horizontal analysis as:

Replication $SS = \dfrac{\Sigma R^2}{ab} - C.F.$

$$= \frac{(84,700)^2 + (100,438)^2 + (100,519)^2}{(6)(3)} - 1,511,109,660$$

$$= 9,220,962$$

A (variety) $SS = \dfrac{\Sigma A^2}{rb} - C.F.$

$$= \frac{(48,755)^2 + \cdots + (28,241)^2}{(3)(3)} - 1,511,109,660$$

$$= 57,100,201$$

$$\text{Error}(a)\ SS = \frac{\Sigma(RA)^2}{b} - C.F. - \text{Replication } SS - A\ SS$$

$$= \frac{(13,703)^2 + \cdots + (10,965)^2}{3} - 1,511,109,660$$

$$- 9,220,962 - 57,100,201$$

$$= 14,922,620$$

Table 3.12   The ReplicationVariety Table of Yield Totals Computed from Data in Table 3.11

Variety	Yield Total (RA)			Variety Total (A)
	Rep. I	Rep. II	Rep. III	
V_1	13,703	17,197	17,855	48,755
V_2	16,690	21,413	18,475	56,578
V_3	14,962	18,508	21,251	54,721
V_4	14,445	20,372	15,304	50,121
V_5	16,876	13,696	16,669	47,241
V_6	8,024	9,252	10,965	28,241
Rep. total (R)	84,700	100,438	100,519	
Grand total (G)				285,657

Table 3.13   The Replication × Nitrogen Table of Yield Totals Computed from Data in Table 3.11

Nitrogen	Yield Total (RB)			Nitrogen Total (B)
	Rep. I	Rep. II	Rep. III	
N_1	18,745	26,891	26,735	72,371
N_2	28,665	35,606	34,337	98,608
N_3	37,290	37,941	39,447	114,678

Table 3.14   The Variety × Nitrogen Table of Yield Totals Computed from Data in Table 3.11

Variety	Yield Total (AB)		
	N_1	N_2	N_3
V_1	10,715	15,396	22,644
V_2	14,803	20,141	21,634
V_3	12,749	18,367	23,605
V_4	12,177	16,661	21,283
V_5	12,305	16,900	18,036
V_6	9,622	11,143	7,476

☐ STEP 5.   Compute the sums of squares for the vertical analysis as:

$$B \text{ (nitrogen) } SS = \frac{\Sigma B^2}{ra} - C.F.$$

$$= \frac{(12,371)^2 + (98,608)^2 + (114,678)^2}{(3)(6)} - 1,511,109,660$$

$$= 50,676,061$$

$$\text{Error}(b) \ SS = \frac{\Sigma (RB)^2}{a} - C.F. - \text{Replication } SS - B \ SS$$

$$= \frac{(18,745)^2 + \cdots + (39,447)^2}{6} - 1,511,109,660$$

$$- 9,220,962 - 50,676,061$$

$$= 2,974,909$$

☐ STEP 6.   Compute the sums of squares for the interaction analysis as:

$$A \times B \text{ (variety} \times \text{nitrogen) } SS = \frac{\Sigma (AB)^2}{r} - C.F. - A \ SS - B \ SS$$

$$= \frac{(10,715)^2 + \cdots + (7,476)^2}{3}$$

$$- 1,511,109,660$$

$$- 57,100,201 - 50,676,061$$

$$= 23,877,980$$

$$\text{Error}(c) \ SS = \text{Total } SS - (\text{the sum of all other } SS)$$

$$= 167,005,649 - (9,220,962 + 57,100,201 + 14,922,620$$

$$+ 50,676,061 + 2,974,909 + 23,877,980)$$

$$= 8,232,916$$

☐ STEP 7.   Compute the mean square for each source of variation by dividing the SS by its *d.f.*:

$$\text{Replication } MS = \frac{9,220,962}{2} = 4,610,481$$

$$A \ MS = \frac{57,100,201}{5} = 11,420,040$$

$$\text{Error}(a)\ MS = \frac{14{,}922{,}620}{10} = 1{,}492{,}262$$

$$B\ MS = \frac{50{,}676{,}061}{2} = 25{,}338{,}031$$

$$\text{Error}(b)\ MS = \frac{2{,}974{,}909}{4} = 743{,}727$$

$$A \times B\ MS = \frac{23{,}877{,}980}{10} = 2{,}387{,}798$$

$$\text{Error}(c)\ MS = \frac{8{,}232{,}916}{20} = 411{,}646$$

☐ STEP 8.   Compute the F values as:

$$F(A) = \frac{A\ MS}{\text{Error}(a)\ MS}$$

$$F(B) = \frac{B\ MS}{\text{Error}(b)\ MS}$$

$$F(A \times B) = \frac{A \times B\ MS}{\text{Error}(c)\ MS}$$

For our example, because the $d.f.$ for error (b) MS is only 4, which is considered inadequate for a reliable estimate of the error variance (see Chapter 2, Section 2.1.2), no test of significance for the main effect of factor B is to be made. Hence, the two other F values are computed as:

$$F(A) = \frac{11{,}420{,}040}{1{,}492{,}262} = 7.65$$

$$F(A \times B) = \frac{2{,}387{,}798}{411{,}646} = 5.80$$

☐ STEP 9.   For each effect whose computed F value is not less than 1, obtain the corresponding tabular F value, from Appendix E, with $f_1 = d.f.$ of the numerator MS and $f_2 = d.f.$ of the denominator MS at the prescribed level of significance.

For our example, the tabular F values corresponding to the computed $F(A \times B)$ value, with $f_1 = 10$ and $f_2 = 20$ degrees of freedom, are 2.35 at the 5% level of significance and 3.37 at the 1% level.

□ STEP 10. Compute the three coefficients of variation corresponding to the three error mean squares as:

$$cv(a) \quad \frac{\sqrt{\text{Error}(a)\ MS}}{\text{Grand Mean}} \times 100$$

$$cv(b) = \frac{\sqrt{\text{Error}(b)\ MS}}{\text{Grand Mean}} \times 100$$

$$cv(c) = \frac{\sqrt{\text{Error}(c)\ MS}}{\text{Grand Mean}} \times 100$$

The $cv(a)$ value indicates the degree of precision associated with the horizontal factor, $cv(b)$ with the vertical factor, and $cv(c)$ with the interaction between the two factors. The value of $cv(c)$ is expected to be the smallest and the precision for measuring the interaction effect is, thus, the highest. For $cv(a)$ and $cv(b)$, however, there is no basis to expect one to be greater or smaller than the other.

For our example, because the $d.f.$ for error(b) MS is inadequate, $cv(b)$ is not computed. The cv values for the two other error terms are computed as:

$$cv(a) = \frac{\sqrt{1,492,262}}{5,290} \times 100 = 23.1\%$$

$$cv(c) = \frac{\sqrt{411,646}}{5,290} \times 100 = 12.1\%$$

Table 3.15   Analysis of Variance of Data in Table 3.11 from a 3 × 6 Factorial Experiment in a Strip-plot Design[a]

Source of Variation	Degree of Freedom	Sum of Squares	Mean Square	Computed F[b]	Tabular F 5%	1%
Replication	2	9,220,962	4,610,481			
Variety (A)	5	57,100,201	11,420,040	7.65**	3.33	5.64
Error(a)	10	14,922,620	1,492,262			
Nitrogen (B)	2	50,676,061	25,338,031	[c]	—	—
Error(b)	4	2,974,909	743,727			
A × B	10	23,877,980	2,387,798	5.80**	2.35	3.37
Error(c)	20	8,232,916	411,646			
Total	53	167,005,649				

[a] $cv(a) = 23.1\%$, $cv(c) = 12.1\%$.
[b] ** = significant at 1% level.
[c] Error(b) $d.f$ is not adequate for valid test of significance.

□ STEP 11. Enter all values computed in steps 3 to 10 in the analysis of variance outline of step 1, as shown in Table 3.15. Compare each computed *F* value with its corresponding tabular *F* values and designate the significant results with the appropriate asterisk notation (see Chapter 2, Section 2.1.2). For our example, both *F* values, one corresponding to the main effect of variety and another to the interaction between variety and nitrogen, are significant. With a significant interaction, caution must be exercised when interpreting the results. See Chapter 5, Section 5.2.4 for appropriate mean comparisons.

3.6 GROUP BALANCED BLOCK IN SPLIT-PLOT DESIGN

The group balanced block design described in Chapter 2, Section 2.5, for single-factor experiments can be used for two-factor experiments. This is done by applying the rules for grouping of treatments (described in Section 2.5) to either one, or both, of the two factors. Thus, the group balanced block design can be superimposed on the split-plot design resulting in what is generally called the group balanced block in split-plot design; or it can be superimposed on the strip-plot design resulting in a group balanced block in strip-plot design.

We limit our discussion to a group balanced block in split-plot design and illustrate it using an experiment with 45 rice varieties and two fertilizer levels. The basic design is a split-plot design in three replications, with fertilizer as the main-plot factor and variety as the subplot factor. The 45 varieties are grouped, according to their growth duration, into group S_1 with less than 105 days, group S_2 with 105 to 115 days, and group S_3 with longer than 115 days. We denote the main-plot factor by A, the subplot factor by B, the level of factor A by a, the level of factor B by b, the number of replications by r, the number of groups in which the b subplot treatments are classified by s, and the group identification by $S_1, S_2, \ldots, S_s$.

3.6.1 Randomization and Layout

The steps in the randomization and layout of the group balanced block in split-plot design are:

□ STEP 1. Divide the experimental area into $r = 3$ replications, each of which is further divided into $a = 2$ main plots. Following the randomization procedure for the standard split-plot design described in Section 3.4.1, randomly assign the two main-plot treatments (F_1 and F_2) to the two main plots in each replication. The result may be as shown in Figure 3.8.

□ STEP 2. Divide each of the six main plots (two main plots for each of the three replications) into three groups of plots, each group consisting of 15

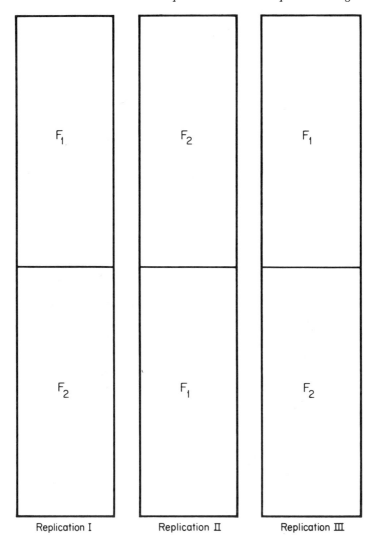

Figure 3.8 Random assignment of two fertilizer rates (main-plot treatments: F_1 and F_2) to the two main plots in each replication, as the first step in the laying out of a group balanced block in split-plot design with three replications.

plots. Using one of the randomization schemes of Chapter 2, Section 2.1.1, randomly assign the three groups of varieties (S_1, S_2, and S_3) to the three groups of plots, separately and independently, for each of the six main plots. The result may be as shown in Figure 3.9.

□ STEP 3. Using the same randomization scheme as in step 2, randomly assign the 15 varieties of each group (i.e., treatments $1, \ldots, 15$ for group S_1, treatments $16, \ldots, 30$ for group S_2, and treatments $31, \ldots, 45$ for group S_3)

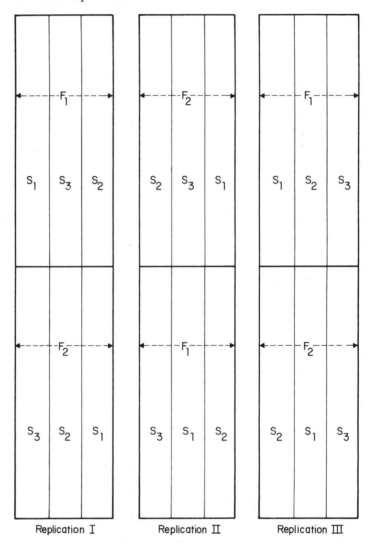

Replication I   Replication II   Replication III

Figure 3.9 A sample layout after the two fertilizer rates (F_1 and F_2) and three groups of varieties (S_1, S_2, and S_3) are assigned, in a group balanced block in split-plot design with three replications.

to the 15 plots in the corresponding group of plots. This process is repeated 18 times (three groups per main plot and a total of six main plots). The final layout may be as shown in Figure 3.10.

3.6.2 Analysis of Variance

For illustration we use data (Table 3.16) from the group balanced block in split-plot design whose layout is shown in Figure 3.10. The computational

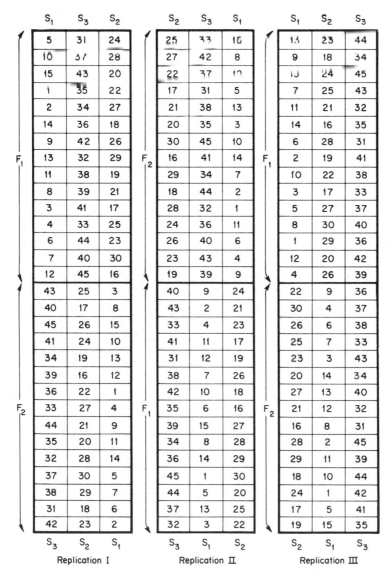

Figure 3.10 A sample layout of a group balanced block in split-plot design with two fertilizer rates (F_1 and F_2) as main-plot treatments and 45 rice varieties (1, 2, ..., 45) grouped in three groups (S_1, S_2, and S_3) as subplot treatments, in three replications.

119

Table 3.16 Grain Yield of 45 Rice Varieties Tested with Two Fertilizer Rates (F_1 and F_2) Using a Group Balanced Block in Split-plot Design with Fertilizer as Main-plot Factor and Variety, Classified in Three Groups,[a] as the Subplot Factor; in Three Replications

Variety Number	Grain Yield, t/ha					
	Rep. I		Rep. II		Rep. III	
	F_1	F_2	F_1	F_2	F_1	F_2
1	4.252	4.331	3.548	5.267	3.114	4.272
2	3.463	3.801	2.720	5.145	2.789	3.914
3	3.228	3.828	2.797	4.498	2.860	4.163
4	4.153	5.082	3.672	5.401	3.738	4.533
5	3.672	4.275	2.781	5.510	2.788	4.481
6	3.337	4.346	2.803	5.874	2.936	4.075
7	3.498	5.557	3.725	4.666	2.627	4.781
8	3.222	4.451	3.142	3.870	2.922	3.721
9	3.161	4.349	3.108	5.293	2.779	4.101
10	3.781	4.603	3.906	4.684	3.295	4.100
11	3.763	5.188	3.709	4.887	3.612	4.798
12	3.177	4.975	3.742	5.021	2.933	4.611
13	3.000	4.643	2.843	5.204	2.776	3.998
14	4.040	4.991	3.251	4.545	3.220	4.253
15	3.790	4.313	3.027	4.742	3.125	4.411
16	3.955	4.311	3.030	4.830	3.000	4.765
17	3.843	4.815	3.207	4.804	3.285	4.263
18	3.558	4.082	3.271	4.817	3.154	4.433
19	3.488	4.140	3.278	4.197	2.784	4.237
20	2.957	5.027	3.284	4.429	2.816	4.415
21	3.237	4.434	2.835	4.030	3.018	3.837
22	3.617	4.570	2.985	4.565	2.958	4.109
23	4.193	5.025	3.639	4.760	3.428	5.225
24	3.611	4.744	3.023	4.221	2.805	3.972
25	3.328	4.274	2.955	4.069	3.031	3.922
26	4.082	4.356	3.089	4.232	2.987	4.181
27	4.063	4.391	3.367	5.069	3.931	4.782
28	3.597	4.494	3.211	4.506	3.238	4.410
29	3.268	4.224	3.913	4.569	3.057	4.377
30	4.030	5.576	3.223	4.229	3.867	5.344
31	3.943	5.056	3.133	4.512	3.357	4.373
32	2.799	3.897	3.184	3.874	2.746	4.499
33	3.479	4.168	3.377	4.036	4.036	4.472
34	3.498	4.502	2.912	4.343	3.479	4.651
35	3.431	5.018	2.879	4.590	3.505	4.510
36	4.140	5.494	4.107	4.856	3.563	4.523
37	4.051	4.600	4.206	4.946	3.563	4.340
38	3.647	4.334	2.863	4.892	2.848	4.509

Table 3.16 (*Continued*)

| Variety Number | Grain Yield, t/ha | | | | | |
| | Rep. I | | Rep. II | | Rep. III | |
	F_1	F_2	F_1	F_2	F_1	F_2
39	4.262	4.852	3.197	4.530	3.680	4.371
40	4.256	5.409	3.091	4.533	3.751	5.134
41	4.501	5.659	3.770	5.050	3.825	4.776
42	4.334	5.121	3.666	5.156	4.222	5.229
43	4.416	4.785	3.824	4.969	3.096	4.870
44	3.578	4.664	3.252	5.582	4.091	4.362
45	4.270	4.993	3.896	5.827	4.312	4.918

[a]Group S_1 (less than 105 days in growth duration) consists of varieties 1 to 15; group S_2 (105 to 115 days in growth duration) consists of varieties 16 to 30; and group S_3 (longer than 115 days in growth duration) consists of varieties 31 to 45.

steps in the analysis of variance are:

□ STEP 1.   Outline the analysis of variance for a group balanced block in split-plot design, with grouping of subplot treatments, as:

Source of Variation	Degree of Freedom	Sum of Squares	Mean Square
Replication	$r - 1 = 2$		
Factor A (A)	$a - 1 = 1$		
Error(a)	$(r - 1)(a - 1) = 2$		
Group (S)	$s - 1 = 2$		
$A \times S$	$(a - 1)(r - 1) = 2$		
Error(b)	$a(s - 1)(r - 1) = 8$		
B within S_1	$(b/s) - 1 = 14$		
B within S_2	$(b/s) - 1 = 14$		
B within S_3	$(b/s) - 1 = 14$		
$A \times (B$ within $S_1)$	$(a - 1)[(b/s) - 1] = 14$		
$A \times (B$ within $S_2)$	$(a - 1)[(b/s) - 1] = 14$		
$A \times (B$ within $S_3)$	$(a - 1)[(b/s) - 1] = 14$		
Error(c)	$as(r - 1)[(b/s) - 1] = 168$		
Total	$rab - 1 = 269$		

□ STEP 2.   Construct three tables of totals:

1. The replication × factor A × group three-way table of totals with replication × factor A totals and replication totals computed. For our example, the replication × fertilizer × group table of totals (RAS), with replication × fertilizer totals (RA) and replication totals (R) computed, is shown in Table 3.17.

2. The factor A × factor B two-way table of totals with factor B totals, factor A totals, group × factor A totals, group totals, and the grand total computed. For our example, the fertilizer × variety table of totals (AB), with variety totals (B), fertilizer totals (A), group × fertilizer totals (SA), group totals (S), and the grand total (G) computed, is shown in Table 3.18.

□ STEP 3.   Compute the correction factor and the various sums of squares in the standard manner as:

$$C.F. = \frac{G^2}{rab}$$

$$= \frac{(1,085.756)^2}{(3)(2)(45)} = 4,366.170707$$

Total $SS = \sum X^2 - C.F.$

$$= \left[(4.252)^2 + \cdots + (4.918)^2 \right] - 4,366.170707$$

$$= 154.171227$$

Table 3.17   The Replication × Fertilizer × Group Table of Totals from Data in Table 3.16

| | Yield Total (RAS) | | | | | |
| | Rep. I | | Rep. II | | Rep.III | |
Group	F_1	F_2	F_1	F_2	F_1	F_2
S_1	53.537	68.733	48.774	74.607	45.514	64.212
S_2	54.827	68.463	48.310	67.327	47.359	66.272
S_3	58.605	72.552	51.357	71.696	54.074	69.537
Total (RA)	166.969	209.748	148.441	213.630	146.947	200.021
Rep. total (R)	376.717		362.071		346.968	

Table 3.18   The Fertilizer × Variety Table of Totals from Data in Table 3.16

Variety Number	Yield Total (AB)		Variety total
	F_1	F_2	(B)
1	10.914	13.870	24.784
2	8.972	12.860	21.832
3	8.885	12.489	21.374
4	11.563	15.016	26.579
5	9.241	14.266	23.507
6	9.076	14.295	23.371
7	9.850	15.004	24.854
8	9.286	12.042	21.328
9	9.048	13.743	22.791
10	10.982	13.387	24.369
11	11.084	14.873	25.957
12	9.852	14.607	24.459
13	8.619	13.845	22.464
14	10.511	13.789	24.300
15	9.942	13.466	23.408
Total (SA)	147.825	207.552	355.377 = S_1
16	9.985	13.906	23.891
17	10.335	13.882	24.217
18	9.983	13.332	23.315
19	9.550	12.574	22.124
20	9.057	13.871	22.928
21	9.090	12.301	21.391
22	9.560	13.244	22.804
23	11.260	15.010	26.270
24	9.439	12.937	22.376
25	9.314	12.265	21.579
26	10.158	12.769	22.927
27	11.361	14.242	25.603
28	10.046	13.410	23.456
29	10.238	13.170	23.408
30	11.120	15.149	26.269
Total (SA)	150.496	202.062	352.558 = S_2
31	10.433	13.941	24.374
32	8.729	12.270	20.999
33	10.892	12.676	23.568
34	9.889	13.496	23.385
35	9.815	14.118	23.933
36	11.810	14.873	26.683
37	11.820	13.886	25.706
38	9.358	13.735	23.093
39	11.139	13.753	24.892
40	11.098	15.076	26.174
41	12.096	15.485	27.581

Table 3.18   (*Continued*)

Variety Number	Yield Total (AB)		Variety total (B)
	F_1	F_2	
42	12.222	15.506	27.728
43	11.336	14.624	25.960
44	10.921	14.608	25.529
45	12.478	15.738	28.216
Total (SA)	164.036	213.785	377.821 = S_3
Fertilizer total (A)	462.357	623.399	
Grand Total (G)			1,085.756

Replication $SS = \dfrac{\Sigma R^2}{ab} - C.F.$

$$= \frac{(376.717)^2 + (362.071)^2 + (346.968)^2}{(2)(45)} - 4{,}366.170707$$

$$= 4.917070$$

A(fertilizer) $SS = \dfrac{\Sigma A^2}{rb} - C.F.$

$$= \frac{(462.357)^2 + (623.399)^2}{(3)(45)} - 4{,}366.170707$$

$$= 96.053799$$

Error(a) $SS = \dfrac{\Sigma(RA)^2}{b} - C.F. -$ Replication $SS - A\ SS$

$$= \frac{(166.969)^2 + \cdots + (200.021)^2}{45} - 4{,}366.170707 - 4.917070$$

$$- 96.053799$$

$$= 2.796179$$

Group $SS = \dfrac{\Sigma S^2}{rab/s} - C.F.$

$$= \frac{(355.377)^2 + (352.558)^2 + (377.821)^2}{(3)(2)(45)/(3)} - 4{,}366.170707$$

$$= 4.258886$$

$$A \times \text{Group } SS = \frac{\Sigma (SA)^2}{rb/s} - C.F. - A\ SS - \text{Group } SS$$

$$= \frac{(147.825)^2 + \cdots + (213.785)^2}{(3)(45)/(3)} - 4{,}366.170707$$

$$- 96.053799 - 4.258886$$

$$= 0.627644$$

$$\text{Error}(b)\ SS = \frac{\Sigma (RAS)^2}{b/s} - C.F. - \text{Replication } SS - A\ SS$$

$$- \text{Error}(a)\ SS - \text{Group } SS - A \times \text{Group } SS$$

$$= \frac{(53.537)^2 + \cdots + (69.537)^2}{45/3} - 4{,}366.170707 - 4.917070$$

$$- 96.053799 - 2.796179 - 4.258886 - 0.627644$$

$$= 2.552576$$

☐ STEP 4. Compute the sums of squares for factor B within the ith group and for its interaction with factor A as:

$$B \text{ within } S_i\ SS = \frac{\Sigma B_i^2}{ra} - \frac{S_i^2}{rab/s}$$

$$A \times (B \text{ within } S_i)\ SS = \frac{\Sigma (AB)_i^2}{r} - \frac{\Sigma (SA)_i^2}{rb/s}$$

$$- B \text{ within } S_i\ SS$$

where the subscript i refers to the ith group and the summation is only over all those totals belonging to the ith group. For example, the summation in the term ΣB_i^2 only covers factor B totals of those levels of factor B belonging to the ith group.

For our example, the computations for each of the three groups are:

• For S_1:

$$\text{Varieties within } S_1 SS = \frac{(24.784)^2 + \cdots + (23.408)^2}{(3)(2)} - \frac{(355.377)^2}{(3)(2)(45)/3}$$

$$= 5.730485$$

$$A \times (\text{varieties within } S_1) \text{ SS} = \frac{(10.914)^2 + \cdots + (13.466)^2}{3}$$

$$- \frac{(147.825)^2 + (207.552)^2}{(3)(45)/(3)} - 5.730485$$

$$= 2.143651$$

- For S_2:

$$\text{Varieties within } S_2 \text{ SS} = \frac{(23.891)^2 + \cdots + (26.269)^2}{(3)(2)} - \frac{(352.558)^2}{(3)(2)(45)/3}$$

$$= 5.484841$$

$$A \times (\text{varieties within } S_2) \text{ SS} = \frac{(9.985)^2 + \cdots + (15.149)^2}{3}$$

$$- \frac{(150.496)^2 + (202.062)^2}{(3)(45)/3} - 5.484841$$

$$= 0.728832$$

- For S_3:

$$\text{Varieties within } S_3 \text{ SS} = \frac{(24.374)^2 + \cdots + (28.216)^2}{(3)(2)} - \frac{(377.821)^2}{(3)(2)(45)/3}$$

$$= 9.278639$$

$$A \times (\text{varieties within } S_3) \text{ SS} = \frac{(10.433)^2 + \cdots + (15.738)^2}{3}$$

$$- \frac{(164.036)^2 + (213.785)^2}{(3)(45)/3} - 9.278639$$

$$= 1.220758$$

□ STEP 5.   Compute the Error(c) SS as:

$$\text{Error}(c) \text{ SS} = \text{Total SS} - (\text{the sum of all other SS})$$

$$= 154.171227 - (4.917070 + 96.053799$$

$$+ 2.796179 + 4.258886 + 0.627644$$

$$+ 2.552576 + 5.730485 + 5.484841$$

$$+ 9.278639 + 2.143651 + 0.728832$$

$$+ 1.220758)$$

$$= 18.377867$$

☐ STEP 6.   Compute the mean square for each source of variation by dividing the SS by its degree of freedom. Then, compute the F value for each effect to be tested by dividing its mean square by the appropriate error mean square.

For our example, for either the replication or the A effect, the divisor is the Error(a) MS. For either the group or the ($A \times$ group) effect, the divisor is the error(b) MS. For all other effects, the divisor is the Error(c) MS. Note that because the Error(a) $d.f.$ is only 2, which is considered inadequate for a reliable estimate of the error variance (see Chapter 2, Section 2.1.2), the F values for testing the replication effect and the A effect are not computed.

☐ STEP 7.   For each computed F value greater than 1, obtain the corresponding tabular F values, from Appendix E, with $f_1 = d.f.$ of the numerator MS and $f_2 = d.f.$ of denominator MS, at the 5% and 1% levels of significance.

☐ STEP 8.   Compute the three coefficients of variation corresponding to the three error terms as:

$$cv(a) = \frac{\sqrt{\text{Error}(a)\ MS}}{\text{Grand mean}} \times 100$$

$$cv(b) = \frac{\sqrt{\text{Error}(b)\ MS}}{\text{Grand mean}} \times 100$$

$$cv(c) = \frac{\sqrt{\text{Error}(c)\ MS}}{\text{Grand mean}} \times 100$$

The $cv(a)$ value indicates the degree of precision attached to the main-plot factor, the $cv(b)$ indicates the degree of precision attached to the group effect and its interaction with the main-plot factor, and the $cv(c)$ value refers to the effects of subplot treatments within the same group and their interactions with the main-plot factor. In the same manner as that of a standard split-plot design, the value of $cv(a)$ is expected to be the largest, followed by $cv(b)$, and finally $cv(c)$.

For our example, because $d.f.$ for Error(a) MS is inadequate, no value of $cv(a)$ is to be computed. The coefficients of variation for the two other error terms are computed as:

$$cv(b) = \frac{\sqrt{0.319072}}{4.021} \times 100 = 14.0\%$$

$$cv(c) = \frac{\sqrt{0.109392}}{4.021} \times 100 = 8.2\%$$

As expected, the $cv(c)$ value is smaller than the $cv(b)$ value. This implies that the degree of precision for mean comparisons involving treatments

Table 3.19 Analysis of Variance of Data in Table 3.16 from a Group Balanced Block in Split-plot Design[a]

Source of Variation	Degree of Freedom	Sum of Squares	Mean Square	Computed F[b]	Tabular F 5%	1%
Replication	2	4.917070	2.458535		—	—
Fertilizer (A)	1	96.053799	96.053799	[c]		
Error(a)	2	2.796179	1.398089			
Group (S)	2	4.258886	2.129443	6.67*	4.46	8.65
$A \times S$	2	0.627644	0.313822	< 1	—	—
Error(b)	8	2.552576	0.319072			
Varieties within S_1	14	5.730485	0.409320	3.74**	1.75	2.19
Varieties within S_2	14	5.484841	0.391774	3.58**	1.75	2.19
Varieties within S_3	14	9.278639	0.662760	6.06**	1.75	2.19
$A \times$ (varieties within S_1)	14	2.143651	0.153118	1.40^{ns}	1.75	2.19
$A \times$ (varieties within S_2)	14	0.728832	0.052059	< 1	—	—
$A \times$ (varieties within S_3)	14	1.220758	0.087197	< 1	—	—
Error(c)	168	18.377867	0.109392			
Total	269	154.171227				

[a]$cv(b) = 14.0\%$, $cv(c) = 8.2\%$.

[b]** = significant at 1% level, * = significant at 5% level, ^{ns} = not significant.

[c]Error(a) $d.f.$ is not adequate for valid test of significance.

belonging to the same group would be higher than that involving treatments of different groups.

☐ STEP 9. Enter all values obtained in steps 3 to 8 in the analysis of variance outline of step 1. Compare each computed F value with its corresponding tabular F values and indicate its significance by appropriate asterisk notation (see Chapter 2, Section 2.1.2). The final result is shown in Table 3.19. The results indicate nonsignificant interaction between variety and fertilizer rate, highly significant differences among varieties within each and all three groups, and a significant difference among the three group means.

CHAPTER 4

Three-or-More-Factor Experiments

A two-factor experiment can be expanded to include a third factor, a three-factor experiment to include a fourth factor, and so on. There are, however, two important consequences when factors are added to an experiment:

1. There is a rapid increase in the number of treatments to be tested, as we illustrated in Chapter 3.
2. There is an increase in the number and type of interaction effects. For example, a three-factor experiment has four interaction effects that can be examined. A four-factor experiment has 10 interaction effects.

Although a large experiment is usually not desirable because of its high cost and complexity, the added information gained from interaction effects among factors can be very valuable. Consequently, the researcher's decision on the number of factors that should be included in a factorial experiment is based on a compromise between the desire to evaluate as many interactions as possible and the need to keep the size of experiment within the limit of available resources.

4.1   INTERACTION BETWEEN THREE OR MORE FACTORS

Building on the definition of a two-factor interaction given in Chapter 3 (Section 3.1), a k-factor interaction (where $k > 2$) may be defined as the difference between the effects of a particular $(k - 1)$-factor interaction over the different levels of the kth factor. For example, a three-factor interaction effect among factors A, B, and C (the $A \times B \times C$ interaction) can be defined in any of the following three ways:

1. The difference between the $A \times B$ interaction effects over the levels of factor C.
2. The difference between the $A \times C$ interaction effects over the levels of factor B.

3. The difference between the $B \times C$ interaction effects over the levels of factor A.

For illustration, consider a $2 \times 2 \times 2$ factorial experiment involving three factors, each with two levels. Two sets of hypothetical data, set(a) showing the presence of the $A \times B \times C$ interaction effect, and set(b) showing its absence, are presented in Table 4.1.

The $A \times B \times C$ interaction effects can be measured by any of three methods:

- Method I is based on the difference in the $A \times B$ interaction effects.

☐ STEP 1.   For each level of factor C, compute the $A \times B$ interaction effect, following the procedure in Chapter 3, Section 3.1:

For set(a) data:

$$\text{At } c_0: A \times B \text{ interaction} = \tfrac{1}{2}(0.5 - 2.0) = -0.75 \text{ t/ha}$$

$$\text{At } c_1: A \times B \text{ interaction} = \tfrac{1}{2}(4.0 - 2.5) = 0.75 \text{ t/ha}$$

For set(b) data:

$$\text{At } c_0: A \times B \text{ interaction} = \tfrac{1}{2}(3.5 - 2.0) = 0.75 \text{ t/ha}$$

$$\text{At } c_1: A \times B \text{ interaction} = \tfrac{1}{2}(3.5 - 2.0) = 0.75 \text{ t/ha}$$

Table 4.1   Two Hypothetical Sets of Data from a $2 \times 2 \times 2$ Factorial Experiment[a]; Set (a) Shows the Presence of the Three-factor Interaction and Set (b) Shows the Absence of the Three-factor Interaction

Level of Factor A	Grain Yield, t/ha					
	c_0			c_1		
	b_0	b_1	$b_1 - b_0$	b_0	b_1	$b_1 - b_0$
	(a) $A \times B \times C$ interaction present					
a_0	2.0	3.0	1.0	2.5	5.0	2.5
a_1	4.0	3.5	-0.5	5.0	9.0	4.0
$a_1 - a_0$	2.0	0.5	-1.5	2.5	4.0	1.5
	(b) $A \times B \times C$ interaction absent					
a_0	2.0	2.5	0.5	3.0	3.5	0.5
a_1	4.0	6.0	2.0	5.0	7.0	2.0
$a_1 - a_0$	2.0	3.5	1.5	2.0	3.5	1.5

[a] Involving three factors A, B, and C, each with two levels; a_0 and a_1 for factor A, b_0 and b_1 for factor B, and c_0 and c_1 for factor C.

☐ STEP 2. Compute the $A \times B \times C$ interaction effect as the difference between the $A \times B$ interaction effects at the two levels of factor C computed in step 1:

For set(a): $A \times B \times C$ interaction $= \frac{1}{2}[0.75 - (-0.75)] = 0.75$ t/ha

For set(b): $A \times B \times C$ interaction $= \frac{1}{2}(0.75 - 0.75) = 0.00$ t/ha

• Method II is based on the difference in the $A \times C$ interaction effects.

☐ STEP 1. For each level of factor B, compute the $A \times C$ interaction effect, following the procedure in Chapter 3, Section 3.1:
For set(a) data:

At b_0: $A \times C$ interaction $= \frac{1}{2}(2.5 - 2.0) = 0.25$ t/ha

At b_1: $A \times C$ interaction $= \frac{1}{2}(4.0 - 0.5) = 1.75$ t/ha

For set(b) data:

At b_0: $A \times C$ interaction $= \frac{1}{2}(2.0 - 2.0) = 0.00$ t/ha

At b_1: $A \times C$ interaction $= \frac{1}{2}(3.5 - 3.5) = 0.00$ t/ha

☐ STEP 2. Compute the $A \times B \times C$ interaction effect as the difference between the $A \times C$ interaction effects at the two levels of factor B computed in step 1:

For set(a): $A \times B \times C$ interaction $= \frac{1}{2}(1.75 - 0.25) = 0.75$ t/ha

For set(b): $A \times B \times C$ interaction $= \frac{1}{2}(0.00 - 0.00) = 0.00$ t/ha

• Method III is based on the difference in the $B \times C$ interaction effects.

☐ STEP 1. For each level of factor A, compute the $B \times C$ interaction effect, following the procedure in Chapter 3, Section 3.1:
For set(a) data:

At a_0: $B \times C$ interaction $= \frac{1}{2}(2.5 - 1.0) = 0.75$ t/ha

At a_1: $B \times C$ interaction $= \frac{1}{2}[(4.0 - (-0.5)] = 2.25$ t/ha

For set(b) data:

$$\text{At } a_0: B \times C \text{ interaction} = \tfrac{1}{2}(0.5 - 0.5) = 0.0 \text{ t/ha}$$

$$\text{At } a_1: B \times C \text{ interaction} = \tfrac{1}{2}(2.0 - 2.0) = 0.0 \text{ t/ha}$$

□ STEP 2. Compute the $A \times B \times C$ interaction effect as the difference between the $B \times C$ interaction effects at the two levels of factor A computed in step 1:

For set(a): $A \times B \times C$ interaction $= \tfrac{1}{2}(2.25 - 0.75) = 0.75$ t/ha

For set(b): $A \times B \times C$ interaction $= \tfrac{1}{2}(0.0 - 0.0) = 0.00$ t/ha

Thus, regardless of computation method, the $A \times B \times C$ interaction effect of set(a) data is 0.75 t/ha and of set(b) data is 0.0 t/ha.

The foregoing procedure for computing the three-factor interaction effect can be easily extended to cover a four-factor interaction, a five-factor interaction, and so on. For example, a four-factor interaction $A \times B \times C \times D$ can be computed in any of the following ways:

- As the difference between the $A \times B \times C$ interaction effects over the levels of factor D
- As the difference between the $A \times B \times D$ interaction effects over the levels of factor C
- As the difference between the $A \times C \times D$ interaction effects over the levels of factor B
- As the difference between the $B \times C \times D$ interaction effects over the levels of factor A.

4.2 ALTERNATIVE DESIGNS

There are many experimental designs that can be considered for use in a three-or-more-factor experiment. For our purpose, these designs can be classified into four categories, namely, the single-factor experimental designs, the two-factor experimental designs, the three-or-more-factor experimental designs, and the fractional factorial designs.

4.2.1 Single-Factor Experimental Designs

All experimental designs for single-factor experiments described in Chapter 2 are applicable to experiments with three or more factors. This is done by treating all the factorial treatment combinations as if they were levels of a single factor.

For illustration, take the case of a three-factor experiment involving two varieties, four nitrogen levels, and three weed-control methods to be tested in three replications. If a randomized complete block design (RCB) is used, the $2 \times 4 \times 3 = 24$ factorial treatment combinations would be assigned completely at random to the 24 experimental plots in each of the three replications. The field layout of such a design may be as shown in Figure 4.1 and the outline of the corresponding analysis of variance shown in Table 4.2. Note that with a RCB design there is only one plot size and only one error variance for testing the significance of all effects (i.e., the three main effects, the three two-factor interaction effects, and one three-factor interaction effect) so that all effects are measured with the same level of precision. Thus, a complete block design, such as RCB, should be used only if:

- All effects (i.e., main effects and interaction effects) are of equal importance and, hence, should be measured with the same level of precision.
- The experimental units are homogeneous enough to achieve a high level of homogeneity within a block.

Because an experiment with three or more factors usually involves a large number of treatments, homogeneity in experimental units within the same block is difficult to achieve and, therefore, the complete block design is not commonly used.

4.2.2 Two-factor Experimental Designs

All experimental designs for two-factor experiments described in Chapter 3 are applicable to experiments with three or more factors. The procedures for applying any of these designs to a three-factor experiment are given below. We illustrate the procedure with the $2 \times 4 \times 3$ factorial experiment described in Section 4.2.1.

☐ STEP 1.   Divide the k factors to be tested into two groups, with k_1 factors in one group and k_2 factors in another group (where $k_1 + k_2 = k$), by putting those factors that are to be measured with the same level of precision in the same group. Each group can contain any number of factors.

For our example with $k = 3$ factors, variety and weed control can be combined in one group and the remaining factor (nitrogen) put in another group. Thus, group I consists of two factors ($k_1 = 2$) and group II consists of one factor ($k_2 = 1$).

☐ STEP 2.   Treat the factorial treatment combinations of k_1 factors in group I as the levels of a single factor called factor A, and the factorial treatment combinations of k_2 factors in group II as the levels of a single factor called

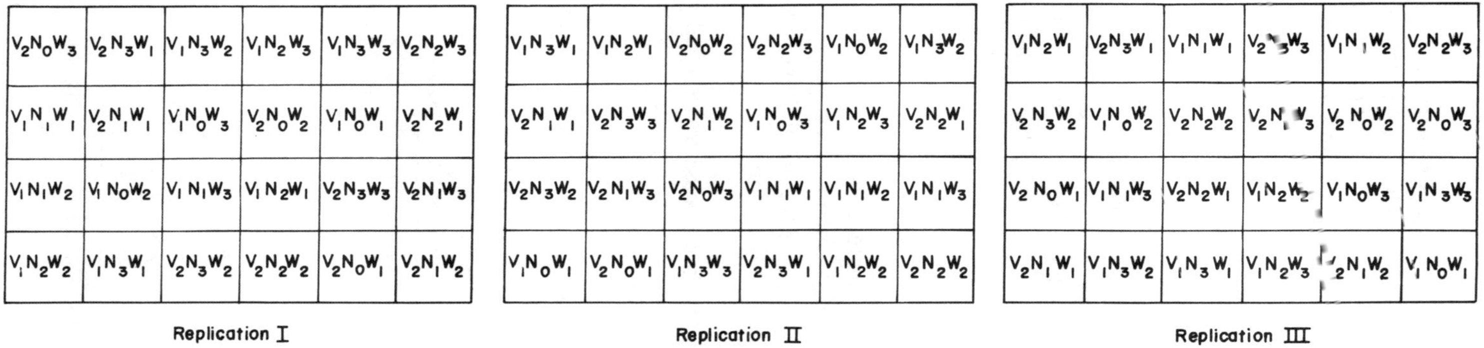
V2N0W3 V2N3W1 V1N3W2 V1N2W3 V1N3W3 V2N2W3
V1N1W1 V2N1W1 V1N0W3 V2N0W2 V1N0W1 V2N2W1
V1N1W2 V1N0W2 V1N1W3 V1N2W1 V2N3W3 V2N1W3
V1N2W2 V1N3W1 V2N3W2 V2N2W2 V2N0W1 V2N1W2
Replication I
V1N3W1 V1N2W1 V2N0W2 V2N2W3 V1N0W2 V1N3W2
V2N1W1 V2N3W3 V2N1W2 V1N0W3 V1N2W3 V2N2W1
V2N3W2 V2N1W3 V2N0W3 V1N1W1 V1N1W2 V1N1W3
V1N0W1 V2N0W1 V1N3W3 V2N3W1 V1N2W2 V2N2W2
Replication II
V1N2W1 V2N3W1 V1N1W1 V2N3W3 V1N1W2 V2N2W3
V2N3W2 V1N0W2 V2N2W2 V2N1W3 V2N0W2 V2N0W3
V2N0W1 V1N1W3 V2N2W1 V1N2W2 V1N0W3 V1N3W3
V2N1W1 V1N3W2 V1N3W1 V1N2W3 V2N1W2 V1N0W1
Replication III

Figure 4.1 A sample layout of a $2 \times 4 \times 3$ factorial experiment involving two varieties (V_1 and V_2), four nitrogen levels (N_0, N_1, N_2, and N_3), and three weed-control methods (W_1, W_2, and W_3) arranged in a randomized complete block design with three replications.

factor B. Thus, the k-factor experiment is now converted to a two-factor experiment involving the two newly created factors A and B.

For our example, the $3 \times 2 = 6$ factorial treatment combinations between variety and weed control in group I are treated as the six levels of the newly created factor A, and the four levels of nitrogen as the levels of the newly created factor B. Thus, the $2 \times 4 \times 3$ factorial experiment can now be viewed as a 6×4 factorial experiment involving A and B.

☐ STEP 3. Select an appropriate experimental design from Chapter 3 and apply it to the simulated two-factor experiment constituted in step 2 by following the corresponding procedure described in Chapter 3.

For our example, if the split-plot design with factor B (nitrogen) as the subplot factor is to be used, the layout of such a design may be as shown in Figure 4.2 and the form of the corresponding analysis of variance shown in Table 4.3.

Note that with this split-plot design, there are two plot sizes and two error mean squares for testing significance of the various effects:

- Error(a) MS for the main effect of variety, the main effect of weed control method, and their interaction effect
- Error(b) MS for the main effect of nitrogen fertilizer and its interaction with the other two variable factors

Because the error(b) MS is expected to be smaller than error(a) MS (see

Table 4.2 Outline of the Analysis of Variance for a 2 × 4 × 3 Factorial Experiment in RCB Design

Source of Variation	Degree of Freedom[a]	Sum of Squares	Mean Square	Computed F	Tabular F 5%	1%
Replication	$r - 1 = 2$					
Treatment	$vnw - 1 = 23$					
Variety (V)	$v - 1 = 1$					
Nitrogen (N)	$n - 1 = 3$					
Weed Control (W)	$w - 1 = 2$					
$V \times N$	$(v - 1)(n - 1) = 3$					
$V \times W$	$(v - 1)(w - 1) = 2$					
$N \times W$	$(n - 1)(w - 1) = 6$					
$V \times N \times W$	$(v - 1)(n - 1)(w - 1) = 6$					
Error	$(r - 1)(vnw - 1) = 46$					
Total	$rvnw - 1 = 71$					

[a] r = number of replications; v, n, and w are levels of the three factors V, N, and W, respectively.

Figure 4.2 A sample layout of a $2 \times 4 \times 3$ factorial experiment involving two varieties (V_1 and V_2), four nitrogen levels (N_0, N_1, N_2, and N_3), and three weed-control methods (W_1, W_2, and W_3) arranged in a split-plot design with nitrogen levels as the subplot treatments, in three replications.

Replication I

$V_1 W_2$	$V_2 W_1$	$V_1 W_3$	$V_2 W_3$	$V_1 W_1$	$V_2 W_2$
N_2	N_3	N_1	N_2	N_3	N_1
N_0	N_2	N_0	N_3	N_0	N_2
N_3	N_1	N_2	N_0	N_2	N_0
N_1	N_0	N_3	N_1	N_1	N_3

Replication II

$V_1 W_2$	$V_2 W_2$	$V_2 W_3$	$V_1 W_1$	$V_1 W_3$	$V_2 W_1$
N_2	N_3	N_0	N_0	N_2	N_3
N_3	N_1	N_1	N_2	N_1	N_1
N_1	N_0	N_3	N_1	N_3	N_0
N_0	N_2	N_2	N_3	N_0	N_2

Replication III

$V_2 W_1$	$V_2 W_2$	$V_1 W_2$	$V_1 W_1$	$V_2 W_3$	$V_1 W_3$
N_2	N_3	N_0	N_1	N_3	N_2
N_3	N_1	N_3	N_2	N_1	N_0
N_0	N_2	N_1	N_3	N_1	N_1
N_1	N_0	N_2	N_0	N_2	N_3

137

Chapter 3, Section 3.4.2) the degree of precision for measuring all effects concerning nitrogen is expected to be higher than that related to either variety or weed control.

Thus, a split-plot design (or a strip-plot design) is appropriate for a three-or-more-factor experiment if both of the following conditions hold:

- The total number of factorial treatment combinations is too large for a complete block design.
- The k factors can be easily divided into two groups with identifiable differences in the desired level of precision attached to each group.

4.2.3 Three-or-More-Factor Experimental Designs

Experimental designs specifically developed for three-or-more-factor experiments commonly used in agricultural research are primarily the extension of either the split-plot or the strip-plot design.

For our example, a split-plot design can be extended to accommodate the third factor through additional subdivision of each subplot into sub-subplots, and further extended to accommodate the fourth factor through additional subdivision of each sub-subplot into sub-sub-subplots, and so on. The resulting designs are referred to as a split-split-plot design, a split-split-split-plot design, and so on. A split-split-plot design, applied to a three-factor experiment, would have the first factor assigned to the main plot, the second factor to the subplot,

Table 4.3 Outline of the Analysis of Variance for a 2 $\times$ 4 $\times$ 3 Factorial Experiment in a Split-plot Design[a]

Source of Variation	Degree of Freedom[b]	Sum of Squares	Mean Square	Computed F	Tabular F 5%	1%
Replication	$r - 1 = 2$					
Main-plot factor:	$vw - 1 = 5$					
Variety (V)	$v - 1 = 1$					
Weed Control (W)	$w - 1 = 2$					
$V \times W$	$(v - 1)(w - 1) = 2$					
Error(a)	$(r - 1)(vw - 1) = 10$					
Subplot factor (N)	$n - 1 = 3$					
Main-plot factor $\times$ subplot factor:	$(vw - 1)(n - 1) = 15$					
$N \times V$	$(n - 1)(v - 1) = 3$					
$N \times W$	$(n - 1)(w - 1) = 6$					
$N \times V \times W$	$(n - 1)(v - 1)(w - 1) = 6$					
Error(b)	$vw(r - 1)(n - 1) = 36$					
Total	$rvwn - 1 = 71$					

[a]Applied to a simulated two-factor experiment with main-plot factor as a combination of two original factors V and W, and subplot factor representing the third original factor N.

[b]r = number of replications; v, n, and w are levels of the three original factors V, N, and W, respectively.

and the third factor to the sub-subplot. In this way, there is no need to combine the three factors into two groups to simulate a two-factor experiment, as is necessary if a split-plot design is applied to a three-factor experiment.

Similarly, the strip-plot design can be extended to incorporate the third factor, the fourth factor, and so on, through the subdivision of each intersection plot into subplots and further subdivision of each subplot into sub-subplots. The resulting designs are referred to as a strip-split-plot design, a strip-split-split-plot design, and so on.

4.2.4   Fractional Factorial Designs

Unlike the designs in Sections 4.2.1 to 4.2.3, where the complete set of factorial treatment combinations is to be included in the test, the fractional factorial design (FFD), as the name implies, includes only a fraction of the complete set of the factorial treatment combinations. The obvious advantage of the FFD is the reduction in size of the experiment, which may be desirable whenever the complete set of factorial treatment combinations is too large for practical implementation. This important advantage is, however, achieved by a reduction in the number of effects that can be estimated.

4.3   SPLIT-SPLIT-PLOT DESIGN

The split-split-plot design is an extension of the split-plot design to accommodate a third factor. It is uniquely suited for a three-factor experiment where three different levels of precision are desired for the various effects. Each level of precision is assigned to the effects associated with each of the three factors. This design is characterized by two important features:

1. There are three plot sizes corresponding to the three factors, namely, the largest plot (main plot) for the main-plot factor, the intermediate plot (subplot) for the subplot factor, and the smallest plot (sub-subplot) for the sub-subplot factor.
2. There are three levels of precision, with the main-plot factor receiving the lowest degree of precision and the sub-subplot factor receiving the highest degree of precision.

We illustrate procedures for randomization, layout, and analysis of variance with a $5 \times 3 \times 3$ factorial experiment with three replications. Treatments are five levels of nitrogen as the main plot, three management practices as the subplot, and three rice varieties as the sub-subplot. We use r to refer to the number of replications; A, B, and C to refer to the main-plot factor, subplot factor, and sub-subplot factor; and a, b, and c to refer to the treatment levels corresponding to factors A, B, and C.

4.3.1   Randomization and Layout

There are three steps in the randomization and layout of a split-split-plot design:

☐ STEP 1.   Divide the experimental area into r replications and each replication into a main plots. Then, randomly assign the a main-plot treatments to the a main plots, separately and independently, for each of the r replications, following any one of the randomization schemes of Chapter 2, Section 2.1.1.

    For our example, the area is divided into three replications and each replication into five main plots. Then, the five nitrogen levels (N_1, N_2, N_3, N_4, and N_5) are assigned at random to the five main plots in each replication. The result may be as shown in Figure 4.3.

☐ STEP 2.   Divide each main plot into b subplots, in which the b subplot treatments are randomly assigned, separately and independently, for each of the $(r)(a)$ main plots.

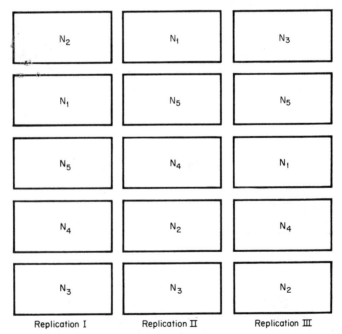

Figure 4.3   Random assignment of five nitrogen levels (N_1, N_2, N_3, N_4, and N_5) to the main plots in each of the three replications as the first step in laying out a split-split-plot design.

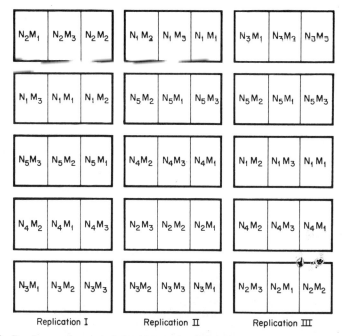

Figure 4.4 Random assignment of three management practices (M_1, M_2, and M_3) to the three subplots in each of the 15 main plots as the second step in laying out a split-split-plot design.

For our example, each main plot is divided into three subplots, into which the three management practices (M_1, M_2, and M_3) are assigned at random. This randomization process is repeated $(r)(a) = 15$ times. The result may be as shown in Figure 4.4.

☐ STEP 3.   Divide each subplot into c sub-subplots, in which the c sub-subplot treatments are randomly assigned, separately and independently, for each of the $(r)(a)(b)$ subplots.

For our example, each subplot is divided into three sub-subplots, into which the three varieties (V_1, V_2, and V_3) are assigned at random. This randomization process is repeated $(r)(a)(b) = 45$ times. The final layout may be as shown in Figure 4.5.

4.3.2   Analysis of Variance

Grain yield data (Table 4.4) from a $5 \times 3 \times 3$ factorial experiment conducted in a split-split-plot design, whose layout is shown in Figure 4.5, is used to illustrate analysis of variance.

☐ STEP 1. Construct an outline of the analysis of variance for the split-split-plot design

Source of Variation	Degree of Freedom	Sum of Squares	Mean Square	Computed F	Tabular F 5% 1%
Main-plot analysis:					
Replication	$r - 1 = 2$				
Main-plot factor (A)	$a - 1 = 4$				
Error(a)	$(r - 1)(a - 1) = 8$				
Subplot analysis:					
Subplot factor (B)	$b - 1 = 2$				
$A \times B$	$(a - 1)(b - 1) = 8$				
Error(b)	$a(r - 1)(b - 1) = 20$				
Sub-subplot analysis:					
Sub-subplot factor (C)	$c - 1 = 2$				
$A \times C$	$(a - 1)(c - 1) = 8$				
$B \times C$	$(b - 1)(c - 1) = 4$				
$A \times B \times C$	$(a - 1)(b - 1)(c - 1) = 16$				
Error(c)	$ab(r - 1)(c - 1) = 60$				
Total	$rabc - 1 = 134$				

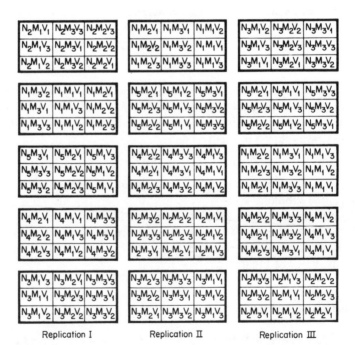

Replication I   Replication II   Replication III

Figure 4.5 A sample layout of a $5 \times 3 \times 3$ factorial experiment arranged in a split-split-plot design with five nitrogen levels (N_1, N_2, N_3, N_4, and N_5) as main-plot treatments, three management practices (M_1, M_2, and M_3) as subplot treatments, and three varieties (V_1, V_2, and V_3) as sub-subplot treatments, in three replications.

Table 4.4 Grain Yields of Three Rice Varieties Grown under Three Management Practices and Five Nitrogen Levels; in a Split-split-plot Design with Nitrogen as Main-plot, Management Practice as Subplot, and Variety as Sub subplot Factors, with Three Replications

	Grain Yield, t/ha								
	V_1			V_2			V_3		
Management	Rep. I	Rep. II	Rep. III	Rep. I	Rep. II	Rep. III	Rep. I	Rep. II	Rep. III
				N_1 (0 kg N/ha)					
M_1(Minimum)	3.320	3.864	4.507	6.101	5.122	4.815	5.355	5.536	5.244
M_2(Optimum)	3.766	4.311	4.875	5.096	4.873	4.166	7.442	6.462	5.584
M_3(Intensive)	4.660	5.915	5.400	6.573	5.495	4.225	7.018	8.020	7.642
				N_2 (50 kg N/ha)					
M_1	3.188	4.752	4.756	5.595	6.780	5.390	6.706	6.546	7.092
M_2	3.625	4.809	5.295	6.357	5.925	5.163	8.592	7.646	7.212
M_3	5.232	5.170	6.046	7.016	7.442	4.478	8.480	9.942	8.714
				N_3 (80 kg N/ha)					
M_1	5.468	5.788	4.422	5.442	5.988	6.509	8.452	6.698	8.650
M_2	5.759	6.130	5.308	6.398	6.533	6.569	8.662	8.526	8.514
M_3	6.215	7.106	6.318	6.953	6.914	7.991	9.112	9.140	9.320
				N_4 (110 kg N/ha)					
M_1	4.246	4.842	4.863	6.209	6.768	5.779	8.042	7.414	6.902
M_2	5.255	5.742	5.345	6.992	7.856	6.164	9.080	9.016	7.778
M_3	6.829	5.869	6.011	7.565	7.626	7.362	9.660	8.966	9.128
				N_5 (140 kg N/ha)					
M_1	3.132	4.375	4.678	6.860	6.894	6.573	9.314	8.508	8.032
M_2	5.389	4.315	5.896	6.857	6.974	7.422	9.224	9.680	9.294
M_3	5.217	5.389	7.309	7.254	7.812	8.950	10.360	9.896	9.712

□ STEP 2. Do a main-plot analysis.

   A. Construct the replication × factor A two-way table of totals and compute the replication totals, the factor A totals, and the grand total. For our example, the replication × nitrogen table of totals (RA), with the nitrogen totals (A) and the grand total (G) computed, is shown in Table 4.5.

   B. Compute the correction factor and the various sums of squares:

$$C.F. = \frac{G^2}{rabc}$$

$$= \frac{(884.846)^2}{(3)(5)(3)(3)} = 5,799.648$$

Table 4.5   The Replication × Nitrogen Table of Yield Totals Computed from Data in Table 4.4

Nitrogen	Yield Total (RA)			Nitrogen Total (A)
	Rep. I	Rep. II	Rep. III	
N_1	49.331	49.598	46.458	145.387
N_2	54.791	59.012	54.146	167.949
N_3	62.461	62.823	63.601	188.885
N_4	63.878	64.099	59.332	187.309
N_5	63.607	63.843	67.866	195.316
Rep. total(R)	294.068	299.375	291.403	
Grand total(G)				884.846

$$\text{Total } SS = \sum X^2 - C.F.$$

$$= \left[(3.320)^2 + \cdots + (9.712)^2 \right] - 5{,}799.648$$

$$= 373.540$$

$$\text{Replication } SS = \frac{\sum R^2}{abc} - C.F.$$

$$= \frac{(294.068)^2 + (299.375)^2 + (291.403)^2}{(5)(3)(3)}$$

$$- 5{,}799.648$$

$$= 0.732$$

$$A \text{ (nitrogen) } SS = \frac{\sum A^2}{rbc} - C.F.$$

$$= \frac{(145.387)^2 + \cdots + (195.316)^2}{(3)(3)(3)}$$

$$- 5{,}799.648$$

$$= 61.641$$

$$\text{Error}(a)\ SS = \frac{\sum (RA)^2}{bc} - C.F. - \text{Replication } SS - A\ SS$$

$$= \frac{(49.331)^2 + \cdots + (67.866)^2}{(3)(3)} - 5{,}799.648$$

$$-0.732 - 61.641$$

$$= 4.451$$

☐ STEP 3. Do a subplot analysis.

A. Construct two tables of totals:

(i) The factor A × factor B two-way table of totals, with the factor B totals computed. For our example, the nitrogen × management table of totals (AB), with the management totals (B) computed, is shown in Table 4.6.

(ii) The replication × factor A × factor B three-way table of totals. For our example, the replication × nitrogen × management table of totals (RAB) is shown in Table 4.7.

B. Compute the various sums of squares:

$$B\ (\text{management})\ SS = \frac{\sum B^2}{rac} - C.F.$$

$$= \frac{(265.517)^2 + (291.877)^2 + (327.452)^2}{(3)(5)(3)}$$

$$-5{,}799.648$$

$$= 42.936$$

Table 4.6   The Nitrogen × Management Table of Yield Totals Computed from Data in Table 4.4

Nitrogen	Yield Total (AB)		
	M_1	M_2	M_3
N_1	43.864	46.575	54.948
N_2	50.805	54.624	62.520
N_3	57.417	62.399	69.069
N_4	55.065	63.228	69.016
N_5	58.366	64.051	71.899
Management total (B)	265.517	291.877	327.452

Table 4.7   The Replication × Nitrogen × Management Table of Yield Totals Computed from Data in Table 4.4

	Yield Total (RAB)		
Management	Rep. I	Rep. II	Rep. III
		$N_1(0\ kg\ N/ha)$	
M_1	14.776	14.522	14.566
M_2	16.304	15.646	14.625
M_3	18.251	19.430	17.267
		$N_2(50\ kg\ N/ha)$	
M_1	15.489	18.078	17.238
M_2	18.574	18.380	17.670
M_3	20.728	22.554	19.238
		$N_3(80\ kg\ N/ha)$	
M_1	19.362	18.474	19.581
M_2	20.819	21.189	20.391
M_3	22.280	23.160	23.629
		$N_4(110\ kg\ N/ha)$	
M_1	18.497	19.024	17.544
M_2	21.327	22.614	19.287
M_3	24.054	22.461	22.501
		$N_5(140\ kg\ N/ha)$	
M_1	19.306	19.777	19.283
M_2	21.470	20.969	22.612
M_3	22.831	23.097	25.971

$$A \times B \text{ (nitrogen × management) } SS = \frac{\sum (AB)^2}{rc} - C.F. - A\ SS - B\ SS$$

$$= \frac{(43.864)^2 + \cdots + (71.899)^2}{(3)(3)} - 5{,}799.648 - 61.641$$

$$-42.936$$

$$= 1.103$$

$$\text{Error}(b)\ SS = \frac{\sum (RAB)^2}{c} - C.F. - \text{Replication } SS - A\ SS$$

$$- \text{Error}(a)\ SS - B\ SS - A \times B\ SS$$

Table 4.8 The Nitrogen × Variety Table of Yield Totals Computed from Data in Table 4.4

Nitrogen	Yield Total (AC)		
	V_1	V_2	V_3
N_1	40.618	46.466	58.303
N_2	42.873	54.146	70.930
N_3	52.514	59.297	77.074
N_4	49.002	62.321	75.986
N_5	45.700	65.596	84.020
Variety total (C)	230.707	287.826	366.313

$$= \frac{(14.776)^2 + \cdots + (25.971)^2}{3} - 5{,}799.648$$

$$-0.732 - 61.641 - 4.451 - 42.936 - 1.103$$

$$= 5.236$$

☐ STEP 4.   Do a sub-subplot analysis.

A.   Construct three tables of totals:

(i)   The factor A × factor C two-way table of totals with the factor C totals computed. For our example, the nitrogen × variety table of totals (AC), with the variety totals (C) computed, is shown in Table 4.8.

(ii)   The factor B × factor C two-way table of totals. For our example, the management × variety table of totals (BC) is shown in Table 4.9.

(iii)   The factor A × factor B × factor C three-way table of totals. For our example, the nitrogen × management × variety table of totals (ABC) is shown in Table 4.10.

Table 4.9 The Management × Variety Table of Yield Totals Computed from Data in Table 4.4

Management	Yield Total (BC)		
	V_1	V_2	V_3
M_1	66.201	90.825	108.491
M_2	75.820	93.345	122.712
M_3	88.686	103.656	135.110

B.   Compute the various sums of squares:

$$C\text{ (variety) }SS = \frac{\sum C^2}{rab} - C.F.$$

$$= \frac{(230.707)^2 + (287.826)^2 + (366.313)^2}{(3)(5)(3)}$$

$$- 5{,}799.648$$

$$= 206.013$$

Table 4.10   The Nitrogen × Management × Variety Table of Yield Totals Computed from Data in Table 4.4

Management	Yield Total (ABC)		
	V_1	V_2	V_3
	$N_1(0\ kg\ N/ha)$		
M_1	11.691	16.038	16.135
M_2	12.952	14.135	19.488
M_3	15.975	16.293	22.680
	$N_2(50\ kg\ N/ha)$		
M_1	12.696	17.765	20.344
M_2	13.729	17.445	23.450
M_3	16.448	18.936	27.136
	$N_3(80\ kg\ N/ha)$		
M_1	15.678	17.939	23.800
M_2	17.197	19.500	25.702
M_3	19.639	21.858	27.572
	$N_4(110\ kg\ N/ha)$		
M_1	13.951	18.756	22.358
M_2	16.342	21.012	25.874
M_3	18.709	22.553	27.754
	$N_5(140\ kg\ N/ha)$		
M_1	12.185	20.327	25.854
M_2	15.600	21.253	28.198
M_3	17.915	24.016	29.968

$$A \times C \ SS = \frac{\sum (AC)^2}{rb} - C.F. - A \ SS - C \ SS$$

$$= \frac{(40.618)^2 + \cdots + (84.020)^2}{(3)(3)} - 5,799.648$$

$$- 61.641 - 206.013$$

$$= 14.144$$

$$B \times C \ SS = \frac{\sum (BC)^2}{ra} - C.F. - B \ SS - C \ SS$$

$$= \frac{(66.201)^2 + \cdots + (135.110)^2}{(3)(5)} - 5,799.648$$

$$- 42.936 - 206.013$$

$$= 3.852$$

$$A \times B \times C \ SS = \frac{\sum (ABC)^2}{r} - C.F. - A \ SS - B \ SS - C \ SS$$

$$- A \times B \ SS - A \times C \ SS - B \times C \ SS$$

$$= \frac{(11.691)^2 + \cdots + (29.968)^2}{3} - 5,799.648$$

$$- 61.641 - 42.936 - 206.013$$

$$- 1.103 - 14.144 - 3.852$$

$$= 3.699$$

$$\text{Error}(c) \ SS = \text{Total } SS - (\text{the sum of all other } SS)$$

$$= 373.540 - (0.732 + 61.641 + 4.451$$

$$+ 42.936 + 1.103 + 5.236 + 206.013$$

$$+ 14.144 + 3.852 + 3.699)$$

$$= 29.733$$

□ STEP 5.   For each source of variation, compute the mean square value by dividing the *SS* by its *d.f.*:

$$\text{Replication } MS = \frac{\text{Replication } SS}{r - 1}$$

$$= \frac{0.732}{2} = 0.3660$$

$$A \text{ } MS = \frac{A \text{ } SS}{a - 1}$$

$$= \frac{61.641}{4} = 15.4102$$

$$\text{Error}(a) \text{ } MS = \frac{\text{Error}(a) \text{ } SS}{(r - 1)(a - 1)}$$

$$= \frac{4.451}{(2)(4)} = 0.5564$$

$$B \text{ } MS = \frac{B \text{ } SS}{b - 1}$$

$$= \frac{42.936}{2} = 21.4680$$

$$A \times B \text{ } MS = \frac{A \times B \text{ } SS}{(a - 1)(b - 1)}$$

$$= \frac{1.103}{(4)(2)} = 0.1379$$

$$\text{Error}(b) \text{ } MS = \frac{\text{Error}(b) \text{ } SS}{a(r - 1)(b - 1)}$$

$$= \frac{5.236}{(5)(2)(2)} = 0.2618$$

$$C \text{ } MS = \frac{C \text{ } SS}{c - 1}$$

$$= \frac{206.013}{2} = 103.0065$$

$$A \times C \ MS = \frac{A \times C \ SS}{(a-1)(r-1)}$$

$$= \frac{14.144}{(4)(2)} = 1.7680$$

$$B \times C \ MS = \frac{B \times C \ SS}{(b-1)(c-1)}$$

$$= \frac{3.852}{(2)(2)} = 0.9630$$

$$A \times B \times C \ MS = \frac{A \times B \times C \ SS}{(a-1)(b-1)(c-1)}$$

$$= \frac{3.699}{(4)(2)(2)} = 0.2312$$

$$Error(c) \ MS = \frac{Error(c) \ SS}{ab(r-1)(c-1)}$$

$$= \frac{29.733}{(5)(3)(2)(2)} = 0.4956$$

□ STEP 6.   Compute the F value for each effect by dividing each mean square by its appropriate error mean square:

$$F(A) = \frac{A \ MS}{Error(a) \ MS}$$

$$= \frac{15.4102}{0.5564} = 27.70$$

$$F(B) = \frac{B \ MS}{Error(b) \ MS}$$

$$= \frac{21.4680}{0.2618} = 82.00$$

$$F(A \times B) = \frac{A \times B \ MS}{Error(b) \ MS}$$

$$= \frac{0.1379}{0.2618} < 1$$

$$F(C) = \frac{C \ MS}{Error(c) \ MS}$$

$$= \frac{103.0065}{0.4956} = 207.84$$

$$F(A \times C) = \frac{A \times C \ MS}{Error(c) \ MS}$$

$$= \frac{1.7680}{0.4956} = 3.57$$

$$F(B \times C) = \frac{B \times C \ MS}{Error(c) \ MS}$$

$$= \frac{0.9630}{0.4956} = 1.94$$

$$F(A \times B \times C) = \frac{A \times B \times C \ MS}{Error(c) \ MS}$$

$$= \frac{0.2312}{0.4956} < 1$$

☐ STEP 7.   For each effect whose computed F value is not less than 1, obtain the corresponding tabular F values from Appendix E, with $f_1 = d.f.$ of the numerator MS and $f_2 = d.f.$ of the denominator MS, at the 5% and 1% levels of significance.

☐ STEP 8.   Compute the three coefficients of variation corresponding to the three error terms:

$$cv(a) = \frac{\sqrt{Error(a) \ MS}}{Grand \ mean} \times 100$$

$$= \frac{\sqrt{0.5564}}{6.55} \times 100 = 11.4\%$$

$$cv(b) = \frac{\sqrt{Error(b) \ MS}}{Grand \ mean} \times 100$$

$$= \frac{\sqrt{0.2618}}{6.55} \times 100 = 7.8\%$$

$$cv(c) = \frac{\sqrt{Error(c) \ MS}}{Grand \ mean} \times 100$$

$$= \frac{\sqrt{0.4956}}{6.55} \times 100 = 10.7\%$$

The $cv(a)$ value indicates the degree of precision associated with the main effect of the main-plot factor, the $cv(b)$ value indicates the degree of precision of the main effect of the subplot factor and of its interaction with the main plot, and the $cv(c)$ value indicates the degree of precision of the main effect of the sub-subplot factor and of all its interactions with the other factors. Normally, the sizes of these three coefficients of variation should decrease from $cv(a)$ to $cv(b)$ and to $cv(c)$.

For our example, the value of $cv(a)$ is the largest as expected, but those of $cv(b)$ and $cv(c)$ do not follow the expected trend. As mentioned in Chapter 3, Section 3.4.2, such unexpected results are occasionally encountered. If they occur frequently, a competent statistician should be consulted.

☐ STEP 9. Enter all values obtained in steps 2 to 8 in the analysis of variance outline of step 1, and compare each computed F value with its corresponding tabular F values, and indicate its significance by the appropriate asterisk notation (see Chapter 2, Section 2.1.2).

For our example, the results, shown in Table 4.11, indicate that the three-factor interaction (nitrogen × management × variety) is not significant, and only one two-factor interaction (nitrogen × variety) is significant. For a proper interpretation of the significant interaction effect and mean comparisons, see appropriate procedures in Chapter 5.

Table 4.11 Analysis of Variance[a] (Split-split- plot Design) of Grain Yield Data in Table 4.4

Source of Variation	Degree of Freedom	Sum of Squares	Mean Square	Computed F[b]	Tabular F 5%	1%
Main-plot analysis						
Replication	2	0.732	0.3660			
Nitrogen (A)	4	61.641	15.4102	27.70**	3.84	7.01
Error(a)	8	4.451	0.5564			
Subplot analysis						
Management (B)	2	42.936	21.4680	82.00**	3.49	5.85
$A \times B$	8	1.103	0.1379	< 1	—	—
Error(b)	20	5.236	0.2618			
Sub-subplot analysis						
Variety (C)	2	206.013	103.0065	207.84**	3.15	4.98
$A \times C$	8	14.144	1.7680	3.57**	2.10	2.82
$B \times C$	4	3.852	0.9630	1.94ns	2.52	3.65
$A \times B \times C$	16	3.699	0.2312	< 1	—	—
Error(c)	60	29.733	0.4956			
Total	134	373.540				

[a] $cv(a) = 11.4\%$, $cv(b) = 7.8\%$, $cv(c) = 10.7\%$.
[b] ** = significant at 1% level, ns = not significant.

4.4   STRIP-SPLIT-PLOT DESIGN

The strip-split-plot design is an extension of the strip-plot design (see Chapter 3, Section 3.5) in which the intersection plot is divided into subplots to accommodate a third factor. The strip-split-plot design is characterized by two main features.

1. There are four plot sizes—the horizontal strip, the vertical strip, the intersection plot, and the subplot.
2. There are four levels of precision with which the effects of the various factors are measured, with the highest level corresponding to the sub-plot factor and its interactions with other factors.

The procedures for randomization, layout, and analysis of variance for the strip-split-plot design are given in the next two sections. We use r as the number of replications; A, B, and C as the vertical, horizontal, and subplot factors; and a, b, and c as the treatment levels corresponding to factors A, B, and C. A three-factor experiment designed to test the effects of two planting methods M_1 and M_2 and three rates of nitrogen application N_1, N_2, and N_3 on the yield of six rice varieties V_1, V_2, V_3, V_4, V_5, and V_6 is used for illustration. This experiment had three replications using nitrogen as the vertical factor, variety as the horizontal factor, and planting method as the subplot factor. Grain yield data are shown in Table 4.12.

4.4.1   Randomization and Layout

The steps involved in the randomization and layout of a strip-split-plot design are:

☐ STEP 1.   Apply the process of randomization and layout for the strip-plot design (Chapter 3, Section 3.5.1) to the vertical factor (nitrogen) and the horizontal factor (variety). The result may be as shown in Figure 4.6.

☐ STEP 2.   Divide each of the $(a)(b)$ intersection plots in each of the r replications into c subplots and, following one of the randomization schemes of Chapter 2, Section 2.1.1, randomly assign the c subplot treatments to the c subplots, separately and independently, in each of the $(r)(a)(b)$ intersection plots.

For our example, each of the $(3)(6) = 18$ intersection plots in each replication is divided into two subplots and the two planting methods P_1 and P_2 are randomly assigned to the subplots, separately and independently, for each of the 54 intersection plots (18 intersection plots per replication and 3 replications). The final layout is shown in Figure 4.7.

Table 4.12 Grain Yields of Six Rice Varieties Tested under Two Planting Methods and Three Nitrogen Rates, in a Strip-split-plot Design with Three Replications

Grain Yield, kg/ha

Variety	P_1 (Broadcast) Rep. I	Rep. II	Rep. III	Total (ABC)	P_2 (Transplanted) Rep. I	Rep. II	Rep. III	Total (ABC)
				N_1 (0 kg N/ha)				
V_1 (IR8)	2,373	3,958	4,384	10,715	2,293	3,528	2,538	8,359
V_2 (IR127-8-1-10)	4,007	5,795	5,001	14,803	4,035	4,885	4,583	13,503
V_3 (IR305-4-12-1-3)	2,620	4,508	5,621	12,749	4,527	4,866	3,628	13,021
V_4 (IR400-2-5-3-3-2)	2,726	5,630	3,821	12,177	5,274	6,200	4,038	15,512
V_5 (IR665-58)	4,447	3,276	4,582	12,305	4,655	2,796	3,739	11,190
V_6 (Peta)	2,572	3,724	3,326	9,622	4,535	5,457	3,537	13,529
				N_2 (60 kg N/ha)				
V_1	4,076	6,431	4,889	15,396	3,085	7,502	4,362	14,949
V_2	5,630	7,334	7,177	20,141	3,728	7,424	5,377	16,529
V_3	4,676	6,672	7,019	18,367	4,946	7,611	6,142	18,699
V_4	4,838	7,007	4,816	16,661	4,878	6,928	4,829	16,635
V_5	5,549	5,340	6,011	16,900	4,646	5,006	4,666	14,318
V_6	3,896	2,822	4,425	11,143	4,627	4,461	4,774	13,852
				N_3 (120 kg N/ha)				
V_1	7,254	6,808	8,582	22,644	6,661	6,353	7,759	20,773
V_2	7,053	8,284	6,297	21,634	6,440	7,648	5,736	19,824
V_3	7,666	7,328	8,611	23,605	8,632	7,101	7,416	22,149
V_4	6,881	7,735	6,667	21,283	6,545	9,838	7,253	22,626
V_5	6,880	5,080	6,076	18,036	6,995	4,486	6,564	18,045
V_6	1,556	2,706	3,214	7,476	5,374	7,218	6,369	18,961

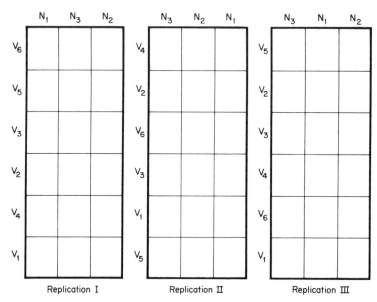

Figure 4.6 Random assignment of six varieties (V_1, V_2, V_3, V_4, V_5, and V_6) to horizontal strips and three nitrogen rates (N_1, N_2, and N_3) to vertical strips, as the first step in the laying out of a strip-split-plot design with three replications.

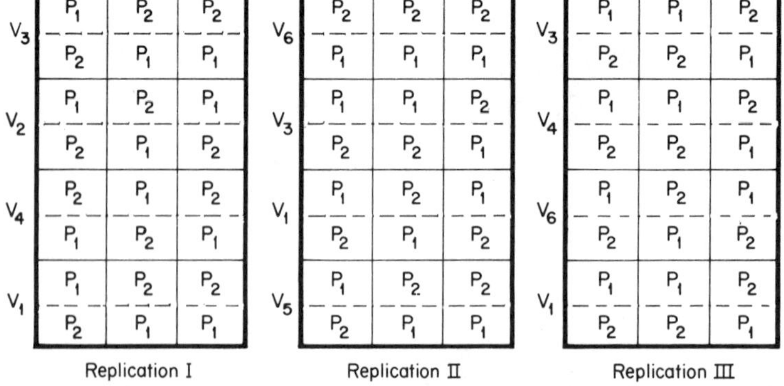

Figure 4.7 A sample layout of a $3 \times 6 \times 2$ factorial experiment arranged in a strip-split-plot design with six varieties (V_1, V_2, V_3, V_4, V_5, and V_6) as horizontal treatments, three nitrogen rates (N_1, N_2, and N_3) as vertical treatments, and two planting methods (P_1 and P_2) as subplot treatments, in three replications.

156

4.4.2 Analysis of Variance

The steps involved in the analysis of variance of a strip split-plot design are:

☐ STEP 1. Construct an outline of the analysis of variance for a strip-split-plot design as follows:

Source of Variation	Degree of Freedom	Sum of Squares	Mean Square	Computed F	Tabular F 5%	1%
Replication	$r - 1 = 2$					
Vertical factor (A)	$a - 1 = 2$					
Error(a)	$(r - 1)(a - 1) = 4$					
Horizontal factor (B)	$b - 1 = 5$					
Error(b)	$(r - 1)(b - 1) = 10$					
$A \times B$	$(a - 1)(b - 1) = 10$					
Error(c)	$(r - 1)(a - 1)(b - 1) = 20$					
Subplot factor (C)	$c - 1 = 1$					
$A \times C$	$(a - 1)(c - 1) = 2$					
$B \times C$	$(b - 1)(c - 1) = 5$					
$A \times B \times C$	$(a - 1)(b - 1)(c - 1) = 10$					
Error(d)	$ab(r - 1)(c - 1) = 36$					
Total	$rabc - 1 = 107$					

☐ STEP 2. Do a vertical analysis.

A. Compute the treatment totals (ABC) as shown in Table 4.12.

B. Construct the replication × vertical factor two-way table of totals, with replication totals, vertical factor totals, and the grand total computed. For our example, the replication × nitrogen table of totals (RA), with replication totals (R), nitrogen totals (A), and grand total (G) computed, is shown in Table 4.13.

C. Compute the correction factor and the various sums of squares:

$$C.F. = \frac{G^2}{rabc}$$

$$= \frac{(580,151)^2}{(3)(3)(6)(2)} = 3,116,436,877$$

$$\text{Total } SS = \sum X^2 - C.F.$$

$$= \left[(2,373)^2 + \cdots + (6,369)^2\right] - 3,116,436,877$$

$$= 307,327,796$$

Table 4.13   The Replication × Nitrogen Table of Yield Totals Computed from Data in Table 4.12

Nitrogen	Yield Total (RA)			Nitrogen Total (A)
	Rep. I	Rep. II	Rep. III	
N_1	44,064	54,623	48,798	147,485
N_2	54,575	74,538	64,487	193,600
N_3	77,937	80,585	80,544	239,066
Rep. Total(R)	176,576	209,746	193,829	
Grand total(G)				580,151

$$\text{Replication } SS = \frac{\sum R^2}{abc} - C.F.$$

$$= \frac{(176{,}576)^2 + (209{,}746)^2 + (193{,}829)^2}{(3)(6)(2)}$$

$$- 3{,}116{,}436{,}877$$

$$= 15{,}289{,}498$$

$$A \text{ (nitrogen) } SS = \frac{\sum A^2}{rbc} - C.F.$$

$$= \frac{(147{,}485)^2 + (193{,}600)^2 + (239{,}066)^2}{(3)(6)(2)}$$

$$- 3{,}116{,}436{,}877$$

$$= 116{,}489{,}164$$

$$\text{Error}(a) \ SS = \frac{\sum (RA)^2}{bc} - C.F. - \text{Replication } SS - A \ SS$$

$$= \frac{(44{,}064)^2 + \cdots + (80{,}544)^2}{(6)(2)} - 3{,}116{,}436{,}877$$

$$- 15{,}289{,}498 - 116{,}489{,}164$$

$$= 6{,}361{,}493$$

□ STEP 3.   Do a horizontal analysis.

A.   Construct the replication × horizontal factor two-way table of totals, with horizontal factor totals computed. For our example, the replica-

Table 4.14   The Replication × Variety Table of Yield Totals Computed from Data In Table 4.12

Variety	Yield Total (RB) Rep. I	Rep. II	Rep. III	Variety Total (B)
V_1	25,742	34,580	32,514	92,836
V_2	30,893	41,370	34,171	106,434
V_3	33,067	38,086	38,437	109,590
V_4	31,142	43,338	31,424	105,904
V_5	33,172	25,984	31,638	90,794
V_6	22,560	26,388	25,645	74,593

tion × variety table of totals (RB) with variety totals (B) computed is shown in Table 4.14.

B.   Compute the various sums of squares:

$$B \text{ (variety) } SS = \frac{\sum B^2}{rac} - C.F.$$

$$= \frac{(92,836)^2 + \cdots + (74,593)^2}{(3)(3)(2)} - 3,166,436,877$$

$$= 49,119,270$$

$$\text{Error}(b) \, SS = \frac{\sum (RB)^2}{ac} - C.F. - \text{Replication } SS - B \, SS$$

$$= \frac{(25,742)^2 + \cdots + (25,645)^2}{(3)(2)} - 3,116,436,877$$

$$- 15,289,498 - 49,119,270$$

$$= 26,721,828$$

☐ STEP 4.   Do an interaction analysis.

A.   Construct two tables of totals.

(i)   The vertical factor × horizontal factor two-way table of totals. For our example, the nitrogen × variety table of totals (AB) is shown in Table 4.15.

(ii)   The replication × vertical factor × horizontal factor three-way table of totals. For our example, the replication × nitrogen × variety table of totals (RAB) is shown in Table 4.16.

Table 4.15   The Nitrogen × Variety Table of Yield Totals Computed from Data in Table 4.12

Variety	Yield Total (AB)		
	N_1	N_2	N_3
V_1	19,074	30,345	43,417
V_2	28,306	36,670	41,458
V_3	25,770	37,066	46,754
V_4	27,689	33,296	44,919
V_5	23,495	31,218	36,081
V_6	23,151	25,005	26,437

Table 4.16   The Nitrogen × Variety × Replication Table of Yield Totals Computed from Data in Table 4.12

Variety	Yield Total (RAB)		
	Rep. I	Rep. II	Rep. III
	N_1 (0 kg N/ha)		
V_1	4,666	7,486	6,922
V_2	8,042	10,680	9,584
V_3	7,147	9,374	9,249
V_4	8,000	11,830	7,859
V_5	9,102	6,072	8,321
V_6	7,107	9,181	6,863
	N_2 (60 kg N/ha)		
V_1	7,161	13,933	9,251
V_2	9,358	14,758	12,554
V_3	9,622	14,283	13,161
V_4	9,716	13,935	9,645
V_5	10,195	10,346	10,677
V_6	8,523	7,283	9,199
	N_3 (120 kg N/ha)		
V_1	13,915	13,161	16,341
V_2	13,493	15,932	12,033
V_3	16,298	14,429	16,027
V_4	13,426	17,573	13,920
V_5	13,875	9,566	12,640
V_6	6,930	9,924	9,583

B. Compute the following sums of squares:

$$A \times B\ SS = \frac{\sum(AB)^2}{rc} - C.F. - A\ SS - B\ SS$$

$$= \frac{(19,074) + \cdots + (26,437)^2}{(3)(2)} - 3,116,436,877$$

$$- 116,489,164 - 49,119,270$$

$$= 24,595,732$$

$$\text{Error}(c)\ SS = \frac{\sum(RAB)^2}{c} - C.F. - \text{Replication } SS - A\ SS$$

$$- \text{Error}(a)\ SS - B\ SS - \text{Error}(b)\ SS - A \times B\ SS$$

$$= \frac{(4,666)^2 + \cdots + (9,583)^2}{2} - 3,116,436,877$$

$$- 15,289,498 - 116,489,164 - 6,361,493$$

$$- 49,119,270 - 26,721,828 - 24,595,732$$

$$= 19,106,732$$

□ STEP 5. Do a subplot analysis.

A. Construct two tables of totals.

(i) The vertical factor × subplot factor two-way table of totals, with subplot factor totals computed. For our example, the nitrogen × planting method table of totals (AC) with planting method totals (C) computed is shown in Table 4.17.

Table 4.17 The Nitrogen × Planting Method Table of Yield Totals Computed from Data in Table 4.12

	Yield Total (AC)	
Nitrogen	P_1	P_2
N_1	72,371	75,114
N_2	98,608	94,992
N_3	114,678	124,388
Planting method total(C)	285,657	294,494

(ii) The horizontal factor × subplot factor two-way table of totals. For our example the variety × planting method (BC) table of totals is shown in Table 4.18.

B. Compute the following sums of squares:

$$C \text{ (planting method) } SS = \frac{\sum C^2}{rab} - C.F.$$

$$= \frac{(285,657)^2 + (294,494)^2}{(3)(3)(6)} - 3,116,436,877$$

$$= 723,078$$

$$A \times C \ SS = \frac{\sum (AC)^2}{rb} - C.F. - A \ SS - C \ SS$$

$$= \frac{(72,371)^2 + \cdots + (124,388)^2}{(3)(6)}$$

$$- 3,116,436,877 - 116,489,164$$

$$- 723,078$$

$$= 2,468,136$$

$$B \times C \ SS = \frac{\sum (BC)^2}{ra} - C.F. - B \ SS - C \ SS$$

$$= \frac{(48,755)^2 + \cdots + (46,352)^2}{(3)(3)}$$

$$- 3,116,436,877 - 49,119,270$$

$$- 723,078$$

$$= 23,761,442$$

Table 4.18 The Variety × Planting Method Table of Yield Totals Computed from Data in Table 4.12

Variety	Yield Total (BC)	
	P_1	P_2
V_1	48,755	44,081
V_2	56,578	49,856
V_3	54,721	54,869
V_4	50,121	55,783
V_5	47,241	43,553
V_6	28,241	46,352

$$A \times B \times C \ SS = \frac{\sum (ABC)^2}{r} - C.F. - A\ SS - B\ SS - C\ SS$$

$$- A \times B\ SS - A \times C\ SS - B \times C\ SS$$

$$= \frac{(10,715)^2 + \cdots + (18,961)^2}{3} - 3,116,436,877$$

$$- 116,489,164 - 49,119,270 - 723,078$$

$$- 24,595,732 - 2,468,136 - 23,761,442$$

$$= 7,512,067$$

$$\text{Error}(d)\ SS = \text{Total}\ SS - (\text{the sum of all other}\ SS)$$

$$= 307,327,796 - (15,289,498 + 116,489,164$$

$$+ 6,361,493 + 49,119,270 + 26,721,828$$

$$+ 24,595,732 + 19,106,732 + 723,078$$

$$+ 2,468,136 + 23,761,442$$

$$+ 7,512,067)$$

$$= 15,179,356$$

☐ STEP 6. For each source of variation, compute the mean square value by dividing the *SS* by its degree of freedom:

$$\text{Replication}\ MS = \frac{\text{Replication}\ SS}{r - 1}$$

$$= \frac{15,289,498}{2} = 7,644,749$$

$$A\ MS = \frac{A\ SS}{a - 1}$$

$$= \frac{116,489,164}{2} = 58,244,582$$

$$\text{Error}(a)\ MS = \frac{\text{Error}(a)\ SS}{(r - 1)(a - 1)}$$

$$= \frac{6,361,493}{(2)(2)} = 1,590,373$$

$$B\ MS = \frac{B\ SS}{b-1}$$

$$= \frac{49,119,270}{5} = 9,823,854$$

$$\text{Error}(b)\ MS = \frac{\text{Error}(b)\ SS}{(r-1)(b-1)}$$

$$= \frac{26,721,828}{(2)(5)} = 2,672,183$$

$$A \times B\ MS = \frac{A \times B\ SS}{(a-1)(b-1)}$$

$$= \frac{24,595,732}{(2)(5)} = 2,459,573$$

$$\text{Error}(c)\ MS = \frac{\text{Error}(c)\ SS}{(r-1)(a-1)(b-1)}$$

$$= \frac{19,106,732}{(2)(2)(5)} = 955,337$$

$$C\ MS = \frac{C\ SS}{c-1}$$

$$= \frac{723,078}{1} = 723,078$$

$$A \times C\ MS = \frac{A \times C\ SS}{(a-1)(c-1)}$$

$$= \frac{2,468,136}{(2)(1)} = 1,234,068$$

$$B \times C\ MS = \frac{B \times C\ SS}{(b-1)(c-1)}$$

$$= \frac{23,761,442}{(5)(1)} = 4,752,288$$

$$A \times B \times C\ MS = \frac{A \times B \times C\ SS}{(a-1)(b-1)(c-1)}$$

$$= \frac{7,512,067}{(2)(5)(1)} = 751,207$$

$$\text{Error}(d) \ MS = \frac{\text{Error}(d) \ SS}{ab(r-1)(c-1)}$$

$$= \frac{15,179,356}{(3)(6)(2)(1)} = 421,649$$

☐ STEP 7. Compute the F value for each effect by dividing each mean square by its appropriate error mean square:

$$F(B) = \frac{B \ MS}{\text{Error}(b) \ MS}$$

$$= \frac{9,823,854}{2,672,183} = 3.68$$

$$F(A \times B) = \frac{A \times B \ MS}{\text{Error}(c) \ MS}$$

$$= \frac{2,459,573}{955,337} = 2.57$$

$$F(C) = \frac{C \ MS}{\text{Error}(d) \ MS}$$

$$= \frac{723,078}{421,649} = 1.71$$

$$F(A \times C) = \frac{A \times C \ MS}{\text{Error}(d) \ MS}$$

$$= \frac{1,234,068}{421,649} = 2.93$$

$$F(B \times C) = \frac{B \times C \ MS}{\text{Error}(d) \ MS}$$

$$= \frac{4,752,288}{421,649} = 11.27$$

$$F(A \times B \times C) = \frac{A \times B \times C \ MS}{\text{Error}(d) \ MS}$$

$$= \frac{751,207}{421,649} = 1.78$$

Note that because of inadequate *d.f.* for error(a) *MS*, the F value for the main effect of factor A is not computed (see Chapter 2, Section 2.1.2.1).

☐ STEP 8.   For each effect whose computed F value is not less than 1, obtain the corresponding tabular F values from Appendix E, with $f_1 = d.f.$ of the numerator MS and $f_2 = d.f.$ of the denominator MS, at the 5% and 1% levels of significance.

☐ STEP 9.   Compute the four coefficients of variation corresponding to the four error mean squares, as follows:

$$cv(a) = \frac{\sqrt{\text{Error}(a)\ MS}}{\text{Grand mean}} \times 100$$

$$cv(b) = \frac{\sqrt{\text{Error}(b)\ MS}}{\text{Grand mean}} \times 100$$

$$cv(c) = \frac{\sqrt{\text{Error}(c)\ MS}}{\text{Grand mean}} \times 100$$

$$cv(d) = \frac{\sqrt{\text{Error}(d)\ MS}}{\text{Grand mean}} \times 100$$

The $cv(a)$ and $cv(b)$ values indicate the degrees of precision associated with the measurement of the effects of vertical and horizontal factors. The $cv(c)$ value indicates the precision of the interaction effect between these two factors and the $cv(d)$ value indicates the precision of all effects concerning the subplot factor. It is normally expected that the values of $cv(a)$ and $cv(b)$ are larger than that of $cv(c)$, which in turn is larger than $cv(d)$.

For our example, the value of $cv(a)$ is not computed because of inadequate error $d.f.$ for error(a) MS (see step 7). The other three cv values are computed as:

$$cv(b) = \frac{\sqrt{2,672,183}}{5,372} \times 100 = 30.4\%$$

$$cv(c) = \frac{\sqrt{955,337}}{5,372} \times 100 = 18.2\%$$

$$cv(d) = \frac{\sqrt{421,649}}{5,372} \times 100 = 12.1\%$$

☐ STEP 10.   Enter all values obtained in steps 2 to 9 in the analysis of variance outline of step 1 and compare each computed F value to its corresponding tabular F values and indicate its significance by the appropriate asterisk notation (see Chapter 2, Section 2.1.2).

Table 4.19 Analysis of Variance[a] (Strip-split-plot Design) of Grain Yield Data in Table 4.12

Source of Variation	Degree of Freedom	Sum of Squares	Mean Square	Computed F^b	Tabular F 5%	Tabular F 1%
Replication	2	15,289,498	7,644,749			
Nitrogen (A)	2	116,489,164	58,244,582	c	—	—
Error(a)	4	6,361,493	1,590,373			
Variety (B)	5	49,119,270	9,823,854	3.68*	3.33	5.64
Error(b)	10	26,721,828	2,672,183			
A × B	10	24,595,732	2,459,573	2.57*	2.35	3.37
Error(c)	20	19,106,732	955,337			
Planting method (C)	1	723,078	723,078	1.71ns	4.11	7.39
A × C	2	2,468,136	1,234,068	2.93ns	3.26	5.25
B × C	5	23,761,442	4,752,288	11.27**	2.48	3.58
A × B × C	10	7,512,067	751,207	1.78ns	2.10	2.86
Error(d)	36	15,179,356	421,649			
Total	107	307,327,796				

[a]$cv(b) = 30.4\%$, $cv(c) = 18.2\%$, $cv(d) = 12.1\%$.
[b]** = significant at 1% level, * = significant at 5% level, ns = not significant.
[c]Error(a) d.f. is not adequate for valid test of significance.

For our example, the results (Table 4.19) show that the three-factor interaction is not significant, and that two of the three two-factor interactions, namely, the nitrogen × variety interaction and the variety × planting method interaction, are significant. These results indicate that the effects of both nitrogen and planting method varied among varieties tested. For a proper interpretation of the significant interactions and appropriate mean comparisons, see appropriate procedures in Chapter 5.

4.5 FRACTIONAL FACTORIAL DESIGN

As the number of factors to be tested increases, the complete set of factorial treatments may become too large to be tested simultaneously in a single experiment. A logical alternative is an experimental design that allows testing of only a fraction of the total number of treatments. A design uniquely suited for experiments involving a large number of factors is the *fractional factorial design* (FFD). It provides a systematic way of selecting and testing only a fraction of the complete set of factorial treatment combinations. In exchange, however, there is loss of information on some preselected effects. Although this information loss may be serious in experiments with one or two factors, such a loss becomes more tolerable with a large number of factors. The number of interaction effects increases rapidly with the number of factors involved, which allows flexibility in the choice of the particular effects to be sacrificed. In fact, in cases where some specific effects are known beforehand to be small or unimportant, use of the FFD results in minimal loss of information.

In practice, the effects that are most commonly sacrificed by use of the FFD are high-order interactions—the four-factor or five-factor interactions and, at times, even the three-factor interaction. In almost all cases, unless the researcher has prior information to indicate otherwise, he should select a set of treatments to be tested so that all main effects and two-factor interactions can be estimated.

In agricultural research, the FFD is most commonly used in exploratory trials where the main objective is to examine the interactions between factors. For such trials, the most appropriate FFD are those that sacrifice only those interactions that involve more than two factors.

With the FFD, the number of effects that can be measured decreases rapidly with the reduction in the number of treatments to be tested. Thus, when the number of effects to be measured is large, the number of treatments to be tested, even with the use of FFD, may still be too large. In such cases, further reduction in the size of the experiment can be achieved by reducing the number of replications. Although use of a FFD without replication is uncommon in agricultural experiments, when FFD is applied to an exploratory trial the number of replications required can be reduced. For example, two replications are commonly used in an exploratory field trial in rice whereas four replications are used for a standard field experiment in rice.

Another desirable feature of FFD is that it allows reduced block size by not requiring a block to contain all treatments to be tested. In this way, the homogeneity of experimental units within the same block can be improved. A reduction in block size is, however, accompanied by loss of information in addition to that already lost through the reduction in number of treatments.

Although the FFD can be tailor-made to fit most factorial experiments, the procedure for doing so is complex and beyond the scope of this book. Thus, we describe only a few selected sets of FFD that are suited for exploratory trials in agricultural research. The major features of these selected designs are that they:

· Apply only to 2^n factorial experiments where n, the number of factors, ranges from 5 to 7.
· Involve only one half of the complete set of factorial treatment combinations (i.e., the number of treatments is $1/2$ of 2^n or 2^{n-1}).
· Have a block size of 16 plots or less.
· Allow all main effects and most, if not all, of the two-factor interactions to be estimated.

The selected plans are given in Appendix M. Each plan provides the list of treatments to be tested and the specific effects that can be estimated. In the designation of the various treatment combinations for all plans, the letters $a, b, c, \ldots$ are used to denote the presence (or use of high level) of factors

$A, B, C, \ldots$. Thus, the treatment combination ab in a 2^5 factorial experiment refers to the treatment combination that contains the high level (or presence) of factors A and B and low level (or absence) of factors $C, D,$ and $E,$ but this same notation (ab) in a 2^6 factorial experiment would refer to the treatment combination that contains the high level of factors A and B and low level of factors $C, D, E,$ and $F.$ In all cases, the treatment combination that consists of the low level of all factors is denoted by the symbol (1).

We illustrate the procedure for randomization, layout, and analysis of variance of a FFD with a field experiment involving six factors $A, B, C, D, E,$ and $F,$ each at two levels (i.e., 2^6 factorial experiment). Only 32 treatments from the total of 64 complete factorial treatment combinations are tested in blocks of 16 plots each. With two replications, the total number of experimental plots is 64.

4.5.1 Randomization and Layout

The steps for randomization and layout are:

☐ STEP 1. Choose an appropriate basic plan of a FFD in Appendix M. The plan should correspond to the number of factors and the number of levels of each factor to be tested. For basic plans that are not given in Appendix M, see Cochran and Cox, 1957.* Our example uses plan 3 of Appendix M.

☐ STEP 2. If there is more than one block per replication, randomly assign the block arrangement in the basic plan to the actual blocks in the field.

For this example, the experimental area is first divided into two replications (Rep. I and Rep. II), each consisting of 32 experimental plots. Each replication is further divided into two blocks (Block 1 and Block 2), each consisting of 16 plots. Following one of the randomization schemes of Chapter 2, Section 2.1.1, randomly reassign the block numbers in the basic plan to the blocks in the field. The result may be as follows:

Block Number in Basic Plan	Block Number Assignment in Field	
	Rep. I	Rep. II
I	2	1
II	1	2

Note that all 16 treatments listed in block I of the basic plan are assigned to block 2 of replication I in the field, all 16 treatments listed in block II of the basic plan are assigned to block 1 of replication I in the field, and so on.

☐ STEP 3. Randomly reassign the treatments in each block of the basic plan to the experimental plots of the reassigned block in the field (from step 2).

---

*W. G. Cochran and G. M. Cox. *Experimental Designs*. New York: Wiley, 1957, pp. 276–292.

For this example, follow the same randomization scheme used in step 2 and randomly assign the 16 treatments of a given block (in the basic plan) to the 16 plots of the corresponding block in the field, separately and independently for each of the four blocks (i.e., two blocks per replication and two replications). The result of the four independent randomization processes may be as follows:

Treatment Number in Basic Plan	Plot Number Assignment in Field			
	Rep. I		Rep. II	
	Block 1	Block 2	Block 1	Block 2
1	6	5	4	11
2	3	4	14	7
3	15	10	3	6
4	12	6	8	1
5	1	12	7	15
6	5	1	11	4
7	13	3	16	14
8	7	8	12	9
9	2	16	9	3
10	10	11	10	5
11	11	15	5	8
12	8	2	6	12
13	4	14	1	16
14	9	9	2	13
15	16	13	15	2
16	14	7	13	10

Note that block 1 of replication I in the field was assigned to receive treatments of block II in the basic plan (step 2); and according to the basic plan used (i.e., plan 3 of Appendix M) treatment 1 of block II is *ae*. Thus, according to the foregoing assignment of treatments, treatment *ae* is assigned to plot 6 in block 1 of replication I. In the same manner, because treatment 2 of block II in the basic plan is *af*, treatment *af* is assigned to plot 3 in block 1 of replication I; and so on. The final layout is shown in Figure 4.8.

4.5.2 Analysis of Variance

The analysis of variance procedures of a FFD, without replication and with replication, are illustrated. We use Yates' method for the computation of sums of squares. This method is suitable for manual computation of large fractional factorial experiments. Other alternative procedures are:

The application of the standard rules for the computation of sums of squares in the analysis of variance (Chapter 3), by constructing two-way tables of totals for two-factor interactions, three-way table of totals for three-factor interactions, and so on.

- The application of the single *d.f.* contrast method (Chapter 5), by specifying a contrast for each of the main effects and interaction effects that are to be estimated.

4.5.2.1 Design without Replication. For illustration, we use data (Table 4.20) from a FFD trial whose layout is shown in Figure 4.8. Here, only data from replication I of Table 4.20 is used. The computational steps in the analysis of variance are:

☐ STEP 1.   Outline the analysis of variance, following that given in Appendix M, corresponding to the basic plan used. For our example, the basic plan is

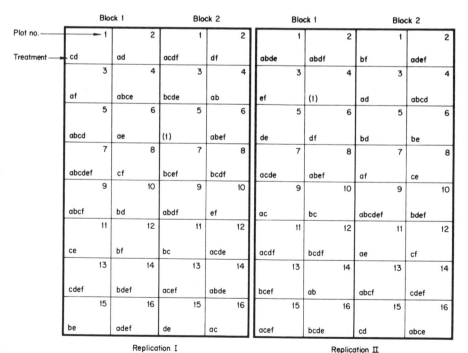

Figure 4.8 A sample layout of a fractional factorial design with two replications: 1/2 of 2^6 factorial treatments arranged in blocks of 16 plots each.

plan 3 of Appendix M and the outline of the analysis of variance is:

Source of Variation	Degree of Freedom	Sum of Squares	Mean Square	Computed F	Tabular F 5%	1%
Block	1					
Main effect	6					
Two-factor interaction	15					
Error	9					
Total	31					

☐ STEP 2.   Determine the number of *real factors* (k) each at two levels, whose complete set of factorial treatments is equal to the number of treatments (t) to be tested (i.e., $2^k = t$). Then select the specific set of k real factors from

Table 4.20 Grain Yield Data from a 2^6 Factorial Experiment Planted in a $\frac{1}{2}$ Fractional Factorial Design in Blocks of 16 Experimental Plots Each, and with Two Replications

Treatment	Grain Yield, t/ha Rep. I	Rep. II	Total	Treatment	Grain Yield, t/ha Rep. I	Rep. II	Total
	Block 1				*Block 2*		
(1)	2.92	2.76	5.68	ad	3.23	3.48	6.71
ab	3.45	3.50	6.95	ae	3.10	3.11	6.21
ac	3.65	3.50	7.15	af	3.52	3.27	6.79
bc	3.16	3.05	6.21	bd	3.29	3.22	6.51
de	3.29	3.03	6.32	be	3.06	3.20	6.26
df	3.34	3.37	6.71	bf	3.27	3.27	6.54
ef	3.28	3.23	6.51	cd	3.68	3.52	7.20
abde	3.88	3.79	7.67	ce	3.08	3.02	6.10
abdf	3.95	4.03	7.98	cf	3.29	3.10	6.39
abef	3.85	3.90	7.75	abcd	3.89	3.99	7.88
acde	4.05	4.18	8.23	abce	3.71	3.80	7.51
acdf	4.37	4.20	8.57	abcf	3.96	3.98	7.94
acef	3.77	3.80	7.57	adef	4.27	3.98	8.25
bcde	4.04	3.87	7.91	bdef	3.69	3.62	7.31
bcdf	4.00	3.76	7.76	cdef	4.29	4.09	8.38
bcef	3.63	3.46	7.09	abcdef	4.80	4.78	9.58
Total (*RB*)	58.63	57.43			58.13	57.43	

the original set of n factors and designate all $(n - k)$ factors not included in the set of k as *dummy factors*.

For our example, the $t = 32$ treatment combinations correspond to a complete set of 2^k factorial treatment combinations, with $k = 5$. For simplicity, the first five factors A, B, C, D, and E are designated as the real factors and F as the dummy factor.

☐ STEP 3.   Arrange the t treatments in a systematic order based on the k real factors:

A.   Treatments with fewer number of letters are listed first. For example, *ab* comes before *abc*, and *abc* comes before *abcde*, and so on. Note that if treatment (1) is present in the set of t treatments, it always appears as the first treatment in the sequence.

B.   Among treatments with the same number of letters, those involving letters corresponding to factors assigned to the lower-order letters come first. For example, *ab* comes before *ac*, *ad* before *bc*, and so on.

C.   All treatment-identification letters corresponding to the dummy factors are ignored in the arrangement process. For our example, factor F is the dummy factor and, thus, *af* is considered simply as *a* and comes before *ab*.

In this example, the systematic arrangement of the 32 treatments is shown in the first column of Table 4.21. Note that:

• The treatments are listed systematically regardless of their block allocation.

• The dummy factor F is placed in parenthesis.

☐ STEP 4.   Compute the t factorial effect totals:

A.   Designate the original data of the t treatments as the initial set or the t_0 values. For our example, the systematically arranged set of 32 t_0 values are listed in the second column of Table 4.21.

B.   Group the t_0 values into $t/2$ successive pairs. For our example, there are 16 successive pairs: the first pair is 2.92 and 3.52, the second pair is 3.27 and 3.45, and the last pair is 4.04 and 4.80.

C.   Add the values of the two treatments in each of the $t/2$ pairs constituted in task 2 to constitute the first half of the second set or the t_1 values. For our example, the first half of the t_1 values are computed as:

$$6.44 = 2.92 + 3.52$$

$$6.72 = 3.27 + 3.45$$

$$\vdots$$

$$8.34 = 4.29 + 4.05$$

$$8.84 = 4.04 + 4.80$$

Table 4.21 Application of Yates' Method for the Computation of Sums of Squares of a 2^6 Factorial Experiment Conducted in a $\frac{1}{2}$ Fractional Factorial Design, without Replication, from Rep. I data in Table 4.20

Treatment Combination	t_0	t_1	t_2	t_3	t_4	t_5	Factorial Effect Identification Preliminary	Final
(1)	2.92	6.44	13.16	27.22	56.97	116.76	(G)	(G)
$a(f)$	3.52	6.72	14.06	29.75	59.79	6.14	A	A
$b(f)$	3.27	6.94	13.81	27.48	3.07	2.50	B	B
ab	3.45	7.12	15.94	32.31	3.07	0.56	AB	AB
$c(f)$	3.29	6.57	13.29	1.94	0.97	5.98	C	C
ac	3.65	7.24	14.19	1.13	1.53	−0.08	AC	AC
bc	3.16	8.05	15.13	1.38	−0.01	−0.48	BC	BC
$abc(f)$	3.96	7.89	17.18	1.69	0.57	−0.50	ABC	ABC(Block)
$d(f)$	3.34	6.38	0.78	0.46	3.03	7.36	D	D
ad	3.23	6.91	1.16	0.51	2.95	−0.50	AD	AD
bd	3.29	6.85	0.55	1.02	0.41	−0.46	BD	BD
$abd(f)$	3.95	7.34	0.58	0.51	−0.49	−0.20	ABD	ABD
cd	3.68	7.56	0.61	0.02	−0.93	2.38	CD	CD
$acd(f)$	4.37	7.57	0.77	−0.03	0.45	−1.16	ACD	ACD
$bcd(f)$	4.00	8.34	1.17	0.36	−0.71	−0.20	BCD	BCD
$abcd$	3.89	8.84	0.52	0.21	0.21	0.94	ABCD	EF
$e(f)$	3.28	0.60	0.28	0.90	2.53	2.82	E	E
ae	3.10	0.18	0.18	2.13	4.83	0.00	AE	AE
be	3.06	0.36	0.67	0.90	−0.81	0.56	BE	BE
$abe(f)$	3.85	0.80	−0.16	2.05	0.31	0.58	ABE	ABE
ce	3.08	−0.11	0.53	0.38	0.05	−0.08	CE	CE
$ace(f)$	3.77	0.66	0.49	0.03	−0.51	−0.90	ACE	ACE
$bce(f)$	3.63	0.69	0.01	0.16	−0.05	1.38	BCE	BCE
$abce$	3.71	−0.11	0.50	−0.65	−0.15	0.92	ABCE	DF
de	3.29	−0.18	−0.42	−0.10	1.23	2.30	DE	DE
$ade(f)$	4.27	0.79	0.44	−0.83	1.15	1.12	ADE	ADE
$bde(f)$	3.69	0.69	0.77	−0.04	−0.35	−0.56	BDE	BDE
$abde$	3.88	0.08	−0.80	0.49	−0.81	−0.10	ABDE	CF
$cde(f)$	4.29	0.98	0.97	0.86	−0.73	−0.08	CDE	CDE
$acde$	4.05	0.19	−0.61	−1.57	0.53	−0.46	ACDE	BF
$bcde$	4.04	−0.24	−0.79	−1.58	−2.43	1.26	BCDE	AF
$abcde(f)$	4.80	0.76	1.00	1.79	3.37	5.80	ABCDE	F

The results of the first 16 t_1 values are shown in the top of the third column of Table 4.21.

D. Subtract the first value from the second in each of the $t/2$ pairs constituted in task 2 to constitute the bottom half of the t_1 values. For our example, the second half of the t_1 values are computed as:

$$0.60 = 3.52 - 2.92$$

$$0.18 = 3.45 - 3.27$$

$$\vdots$$

$$-0.24 = 4.05 - 4.29$$

$$0.76 = 4.80 - 4.04$$

The results of the last 16 t_1 values are shown in the bottom half of the third column of Table 4.21.

E. Reapply tasks B to D using the values of t_1 instead of t_0 to derive the third set or the t_2 values. For our example, tasks B to D are reapplied to t_1 values to arrive at the t_2 values shown in the fourth column of Table 4.21.

F. Repeat task E, $(k - 2)$ times. Each time use the newly derived values of t. For our example, task E is repeated three more times to derive t_3 values, t_4 values, and t_5 values as shown in the fifth, sixth, and seventh columns of Table 4.21.

☐ STEP 5. Identify the specific factorial effect that is represented by each of the values of the last set (commonly referred to as the factorial effect totals) derived in step 4. Use the following guidelines:

A. The first value represents the grand total (G).

B. For the remaining $(t - 1)$ values, assign the *preliminary* factorial effects according to the letters of the corresponding treatments, with the dummy factors ignored. For our example, the second t_5 value corresponds to treatment combination $a(f)$ and, hence, is assigned to the A main effect. The fourth t_5 value corresponds to treatment ab and is assigned to the $A \times B$ interaction effect, and so on. The results for all 32 treatments are shown in the eighth column of Table 4.21.

C. For treatments involving the dummy factor (or factors) adjust the preliminary factorial effects derived in task B as follows:

  • Based on the *conditions* stated in the basic plan of Appendix M, identify all effects involving the dummy factor that are estimable (i.e., that can be estimated). For our example, the estimable effects involving the dummy factor F consist of the main effect of F and all its two-factor interactions AF, BF, CF, DF, and EF.

- Identify the *aliases* of all effects listed immediately above. The alias of any effect is defined as its *generalized interaction* with the defining contrast. The generalized interaction between any two factorial effects is obtained by combining all the letters that appear in the two effects and canceling all letters that enter twice. For example, the generalized interaction between *ABC* and *AB* is *AABBC* or *C*.

  For our example, because the defining contrast is *ABCDEF* (see plan 3 of Appendix M) the aliases of the six effects involving the dummy factor *F* are: *F = ABCDE, AF = BCDE, BF = ACDE, CF = ABDE, DF = ABCE,* and *EF = ABCD.*

  The two factorial effects involved in each pair of aliases (one to the left and another to the right of the equal sign) are not separable (i.e., can not be estimated separately). For example, for the first pair, *F* and *ABCDE*, the main effect of factor *F* cannot be separated from the $A \times B \times C \times D \times E$ interaction effect and, hence, unless one of the pair is known to be absent there is no way to know which of the pairs is the contributor to the estimate obtained.

- Replace all preliminary factorial effects that are aliases of the estimable effects involving the dummy factors by the latter. For example, because *ABCDE* (corresponding to the last treatment in Table 4.21) is the alias of *F*, it is replaced by *F*. In the same manner, *BCDE* is replaced by *AF*, *ACDE* by *BF*, *ABDE* by *CF*, *ABCE* by *DF*, and *ABCD* by *EF*.

- When blocking is used, identify the factorial effects that are confounded with blocks. Such effects are stated for each plan of Appendix M. For our example, *ABC* is confounded with block (see plan 3 of Appendix M) and the preliminary factorial effect *ABC* is, therefore, replaced by the *block effect*. That means that the estimate of the *ABC* effect becomes the measure of the block effect.

  The final results of the factorial effect identification are shown in the last column of Table 4.21.

☐ STEP 6.   For each source of variation in the analysis of variance (step 1) identify the corresponding factorial effects. For our example, there is only one factorial effect (i.e., *ABC*) corresponding to the first source of variation of block. For the second source of variation (main effects) there are six factorial effects corresponding to the six main effects (*A, B, C, D, E,* and *F*). And, for the third source of variation (two-factor interactions) there are 15 factorial effects (i.e., all 15 possible two-factor interaction effects among the six factors). All the remaining nine factorial effects correspond to the last source of variation (error).

☐ STEP 7.   For each source of variation in the analysis of variance of step 1, compute its *SS* as the sum of the squares of the factorial effect totals of the corresponding factorial effects (identified in step 6) divided by the total number of treatments tested in the experiment. For our example, the various *SS* are computed as:

$$\text{Block } SS = \frac{(ABC)^2}{32}$$

$$= \frac{(-0.50)^2}{32} = 0.007812$$

$$\text{Main effect } SS = \frac{(A)^2 + (B)^2 + (C)^2 + (D)^2 + (E)^2 + (F)^2}{32}$$

$$= \left[(6.14)^2 + (2.50)^2 + (5.98)^2 + (7.36)^2\right.$$

$$\left. + (2.82)^2 + (5.80)^2\right]/32$$

$$= 5.483500$$

$$\text{Two-factor interaction } SS = \left[(AB)^2 + (AC)^2 + (BC)^2 + \cdots + (CF)^2\right.$$

$$\left. + (BF)^2 + (AF)^2\right]/32$$

$$= \left[(0.56)^2 + (-0.08)^2 + (-0.48)^2\right.$$

$$+ \cdots + (-0.10)^2 + (-0.46)^2 + (1.26)^2\Big]/32$$

$$= 0.494550$$

$$\text{Error } SS = \left[(ABD)^2 + (ACD)^2 + (BCD)^2 + \cdots\right.$$

$$\left. + (ADE)^2 + (BDE)^2 + (CDE)^2\right]/32$$

$$= \left[(-0.20)^2 + (-1.16)^2 + (-0.20)^2 + \cdots + (1.12)^2\right.$$

$$\left. + (-0.56)^2 + (-0.08)^2\right]/32$$

$$= 0.189088$$

Note that the error *SS* can also be computed as the difference between the total *SS* and the sum of all other *SS*, where the total *SS* is computed from all factorial effect totals. For our example, the total SS and the error SS are:

$$\text{Total } SS = \frac{(A)^2 + (B)^2 + (AB)^2 + \cdots + (BF)^2 + (AF)^2 + (F)^2}{32}$$

$$= \left[(6.14)^2 + (2.50)^2 + (0.56)^2 + \cdots + (-0.46)^2 \right.$$

$$\left. + (1.26)^2 + (5.80)^2\right] / 32$$

$$= 6.174950$$

$$\text{Error } SS = \text{Total } SS - \text{Main effect } SS - \text{Two-factor interaction } SS$$

$$- \text{Block } SS$$

$$= 6.174950 - 5.483500 - 0.494550 - 0.007812$$

$$= 0.189088$$

☐ STEP 8.   Determine the degree of freedom for each SS as the number of factorial effect totals used in its computation. For example, the computation of the block SS involves only one effect, namely ABC; hence, its $d.f.$ is 1. On the other hand, there are six effect totals involved in the computation of the main effect SS; hence, its $d.f.$ is 6. The results are shown in the second column of Table 4.22.

☐ STEP 9.   Compute the mean square for each source of variation by dividing each SS by its $d.f.$:

$$\text{Block } MS = \frac{\text{Block } SS}{1}$$

$$= 0.007812$$

Table 4.22   Analysis of Variance of Data from a Fractional Factorial Design: $\frac{1}{2}$ of a 2^6 Factorial Experiment without Replication[a]

Source of Variation	Degree of Freedom	Sum of Squares	Mean Square	Computed F[b]	Tabular F 5%	1%
Block	1	0.007812	0.007812	< 1	—	—
Main effect	6	5.483500	0.913917	43.50**	3.37	5.80
Two-factor interaction	15	0.494550	0.032970	1.57[ns]	3.00	4.96
Error	9	0.189088	0.021010			
Total	31	6.174950				

[a]Source of data: Rep. I data of Table 4.20.
[b]** = significant at 1% level, [ns] = not significant.

$$\text{Main effect } MS = \frac{\text{Main effect } SS}{6}$$

$$= \frac{5.483500}{6} = 0.913917$$

$$\text{Two-factor interaction } MS = \frac{\text{Two-factor interaction } SS}{15}$$

$$= \frac{0.494550}{15} = 0.032970$$

$$\text{Error } MS = \frac{\text{Error } SS}{9}$$

$$= \frac{0.189088}{9} = 0.021010$$

☐ STEP 10.   Compute the F value for each effect by dividing its MS by the error MS:

$$F(\text{block}) = \frac{\text{Block } MS}{\text{Error } MS}$$

$$= \frac{0.007812}{0.021010} < 1$$

$$F(\text{main effect}) = \frac{\text{Main effect } MS}{\text{Error } MS}$$

$$= \frac{0.913917}{0.021010} = 43.50$$

$$F(\text{two-factor interaction}) = \frac{\text{Two-factor interaction } MS}{\text{Error } MS}$$

$$= \frac{0.032970}{0.021010} = 1.57$$

☐ STEP 11.   Compare each computed F value with the corresponding tabular F values, from Appendix E, with $f_1 = d.f.$ of the numerator MS and $f_2 = $ error $d.f.$

The final analysis of variance is shown in Table 4.22. The results indicate a highly significant main effect but not the two-factor interaction effect.

4.5.2.2   *Design with Replication.* We show the computations involved in the analysis of variance of a FFD with data from both replications in Table 4.20.

☐ STEP 1.   Outline the analysis of variance, following that given in plan 3 of Appendix M:

Source of Variation	Degree of Freedom	Sum of Squares	Mean Square	Computed F	Tabular F 5%	1%
Replication	1					
Block	1					
Block × Replication	1					
Main effect	6					
Two-factor interaction	15					
Three-factor interaction	9					
Error	30					
Total	63					

☐ STEP 2.   Compute the replication × block totals (RB) as shown in Table 4.20. Then compute the replication total for each of the two replications (R), the block totals for each of the two blocks (B), and the grand total (G) as:

$$R_1 = 58.63 + 58.13 = 116.76$$

$$R_2 = 57.43 + 57.43 = 114.86$$

$$B_1 = 58.63 + 57.43 = 116.06$$

$$B_2 = 58.13 + 57.43 = 115.56$$

$$G = 116.76 + 114.86$$

$$= 116.06 + 115.56 = 231.62$$

☐ STEP 3.   Let r denote the number of replications, p the number of blocks in each replication, and t the total number of treatments tested. Compute the correction factor, total SS, replication SS, block SS, and block × replication SS as:

$$C.F. = \frac{G^2}{rt}$$

$$= \frac{(231.62)^2}{(2)(32)} = 838.247256$$

Total $SS = \sum X^2 - C.F.$

$$- \left[(7.92)^2 + \cdots + (4.78)^2 \right] - 838.247256$$

$$= 12.419344$$

Replication $SS = \dfrac{\sum R^2}{t} - C.F.$

$$= \dfrac{(116.76)^2 + (114.86)^2}{32} - 838.247256$$

$$= 0.056406$$

Block $SS = \dfrac{\sum B^2}{t} - C.F.$

$$= \dfrac{(116.06)^2 + (115.56)^2}{32} - 838.247256$$

$$= 0.003906$$

Block $\times$ Replication $SS = \dfrac{\sum (RB)^2}{t/p} - C.F. -$ Replication $SS -$ Block SS

$$= \dfrac{(58.63)^2 + (57.43)^2 + (58.13)^2 + (57.43)^2}{32/2}$$

$$- 838.247256 - 0.056406 - 0.003906$$

$$= 0.003907$$

□ STEP 4. Follow steps 2 to 7 of Section 4.5.2.1; with one modification, namely that the grain yield data in the second column of Table 4.21 is replaced by the yield totals over two replications as shown in Table 4.23. Then compute the various SS as follows:

Main effect $SS = \dfrac{(A)^2 + (B)^2 + (C)^2 + (D)^2 + (E)^2 + (F)^2}{(r)(2^k)}$

$$= \left[(13.86)^2 + (6.08)^2 + (11.32)^2 + (14.32)^2 \right.$$

$$\left. + (5.68)^2 + (10.62)^2 \right] / (2)(32)$$

$$= 11.051838$$

Two-factor interaction $SS = \left[(AB)^2 + (AC)^2 + (BC)^2 + \cdots + (CF)^2\right.$

$$+ (BF)^2 + (AF)^2\Big]/(r)(2^k)$$

$$= \left[(1.48)^2 + (0.92)^2 + (-1.50)^2 + \cdots + (-0.44)^2\right.$$

$$+ (-0.52)^2 + (1.62)^2\Big]/(2)(32)$$

$$= 0.787594$$

Three-factor interaction $SS = \left[(ABD)^2 + (ACD)^2 + (BCD)^2 + \cdots\right.$

$$+ (ADE)^2 + (BDE)^2 + (CDE)^2\Big]/(r)(2^k)$$

$$= \left[(-0.54)^2 + (-2.42)^2 + (0.04)^2 + \cdots\right.$$

$$+ (1.78)^2 + (-0.24)^2 + (1.24)^2\Big]/(2)(32)$$

$$= 0.238206$$

Error SS = Total SS − (the sum of all other SS)

$$= 12.419344 - (0.056406 + 0.003906 + 0.003907$$

$$+ 11.051838 + 0.787594 + 0.238206)$$

$$= 0.277487$$

☐ STEP 5.   Compute the mean square for each source of variation, by dividing the SS by its $d.f.$ (see step 8 of Section 4.5.2.1 for the determination of $d.f$) as:

$$\text{Replication } MS = \frac{\text{Replication } SS}{1}$$

$$= \frac{0.056406}{1} = 0.056406$$

$$\text{Block } MS = \frac{\text{Block } SS}{1}$$

$$= \frac{0.003906}{1} = 0.003906$$

$$\text{Block} \times \text{Replication } MS = \frac{\text{Block} \times \text{Replication } SS}{1}$$

$$= \frac{0.003907}{1} = 0.003907$$

Table 4.23 Application of Yates' Method for the Computation of Sums of Squares of a 2^6 Factorial Experiment Conducted in a $\frac{1}{2}$ Fractional Factorial Design, with Two Replications; from Data in Table 4.20

Treatment Combination	t_0	t_1	t_2	t_3	t_4	t_5	Factorial Effect Identification Preliminary	Final
(1)	5.68	12.47	25.96	53.65	112.97	231.62	(G)	(G)
$a(f)$	6.79	13.49	27.69	59.32	118.65	13.86	A	A
$b(f)$	6.54	13.54	27.91	55.00	6.97	6.08	B	B
ab	6.95	14.15	31.41	63.65	6.89	1.48	AB	AB
$c(f)$	6.39	13.42	26.73	4.01	2.57	11.32	C	C
ac	7.15	14.49	28.27	2.96	3.51	0.92	AC	AC
bc	6.21	15.77	29.55	3.08	0.49	−1.50	BC	BC
$abc(f)$	7.94	15.64	34.10	3.81	0.99	−0.50	ABC	ABC(Block)
$d(f)$	6.71	12.72	1.52	1.63	5.23	14.32	D	D
ad	6.71	14.01	2.49	0.94	6.09	−0.32	AD	AD
bd	6.51	13.67	1.47	2.22	0.99	−1.62	BD	BD
$abd(f)$	7.98	14.60	1.49	1.29	−0.07	−0.54	ABD	ABD
cd	7.20	14.57	1.19	0.27	−1.61	4.78	CD	CD
$acd(f)$	8.57	14.98	1.89	0.22	0.11	−2.42	ACD	ACD
$bcd(f)$	7.76	16.61	2.29	0.74	−1.05	0.04	BCD	BCD
$abcd$	7.88	17.49	1.52	0.25	0.55	1.84	ABCD	EF
$e(f)$	6.51	1.11	1.02	1.73	5.67	5.68	E	E
ae	6.21	0.41	0.61	3.50	8.65	−0.08	AE	AE
be	6.26	0.76	1.07	1.54	−1.05	0.94	BE	BE
$abe(f)$	7.75	1.73	−0.13	4.55	0.73	0.50	ABE	ABE
ce	6.10	0.00	1.29	0.97	−0.69	0.86	CE	CE
$ace(f)$	7.57	1.47	0.93	0.02	−0.93	−1.06	ACE	ACE
$bce(f)$	7.09	1.37	0.41	0.70	−0.05	1.72	BCE	BCE
$abce$	7.51	0.12	0.88	−0.77	−0.49	1.60	ABCE	DF
de	6.32	−0.30	−0.70	−0.41	1.77	2.98	DE	DE
$ade(f)$	8.25	1.49	0.97	−1.20	3.01	1.78	ADE	ADE
$bde(f)$	7.31	1.47	1.47	−0.36	−0.95	−0.24	BDE	BDE
$abde$	7.67	0.42	−1.25	0.47	−1.47	−0.44	ABDE	CF
$cde(f)$	8.38	1.93	1.79	1.67	−0.79	1.24	CDE	CDE
$acde$	8.23	0.36	−1.05	−2.72	0.83	−0.52	ACDE	BF
$bcde$	7.91	−0.15	−1.57	−2.84	−4.39	1.62	BCDE	AF
$abcde(f)$	9.58	1.67	1.82	3.39	6.23	10.62	ABCDE	F

$$\text{Main effect } MS = \frac{\text{Main effect } SS}{6}$$

$$= \frac{11.051838}{6} = 1.841973$$

$$\text{Two-factor interaction } MS = \frac{\text{Two-factor interaction } SS}{15}$$

$$= \frac{0.787594}{15} = 0.052506$$

$$\text{Three-factor interaction } MS = \frac{\text{Three-factor interaction } SS}{9}$$

$$= \frac{0.238206}{9} = 0.026467$$

$$\text{Error } MS = \frac{\text{Error } SS}{30}$$

$$= \frac{0.277487}{30} = 0.009250$$

□ STEP 6. Compute the F value for each effect, by dividing its MS by the error MS as:

$$F(\text{replication}) = \frac{\text{Replication } MS}{\text{Error } MS}$$

$$= \frac{0.056406}{0.009250} = 6.10$$

$$F(\text{block}) = \frac{\text{Block } MS}{\text{Error } MS}$$

$$= \frac{0.003906}{0.009250} < 1$$

$$F(\text{block} \times \text{replication}) = \frac{\text{Block} \times \text{replication } MS}{\text{Error } MS}$$

$$= \frac{0.003907}{0.009250} < 1$$

$$F(\text{main effect}) = \frac{\text{Main effect } MS}{\text{Error } MS}$$

$$= \frac{1.841973}{0.009250} = 199.13$$

$$F(\text{two-factor interaction}) = \frac{\text{Two-factor interaction } MS}{\text{Error } MS}$$

$$= \frac{0.052506}{0.009250} = 5.68$$

$$F(\text{three-factor interaction}) = \frac{\text{Three-factor interaction } MS}{\text{Error } MS}$$

$$= \frac{0.026467}{0.009250} = 2.86$$

□ STEP 7. Compare each computed F value with the corresponding tabular F values, from Appendix E, with $f_1 = d.f.$ of the numerator MS and $f_2 = $ error $d.f.$ The results indicate that the main effects, the two-factor interactions, and the three-factor interactions are all significant.

The final analysis of variance is shown in Table 4.24. There are two important points that should be noted in the results of this analysis of variance obtained from two replications as compared to that without replication (Table 4.22):

- The effect of the three-factor interactions can be estimated only when there is replication.

Table 4.24 Analysis of Variance of Grain Yield Data in Table 4.20, from a Fractional Factorial Design: $\frac{1}{2}$ of a 2^6 Factorial Experiment with Two Replications

Source of Variation	Degree of Freedom	Sum of Squares	Mean Square	Computed F^a	Tabular F 5%	1%
Replication	1	0.056406	0.056406	6.10*	4.17	7.56
Block	1	0.003906	0.003906	< 1	—	—
Block × replication	1	0.003907	0.003907	< 1	—	—
Main effect	6	11.051838	1.841973	199.13**	2.42	3.47
Two-factor interaction	15	0.787594	0.052506	5.68**	2.02	2.70
Three-factor interaction	9	0.238206	0.026467	2.86*	2.21	3.06
Error	30	0.277487	0.009250			
Total	63	12.419344				

^{a} ** = significant at 1% level, * = significant at 5% level.

Without replication, the error term is estimated as the values of the three-factor interaction effects; whereas, with replication, an independent estimate of error is available. Thus, when the three-factor interaction effect is large and significant (as is the present case) the error term is highly overestimated and the sensitivity of the F test greatly reduced. This is clearly shown in our example in which the significance of the two-factor interaction cannot be detected in the case without replication.

CHAPTER 5

Comparison Between Treatment Means

There are many ways to compare the means of treatments tested in an experiment. Only those comparisons helpful in answering the experimental objectives should be thoroughly examined. Consider an experiment in rice weed control with 15 treatments—4 with hand weeding, 10 with herbicides, and 1 with no weeding (control). The probable questions that may be raised, and the specific mean comparisons that can provide their answers, may be:

- Is any treatment effective in controlling weeds? This could be answered simply by comparing the mean of the nonweeded treatment with the mean of each of the 14 weed-control treatments.
- Are there differences between the 14 weed-control treatments? If so, which is effective and which is not? Among the effective treatments, are there differences in levels of effectivity? If so, which is the best? To answer these questions, the mean of each of the 14 weed-control treatments is compared to the control's mean and those that are significantly better than the control are selected. In addition, the selected treatments are compared to identify the best among them.
- Is there any difference between the group of hand-weeding treatments and the group of herbicide treatments? To answer this question, the means of the four hand-weeding treatments are averaged and compared with the averaged means of the 10 herbicide treatments.
- Are there differences between the four hand-weeding treatments? If so, which treatment is best? To answer these questions, the four hand-weeding treatment means are compared to detect any significant difference among them and the best treatments are identified.
- Are there differences among the 10 herbicide treatments? If so, which treatment is best? Or, which herbicide gave better performance than the 2, 4-D treatment, the current leading herbicide for rice? The comparisons needed for the first two questions are similar to those described previously,

except that four hand-weeding treatments replaced the 10 herbicide treatments. For the third question, the mean of each herbicide treatment is compared with the mean of the 2, 4-D treatment to identify those herbicides giving significantly better performance than 2, 4-D.

· If five of the 10 herbicide treatments represent five different rates of a single herbicide, are there grain yield differences among the rates of application? To answer this question, a functional relationship between the response and the treatment (i.e., rate of herbicide application) is evaluated to characterize the change in grain yield for every change in the amount of herbicide applied (see Chapter 9 on regression analysis).

This weed-control experiment illustrates the diversity in the types of mean comparison. These different types can, however, be classified either as *pair comparison* or *group comparison*. In this chapter we focus on the statistical procedures for making these two types of comparison.

5.1   PAIR COMPARISON

Pair comparison is the simplest and most commonly used comparison in agricultural research. There are two types:

· *Planned pair comparison*, in which the specific pair of treatments to be compared was identified before the start of the experiment. A common example is comparison of the control treatment with each of the other treatments.
· *Unplanned pair comparison*, in which no specific comparison is chosen in advance. Instead, every possible pair of treatment means is compared to identify pairs of treatments that are significantly different.

The two most commonly used test procedures for pair comparisons in agricultural research are the *least significant difference* (LSD) *test* which is suited for a planned pair comparison, and *Duncan's multiple range test* (DMRT) which is applicable to an unplanned pair comparison. Other test procedures, such as the honestly significant difference (HSD) test and the Student-Newman-Keuls' multiple range test, can be found in Steel and Torrie, 1980,[*] and Snedecor and Cochran, 1980.[†]

5.1.1   Least Significant Difference Test

The least significant difference (LSD) test is the simplest and the most commonly used procedure for making pair comparisons. The procedure pro-

[*]R. G. D. Steel and J. A. Torrie, *Principles and Procedures of Statistics*, 2nd ed., USA: McGraw-Hill, 1980. pp. 183–193.
[†]G. W. Snedecor and W. G. Cochran. *Statistical Methods*. USA: The Iowa State University Press, 1980. pp. 232–237.

vides for a single LSD value, at a prescribed level of significance, which serves as the boundary between significant and nonsignificant differences between any pair of treatment means. That is, two treatments are declared significantly different at a prescribed level of significance if their difference exceeds the computed LSD value; otherwise they are not significantly different.

The LSD test is most appropriate for making planned pair comparisons but, strictly speaking, is not valid for comparing all possible pairs of means, especially when the number of treatments is large. This is so because the number of possible pairs of treatment means increases rapidly as the number of treatments increases—10 possible pairs of means with 5 treatments, 45 pairs with 10 treatments, and 105 pairs with 15 treatments. The probability that, due to chance alone, at least one pair will have a difference that exceeds the LSD value increases with the number of treatments being tested. For example, in experiments where no real difference exists among all treatments, it can be shown that the numerical difference between the largest and the smallest treatment means is expected to exceed the LSD value at the 5% level of significance 29% of the time when 5 treatments are involved, 63% of the time when 10 treatments are involved, and 83% of the time when 15 treatments are involved. Thus avoid use of the LSD test for comparisons of all possible pairs of means. If the LSD test must be used, apply it only when the F test for treatment effect is significant and the number of treatments is not too large—less than six.

The procedure for applying the LSD test to compare any two treatments, say the ith and the jth treatments, involve these steps:

□ STEP 1. Compute the mean difference between the ith and the jth treatment as:

$$d_{ij} = \overline{X}_i - \overline{X}_j$$

where $\overline{X}_i$ and $\overline{X}_j$ are the means of the ith and the jth treatments.

□ STEP 2. Compute the LSD value at α level of significance as:

$$\text{LSD}_\alpha = (t_\alpha)(s_{\bar{d}})$$

where $s_{\bar{d}}$ is the standard error of the mean difference and t_α is the tabular t value, from Appendix C, at α level of significance and with n = error degree of freedom.

□ STEP 3. Compare the mean difference computed in step 1 to the LSD value computed in step 2 and declare the ith and jth treatments to be significantly different at the α level of significance, if the absolute value of d_{ij} is greater than the LSD value, otherwise it is not significantly different.

In applying the foregoing procedure, it is important that the appropriate standard error of the mean difference for the treatment pair being compared is correctly identified. This task is affected by the experimental design used, the number of replications of the two treatments being compared, and the specific type of means to be compared. Thus in the succeeding sections we illustrate the procedure for implementing an LSD test for various experimental designs. Emphasis is on how to compute the $s_{\bar{d}}$ value to be used and other special modifications that may be required.

5.1.1.1 Complete Block Design.

For a complete block design where only one error term is involved, such as completely randomized, randomized complete block, or latin square, the standard error of the mean difference for any pair of treatment means is computed as:

$$s_{\bar{d}} = \sqrt{\frac{2s^2}{r}}$$

where r is the number of replications that is common to both treatments in the pair and s^2 is the error mean square in the analysis of variance.

When the two treatments do not have the same number of replications, $s_{\bar{d}}$ is computed as:

$$s_{\bar{d}} = \sqrt{s^2\left(\frac{1}{r_i} + \frac{1}{r_j}\right)}$$

where r_i and r_j are the number of replications of the ith and the jth treatments.

In a factorial experiment, there are several types of treatment mean. For example, a 2×3 factorial experiment, involving factor A with two levels and factor B with three levels, has four types of mean that can be compared:

1. The two A means, averaged over all three levels of factor B
2. The three B means, averaged over both levels of factor A
3. The six A means, two means at each of the three levels of factor B
4. The six B means, three means at each of the two levels of factor A

The type-1 mean is an average of $3r$ observations; the type 2 is an average of $2r$ observations; and the type 3 or type 4 is an average of r observations.

Thus, the formula $s_{\bar{d}} = (2s^2/r)^{1/2}$ is appropriate only for the mean difference involving either type-3 or type-4 mean. For type 1 and type 2, the divisor r in the formula should be replaced by $3r$ and $2r$. That is, to compare two A means averaged over all levels of factor B, the $s_{\bar{d}}$ value is computed as $(2s^2/3r)^{1/2}$ and to compare any pair of B means averaged over all levels of factor A, the $s_{\bar{d}}$ value is computed as $(2s^2/2r)^{1/2}$ or simply $(s^2/r)^{1/2}$.

We illustrate the LSD test procedure with two examples. One is a case with equal replication; the other a case with unequal replication.

5.1.1.1.1 Equal Replication. Data from a completely randomized design experiment with seven treatments (six insecticide treatments and one control treatment) were tested in four replications (Chapter 2, Table 2.1). Assume that the primary objective of the experiment is to identify one or more of the six insecticide treatments that is better than the control treatment. For this example, the appropriate comparison is the series of planned pair comparisons in which each of the six insecticide treatment means is compared to the control mean. The steps involved in applying the LSD test to each of the six pair comparisons are:

☐ STEP 1. Compute the mean difference between the control treatment and each of the six insecticide treatments, as shown in Table 5.1.

☐ STEP 2. Compute the LSD value at α level of significance as:

$$\text{LSD}_\alpha = t_\alpha \sqrt{\frac{2s^2}{r}}$$

For our example, the error mean square s^2 is 94,773, the error degree of freedom is 21, and the number of replications is four. The tabular t values (Appendix C), with $n = 21$ degrees of freedom, are 2.080 at the 5% level of

Table 5.1 Comparison between Mean Yields of a Control and Each of the Six Insecticide Treatments, Using the LSD Test (Data in Table 2.1)

Treatment	Mean Yield,[a] kg/ha	Difference from Control,[b] kg/ha
Dol-Mix (1 kg)	2,127	811**
Dol-Mix (2 kg)	2,678	1,362**
DDT + γ-BHC	2,552	1,236**
Azodrin	2,128	812**
Dimecron-Boom	1,796	480*
Dimecron-Knap	1,681	365[ns]
Control	1,316	—

[a]Average of four replications.
[b]** = significant at 1% level, * = significant at 5% level, ns = not significant.

significance and 2.831 at the 1% level. The LSD values are computed:

$$LSD_{.05} = 2.080 \sqrt{\frac{2(94,773)}{4}} = 453 \text{ kg/ha}$$

$$LSD_{.01} = 2.831 \sqrt{\frac{2(94,773)}{4}} = 616 \text{ kg/ha}$$

☐ STEP 3.   Compare each of the mean differences computed in step 1 to the LSD values computed in step 2 and indicate its significance with the appropriate asterisk notation (see Chapter 2, Section 2.1.2). For our example, the mean difference between the first treatment and the control of 811 kg/ha (Table 5.1) exceeds both computed LSD values and, thus, receives two asterisks to indicate that the two treatments are significantly different at the 1% level of significance. The results for the six pair comparisons show that, except for Dimecron-Knap, all insecticide treatments gave yields that were significantly higher than that of control.

5.1.1.1.2   Unequal Replication.   Using data from a completely randomized design experiment with 11 treatments (10 weed-control treatments and a control) and unequal replications (Chapter 2, Table 2.3), a researcher wishes to determine whether any of the 10 weed-control treatments is better than the control treatment. For this example, the appropriate comparison is the planned pair comparisons in which each of the 10 weed-control treatment means is compared to the control mean. The steps involved in applying the LSD test to each of these 10 pair comparisons are:

☐ STEP 1.   Compute the mean difference between the control treatment and each of the 10 weed-control treatments, as shown in column 4 of Table 5.2.

☐ STEP 2.   Compute the LSD value at α level of significance. Because some treatments have four replications and others have three, two sets of LSD values must be computed. Using the error mean square s^2 of 176,532 (Chapter 2, Table 2.4), the error degree of freedom of 29, and the tabular t values with 29 degrees of freedom of 2.045 at the 5% level of significance and 2.756 at the 1% level, the two sets of LSD values are computed:

• For comparing the control mean (with four replications) and each weed-control treatment having four replications, compute the LSD values by the same formula as in Section 5.1.1.1.1, step 2:

$$LSD_{.05} = 2.045 \sqrt{\frac{2(176,532)}{4}} = 608 \text{ kg/ha}$$

$$LSD_{.01} = 2.756 \sqrt{\frac{2(176,532)}{4}} = 819 \text{ kg/ha}$$

Table 5.2 Comparison between Mean Yields of Each of the 10 Treatments and the Control Treatment, Using the LSD Test with Unequal Replication (Data in Table 2.3)

Treatment Number	Replications, no.	Mean Yield, kg/ha	Difference from Control,[a] kg/ha	LSD Values 5%	1%
1	4	3,644	2,407**	608	819
2	3	3,013	1,776**	656	884
3	4	2,948	1,711**	608	819
4	4	2,910	1,673**	608	819
5	3	2,568	1,331**	656	884
6	3	2,565	1,328**	656	884
7	4	2,484	1,247**	608	819
8	3	2,206	969**	656	884
9	4	2,041	804*	608	819
10	4	2,798	1,561**	608	819
11(Control)	4	1,237	—	—	—

[a]** = significant at 1% level, * = significant at 5% level.

- For comparing the control mean (with four replications) and each weed-control treatment having three replications, compute the LSD values following the formula:

$$\text{LSD}_\alpha = (t_\alpha)(s_{\bar{d}})$$

where

$$s_{\bar{d}} = \sqrt{s^2\left(\frac{1}{r_i} + \frac{1}{r_j}\right)}$$

Thus,

$$\text{LSD}_{.05} = 2.045\sqrt{176,532\,(1/3 + 1/4)}$$

$$= 656 \text{ kg/ha}$$

$$\text{LSD}_{.01} = 2.756\sqrt{176,532\,(1/3 + 1/4)}$$

$$= 884 \text{ kg/ha}$$

☐ STEP 3. Compare each of the mean differences computed in step 1 to its corresponding LSD values computed in step 2 and place the appropriate asterisk notation (see Chapter 2, Section 2.1.2). The mean difference between the first treatment (four replications) and the control (four replications) is 2,407 kg/ha. Compare it to the first set of LSD values in step 2—608 kg/ha and 819 kg/ha. Because the mean difference is higher than the corresponding LSD value at the 1% level of significance, it is declared significant at the 1% level of significance and is indicated with two asterisks.

On the other hand, the mean difference between the second treatment (with three replications) and the control (with four replications) is 1,776 kg/ha. Compare it to the second set of LSD values in step 2—656 kg/ha and 884 kg/ha. Because the mean difference is also higher than the corresponding LSD value at the 1% level of significance, it is declared significant at the 1% level of significance and is indicated with two asterisks.

The test results for all pairs, shown in Table 5.2, indicate that all weed-control treatments gave significantly higher yields than that of the control treatment.

5.1.1.2 Balanced lattice Design.   The application of the LSD test to data from a balanced lattice design involves two important adjustments:

- The adjusted treatment mean is used in computing the mean difference.
- The effective error mean square is used in computing the standard error of the mean difference.

For illustration, consider the 4×4 balanced lattice design described in Chapter 2 and the corresponding data in Table 2.10. Assume that one of the 16 treatments (treatment 10) is the no-fertilizer control treatment and that the researcher wishes to determine whether there is any significant response to each of the 15 fertilizer treatments. For this purpose, the appropriate mean comparison is the planned pair comparisons in which each of the 15 fertilizer treatments is compared with the control treatment. The steps involved in applying the LSD test are:

☐ STEP 1.   Compute the mean difference between the control (treatment 10) and each of the 15 fertilizer treatments, using the adjusted treatment means (see Chapter 2, Section 2.4.1.2) shown in Table 5.3.

☐ STEP 2.   Compute the LSD value at α level of significance as:

$$\text{LSD}_\alpha = t_\alpha \sqrt{\frac{2(\text{effective error } MS)}{r}}$$

where effective error MS is as defined in Chapter 2, Section 2.4.1.2, step 13.

For our example, each treatment is replicated five times and the effective error MS with 45 degrees of freedom is 369. The tabular t values (Appendix C) with $n = 45$ degrees of freedom are 2.016 at the 5% level of significance

Table 5.3 Comparison between Mean Tiller Count of Each of the 15 Fertilizer Treatments with That of Control, Using the LSD Test (Data in Table 2.10)

Treatment Number	Adjusted Mean, no./m^2	Difference from Control,[a] no./m^2
1	166	47**
2	161	42**
3	184	65**
4	176	57**
5	163	44**
6	174	55**
7	168	49**
8	177	58**
9	163	44**
10(control)	119	—
11	188	69**
12	191	72**
13	170	51**
14	197	78**
15	186	67**
16	168	49**

[a]** = significant at 1% level.

and 2.693 at the 1% level. Thus, the LSD values are:

$$LSD_{.05} = 2.016\sqrt{\frac{2(369)}{5}} = 24/m^2$$

$$LSD_{.01} = 2.693\sqrt{\frac{2(369)}{5}} = 33/m^2$$

☐ STEP 3.   Compare the mean difference of each pair of treatments computed in step 1 to the LSD values computed in step 2 and place the appropriate asterisk notation (see Chapter 2, Section 2.1.2). For example, the mean difference between treatment 1 and the control treatment of 47 tillers/m^2 exceeds both computed LSD values and is indicated with two asterisks. The results for all 15 pairs are highly significant (Table 5.3).

5.1.1.3 Partially Balanced Lattice Design.   As in the case of the balanced lattice design (Section 5.1.1.2), the application of the LSD test to data from a partially balanced lattice design involves two important adjustments:

• The adjusted treatment mean is used in the computation of the mean difference.

- The appropriate effective error mean square is used in the computation of the standard error of the mean difference.

But unlike the balanced lattice design where there is only one standard error of the mean difference, there are two standard errors of the mean difference for a partially balanced lattice design—one corresponds to treatment pairs that were tested in the same incomplete block and another corresponds to treatment pairs that never appeared together in the same incomplete block. For the first set, the effective error $MS(1)$ is used; for the second, the effective error $MS(2)$ is used (see Chapter 2, Section 2.4.2.2.1, step 17 for formulas).

Consider the varietal test in a 9×9 triple lattice design, as described in Chapter 2, Section 2.4.2, with the data shown in Table 2.13. Assume that the researcher wishes to identify varieties that significantly outyielded the local variety (variety no. 2). For this purpose, the appropriate mean comparison is the planned pair comparisons in which each of the 80 test varieties is compared to the local variety. The steps involved in applying the LSD test are:

☐ STEP 1.   Compute the mean difference between the local variety and each of the 80 test varieties based on the adjusted treatment means, and indicate whether each pair was or was not tested together in the same incomplete block, as shown in Table 5.4.

☐ STEP 2.   Compute the two sets of LSD values:
- For comparing two treatments that were tested together in an incomplete block:

$$\text{LSD}_\alpha = t_\alpha \sqrt{\frac{2[\text{effective error } MS(1)]}{r}}$$

where effective error $MS(1)$ is as defined in Chapter 2.
- For comparing two treatments that were not tested together in an incomplete block:

$$\text{LSD}_\alpha = t_\alpha \sqrt{\frac{2[\text{effective error } MS(2)]}{r}}$$

where effective error $MS(2)$ is as defined in Chapter 2.

For our example, each variety is replicated three times, the effective error $MS(1)$ is 0.2786, and the effective error $MS(2)$ is 0.2856. The tabular t values (Appendix C) with $n = 136$ degrees of freedom are 1.975 at the 5% level of significance and 2.606 at the 1% level. Thus, the two sets of LSD values are

Table 5.4   Comparison between Adjusted Mean Yield of 80 Rice Varieties and That of Local Variety, Using the LSD Test (Data in Table 2.13)

Variety Number	Adjusted Mean,[a] t/ha	Difference from Local Variety,[b] t/ha	Variety Number	Adjusted Mean,[a] t/ha	Difference from Local Variety,[b] t/ha	Variety Number	Adjusted Mean,[a] t/ha	Difference from Local Variety,[b] t/ha
1	3.28	1.86**	29	4.07	2.65**	56	4.14	2.72**
3	4.41	2.99**	30	5.16	3.74**	57	2.82	1.40**
4	3.00	1.58**	31	3.75	2.33**	58	3.84	2.42**
5	3.57	2.15**	32	3.78	2.36**	59	3.26	1.84**
6	3.41	1.99**	33	3.59	2.17**	60	4.01	2.59**
7	2.35	0.93*	34	3.71	2.29**	61	4.20	2.78**
8	4.60	3.18**	35	4.22	2.80**	62	3.98	2.56**
9	4.66	3.24**	36	4.99	3.57**	63	3.07	1.65**
10	3.38	1.96**	37	3.80	2.38**	64	4.59	3.17**
11	5.19	3.77**	38	4.27	2.85**	65	3.06	1.64**
12	3.17	1.75**	39	5.19	3.77**	66	2.10	0.68^{ns}
13	1.66	0.24^{ns}	40	4.80	3.38**	67	5.26	3.84**
14	3.62	2.20**	41	4.94	3.52**	68	4.56	3.14**
15	4.06	2.64**	42	4.16	2.74**	69	5.09	3.67**
16	1.78	0.36^{ns}	43	3.39	1.97**	70	3.29	1.87**
17	2.81	1.39**	44	3.70	2.28**	71	3.74	2.32**
18	2.93	1.51**	45	3.74	2.32**	72	4.73	3.31**
19	3.95	2.53**	46	3.99	2.57**	73	3.56	2.14**
20	3.18	1.76**	47	2.60	1.18**	74	5.52	4.10**
21	5.23	3.81**	48	2.58	1.16**	75	3.69	2.27**
22	5.50	4.08**	49	2.75	1.33**	76	3.65	2.23**
23	1.81	0.39^{ns}	50	3.29	1.87**	77	4.02	2.60**
24	3.74	2.32**	51	3.73	2.31**	78	6.36	4.94**
25	5.26	3.84**	52	1.32	−0.10^{ns}	79	5.40	3.98**
26	5.50	4.08**	53	3.16	1.74**	80	2.72	1.30**
27	4.13	2.71**	54	4.04	2.62**	81	4.45	3.03**
28	3.73	2.31**	55	3.16	1.74**			

[a] Italicized means are from varieties that were tested together with the local variety in the same incomplete block.
[b] ** = significant at 1% level, * = significant at 5% level, ns = not significant.

computed as:

- For comparing two varieties that were tested together in an incomplete block, the LSD values are computed following the first formula given:

$$LSD(1)_{.05} = 1.975\sqrt{\frac{2(0.2786)}{3}} = 0.85 \text{ t/ha}$$

$$LSD(1)_{.01} = 2.606\sqrt{\frac{2(0.2786)}{3}} = 1.12 \text{ t/ha}$$

- For comparing two varieties that were not tested together in an incomplete block, the LSD values are computed following the second formula given:

$$\text{LSD}(2)_{.05} = 1.975\sqrt{\frac{2(0.2856)}{3}} = 0.86 \text{ t/ha}$$

$$\text{LSD}(2)_{.01} = 2.606\sqrt{\frac{2(0.2856)}{3}} = 1.14 \text{ t/ha}$$

Note that whenever the two effective error *MS* do not differ much, the use of the average error *MS* (see Chapter 2, Section 2.4.2.2.1, step 17) is appropriate, and only one set of LSD values needs to be computed as:

$$\text{LSD}_\alpha = t_\alpha\sqrt{\frac{2(\text{av. effective error } MS)}{r}}$$

This LSD value can be used for comparing any pair of treatments regardless of their block configuration. In this example, the use of the average effective error *MS* of 0.2835 is applicable. Hence, the only set of LSD values needed is computed as:

$$\text{LSD}_{.05} = 1.975\sqrt{\frac{2(0.2835)}{3}} = 0.86 \text{ t/ha}$$

$$\text{LSD}_{.01} = 2.606\sqrt{\frac{2(0.2835)}{3}} = 1.13 \text{ t/ha}$$

Although this set of LSD values is applicable to the comparison of any pair of treatment means, for illustration purposes, in succeeding steps, we will use the two sets of LSD values computed from effective error *MS*(1) and effective error *MS*(2).

☐ STEP 3.   Compare the mean difference of each pair of varieties computed in step 1 to the appropriate set of LSD values computed in step 2. Use LSD(1) values for pairs that were tested together in an incomplete block, otherwise use LSD(2) values. For example, because variety no. 1 and the local variety were tested together in an incomplete block, their mean difference of 1.86 t/ha is compared with the LSD(1) values of 0.85 and 1.12 t/ha. The result indicates a highly significant difference between variety no. 1 and local variety. On the other hand, because variety no. 12 was not tested together with the local variety in any block, their mean difference of 1.75 t/ha is compared with the LSD(2) values of 0.86 and 1.14 t/ha. The results for all pairs are shown in Table 5.4.

5.1.1.4 Split-Plot Design. In a split-plot design, with two variable factors and two error terms, there are four different types of pair comparison. Each requires its own set of LSD values. These comparisons are:

- Comparison between two main-plot treatment means averaged over all subplot treatments.
- Comparison between two subplot treatment means averaged over all main-plot treatments.
- Comparison between two subplot treatment means at the same main-plot treatment.
- Comparison between two main-plot treatment means at the same or different subplot treatments (i.e., means of any two treatment combinations).

Table 5.5 gives the formula for computing the appropriate standard error of the mean difference $s_{\bar{d}}$ for each of these types of pair comparison. When the computation of $s_{\bar{d}}$ involves more than one error term, such as in comparison type 4, the standard tabular t values from Appendix C cannot be used directly and the weighted tabular t values need to be computed. The formulas for weighted tabular t values are given in Table 5.6.

Consider the 6×4 factorial experiment whose data are shown in Tables 3.7 through 3.9 of Chapter 3. The analysis of variance (Table 3.10) shows a highly

Table 5.5 Standard Error of the Mean Difference for Each of the Four Types of Pair Comparison in a Split-plot Design

	Type of Pair Comparison	
Number	Between	$s_{\bar{d}}{}^a$
1	Two main-plot means (averaged over all subplot treatments)	$\sqrt{\dfrac{2E_a}{rb}}$
2	Two subplot means (averaged over all main-plot treatments)	$\sqrt{\dfrac{2E_b}{ra}}$
3	Two subplot means at the same main-plot treatment	$\sqrt{\dfrac{2E_b}{r}}$
4	Two main-plot means at the same or different subplot treatments	$\sqrt{\dfrac{2[(b-1)E_b + E_a]}{rb}}$

$^a E_a$ = error(a) MS, E_b = error(b) MS, r = no. of replications, a = no. of main-plot treatments, and b = no. of subplot treatments.

Table 5.6 The Weighted Tabular *t* Values Associated with the Different Mean Comparisons in Tables 5.5, 5.8, and 5.10, Whose Standard Error of the Mean Difference Involves More Than One Error Mean Square

	Treatment Comparison		
		Source	Weighted Tabular
Number	Table Number	Comparison Number	*t* Value[a]
1	5.5	4	$\dfrac{(b-1)E_b t_b + E_a t_a}{(b-1)E_b + E_a}$
2	5.8	3	$\dfrac{(b-1)E_c t_c + E_a t_a}{(b-1)E_c + E_a}$
3	5.8	4	$\dfrac{(a-1)E_c t_c + E_b t_b}{(a-1)E_c + E_b}$
4	5.10	8	$\dfrac{(b-1)E_b t_b + E_a t_a}{(b-1)E_b + E_a}$
5	5.10	9	$\dfrac{(c-1)E_c t_c + E_b t_b}{(c-1)E_c + E_b}$
6	5.10	10	$\dfrac{(c-1)E_c t_c + E_b t_b}{(c-1)E_c + E_b}$
7	5.10	11	$\dfrac{(c-1)E_c t_c + E_a t_a}{(c-1)E_c + E_a}$
8	5.10	12	$\dfrac{b(c-1)E_c t_c + (b-1)E_b t_b + E_a t_a}{b(c-1)E_c + (b-1)E_b + E_a}$

[a] For definitions of a, b, c, E_a, E_b, and E_c, see Tables 5.5, 5.8, and 5.10; t_a, t_b, and t_c are the tabular t values from Appendix C with $n = d.f.$ corresponding to E_a, E_b, and E_c, respectively.

significant interaction between nitrogen and variety, indicating that varietal effects varied with the rate of nitrogen applied. Hence, comparison between nitrogen means averaged over all varieties or between variety means averaged over all nitrogen rates is not useful (see Chapter 3, Section 3.1).

The more appropriate mean comparisons are those between variety means under the same nitrogen rate or between nitrogen-rate means of the same variety. However, because pair comparison between nitrogen-rate means of the same variety is not appropriate because of the quantitative nature of the nitrogen-rate treatments, only the comparison between variety means with the same nitrogen rate is illustrated.

The steps involved in the computation of the LSD test for comparing two variety means with the same nitrogen rate (i.e., two subplot means at the same main-plot treatment) are:

□ STEP 1. Compute the standard error of the mean difference following the formula for comparison type 3 of Table 5.5;

$$s_{\bar{d}} = \sqrt{\frac{2E_b}{r}}$$

$$= \sqrt{\frac{2(349,580)}{3}} = 482.8 \text{ kg/ha}$$

where the E_b value of 349,580 is obtained from the error(b) MS in the analysis of variance of Table 3.10.

□ STEP 2. From Appendix C, obtain the tabular t values with $n = $ error(b) $d.f. = 36$ degrees of freedom as 2.029 at the 5% level of significance and 2.722 at the 1% level.

□ STEP 3. Following the formula $LSD_\alpha = (t_\alpha)(s_{\bar{d}})$ compute the LSD values at the 5% and 1% levels of significance:

$$LSD_{.05} = (2.029)(482.8) = 980 \text{ kg/ha}$$

$$LSD_{.01} = (2.722)(482.8) = 1,314 \text{ kg/ha}$$

□ STEP 4. Construct the variety × nitrogen two-way table of means with the LSD values for comparing two variety means at the same nitrogen rate as shown in Table 5.7. For each pair of varieties (with the same nitrogen rate)

Table 5.7 **Mean Yields of Four Rice Varieties Tested with Six Rates of Nitrogen in a Split-plot Design (Data in Table 3.7)**

Nitrogen Rate, kg/ha	Mean Yield,[a] kg/ha			
	IR8	IR5	C4-63	Peta
0	4,253	4,306	3,183	4,481
60	5,672	5,982	5,443	4,816
90	6,400	6,259	5,994	4,812
120	6,733	6,895	6,014	3,816
150	7,563	6,951	6,687	2,047
180	8,701	6,540	6,065	1,881

[a]Average of three replications. The LSD values for comparing two varieties under the same nitrogen rate are 980 kg/ha at the 5% level of significance and 1,314 kg/ha at the 1% level.

to be compared, compute the mean difference and compare it to the LSD values. For example, one mean difference of interest may be between Peta and IR8 at 0 kg N/ha, which is computed as $4{,}481 - 4{,}253 = 228$ kg/ha. Because this mean difference is smaller than the LSD value at the 5% level of significance, it is not significant.

5.1.1.5   Strip-Plot Design.   As in the case of the split-plot design (Section 5.1.1.4) a strip-plot design has four types of pair comparison, each requiring its own set of LSD values. These four types and the appropriate formulas for the computation of the corresponding $s_{\bar{d}}$ values are shown in Table 5.8.

The procedure for applying the LSD test to pair comparison in a strip-plot design is illustrated with a 6×3 factorial experiment. Data and analysis of variance are given in Tables 3.11 through 3.15 of Chapter 3. Because the interaction effect between variety and nitrogen is significant (Table 3.15), the only appropriate type of pair comparison is that among varieties under the same nitrogen rate (see related discussion in Section 5.1.1.4).

The steps in the computation of the LSD values for comparing two varieties grown with the same nitrogen rate are:

☐ STEP 1.   Compute the $s_{\bar{d}}$ value, following the formula for comparison type 3 of Table 5.8:

$$s_{\bar{d}} = \sqrt{\frac{2[(b-1)E_c + E_a]}{rb}}$$

$$= \sqrt{\frac{2[(2)(411{,}646) + 1{,}492{,}262]}{(3)(3)}}$$

$$= 717.3 \text{ kg/ha}$$

☐ STEP 2.   Because there are two error terms (E_a and E_c) involved in the formula used in step 1, compute the weighted tabular t values as:

- From Appendix C, obtain the tabular t values corresponding to E_a with $n = 10$ d.f. (i.e., t_a) and the tabular t values corresponding to E_c with $n = 20$ d.f. (i.e., t_c) at the 5% and 1% levels of significance:

$$t_a(.05) = 2.228 \text{ and } t_a(.01) = 3.169$$

$$t_c(.05) = 2.086 \text{ and } t_c(.01) = 2.845$$

- Compute the weighted tabular t values, following the corresponding formula given in Table 5.6 (i.e., formula 2):

$$t' = \frac{(b-1)E_c t_c + E_a t_a}{(b-1)E_c + E_a}$$

Table 5.8   Standard Error of the Mean Difference for Each of the Four Types of Pair Comparisons in a Strip-plot Design

Number	Between	$s_{\bar{d}}$ [a]
	Type of Pair Comparison	
1	Two horizontal means (averaged over all vertical treatments)	$\sqrt{\dfrac{2E_a}{rb}}$
2	Two vertical means (averaged over all horizontal treatments)	$\sqrt{\dfrac{2E_b}{ra}}$
3	Two horizontal means at the same level of vertical factor	$\sqrt{\dfrac{2[(b-1)E_c + E_a]}{rb}}$
4	Two vertical means at the same level of horizontal factor	$\sqrt{\dfrac{2[(a-1)E_c + E_b]}{ra}}$

[a] E_a = error(a) MS, E_b = error(b) MS, E_c = error(c) MS, r = no. of replications, a = levels of horizontal-strip factor, and b = levels of vertical-strip factor.

$$t'(.05) = \frac{(2)(411{,}646)(2.086) + (1{,}492{,}262)(2.228)}{(2)(411{,}646) + 1{,}492{,}262}$$

$$= 2.178$$

$$t'(.01) = \frac{(2)(411{,}646)(2.845) + (1{,}492{,}262)(3.169)}{(2)(411{,}646) + 1{,}492{,}262}$$

$$= 3.054$$

☐ STEP 3.   Compute the LSD values at the 5% and 1% levels of significance:

$$\mathrm{LSD}_{.05} = t'(.05)(s_{\bar{d}})$$

$$= (2.178)(717.3) = 1{,}562 \text{ kg/ha}$$

$$\mathrm{LSD}_{.01} = t'(.01)(s_{\bar{d}})$$

$$= (3.054)(717.3) = 2{,}191 \text{ kg/ha}$$

☐ STEP 4.   Construct the variety × nitrogen two-way table of means, with the LSD values (computed in step 3) indicated, as shown in Table 5.9. For example, to determine whether the mean yield of IR8 is significantly different from that of Peta at a rate of 0 kg N/ha, their mean difference $(3{,}572 - 3{,}207 = 365 \text{ kg/ha})$ is compared to the computed LSD values of

Table 5.9 Mean Yields of Six Rice Varieties Tested with Three Nitrogen Rates in a Strip-plot Design (Data in Table 3.11)

Variety	Mean Yield,[a] kg/ha		
	0 kg N/ha	60 kg N/ha	120 kg N/ha
IR8	3,572	5,132	7,548
IR127-80	4,934	6,714	7,211
IR305-4-12	4,250	6,122	7,868
IR400-2-5	4,059	5,554	7.094
IR665-58	4,102	5,633	6,012
Peta	3,207	3,714	2,492

[a]Average of three replications. The LSD values for comparing two varieties with the same nitrogen rate are 1,562 kg/ha at the 5% level of significance and 2,191 kg/ha at the 1% level.

1,562 and 2,191 kg/ha. Because the mean difference is smaller than the LSD value at the 5% level of significance, the mean yields of Peta and IR8 at 0 kg N/ha are not significantly different.

5.1.1.6 Split-Split-Plot Design. For a split-split-plot design, there are 12 types of pair comparison, each requiring its own set of LSD values. These pair comparisons, together with the appropriate formulas for computing the corresponding $s_{\bar{d}}$ values, are shown in Table 5.10.

The procedure for applying the LSD test to pair comparison in a split-split-plot design is illustrated with a $5 \times 3 \times 3$ factorial experiment. The data and analysis are given in Tables 4.4 through 4.11 of Chapter 4. From the analysis of variance (Table 4.11), all three main effects and one interaction effect between nitrogen and variety are significant. Consequently, only the following types of mean comparison should be tested:

1. Comparison between the three management practices averaged over all varieties and nitrogen rates— because none of the interaction effects involving management practices is significant.

2. Comparison between the three varieties averaged over all management practices but at the same nitrogen rate—because the nitrogen $\times$ variety interaction is significant.

3. Comparison between the five nitrogen rates averaged over all management practices but with the same variety—because the nitrogen $\times$ variety interaction is significant.

Because pair comparison is appropriate only for comparisons 1 and 2 and not for comparison 3 where the treatments involved (i.e., nitrogen rates) are quantitative, we give the LSD-test procedures only for comparisons 1 and 2.

Table 5.10 Standard Error of the Mean Difference for Each of the 12 Types of Pair Comparison in a Split-split-plot design

Type of Pair Comparison		
Number	Between	$s_{\bar{d}}^{a}$
1	Two main-plot means (averaged over all subplot and sub-subplot treatments)	$\sqrt{\dfrac{2E_a}{rbc}}$
2	Two subplot means (averaged over all main-plot and sub-subplot treatments)	$\sqrt{\dfrac{2E_b}{rac}}$
3	Two subplot means (averaged over all sub-subplot treatments) at the same or different levels of main-plot factor	$\sqrt{\dfrac{2E_b}{rc}}$
4	Two sub-subplot means (averaged over all main-plot and subplot treatments)	$\sqrt{\dfrac{2E_c}{rab}}$
5	Two sub-subplot means at the same level of main-plot factor (averaged over all subplot treatments)	$\sqrt{\dfrac{2E_c}{rb}}$
6	Two sub-subplot means at the same level of subplot factor (averaged over all main-plot treatments)	$\sqrt{\dfrac{2E_c}{ra}}$
7	Two sub-subplot means at the same combination of main-plot and subplot treatments	$\sqrt{\dfrac{2E_c}{r}}$
8	Two main-plot means (averaged over all sub-subplot treatments) at the same or different levels of subplot factor	$\sqrt{\dfrac{2[(b-1)E_b + E_a]}{rbc}}$
9	Two subplot means (averaged over all main-plot treatments) at the same or different levels of sub-subplot factor	$\sqrt{\dfrac{2[(c-1)E_c + E_b]}{rac}}$
10	Two subplot means at the same combination of main-plot and sub-subplot treatments	$\sqrt{\dfrac{2[(c-1)E_c + E_b]}{rc}}$
11	Two main-plot means (averaged over all subplot treatments) at the same or different levels of sub-subplot factor	$\sqrt{\dfrac{2[(c-1)E_c + E_a]}{rbc}}$
12	Two main-plot means at the same combination of subplot and sub-subplot treatments	$\sqrt{\dfrac{2[b(c-1)E_c + (b-1)E_b + E_a]}{rbc}}$

[a] E_a = error(a) MS, E_b = error(b) MS, E_c = error(c) MS, r = number of replications, a = number of main-plot treatments, b = number of subplot treatments, and c = number of sub-subplot treatments.

For comparison 1 (comparison between the three management practices averaged over all varieties and nitrogen rates) the steps for applying the LSD test are:

☐ STEP 1.   Compute the $s_{\bar{d}}$ value, following the formula for comparison type 2 of Table 5.10:

$$s_{\bar{d}} = \sqrt{\frac{2E_b}{rac}}$$

$$= \sqrt{\frac{2(0.2618)}{(3)(5)(3)}} = 0.108 \text{ t/ha}$$

☐ STEP 2.   From Appendix C, obtain the tabular t values with $n = \text{error}(b)$ $d.f. = 20\ d.f.$ as 2.086 at the 5% level of significance and 2.845 at the 1% level.

☐ STEP 3.   Compute the LSD values at the 5% and 1% levels of significance:

$$\text{LSD}_\alpha = (t_\alpha)(s_{\bar{d}})$$

$$\text{LSD}_{.05} = (2.086)(0.108) = 0.225 \text{ t/ha}$$

$$\text{LSD}_{.01} = (2.845)(0.108) = 0.307 \text{ t/ha}$$

☐ STEP 4.   Compute the mean yields of the three management practices averaged over all nitrogen rates and varieties:

Management Practice	Mean Yield, t/ha
M_1	5.900
M_2	6.486
M_3	7.277

☐ STEP 5.   Using the mean yields computed in step 4, compute the mean difference for any pair of management practices of interest and compare it with the LSD values computed in step 3. For example, to compare M_1 and M_2 the mean difference is computed as $6.486 - 5.900 = 0.586$ t/ha. Because the computed mean difference is higher than the LSD value at the 1% level of significance, the difference between M_1 and M_2 is declared highly significant.

For comparison 2 (comparison between the three varieties averaged over all management practices but at the same nitrogen rate) the step-by-step procedures for applying the LSD test are:

☐ STEP 1.   Compute the $s_{\bar{d}}$ value, following the formula for comparison type 5 of Table 5.10:

$$s_{\bar{d}} = \sqrt{\frac{2E_c}{rb}}$$

$$= \sqrt{\frac{2(0.4956)}{(3)(3)}} = 0.332 \text{ t/ha}$$

☐ STEP 2.   From Appendix C, obtain the tabular t values with $n = $ error(c) $d.f. = 60$ $d.f.$ as 2.000 at the 5% level of significance and 2.660 at the 1% level.

☐ STEP 3.   Compute the LSD value, at the 5% and 1% levels of significance:

$$\text{LSD}_\alpha = (t_\alpha)(s_{\bar{d}})$$

$$\text{LSD}_{.05} = (2.000)(0.332) = 0.664 \text{ t/ha}$$

$$\text{LSD}_{.01} = (2.660)(0.332) = 0.883 \text{ t/ha}$$

☐ STEP 4.   Construct the variety × nitrogen two-way table of means averaged over the three management practices, as shown in Table 5.11. To compare any pair of variety means at the same nitrogen rate, compute the mean difference and compare it to the LSD values computed in step 3. For example, to compare V_1 and V_2 at 140 kg N/ha, the mean difference is $7.288 - 5.078 = 2.210$ t/ha. Because this mean difference is higher than the LSD value at the 1% level of significance, the mean yields of V_1 and V_2 at 140 kg N/ha are declared highly significantly different.

5.1.2   Duncan's Multiple Range Test

For experiments that require the evaluation of all possible pairs of treatment means, the LSD test is usually not suitable. This is especially true when the total number of treatments is large (see Section 5.1.1). In such cases, Duncan's multiple range test (DMRT) is useful.

The procedure for applying the DMRT is similar to that for the LSD test; DMRT involves the computation of numerical boundaries that allow for the

Table 5.11   Mean Yields of Three Rice Varieties Grown with Five Nitrogen Rates in a Split-split-plot Design (Data in Table 4.7)

Nitrogen Rate, kg/ha	Mean Yield, t/ha[a]		
	V_1	V_2	V_3
0	4.513	5.163	6.478
50	4.764	6.016	7.881
80	5.835	6.589	8.564
110	5.445	6.925	8.443
140	5.078	7.288	9.336

[a]Average of three management practices, each replicated three times. The LSD values for comparing two varieties with the same nitrogen rates are 0.664 t/ha at the 5% level of significance and 0.883 t/ha at the 1% level.

classification of the difference between any two treatment means as significant or nonsignificant. However, unlike the LSD test in which only a single value is required for any pair comparison at a prescribed level of significance, the DMRT requires computation of a series of values, each corresponding to a specific set of pair comparisons.

The procedure for computing the DMRT values, as for the LSD test, depends primarily on the specific $s_{\bar{d}}$ of the pair of treatments being compared. Because the procedures for computing the appropriate $s_{\bar{d}}$ value for the various experimental designs are already discussed for the LSD test in Section 5.1.1, we illustrate the procedure for applying the DMRT for only one case—a single-factor experiment in a completely randomized design.

The steps for computation of the DMRT values for comparing all possible pairs of means are given for a completely randomized design experiment testing seven insecticide treatments in four replications. The data and analysis of variance are given in Tables 2.1 and 2.2 of Chapter 2.

☐ STEP 1.   Rank all the treatment means in decreasing (or increasing) order. It is customary to rank the treatment means according to the order of preference. For yield data, means are usually ranked from the highest-yielding treatment to the lowest-yielding treatment. For data on pest incidence, means are usually ranked from the least-infested treatment to the most severely infested treatment.

For our example, the seven treatment means arranged in decreasing order of yield are:

Treatment	Mean Yield, kg/ha	Rank
T_2: Dol-Mix (2 kg)	2,678	1
T_3: DDT + γ-BHC	2,552	2
T_4: Azodrin	2,128	3
T_1: Dol-Mix (1 kg)	2,127	4
T_5: Dimecron-Boom	1,796	5
T_6: Dimecron-Knap	1,681	6
T_7: Control	1,316	7

☐ STEP 2. Compute the $s_{\bar{d}}$ value following the appropriate procedures for specific designs described in Section 5.1.1. For our example, $s_{\bar{d}}$ is computed as:

$$s_{\bar{d}} = \sqrt{\frac{2s^2}{r}}$$

$$= \sqrt{\frac{2(94,773)}{4}} = 217.68 \text{ kg/ha}$$

☐ STEP 3. Compute the $(t-1)$ values of the *shortest significant ranges* as:

$$R_p = \frac{(r_p)(s_{\bar{d}})}{\sqrt{2}} \quad \text{for } p = 2, 3, \ldots, t$$

where t is the total number of treatments, $s_{\bar{d}}$ is the standard error of the mean difference computed in step 2, r_p values are the tabular values of the *significant studentized ranges* obtained from Appendix F, and p is the distance in rank between the pairs of treatment means to be compared (i.e., $p = 2$ for the two means with consecutive rankings and $p = t$ for the highest and lowest means).

For our example, the r_p values with error $d.f.$ of 21 and at the 5% level of significance are obtained from Appendix F as:

p	$r_p(.05)$
2	2.94
3	3.09
4	3.18
5	3.24
6	3.30
7	3.33

The $(t - 1) = 6 \; R_p$ values are then computed:

p	$R_{\mathrm{p}} = \dfrac{(r_p)(s_{\bar{d}})}{\sqrt{2}}$
2	$\dfrac{(2.94)(217.68)}{\sqrt{2}} = 453$
3	$\dfrac{(3.09)(217.68)}{\sqrt{2}} = 476$
4	$\dfrac{(3.18)(217.68)}{\sqrt{2}} = 489$
5	$\dfrac{(3.24)(217.68)}{\sqrt{2}} = 499$
6	$\dfrac{(3.30)(217.68)}{\sqrt{2}} = 508$
7	$\dfrac{(3.33)(217.68)}{\sqrt{2}} = 513$

☐ STEP 4.   Identify and group together all treatment means that do not differ significantly from each other:

A.   Compute the difference between the largest treatment mean and the largest R_p value (the R_p value at $p = t$) computed in step 3, and declare all treatment means whose values are less than the computed difference as significantly different from the largest treatment mean.

Next, compute the range between the remaining treatment means (i.e., those means whose values are larger than or equal to the difference between the largest mean and the largest R_p value) and compare this range with the value of R_p at $p = m$ where m is the number of treatments in the group. If the computed range is smaller than the corresponding R_p value, all the m treatment means in the group are declared not significantly different from each other.

Finally, in the array of means in step 1, draw a vertical line connecting all means that have been declared not significantly different from each other.

For our example, the difference between the largest R_p value (the R_p value at $p = 7$) of 513 and the largest treatment mean (T_2 mean) of 2,678 is $2,678 - 513 = 2,165$ kg/ha. From the array of means obtained in step 1, all treatment means, except that of T_3, are less than the

computed difference of 2,165 kg/ha. Hence, they are declared significantly different from T_1.

From the $m = 2$ remaining treatment means (T_2 and T_3) whose
values are larger than the computed difference of 2,165 kg/ha, compute
the range as $2,678 - 2,552 = 126$ kg/ha and compare it to the R_p
value at $p = m = 2$ of 453. Because the computed difference is smaller
than the R_p value at $p = 2$, T_2 mean and T_3 mean are declared not
significantly different from each other. A vertical line is then drawn to
connect these two means in the array of means, as shown following:

Treatment	Mean Yield, kg/ha	
T_2	2,678	⎤
T_3	2,552	⎦
T_4	2,128	
T_1	2,127	
T_5	1,796	
T_6	1,681	
T_7	1,316	

B. Compute the difference between the second largest treatment mean and
the second largest R_p value (the R_p value at $p = t - 1$) computed in
step 3, and declare all treatment means whose values are less than this
difference as significantly different from the second largest treatment
mean. For the m_1 remaining treatment means whose values are larger
than or equal to the computed difference, compute its range and
compare it with the appropriate R_p value (R_p at $p = m_1$). Declare all
treatments within the range not significantly different from each other
if the range is smaller than the corresponding R_p value.

For our example, the difference between the second largest R_p value
(the R_p value at $p = 6$) and the second largest treatment mean (T_3
mean) is computed as $2,552 - 508 = 2,044$ kg/ha. Because the means
of treatments T_5, T_6, and T_7 are less than 2,044 kg/ha, they are
declared significantly different from the mean of T_3.

The $m_1 = 3$ remaining treatment means, which have not been declared significantly different, are T_3, T_4, and T_1. Its range is computed
as $T_3 - T_1 = 2,552 - 2,127 = 425$ kg/ha, which is compared to the
corresponding R_p value at $p = m_1 = 3$ of 476 kg/ha. Because the range
is smaller than the R_p value at $p = 3$, the three remaining means are
not significantly different from each other. A vertical line is then drawn

to connect the means of T_3, T_4, and T_1, as shown following:

Treatment	Mean Yield, kg/ha	
T_2	2,678	
T_3	2,552	
T_4	2,128	
T_1	2,127	
T_5	1,796	
T_6	1,681	
T_7	1,316	

C. Continue the process with the third largest treatment mean, then the fourth, and so on, until all treatment means have been properly compared. For our example, the process is continued with the third largest treatment mean. The difference between the third largest treatment mean (T_4 mean) and the third largest R_p value (R_p value at $p = 5$) is computed as $2,128 - 499 = 1,629$ kg/ha. Only the mean of treatment T_7 is less than the computed difference of 1,629 kg/ha. Thus, T_7 is declared significantly different from T_4. The four remaining treatments, which have not been declared significantly different, are T_4, T_1, T_5, and T_6. Its range is computed as $T_4 - T_6 = 2,128 - 1,681 = 447$ kg/ha. Because the computed range is less than the corresponding R_p value at $p = 4$ of 489 kg/ha, all the four remaining means are declared not significantly different from each other. A vertical line is then drawn to connect the means of T_4, T_1, T_5, and T_6, as shown following:

Treatment	Mean Yield, kg/ha		
T_2	2,678		
T_3	2,552		
T_4	2,128		
T_1	2,127		
T_5	1,796		
T_6	1,681		
T_7	1,316		

At this point, the same process can be continued with the fourth largest treatment mean, and so on. However, because the mean of T_7 is the only one outside the groupings already made, it is simpler just to compare the T_7 mean, using the appropriate R_p values, with the rest of the means (namely: T_1, T_5, and T_6). These comparisons are made as follows.

T_1 vs. T_7: $2{,}127 - 1{,}316 = 811 > R_p$ (at $p = 4$) of 489

T_5 vs. T_7: $1{,}796 - 1{,}316 = 480 > R_p$ (at $p = 3$) of 476

T_6 vs. T_7: $1{,}681 - 1{,}316 = 365 < R_p$ (at $p = 2$) of 453

Of the three comparisons, the only one whose difference is less than the corresponding R_p value is that between T_6 and T_7. Thus, T_6 and T_7 are declared not significantly different from each other. A vertical line is then drawn to connect the means of T_6 and T_7, as shown following:

Treatment	Mean Yield, kg/ha
T_2	2,678
T_3	2,552
T_4	2,128
T_1	2,127
T_5	1,796
T_6	1,681
T_7	1,316

Because the last treatment in the array (T_7) has been reached, the process of grouping together all treatment means that do not differ significantly from each other is completed.

☐ STEP 5. Present the test results in one of the two following ways:

1. Use the *line notation* if the sequence of the treatments in the presentation of the results can be arranged according to their ranks, as shown in Table 5.12.

2. Use the *alphabet notation* if the desired sequence of the treatments in the presentation of the results is not to be based on their ranks. The alphabet notation can be derived from the line notation simply by assigning the same alphabet to all treatment means connected by the same vertical line. It is customary to use *a* for the first line, *b* for the second, *c* for the third, and so on. For our example, *a*, *b*, *c*, and *d* are

Table 5.12 DMRT for Comparing All Possible Pairs of Treatment Means, from a CRD Experiment Involving Seven Treatments, Using the Line Notation (Data in Table 2.1)

Treatment	Mean Yield, kg/ha[a]	DMRT[b]
T_2	2,678	
T_3	2,552	
T_4	2,128	
T_1	2,127	
T_5	1,796	
T_6	1,681	
T_7	1,316	

[a]Average of four replications.
[b]Any two means connected by the same vertical line are not significantly different at the 5% level of significance.

Table 5.13 DMRT for Comparing All Possible Pairs of Treatment Means, from a CRD Experiment Involving Seven Treatments, Using the Alphabet Notation (Data in Table 2.1)

Treatment	Mean Yield, kg/ha[a]	DMRT[b]
T_1	2,127	bc
T_2	2,678	a
T_3	2,552	ab
T_4	2,128	bc
T_5	1,796	c
T_6	1,681	cd
T_7	1,316	d

[a]Average of four replications.
[b]Any two means having a common letter are not significantly different at the 5% level of significance.

assigned to the four vertical lines in Table 5.12 as follows:

Treatment	Mean Yield, kg/ha
T_2	2,678 *a*
T_3	2,552 *b*
T_4	2,128 *c*
T_1	2,127
T_5	1,796
T_6	1,681 *d*
T_7	1,316

The final presentation using the alphabet notation is shown in Table 5.13. Note that one or more letters can be assigned to each treatment. For example, only one letter, *a*, is assigned to T_2 while two letters, *ab*, are assigned to T_3.

5.2 GROUP COMPARISON

For group comparison, more than two treatments are involved in each comparison. There are four types of comparison:

- *Between-group comparison*, in which treatments are classified into *s* (where *s* > 2) meaningful groups, each group consisting of one or more treatments, and the aggregate mean of each group is compared to that of the others.
- *Within-group comparison*, which is primarily designed to compare treatments belonging to a subset of all the treatments tested. This subset generally corresponds to a group of treatments used in the between-group comparison. In some instances, the subset of the treatments in which the within-group comparison is to be made may be selected independently of the between-group comparison.
- *Trend comparison*, which is designed to examine the functional relationship between treatment levels and treatment means. Consequently, it is applicable only to treatments that are quantitative, such as rate of herbicide application, rate of fertilizer application, and distance of planting.
- *Factorial comparison*, which, as the name implies, is applicable only to factorial treatments in which specific sets of treatment means are compared to investigate the main effects of the factors tested and, in particular, the nature of their interaction.

The most commonly used test procedure for making a group comparison is to partition the treatment sum of squares into meaningful components. The procedures is similar to that of the analysis of variance where the total sum of squares is partitioned into a fixed set of components directed by the experimental design used. For example, the total SS in the RCB design has three components, namely, replication, treatment, and experimental error. With further partitioning of the treatment SS into one or more components, specific causes of the difference between treatment means can be determined and the most important ones readily identified.

The procedure for partitioning the treatment SS consists essentially of:

· Selecting a desired set of group comparisons to clearly meet the experimental objective. The relationship between an experimental objective and the selected set of group comparisons is clearly illustrated by the weed-control experiment described in Section 5.1.

· Computing the SS for each desired group comparison and testing the significance of each comparison by an F test.

Each component of a partitioned treatment SS can be either a single $d.f.$ or a multiple $d.f.$ contrast.

A *single d.f. contrast* is a linear function of the treatment totals:

$$L = c_1 T_1 + c_2 T_2 + \cdots + c_t T_t = \sum_{i=1}^{t} c_i T_i$$

where T_i is the treatment total of the ith treatment, t is the total number of treatments, and c_i is the contrast coefficient associated with the ith treatment. The sum of the contrast coefficients is equal to zero:

$$\sum_{i=1}^{t} c_i = 0$$

The SS for the single $d.f.$ contrast L is computed as:

$$SS(L) = \frac{L^2}{r(\Sigma c^2)}$$

Two single $d.f.$ contrasts are said to be *orthogonal* if the sum of cross products of their coefficients equals zero. That is, the two single $d.f.$ contrasts:

$$L_1 = c_{11} T_1 + c_{12} T_2 + \cdots + c_{1t} T_t$$

$$L_2 = c_{21} T_1 + c_{22} T_2 + \cdots + c_{2t} T_t$$

are said to be orthogonal if the following condition holds:

$$\sum_{i=1}^{t} v_{1i}c_{2i} = c_{11}c_{21} + c_{12}c_{22} + \cdots + c_{1t}c_{2t} = 0$$

A group of p single $d.f.$ contrasts, where $p > 2$, is said to be *mutually orthogonal* if each pair, and all pairs, of the contrasts in the group are orthogonal. For an experiment with t treatments, the maximum number of mutually orthogonal single $d.f.$ contrasts that can be constructed is $(t - 1)$ or the $d.f.$ for the treatment SS. Also, for any set of $(t - 1)$ mutually orthogonal single $d.f.$ contrasts, the sum of their SS equals the treatment SS. That is:

$$SS(L_1) + SS(L_2) + \cdots + SS(L_{t-1}) = \text{Treatment } SS$$

where $L_1, L_2, \ldots, L_{t-1}$ are $(t - 1)$ mutually orthogonal single $d.f.$ contrasts.

The single $d.f.$ contrast method is applicable to all four types of group comparison defined earlier, and any group comparison can be represented by one or more single $d.f.$ contrasts.

A *multiple $d.f.$ contrast* represents a group of single $d.f.$ contrasts, and is usually defined in terms of a between-group comparison as:

$$M = g_1 \text{ vs. } g_2 \text{ vs. } g_3 \text{ vs. } \cdots \text{ vs. } g_s$$

where g_i is the ith group consisting of m_i treatments and there is no overlapping of treatments among the s groups (i.e., no single treatment appears in more than one group).

The SS for the multiple $d.f.$ contrast M with $(s - 1)$ $d.f.$ is computed as:

$$SS(M) = \frac{1}{r} \sum_{i=1}^{r} \frac{G_i^2}{m_i} - \frac{\left(\sum_{i=1}^{s} G_i\right)^2}{r \sum_{i=1}^{s} m_i}$$

where G_i is the sum of the treatment totals over the m_i treatments in the g_i group, and r is the number of replications.

We describe the partitioning of the treatment SS, either by the single $d.f.$ contrast method or the multiple $d.f.$ contrast method, to implement each of the four types of group comparison: between-group comparison, within-group comparison, trend comparison, and factorial comparison.

5.2.1 Between-Group Comparison

A between-group comparison involving s groups can be represented by a multiple $d.f.$ contrast or as a set of $(s - 1)$ mutually orthogonal single $d.f.$

contrasts. The basis for choosing between those is the total number of groups involved (the size of s) and the additional comparisons needed. With a large s, the set of mutually orthogonal single $d.f.$ contrasts would be large and the computational procedure would become lengthy. However, because many of these contrasts may be useful for subsequent within-group comparisons, the additional computation could be justified.

5.2.1.1 Single d.f. Contrast Method.

The primary task in the use of single $d.f.$ contrast method is the construction of an appropriate set of $(s - 1)$ mutually orthogonal single $d.f.$ contrasts involving the s treatment groups. Although there are several ways in which such a set of $(s - 1)$ mutually orthogonal contrasts can be derived, we illustrate a simplified procedure using an example of five treatment groups ($s = 5$). Let $g_1, g_2, g_3, g_4,$ and g_5 represent the original five treatment groups; and $m_1, m_2, m_3, m_4,$ and m_5 represent the corresponding number of treatments in each group.

☐ STEP 1. From the s original treatment groups, identify a set of $(s - 1)$ between-group comparisons each of which involves only two groups of treatments.

A. Place the original treatment groups first into two sets in any manner desired. For our example, $g_1, g_2,$ and g_3 may be placed in the first set and g_4 and g_5 in the second set. These two sets comprise the first newly created between-group comparison (i.e., comparison 1 of Table 5.14).

B. Examine each of the two sets derived in step 1A to see if either, or both, has more than two original treatment groups. If any set does, further subdivide that set until all sets contain no more than two original treatment groups per set. For our example, because set 2 of comparison 1 contains only two original treatment groups (g_4 and g_5) no further regrouping is required. On the other hand, set 1 contains three original

Table 5.14 The Construction of a Set of Four Mutually Orthogonal Single *d.f.* Contrasts to Represent a Between-group Comparison Involving Five Groups of Treatments (g_1, g_2, g_3, g_4, and g_5) Each Consisting of Two Treatments

Between-group Comparison			
Number	Set 1	Set 2	Single $d.f.$ Contrast[a]
1	g_1, g_2, g_3	g_4, g_5	$2(G_1 + G_2 + G_3) - 3(G_4 + G_5)$
2	g_1	g_2, g_3	$2G_1 - (G_2 + G_3)$
3	g_2	g_3	$G_2 - G_3$
4	g_4	g_5	$G_4 - G_5$

[a]G_i is the sum of treatment totals over all treatments in group g_i.

treatment groups (g_1, g_2, and g_3) and should thus be further regrouped. We placed g_1 in one set and g_2 and g_3 in another, resulting in the second newly created between-group comparison (i.e. comparison 2 of Table 5.14). Because neither of these two new sets in comparison 2 consists of more than two original treatment groups, the regrouping process is terminated.

C. Split each set that involves two original treatment groups into two sets of one original treatment group each. For example, there are at this point two sets that have two original treatment groups: (g_4, g_5) of comparison 1 and (g_2, g_3) of comparison 2. Hence, split each of these two sets into new sets, each consisting of one original treatment group (i.e., comparisons 3 and 4 of Table 5.14).

☐ STEP 2.   Represent each of the ($s - 1$) between-group comparisons derived in step 1 by its corresponding single $d.f.$ contrast. A between-group comparison involving two treatment groups ($s = 2$) can always be represented by the following single $d.f$ contrast:

$$L = c_1 G_1 - c_2 G_2$$

where G_1 is the sum of the treatment totals of all m_1 treatments belonging to the first group, G_2 is the sum of the treatment totals of all m_2 treatments belonging to the second group, and c_1 and c_2 are the contrast coefficients that satisfy the condition:

$$m_1 c_1 = m_2 c_2$$

For example, comparison 1 of Table 5.14 with set 1 consisting of the first three original treatment groups (g_1, g_2, g_3) and set 2 consisting of the last two (g_4, g_5) can be represented by the single $d.f.$ contrast:

$$L = a_1(G_1 + G_2 + G_3) - a_2(G_4 + G_5)$$

where G_i is the sum of the treatment totals of m_i treatments in group g_i ($i = 1, \ldots, 5$) and a_1 and a_2 are constants such that:

$$(m_1 + m_2 + m_3)a_1 = (m_4 + m_5)a_2$$

For example, if each $m_i = 2$, then the single $d.f.$ contrast for comparison 1 of Table 5.14 would be:

$$L = 2(G_1 + G_2 + G_3) - 3(G_4 + G_5)$$

The single $d.f.$ contrast for all four between-group comparisons involving five groups of treatments created in step 1, assuming that $m_i = 2$, are shown in Table 5.14.

Once the appropriate set of single *d.f.* contrasts associated with the desired between-group comparison is specified, the corresponding computation of its *SS* and the test for its significance are fairly straightforward.

We illustrate the procedure for doing this with data from the RCB experiment with four replications and six rates (kg/ha) of seeding ($T_1 = 25$, $T_2 = 50$, $T_3 = 75$, $T_4 = 100$, $T_5 = 125$, and $T_6 = 150$) shown in Table 2.5 of Chapter 2. It is assumed that the researcher wishes to compare between three groups of treatments, with T_1 and T_2 in the first group (g_1), T_3 and T_4 in the second group (g_2), and T_5 and T_6 in the third group (g_3). Thus:

$$M = g_1 \text{ vs. } g_2 \text{ vs. } g_3$$

Following the procedure just outlined for constructing a set of $(s - 1)$ single *d.f.* contrasts, the appropriate set of two single *d.f.* contrasts for our example may be:

$$L_1 = 2g_1 - (g_2 + g_3)$$

$$= 2(T_1 + T_2) - (T_3 + T_4 + T_5 + T_6)$$

$$L_2 = g_2 - g_3$$

$$= (T_3 + T_4) - (T_5 + T_6)$$

The step-by-step procedures for computing the *SS* and for testing its significance are:

☐ STEP 1. Verify the orthogonality among the $(s - 1)$ single *d.f.* contrasts. For our example, orthogonality of the two single *d.f.* contrasts L_1 and L_2 is verified because the sum of the cross products of their contrast coefficients ($\sum_{i=1}^{t} c_{1i} c_{2i}$) is zero [i.e., $(2)(0) + (2)(0) + (-1)(1) + (-1)(1) + (-1)(-1) + (-1)(-1) = 0$].

☐ STEP 2. Compute the *SS* for each of the $(s - 1)$ single *d.f.* contrasts. For our example, with $r = 4$ and the values of treatment total taken from Table 2.5 of Chapter 2, the *SS* for L_1 and L_2 are computed as:

$$SS(L_1) = \frac{[2(20{,}496 + 20{,}281) - (21{,}217 + 19{,}391 + 18{,}832 + 18{,}813)]^2}{4(4 + 4 + 1 + 1 + 1 + 1)}$$

$$= 227{,}013$$

$$SS(L_2) = \frac{[(21{,}217 + 19{,}391) - (18{,}832 + 18{,}813)]^2}{4(1 + 1 + 1 + 1)}$$

$$= 548{,}711$$

☐ STEP 3. Compute the SS for the original between-group comparison involving s groups as the sum of the $(s - 1)$ SS computed in step 2. For our example, the SS for the original between-group comparison M involving three groups is computed as:

$$SS(M) = SS(L_1) + SS(L_2) = 227{,}013 + 548{,}711$$

$$= 775{,}724$$

☐ STEP 4. Compute the F value as:

$$F = \frac{\dfrac{SS(M)}{(s - 1)}}{\text{Error } MS}$$

where error MS is from the analysis of variance. For our example, with error MS of 110,558 (Table 2.6), the F value is computed as:

$$F = \frac{\dfrac{775{,}724}{2}}{110{,}558} = 3.51$$

☐ STEP 5. Compare the computed F value with the tabular F values (Appendix E) with $f_1 = (s - 1)$ and $f_2 = $ error $d.f.$ For our example, the tabular F values with $f_1 = 2$ and $f_2 = 15$ $d.f.$ are 3.68 at the 5% level of significance and 6.36 at the 1% level. Because the computed F value is smaller than the tabular F value at the 5% level of significance, the means of the three groups of treatment do not differ significantly from each other.

5.2.1.2 Multiple d.f. Contrast Method. To illustrate the procedure for using a multiple $d.f.$ contrast to make between-group comparison, we use the same set of data and same group comparison that was used to illustrate the single $d.f.$ contrast method in Section 5.2.1.1. The steps are:

☐ STEP 1. For each of the s groups, compute the sum of treatment totals of all treatments in the group. For our example, the total for each of the three groups is computed as:

Group Number	Number of Treatments	Treatment Total
1	2	$G_1 = T_1 + T_2 = 40{,}777$
2	2	$G_2 = T_3 + T_4 = 40{,}608$
3	2	$G_3 = T_5 + T_6 = 37{,}645$
Total	6	$G_1 + G_2 + G_3 = 119{,}030$

☐ STEP 2. Compute the *SS* of the between-group comparison involving *s* groups as:

$$SS(M) = \frac{1}{r} \sum_{i=1}^{s} \frac{G_i}{m_i} - \frac{\left(\sum_{i=1}^{s} G_i \right)^2}{r \sum_{i=1}^{s} m_i}$$

For our example, the *SS* of the between-group comparison involving three groups is computed as:

$$SS(M) = \frac{G_1^2 + G_2^2 + G_3^2}{(4)(2)} - \frac{(G_1 + G_2 + G_3)^2}{(4)(6)}$$

$$= \frac{(40{,}777)^2 + (40{,}608)^2 + (37{,}645)^2}{8}$$

$$- \frac{(119{,}030)^2}{24}$$

$$= 591{,}114{,}927 - 590{,}339{,}204$$

$$= 775{,}723$$

Note that this *SS*, except for rounding error, is equal to the *SS* computed by the single *d.f.* contrast method in Section 5.2.1.1.

☐ STEP 3. Follow steps 4 and 5 of the single *d.f.* contrast method in Section 5.2.1.1.

5.2.2 Within-Group Comparison

Although both the single *d.f.* and the multiple *d.f.* contrast methods are applicable to a within-group comparison, the multiple *d.f.* is simpler and is preferred in practice. We illustrate both methods using the maize yield data in Table 2.7 of Chapter 2, assuming that the researcher wishes to determine the significance of the difference among the three hybrids *A*, *B*, and *D*.

5.2.2.1 Single d.f. Contrast Method. The procedure for applying the single *d.f.* contrast method to the within-group comparison is essentially the same as that for the between-group comparison described in Section 5.2.1. The *s* treatments in the within-group comparison are treated as if they are *s* groups in the between-group comparison. The procedures are:

☐ STEP 1. Construct a set of two orthogonal single *d.f.* contrasts to represent the desired within-group comparison involving three treatments (*W* = *A* vs. *B* vs. *D*) as:

$$L_1 = T_a + T_b - 2T_d$$

$$L_2 = T_a - T_b$$

where T_a, T_b, and T_d are the treatment totals of the three hybrids *A*, *B*, and *D*.

☐ STEP 2. Verify the orthogonality of the two single *d.f.* contrasts L_1 and L_2 constructed in step 1. Because the sum of cross products of the contrast coefficients of L_1 and L_2 is zero [i.e., $(1)(1) + (1)(-1) + (-2)(0) = 0$] their orthogonality is verified.

☐ STEP 3. Compute the *SS* for each single *d.f.* contrast using the formula $SS(L) = L^2/r(\Sigma c^2)$ as:

$$SS(L_1) = \frac{[5.855 + 5.885 - 2(5.355)]^2}{(4)(1 + 1 + 4)} = 0.044204$$

$$SS(L_2) = \frac{(5.855 - 5.885)^2}{(4)(1 + 1)} = 0.000112$$

☐ STEP 4. Compute the *SS* for the desired within-group comparison (*W*), with 2 *d.f.*, as the sum of the two *SS* computed in step 3:

$$SS(W) = SS(L_1) + SS(L_2)$$

$$= 0.044204 + 0.000112 = 0.044316$$

☐ STEP 5.   Compute the F value as:

$$F = \frac{\dfrac{SS(W)}{(s-1)}}{\text{Error } MS}$$

$$= \frac{\dfrac{0.044316}{2}}{0.021598} = 1.03$$

☐ STEP 6.   Because the computed F value is smaller than the corresponding tabular F value (Appendix E) with $f_1 = 2$ and $f_2 = 6$ degrees of freedom and at the 5% level of significance of 5.14, there is no significant difference in mean yield among the three maize hybrids.

5.2.2.2 *Multiple d.f. Contrast Method.* The procedures for applying the multiple $d.f.$ contrast method to the within-group comparison are:

☐ STEP 1.   Compute the SS of a within-group comparison involving s treatments as:

$$SS(W) = \frac{\sum\limits_{i=1}^{s} T_i^2}{r} - \frac{\left(\sum\limits_{i=1}^{s} T_i\right)^2}{rs}$$

where T_i is the total of the ith treatment and r is the number of replications common to all s treatments. This SS has $(s-1)$ degrees of freedom.

For our example, the SS of the within-group comparison involving three treatments ($W = A$ vs. B vs. D) is computed as:

$$SS(W) = \frac{T_a^2 + T_b^2 + T_d^2}{4} - \frac{(T_a + T_b + T_d)^2}{(4)(3)}$$

$$= \frac{(5.855)^2 + (5.885)^2 + (5.355)^2}{4}$$

$$- \frac{(5.855 + 5.885 + 5.355)^2}{12}$$

$$= 0.044317$$

Note that this SS, except for rounding error, is the same as that computed earlier through the single $d.f.$ contrast method.

☐ STEP 2.   Follow steps 5 and 6 of Section 5.2.2.1.

5.2.3   Trend Comparison

With quantitative treatments, such as plant density or rate of fertilizer applied, there is continuity from one treatment level to another and the number of possible treatment levels that could be tested is infinite. Although only a finite number of treatment levels can be tested in a trial, the researcher's interest usually covers the whole range of treatments. Consequently, the types of mean comparison that focus on the specific treatments tested are not adequate. A more appropriate approach is to examine the functional relationship between response and treatment that covers the whole range of the treatment levels tested.

For example, in a rice fertilizer trial where nitrogen rates of 0, 30, 60, 90, and 120 kg N/ha are tested, a researcher is not interested simply in establishing that grain yield at 30 kg N/ha is higher than that at 0 kg N/ha, and that grain yield at 60 kg N/ha is still higher than that at 30 kg N/ha, and so on. Instead, the interest is that of describing yield response over the whole range of nitrogen rates tested. Even though a certain specific nitrogen rate, for example 45 kg N/ha, was not actually tested, it is desirable to estimate what the yield would have been if 45 kg N/ha had been tested. This is achieved by examining a *nitrogen response function* that can describe the change in yield for every change in the rate of nitrogen applied. This type of analysis is referred to as *trend comparison*.

Although trend comparison can be made for any prescribed functional relationship, the simplest and most commonly used is the one based on *polynomials* (see Chapter 9 for more information on polynomials and other types of functional relationship). An nth degree polynomial describing the relationship between a dependent variable Y and an independent variable X is represented by:

$$Y = \alpha + \beta_1 X + \beta_2 X^2 + \cdots + \beta_n X^n$$

where α is the intercept and β_i ($i = 1, \ldots, n$) is the partial regression coefficient associated with the ith degree polynomial.

The trend comparison procedure based on polynomials, usually referred to as the *method of orthogonal polynomials*, seeks the lowest degree polynomial that can adequately represent the relationship between a dependent variable Y (usually represented by crop or non-crop response) and an independent variable X (usually represented by the treatment level). The procedure consists of:

1. Construction of a set of mutually orthogonal single $d.f.$ contrasts, with the first contrast representing the first degree polynomial (linear), the second contrast representing the second degree polynomial (quadratic), and so on. The number of polynomials that can be examined depends

on the number of paired observations (n) or, generally, the number of treatments tested (t). In fact, the highest degree polynomial that can be examined is equal to ($n - 1$) or ($t - 1$).

2. Computation of the SS, and the test of significance, for each contrast.
3. Selection of the specific degree polynomial that best describes the relationship between the treatment and the response. For example, in the polynomial equation given, if only β_1 is significant, then the relationship is linear; and, if only β_1 and β_2, or only β_2, is significant, then the relationship is quadratic; etc.

We illustrate the method of orthogonal polynomials for two cases. One has treatments of equal intervals; the other has treatments of unequal intervals.

5.2.3.1 Treatments with Equal Intervals.
For *treatments with equal intervals*, we use rice yield data from a RCB experiment where six rates (kg/ha) of seeding ($T_1 = 25$, $T_2 = 50$, $T_3 = 75$, $T_4 = 100$, $T_5 = 125$, and $T_6 = 150$) were tested in four replications (Table 2.5 of Chapter 2). Note that the treatments have an equal interval of 25 kg seed/ha. The steps involved in applying the orthogonal polynomial method to compare the trends among the six treatment means follow:

☐ STEP 1.   From Appendix G, obtain the set of ($t - 1$) single $d.f.$ contrasts representing the orthogonal polynomials, where t is the number of treatments tested. For our example, the five single $d.f.$ contrasts representing the orthogonal polynomials are listed in terms of its contrast coefficients and the corresponding sum of squares of the coefficients:

Degree of Polynomial	Orthogonal Polynomial Coefficient (c)						Sum of Squares (Σc^2)
	T_1	T_2	T_3	T_4	T_5	T_6	
Linear (1st)	-5	-3	-1	$+1$	$+3$	$+5$	70
Quadratic (2nd)	$+5$	-1	-4	-4	-1	$+5$	84
Cubic (3rd)	-5	$+7$	$+4$	-4	-7	$+5$	180
Quartic (4th)	$+1$	-3	$+2$	$+2$	-3	$+1$	28
Quintic (5th)	-1	$+5$	-10	$+10$	-5	$+1$	252

☐ STEP 2.   Compute the SS for each single $d.f.$ contrast, or each orthogonal polynomial, derived in step 1. For our example, with $r = 4$ and using the treatment totals in Table 2.5 and the formula $SS(L) = \Sigma L^2 / r(\Sigma c^2)$, the SS

for each degree of polynomial is computed as:

$$SS_1 = [(\quad 5)(20{,}496) + (-3)(20{,}281) + (-1)(21{,}217)$$

$$+ (1)(19{,}391) + (3)(18{,}832) + (5)(18{,}813)]^2/(4)(70)$$

$$= 760{,}035$$

$$SS_2 = [(5)(20{,}496) + (-1)(20{,}281) + (-4)(21{,}217)$$

$$+ (-4)(19{,}391) + (-1)(18{,}832) + (5)(18{,}813)]^2/(4)(84)$$

$$= 74{,}405$$

$$SS_3 = [(-5)(20{,}496) + (7)(20{,}281) + (4)(21{,}217)$$

$$+ (-4)(19{,}391) + (-7)(18{,}832) + (5)(18{,}813)]^2/(4)(180)$$

$$= 113{,}301$$

$$SS_4 = [(1)(20{,}496) + (-3)(20{,}281) + (2)(21{,}217)$$

$$+ (2)(19{,}391) + (-3)(18{,}832) + (1)(18{,}813)]^2/(4)(28)$$

$$= 90{,}630$$

$$SS_5 = [(-1)(20{,}496) + (5)(20{,}281) + (-10)(21{,}217)$$

$$+ (10)(19{,}391) + (-5)(18{,}832) + (1)(18{,}813)]^2/(4)(252)$$

$$= 159{,}960$$

where the subscripts 1, 2, 3, 4, and 5 of the SS refer to the first, second, third, fourth, and fifth degree polynomial, respectively.

☐ STEP 3. Compute the F value for each degree polynomial by dividing each SS computed in step 2 by the error mean square from the analysis of variance. With the error MS of 110,558 from Table 2.6, the F value corresponding to each SS computed in step 2 is:

$$F_1 = \frac{SS_1}{\text{Error } MS}$$

$$= \frac{760{,}035}{110{,}558} = 6.87$$

$$F_2 = \frac{SS_2}{\text{Error } MS}$$

$$= \frac{74{,}405}{110{,}558} = 0.67$$

$$F_3 = \frac{SS_3}{\text{Error } MS}$$

$$= \frac{113{,}301}{110{,}558} = 1.02$$

$$F_4 = \frac{SS_4}{\text{Error } MS}$$

$$= \frac{90{,}630}{110{,}558} = 0.82$$

$$F_5 = \frac{SS_5}{\text{Error } MS}$$

$$= \frac{159{,}960}{110{,}558} = 1.45$$

☐ STEP 4. Compare each computed F value with the tabular F value (Appendix E) with $f_1 = 1$ and $f_2 = $ error $d.f.$ at the prescribed level of significance. The tabular F values with $f_1 = 1$ and $f_2 = 15$ degrees of freedom are 4.54 at the 5% level of significance and 8.68 at the 1% level. Except for F_1, all other computed F values are smaller than the tabular F value at the 5% level of significance. Thus, the results indicate that only the first degree polynomial is significant, or that the relationship between yield and seeding rate is linear within the range of the seeding rates tested.

☐ STEP 5. Pool the SS over all polynomials that are at least two degrees higher than the highest significant polynomial. This pooled SS value is usually referred to as the residual SS.

For our example, because all degree polynomials, except the first, are not significant, the third, fourth, and fifth degree polynomials are pooled. That is, the residual SS, with three $d.f.$, is computed as the sum of the SS corresponding to the third, fourth, and fifth degree polynomials:

$$\text{Residual } SS = SS_3 + SS_4 + SS_5$$

$$= 113{,}301 + 90{,}630 + 159{,}960$$

$$= 363{,}891$$

The residual $d.f.$ is equal to the number of SS pooled; three in this case. The residual mean square and the corresponding F value can be computed, in the usual manner:

$$\text{Residual } MS = \frac{\text{Residual } SS}{\text{Residual } d.f.}$$

$$= \frac{363,891}{3} = 121,297$$

$$F = \frac{\text{Residual } MS}{\text{Error } MS}$$

$$= \frac{121,297}{110,558} = 1.10$$

The computed F value can be compared with the tabular F value with f_1 = residual $d.f.$ and f_2 = error $d.f.$ at the prescribed level of significance. For our example, the tabular F values with $f_1 = 3$ and $f_2 = 15$ $d.f.$ are 3.29 at the 5% level of significance and 5.42 at the 1% level. As expected, the combined effects of the third, fourth, and fifth degree polynomials are not significant at the 5% level of significance.

☐ STEP 6.   Enter all values obtained in steps 2 to 5 in the analysis of variance table. The final results are shown in Table 5.15.

5.2.3.2   Treatments with Unequal Intervals.   In the orthogonal polynomial method, the only difference between the case of equal intervals and that of unequal intervals is in the derivation of the appropriate set of mutually orthogonal single *d.f.* contrasts to represent the orthogonal polynomials. Instead of obtaining the contrast coefficients directly from a standardized table, such as Appendix G, the contrast coefficients must be derived for each case of unequal treatment intervals. However, once the contrast coefficients are

Table 5.15   Analysis of Variance with the Treatment Sum of Squares Partitioned Following the Procedure of Trend Comparison (Data in Table 2.5)

Source of Variation	Degree of Freedom	Sum of Squares	Mean Square	Computed F^a	Tabular F 5%	Tabular F 1%
Replication	3	1,944,361				
Seeding rate	5	1,198,331	239,666	2.17^{ns}	2.90	4.56
Linear	(1)	760,035	760,035	6.87*	4.54	8.68
Quadratic	(1)	74,405	74,405	< 1	—	—
Residual	(3)	363,891	121,297	1.10^{ns}	3.29	5.42
Error	15	1,658,376	110,558			
Total	23	4,801,068				

[a]* = significant at 5% level, ns = not significant.

specified, the computational procedures are the same for both cases. Thus, we focus mainly on the procedure for deriving the orthogonal polynomial coefficients for the case of unequal intervals.

The procedures for deriving the orthogonal polynomial coefficients for treatments with unequal intervals are complex, especially when higher-degree polynomials are involved. For simplicity, we discuss only the derivation of orthogonal polynomial coefficients of up to the third degree, but for any number of treatments. This limitation is not too restrictive in agricultural research because most biological responses to environmental factors can be adequately described by polynomials that are no higher than the third degree.

Consider a trial where the treatments consist of four nitrogen rates—0, 60, 90, and 120 kg N/ha. Note that the intervals between successive treatments are not the same. The steps involved in the derivation of the three sets of orthogonal polynomial coefficients (i.e., first, second, and third degree polynomials) are:

☐ STEP 1.   Code the treatments to the smallest integers. For our example, the codes X_1, X_2, X_3, and X_4, corresponding to the four nitrogen rates, are obtained by dividing each nitrogen rate by 30:

Nitrogen Rate, kg/ha	Code (X)
0	0
60	2
90	3
120	4

☐ STEP 2.   Compute the three sets of orthogonal polynomial coefficients, corresponding to the first (linear), second (quadratic), and third (cubic) degree polynomials as:

$$L_i = a + X_i$$

$$Q_i = b + cX_i + X_i^2$$

$$C_i = d + eX_i + fX_i^2 + X_i^3$$

where L_i, Q_i, and C_i ($i = 1, \ldots, t$) are the coefficients of the ith treatment corresponding to linear, quadratic, and cubic, respectively; t is the number of treatments; and a, b, c, d, e, and f are the parameters that need to be estimated from the following six equations:*

$$\sum_{i=1}^{t} L_i - la + \sum_{i=1}^{t} X_i = 0$$

$$\sum_{i=1}^{t} Q_i = tb + c\sum_{i=1}^{t} X_i + \sum_{i=1}^{t} X_i^2 = 0$$

$$\sum_{i=1}^{t} C_i = td + e\sum_{i=1}^{t} X_i + f\sum_{i=1}^{t} X_i^2 + \sum_{i=1}^{t} X_i^3 = 0$$

$$\sum_{i=1}^{t} L_i Q_i = \sum_{i=1}^{t} (a + X_i)(b + cX_i + X_i^2) = 0$$

$$\sum_{i=1}^{t} L_i C_i = \sum_{i=1}^{t} (a + X_i)(d + eX_i + fX_i^2 + X_i^3) = 0$$

$$\sum_{i=1}^{t} Q_i C_i = \sum_{i=1}^{t} (b + cX_i + X_i^2)(d + eX_i + fX_i^2 + X_i^3) = 0$$

The general solution of the six parameters and the computation of their particular values, for our example, are:

$$a = -\frac{\Sigma X}{t}$$

$$= \frac{-9}{4}$$

$$b = \frac{(\Sigma X)(\Sigma X^3) - (\Sigma X^2)^2}{t(\Sigma X^2) - (\Sigma X)^2}$$

$$= \frac{(9)(99) - (29)^2}{4(29) - (9)^2} = \frac{10}{7}$$

$$c = \frac{(\Sigma X)(\Sigma X^2) - t(\Sigma X^3)}{t(\Sigma X^2) - (\Sigma X)^2}$$

$$= \frac{(9)(29) - 4(99)}{4(29) - (9)^2} = -\frac{27}{7}$$

*The first three equations are derived from the definition of single *d.f* contrast that the sum of its coefficients must be zero. The last three equations are derived from the orthogonality conditions that the sum of the cross products of coefficients for each pair of contrasts must be zero.

$$f = \left\{ \Sigma X^5 \left[(\Sigma X)^2 - t(\Sigma X^2) \right] + \Sigma X^4 \left[t\Sigma X^3 - (\Sigma X)(\Sigma X^2) \right] \right.$$

$$\left. + \Sigma X^3 \left[(\Sigma X^2)^2 - (\Sigma X)(\Sigma X^3) \right] \right\}$$

$$\left/ \left\{ \Sigma X^4 \left[t\Sigma X^2 - (\Sigma X)^2 \right] + \Sigma X^3 \left[(\Sigma X)(\Sigma X^2) - t\Sigma X^3 \right] \right. \right.$$

$$\left. + \Sigma X^2 \left[(\Sigma X)(\Sigma X^3) - (\Sigma X^2)^2 \right] \right\}$$

$$= \frac{1{,}299 \left[(9)^2 - 4(29) \right] + 353[4(99) - 9(29)] + 99 \left[(29)^2 - 9(99) \right]}{353 \left[4(29) - (9)^2 \right] + 99[(9)(29) - 4(99)] + 29 \left[(9)(99) - (29)^2 \right]}$$

$$= -\frac{69}{11}$$

$$e = \frac{f \left[(\Sigma X_i)(\Sigma X_i^2) - t(\Sigma X_i^3) \right] + \left[(\Sigma X_i)(\Sigma X_i^3) - t(\Sigma X_i^4) \right]}{t(\Sigma X_i^2) - (\Sigma X_i)^2}$$

$$= \frac{\left(-\dfrac{69}{11} \right) [(9)(29) - 4(99)] + [(9)(99) - 4(353)]}{4(29) - (9)^2}$$

$$= \frac{512}{55}$$

$$d = -\frac{e(\Sigma X_i) + f(\Sigma X_i^2) + \Sigma X_i^3}{t}$$

$$= -\frac{\left(\dfrac{512}{55} \right)(9) + \left(-\dfrac{69}{11} \right)(29) + 99}{4}$$

$$= -\frac{12}{55}$$

The values of the parameters a, b, c, d, e, and f computed are then used in the equations in step 2 to compute the values of L_i, Q_i, and C_i for each nitrogen rate.

For example, for 0 kg N/ha, the three coefficients are computed as:

$$L_1 = a + X_1$$

$$= -\frac{9}{4} + 0 = -\frac{9}{4}$$

$$Q_1 = b + cX_1 + X_1^2$$

$$= \frac{10}{7} - \frac{27}{7}(0) + 0 = \frac{10}{7}$$

$$C_1 = d + eX_1 + fX_1^2 + X_1^3$$

$$= -\frac{12}{55} + \frac{512}{55}(0) + 0 = -\frac{12}{55}$$

And, for 60 kg N/ha, they are:

$$L_2 = a + X_2$$

$$= -\frac{9}{4} + 2 = -\frac{1}{4}$$

$$Q_2 = b + cX_2 + X_2^2$$

$$= \frac{10}{7} - \frac{27}{7}(2) + (2)^2 = -\frac{16}{7}$$

$$C_2 = d + eX_2^2 + fX_2^2 + X_2^3$$

$$= -\frac{12}{55} + \frac{512}{55}(2) - \frac{69}{11}(2)^2 + (2)^3 = \frac{72}{55}$$

The results for all four nitrogen rates are:

Treatment		Orthogonal Polynomial Coefficient		
Nitrogen Rate, kg/ha	Code (X)	Linear	Quadratic	Cubic
0	0	−9	5	−1
60	2	−1	−8	6
90	3	3	−4	−8
120	4	7	7	3

Note that the common denominator for all coefficients of the same degree polynomial is removed. For example, the four coefficients of the linear contrast, namely, $-\frac{9}{4}$, $-\frac{1}{4}$, $\frac{3}{4}$, and $\frac{7}{4}$, are simply shown as −9, −1, 3, and 7.

5.2.4 Factorial Comparison

For a factorial experiment, the partitioning of the treatment *SS* into components associated with the main effects of the factors tested, and their interaction effects, is a standard part of the analysis of variance (see Chapter 3, Section 3.3). Each effect, however, may be further partitioned into several subcomponents. In fact, the procedures described in Sections 5.2.1, 5.2.2, and 5.2.3 can be applied directly to partition the main-effect SS. For example, consider the 3×5 factorial experiment whose analysis of variance is shown in Table 3.6 of Chapter 3. If the researcher wishes to make a trend comparison between the five nitrogen rates (i.e., partition the nitrogen SS into linear component, quadratic component, etc.) the procedure described in Section 5.2.3 is directly applicable. Or if the researcher wishes to make a between-group comparison to compare the mean of the three varieties (V_1, V_2, and V_3) he can apply one of the methods described in Section 5.2.1. Thus only the procedure for partitioning the interaction effect SS is so far unspecified.

We illustrate the computational procedures for partitioning a two-factor interaction SS with data from a 6×3 factorial experiment involving six rice varieties and three nitrogen rates, as shown in Table 3.11 of Chapter 3. Note that the analysis of variance (Table 3.15) showed a highly significant interaction effect between variety and nitrogen, indicating that varietal differences are not the same at different rates of nitrogen and, similarly, that nitrogen responses differ among varieties tested. Thus, further partitioning of the interaction SS could be useful in understanding the nature of the interaction between variety and fertilizer rate. The step-by-step procedures for partitioning a two-factor interaction SS are:

□ STEP 1.   Construct a set of mutually orthogonal contrasts (see Section 5.2.1) for one of the factors, say factor A (to be referred to as the *primary factor*), corresponding to the objective of the trial. This set of contrasts could be composed of either single $d.f.$ or multiple $d.f.$ contrasts or a mixture of the two. To minimize misinterpretation of results, the contrasts should be selected so they are mutually orthogonal.

For our example, nitrogen factor may be considered as the primary factor. In such a case, the trend comparison would be an appropriate set of contrasts to examine. With three rates of nitrogen (0, 60, and 120 kg/ha), two orthogonal single $d.f.$ contrasts A_1 and A_2 representing the linear and quadratic polynomials can be constructed. Because the three nitrogen rates are of equal intervals (see Section 5.2.3.1) the two sets of orthogonal polynomial coefficients are obtained directly from Appendix G as:

Nitrogen Rate, kg/ha	Orthogonal Polynomial Coefficient	
	Linear (A_1)	Quadratic (A_2)
0	−1	+1
60	0	−2
120	+1	+1

☐ STEP 2. Compute the *SS* for each of the contrasts constructed in step 1, based on the *A* totals over all levels of factor *B*, following the appropriate procedure described in Sections 5.2.1, 5.2.2, or 5.2.3.

For our example, the *SS* for each of the two single *d.f.* contrasts is computed following the procedure of Section 5.2.3.1:

$$A_1 \; SS = N_L \; SS = \frac{[(-1)(N_1) + (0)(N_2) + (1)(N_3)]^2}{(r)(a)[(-1)^2 + (0)^2 + (1)^2]}$$

$$A_2 \; SS = N_Q \; SS = \frac{[(1)(N_1) + (-2)(N_2) + (1)(N_3)]^2}{(r)(a)[(1)^2 + (-2)^2 + (1)^2]}$$

where N_1, N_2, and N_3 are nitrogen totals for the first, second, and third level of nitrogen, respectively; *r* is the number of replications; and *a* is the number of varieties. Note that the quantity $(r)(a)$ is used in the divisor instead of *r* because the treatment totals (N_1, N_2, and N_3) used in the computation of the *SS* are summed over $(r)(a)$ observations. Note further that because there are only two orthogonal single *d.f.* contrasts, only one of the two *SS* needs to be computed directly; the other one can be obtained by subtraction. That is, if N_L *SS* is computed directly, then N_Q *SS* can be derived simply as:

$$N_Q \; SS = \text{Nitrogen } SS - N_L \; SS$$

Substituting the nitrogen totals from Table 3.13, the values of the two *SS* of the first contrast is computed as:

$$A_1 \; SS = N_L \; SS = \frac{[(-1)(72,371) + (1)(114,678)]^2}{(3)(6)(2)}$$

$$= 49,718,951$$

The *SS* of the second contrast is either computed directly as:

$$A_2 \; SS = N_Q \; SS = \frac{[(1)(72,371) + (-2)(98,608) + (1)(114,678)]^2}{(18)(6)}$$

$$= 957,110$$

or computed through subtraction as:

$$A_2 \; SS = N_Q \; SS = \text{Nitrogen } SS - N_L \; SS$$

$$= 50,676,061 - 49,718,951 = 957,110$$

☐ STEP 3. Following the procedure of step 2, compute the SS for each of the contrasts constructed in step 1 but based on the A totals at each level of factor B (instead of the A totals over all levels of factor B used in step 2). Note that for each contrast, the number of SS to be computed is equal to b, the levels of factor B.

For our example, there are six SS for the linear component of nitrogen SS ($N_L SS$) and six SS for the quadratic component $N_Q SS$; with each SS corresponding to each of the six varieties. The computation of these SS is shown in Table 5.16. For example, the $N_L SS$ and $N_Q SS$ for V_1 are computed as:

$$N_L SS = \frac{[(-1)(10{,}715) + (1)(22{,}644)]^2}{(3)(2)}$$

$$= 23{,}716{,}840$$

$$N_Q SS = \frac{[(1)(10{,}715) + (-2)(15{,}396) + (1)(22{,}644)]^2}{(3)(6)}$$

$$= 366{,}083$$

☐ STEP 4. Compute the components of the $A \times B\ SS$, corresponding to the set of mutually orthogonal contrasts constructed for factor A in step 1 as:

$$A_i \times B\ SS = \sum_{j=1}^{b} (A_i\ SS)_j - A_i\ SS$$

Table 5.16 Computational Procedure for the Partitioning of Nitrogen × Variety Interaction SS in Table 3.15 into Two Components, Based on the Linear and Quadratic Components of Nitrogen SS

| Variety | Treatment Total | | | Sum of Squares[a] | |
	N_1	N_2	N_3	Linear	Quadratic
V_1	10,715	15,396	22,644	23,716,840	366,083
V_2	14,803	20,141	21,634	7,777,093	821,335
V_3	12,749	18,367	23,605	19,642,123	8,022
V_4	12,177	16,661	21,283	13,819,873	1,058
V_5	12,305	16,900	18,036	5,474,060	664,704
V_6	9,622	11,143	7,476	767,553	1,495,297
Total				71,197,542	3,356,499

[a] The linear $SS = (-N_1 + N_3)^2/6$ and the quadratic $SS = (N_1 - 2N_2 + N_3)^2/[(3)(6)]$.

where $A_i \times B\ SS$ is the ith component of the $A \times B\ SS$, $A_i\ SS$ is the ith component of the main-effect SS as computed in step 2, $(A_i\ SS)_j$ is the SS for the A_i contrast corresponding to the jth level of factor D as computed in step 3, and b is the number of levels of factor B.

For our example, the two components of the $N \times V\ SS$ are computed as.

$$N_L \times V\ SS = \sum_{j=1}^{6} (N_L\ SS)_j - N_L SS$$

$$N_Q \times V\ SS = \sum_{j=1}^{6} (N_Q\ SS)_j - N_Q SS$$

where $(N_L\ SS)_j$ and $(N_Q\ SS)_j$ are the SS, computed in step 3, associated with the jth variety for the linear and quadratic components of nitrogen SS, respectively. $N_L\ SS$ and $N_Q\ SS$ are similarly defined SS computed in step 2.

Substituting the values obtained in steps 2 and 3 in the two preceding equations, the following values are obtained:

$$N_L \times V\ SS = (23,716,840 + \cdots + 767,553) - 49,718,951$$

$$= 71,197,543 - 49,718,951 = 21,478,591$$

$$N_Q \times V\ SS = (366,083 + \cdots + 1,495,297) - 957,110$$

$$= 3,356,499 - 957,110 = 2,399,389$$

Note that the $N_Q \times V\ SS$ can also be computed by subtraction:

$$N_Q \times V\ SS = N \times V\ SS - N_L \times V\ SS$$

where $N \times V\ SS$ is the interaction SS computed in the standard analysis of variance as shown in Table 3.15. Thus,

$$N_Q \times V\ SS = 23,877,980 - 21,478,591 = 2,399,389$$

☐ STEP 5. Enter all values of the SS computed in steps 2 to 4 in the original analysis of variance. For our example, the values of the $N_L\ SS$, $N_Q\ SS$, $N_L \times V\ SS$, and $N_Q \times V\ SS$ are entered in the analysis of variance of Table 3.15, as shown in Table 5.17. The result indicates that the existence of the variety × nitrogen interaction is mainly due to the difference in the linear part of the yield responses to nitrogen rates of the different varieties.

Note that, at this point, the partitioning of the nitrogen × variety interaction SS, based on the prescribed trend comparison of nitrogen means, is completed. The researcher should decide at this point if the information obtained so far is adequate to answer the experimental objectives. If the objectives are answered, the procedure can be terminated. Otherwise, additional analyses may be needed. In general, further analysis may be required when one or both of the following cases occur:

- When the primary factor or the specific set of contrasts originally selected for the primary factor is shown to be not appropriate. In our example, it may be suspected that the use of variety instead of nitrogen as the primary factor for partitioning the interaction *SS* could provide better answers to the experimental objectives. Or, instead of choosing the set of orthogonal polynomials as the basis for partitioning the nitrogen *SS*, it is suspected that some between-group comparisons may be more useful.
- The findings have generated additional questions. For example, in the illustration we used here, the result obtained so far leads to the question: what are the varieties that contribute to the differences in the rates of nitrogen response? To answer this question, further partitioning of the linear component of the interaction *SS* (i.e., $N_L \times V$ *SS*) is needed. The following additional steps are required:

☐ STEP 6.   Make visual observations of the data for probable answers to the new question raised. For our example, the nitrogen responses of the different varieties can be easily examined through a freehand graphical representation of the responses, as shown in Figure 5.1. It can be seen that:

- V_6 is the only variety with a negative response.
- Among the remaining five varieties, V_2 and V_5 seem to have a declining response within the range of nitrogen rates tested while the others (V_1,

Table 5.17   Analysis of Variance with Partitioning of Main Effect *SS* and Corresponding Interaction *SS* (Original Analysis of Variance in Table 3.15)

Source of Variation	Degree of Freedom	Sum of Squares	Mean Square	Computed F^a
Replication	2	9,220,962	4,610,481	
Variety (V)	5	57,100,201	11,420,040	7.65**
Error(a)	10	14,922,620	1,492,262	
Nitrogen (N)	2	50,676,061	25,338,031	b
Linear (N_L)	(1)	49,718,951	49,718,951	b
Quadratic (N_Q)	(1)	957,110	957,110	b
Error(b)	4	2,974,909	743,727	
$N \times V$	10	23,877,980	2,387,798	5.80**
$N_L \times V$	(5)	21,478,591	4,295,718	10.44**
$N_Q \times V$	(5)	2,399,389	479,878	1.17^{ns}
Error(c)	20	8,232,916	411,646	
Total	53	167,005,649		

^{a}** = significant at 1% level, * = significant at 5% level, ^{ns} = not significant.
^{b}Error(b) $d.f.$ is not adequate for valid test of significance.

V_3, and V_4) do not.
- There seems to be no appreciable difference in the responses either between V_2 and V_5 or among V_1, V_3, and V_4.

Thus, the six varieties can be classified into three groups according to their nitrogen responses: group 1 composed of V_6 with a negative response, group 2 composed of V_2 and V_5 with declining responses, and group 3 composed of V_1, V_3, and V_4 with linear positive responses.

□ STEP 7. Confirm the visual observation made in step 6 through partitioning of the interaction SS into appropriate components. For our example, the $N_L \times V$ SS can be further partitioned into the following four components:

Component		Degree of Freedom	Question to be Answered
Number	Definition		
1	$N_L \times (V_6$ vs. others$)$	1	Does the linear response of V_6 differ from that of the other varieties?
2	$N_L \times [(V_2, V_5)$ vs. $(V_1, V_3, V_4)]$	1	Does the mean linear response of V_2 and V_5 differ from that of V_1, V_3, and V_4?
3	$N_L \times (V_2$ vs. $V_5)$	1	Does the linear response of V_2 differ from that of V_5?
4	$N_L \times (V_1$ vs. V_3 vs. $V_4)$	2	Is there any difference in the linear responses of V_1, V_3, and V_4?

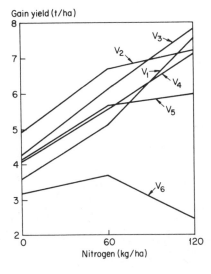

Figure 5.1 Mean yield of 6 rice varieties with 3 nitrogen rates (data in Table 3.11).

☐ STEP 8.   Compute the *SS* for each of the components of the $N_L \times V$ *SS* constructed in step 7. For our example, the computation of the *SS* for each of the four components of the $N_L \times V$ *SS* follows the first formula in Section 5.2.4, step 4. The final results are shown in Table 5.18.

Results of the partitioning of $N_L \times V$ *SS* confirmed the visual observation made in step 6, as follows:

- Component 1 is significant, indicating that there is a significant difference between the linear response of V_6 and the mean linear response of the other five varieties.
- Component 2 is significant, indicating that among the five varieties, there is a significant difference between the mean linear response of V_2 and V_5 and that of V_1, V_3, and V_4.
- Components 3 and 4 are both nonsignificant, indicating that there is no significant difference in the linear responses between V_2 and V_5 or between V_1, V_3, and V_4.

Table 5.18 Additional Partitioning of the $N_L \times V$ *SS* in Table 5.17, to Support the Visual Observation in Figure 5.1

Source of Variation	Degree of Freedom	Sum of Squares	Mean Square	Computed F^a	Tabular F 5%	Tabular F 1%
$N_L \times V$	5	21,478,591	4,295,718	10.44**	2.71	4.10
$N_L \times (V_6$ vs. others)	(1)	16,917,575	16,917,575	41.10**	4.35	8.10
$N_L \times [(V_2, V_5)$ vs. $(V_1, V_3, V_4)]$	(1)	3,783,340	3,783,340	9.19**	4.35	8.10
$N_L \times (V_2$ vs. $V_5)$	(1)	100,834	100,834	< 1	—	—
$N_L \times (V_1$ vs. V_3 vs. $V_4)$	(2)	676,842	338,421	< 1	—	—
Error(c)	20	8,232,916	411,646			

^{a}** = significant at 1% level.

CHAPTER 6

Analysis of Multiobservation Data

When a single character from the same experimental unit is measured more than once, the data is called *multiobservation* data. There are two types:

- Data from plot sampling in which s sampling units are measured from each plot, as in the measurement of plant height in transplanted rice where 10 hills may be measured in each plot.
- Data from measurements made over time in which the same character is measured at different growth stages of the crop, as in plant height, tiller number, and dry matter production, which may be measured every 20 days.

Standard analysis of variance (Chapters 2 to 4), which requires that there is a single observation per character per experimental unit, is not directly applicable to multiobservation data. It can be applied only to the average of all samples from a plot, or to the average of all measurements made over time, for each plot. We focus on the appropriate procedures for directly analyzing multiobservation data.

6.1 DATA FROM PLOT SAMPLING

For data from plot sampling, an additional source of variation can be measured: that due to sampling variation, which is commonly referred to as sampling error. The formats for the analysis of variance for data from plot sampling of a completely randomized design (CRD) and a randomized complete block (RCB) design, with t treatments and r replications; and of a split-plot design with a main-plot treatments, b subplot treatments, and r replications, are shown in Tables 6.1, 6.2, and 6.3.

Because the only distinct feature of the analysis of variance for data from plot sampling is the part involving sampling error, we illustrate the computational procedure with one example from an experiment with a RCB design, and another example from an experiment with a split-plot design.

Table 6.1 Format for the Analysis of Variance of Data from Plot Sampling in a CRD with *t* Treatments, *r* Replications, and *s* Sampling Units per Plot

Source of Variation	Degree of Freedom	Sum of Squares	Mean Squares	Computed F	Tabular F 5%	1%
Treatment	$t - 1$					
Experimental error	$t(r - 1)$					
Sampling error	$rt(s - 1)$					
Total	$rts - 1$					

Table 6.2 Format for the Analysis of Variance of Data from Plot Sampling in a RCB Design with *t* Treatments, *r* Replications, and *s* Sampling Units per Plot

Source of Variation	Degree of Freedom	Sum of Squares	Mean Square	Computed F	Tabular F 5%	1%
Replication	$r - 1$					
Treatment	$t - 1$					
Experimental error	$(r - 1)(t - 1)$					
Sampling error	$rt(s - 1)$					
Total	$rts - 1$					

Table 6.3 Format for the Analysis of Variance of Data from Plot Sampling in a Split-Plot Design with *a* Main-plot Treatments, *b* Subplot Treatments, *r* Replications, and *s* Sampling Units per Plot

Source of Variation	Degree of Freedom	Sum of Squares	Mean Square	Computed F	Tabular F 5%	1%
Replication	$r - 1$					
Main-plot treatment (A)	$a - 1$					
Error(a)	$(r - 1)(a - 1)$					
Subplot treatment (B)	$b - 1$					
$A \times B$	$(a - 1)(b - 1)$					
Error(b)	$a(r - 1)(b - 1)$					
Sampling error	$abr(s - 1)$					
Total	$rabs - 1$					

6.1.1 RCB Design

The computations we show here use data from a RCB experiment to evaluate forms of urea and their mode of application in wetland rice. In an experiment with nine treatments in four replications, data on tiller count, collected from four randomly selected 2×2-hill sampling units per plot, is shown in Table 6.4.

Let t denote the number of treatments, r the number of replications, and s the number of sampling units selected per plot. The steps to compute analysis of variance are:

☐ STEP 1. Construct an appropriate outline of the analysis of variance of data from plot sampling based on the experimental design used. For this example, the form of the analysis of variance is shown in Table 6.2.

☐ STEP 2. Construct the replication × treatment table of totals (RT) and compute the replication totals (R), the treatment totals (T), and the grand total (G). For our example, such a table is shown in Table 6.5.

☐ STEP 3. Compute the correction factor and the sums of squares as:

$$C.F. = \frac{G^2}{trs}$$

$$= \frac{(7,723)^2}{(9)(4)(4)} = 414,199.51$$

Table 6.4 Tiller Count (no./4 hills) of Rice Variety IR729-67-3, Tested under Nine Fertilizer Treatments in a RCB Experiment with Four Replications and Four Sampling Units (S_1, S_2, S_3, and S_4)

Treatment Number	Rep. I				Rep. II				Rep. III				Rep. IV			
	S_1	S_2	S_3	S_4	S_1	S_2	S_3	S_4	S_1	S_2	S_3	S_4	S_1	S_2	S_3	S_4
1	30	23	27	22	22	26	25	32	34	26	30	24	40	42	37	26
2	48	46	33	42	57	60	38	50	67	64	63	58	40	57	36	60
3	52	47	61	46	49	41	43	70	52	48	54	56	50	61	58	74
4	45	51	73	55	65	62	79	54	75	56	75	75	58	41	47	58
5	52	62	56	52	50	72	51	51	56	39	49	59	53	53	40	72
6	62	63	56	43	52	48	54	56	74	58	48	51	63	59	46	52
7	58	46	63	55	47	50	70	53	75	48	73	52	66	76	72	74
8	63	56	59	49	47	53	60	68	47	58	65	78	63	70	80	68
9	70	72	72	49	55	44	42	52	69	55	56	59	53	52	44	49

Table 6.5 The Replication × Treatment Table of Totals Computed from Data in Table 6.4

Treatment Number	Tiller Count Total (RT)				Treatment Total (T)
	Rep. I	Rep. II	Rep. III	Rep. IV	
1	102	105	114	145	466
2	169	205	252	193	819
3	206	203	210	243	862
4	224	260	281	204	969
5	222	224	203	218	867
6	224	210	231	220	885
7	222	220	248	288	978
8	227	228	248	281	984
9	263	193	239	198	893
Rep. total (R)	1,859	1,848	2,026	1,990	
Grand total (G)					7,723

Total $SS = \sum X^2 - C.F.$

$$= \left[(30)^2 + (23)^2 + \cdots + (49)^2 \right] - 414{,}199.51$$

$$= 25{,}323.49$$

Replication $SS = \dfrac{\sum R^2}{ts} - C.F.$

$$= \frac{(1{,}859)^2 + (1{,}848)^2 + (2{,}026)^2 + (1{,}990)^2}{(9)(4)} - 414{,}199.51$$

$$= 682.74$$

Treatment $SS = \dfrac{\sum T^2}{rs} - C.F.$

$$= \frac{(466)^2 + (819)^2 + \cdots + (893)^2}{(4)(4)} - 414{,}199.51$$

$$= 12{,}489.55$$

Experimental error $SS = \dfrac{\sum (RT)^2}{s}$

$\qquad \qquad$ (.F. - Replication SS - Treatment SS

$\qquad = \dfrac{(102)^2 + (105)^2 + (114)^2 + \cdots + (198)^2}{4}$

$\qquad \qquad - 414{,}199.51 - 682.74 - 12{,}489.55$

$\qquad = 3{,}882.95$

Sampling error $SS = $ Total $SS - ($ sum of all other $SS)$

$\qquad \qquad = 25{,}323.49 - (682.74 + 12{,}489.55 + 3{,}882.95)$

$\qquad \qquad = 8{,}268.25$

☐ STEP 4.   For each source of variation, compute the mean square by dividing the SS by its corresponding $d.f.$:

$$\text{Replication } MS = \frac{\text{Replication } SS}{r - 1}$$

$$= \frac{682.74}{3} = 227.58$$

$$\text{Treatment } MS = \frac{\text{Treatment } SS}{t - 1}$$

$$= \frac{12{,}489.55}{8} = 1{,}561.19$$

$$\text{Experimental error } MS = \frac{\text{Experimental error } SS}{(r - 1)(t - 1)}$$

$$= \frac{3{,}882.95}{(3)(8)} = 161.79$$

$$\text{Sampling error } MS = \frac{\text{Sampling error } SS}{tr(s - 1)}$$

$$= \frac{8{,}268.25}{(9)(4)(3)} = 76.56$$

☐ STEP 5.   To test the significance of the treatment effect, compute the F value as:

$$F = \frac{\text{Treatment } MS}{\text{Experimental error } MS}$$

$$= \frac{1{,}561.19}{161.79} = 9.65$$

and compare it with the tabular F value (Appendix E) with $f_1 = (t - 1) = 8$ and $f_2 = (r - 1)(t - 1) = 24$ degrees of freedom, at the prescribed level of significance.

For our example, the computed F value of 9.65 is greater than the tabular F value with $f_1 = 8$ and $f_2 = 24$ degrees of freedom at the 1% level of significance of 3.36. Hence, the treatment difference is significant at the 1% level of significance.

☐ STEP 6.   Enter all values obtained in steps 2 to 5 in the analysis of variance outline of step 1. The final result is shown in Table 6.6.

☐ STEP 7.   For mean comparison, compute the standard error of the difference between the ith and jth treatments as:

$$s_{\bar{d}} = \sqrt{\frac{2(MS_2)}{rs}}$$

where MS_2 is the experimental error MS in the analysis of variance. For our example, the standard error of the difference between any pair of treatments is:

$$s_{\bar{d}} = \sqrt{\frac{2(161.79)}{(4)(4)}} = 4.50$$

Table 6.6   Analysis of Variance (RCB with Data from Plot Sampling) of Data in Table 6.4[a]

Source of Variation	Degree of Freedom	Sum of Squares	Mean Square	Computed F[b]	Tabular F 5%	1%
Replication	3	682.74	227.58			
Treatment	8	12,489.55	1,561.19	9.65**	2.36	3.36
Experimental error	24	3,882.95	161.79			
Sampling error	108	8,268.25	76.56			
Total	143	25,323.49				

[a]$cv = 23.6\%$.
[b]** = significant at 1% level.

☐ STEP 8. Compute the estimates of the sampling error variance and of the experimental error variance as;

$$s_S^2 = MS_1$$

$$s_E^2 = \frac{MS_2 - MS_1}{s}$$

where MS_1 is the sampling error MS in the analysis of variance. For our example, the two variance estimates and their corresponding cv values are:

$$s_S^2 = 76.56$$

$$s_E^2 = \frac{161.79 - 76.56}{4} = 21.31$$

$$cv(S) = \frac{\sqrt{76.56}}{54} \times 100 = 16.2\%$$

$$cv(E) = \frac{\sqrt{21.31}}{54} \times 100 = 8.5\%$$

For examples on the use of variance estimates in the development of sampling techniques, see Chapter 15, Section 15.2.

6.1.2 Split-Plot Design

The computational procedure we show below uses data from a split-plot experiment involving eight management levels as main-plot treatments and four times of nitrogen application as subplot treatments. There are three replications. The data on plant height, measured on two single-hill sampling units per plot, is shown in Table 6.7.

We denote the main-plot factor by A, the subplot factor by B, the levels of factor A by a, the levels of factor B by b, the number of replications by r, and the number of sampling units per plot by s. The computational procedures are:

☐ STEP 1. Construct the outline of an appropriate analysis of variance of data from plot sampling based on the experimental design used. For our example, the form of the analysis of variance is shown in Table 6.3.

☐ STEP 2. Construct three tables of totals:
- The replication × factor A two-way table of totals (RA), including the replication totals (R), factor A totals (A), and the grand total (G). For our example, this is shown in Table 6.8.

Table 6.7 Height of Rice Plants Measured on Two Sampling Units (S_1 and S_2) per Plot, from a Split-Plot Experiment Involving Eight Management Levels ($M_1, M_2,..., M_8$) and Four Times of Nitrogen Application ($T_1, T_2, T_3,$ and T_4) with Three Replications

Treatment Combination		Plant Height, cm					
Time of Application	Management Level	Rep. I		Rep. II		Rep. III	
		S_1	S_2	S_1	S_2	S_1	S_2
T_1	M_1	104.5	106.5	112.3	109.0	109.2	106.7
	M_2	92.3	92.0	113.3	109.6	108.0	106.3
	M_3	96.8	95.5	108.3	110.2	102.4	103.2
	M_4	94.7	94.4	108.1	107.0	102.5	104.4
	M_5	105.7	103.0	104.9	102.4	100.8	101.3
	M_6	100.5	102.0	106.3	104.5	106.0	108.4
	M_7	86.0	89.0	105.0	102.0	104.0	103.7
	M_8	105.9	104.6	108.9	105.8	95.8	99.2
T_2	M_1	109.7	112.2	110.3	108.0	113.6	113.5
	M_2	100.5	100.0	113.5	112.5	103.6	102.0
	M_3	91.4	92.0	109.2	106.2	113.0	111.9
	M_4	100.8	103.2	115.0	112.0	109.6	108.2
	M_5	97.0	96.1	105.1	102.3	116.3	114.3
	M_6	102.3	100.0	105.2	108.2	115.5	118.8
	M_7	100.3	100.8	97.5	96.3	100.0	102.3
	M_8	102.7	102.5	104.3	107.5	106.8	107.6
T_3	M_1	97.5	95.2	107.6	106.2	113.2	115.0
	M_2	95.0	96.2	102.5	105.8	106.7	104.6
	M_3	86.6	85.5	104.1	102.3	105.0	105.3
	M_4	91.2	90.0	108.1	105.6	103.8	104.3
	M_5	100.0	100.0	99.2	101.3	100.1	98.6
	M_6	94.4	93.5	96.2	96.0	106.1	104.4
	M_7	92.3	93.4	98.1	96.2	102.0	100.6
	M_8	101.9	103.0	104.3	106.4	94.2	92.0
T_4	M_1	103.8	105.0	110.1	109.5	115.0	112.5
	M_2	93.2	92.5	111.0	111.3	96.0	97.2
	M_3	95.0	95.2	108.1	106.0	107.2	107.4
	M_4	103.9	103.6	112.0	109.3	117.6	119.5
	M_5	96.0	93.5	102.5	103.8	108.0	107.2
	M_6	102.3	102.8	111.7	110.5	107.5	107.3
	M_7	91.2	93.0	99.5	97.8	104.3	103.3
	M_8	106.0	106.4	100.0	103.0	104.9	106.4

Table 6.8 The Replication × Management Level Table of Totals Computed from Data in Table 6.7

Management Level	Plant Height Total (*RA*)			Management Total (*A*)
	Rep. I	Rep. II	Rep. III	
M_1	834.4	873.0	898.7	2,606.1
M_2	761.7	879.5	824.4	2,465.6
M_3	738.0	854.4	855.4	2,447.8
M_4	781.8	877.1	869.9	2,528.8
M_5	791.3	821.5	846.6	2,459.4
M_6	797.8	838.6	874.0	2,510.4
M_7	746.0	792.4	820.2	2,358.6
M_8	833.0	840.2	806.9	2,480.1
Rep. total (*R*)	6,284.0	6,776.7	6,796.1	
Grand total (*G*)				19,856.8

- The factor A × factor B two-way table of totals (AB) including factor B totals (B). For our example, this is shown in Table 6.9.
- The replication × factor A × factor B three-way table of totals (RAB). For our example, this is shown in Table 6.10.

☐ STEP 3. Compute the correction factor and the sums of squares:

$$C.F. = \frac{G^2}{rabs}$$

$$= \frac{(19,856.8)^2}{(3)(8)(4)(2)} = 2,053,606.803$$

Table 6.9 The Management Level × Time of Nitrogen Application Table of Totals Computed from Data in Table 6.7

Management Level	Plant Height Total (*AB*)			
	T_1	T_2	T_3	T_4
M_1	648.2	667.3	634.7	655.9
M_2	621.5	632.1	610.8	601.2
M_3	616.4	623.7	588.8	618.9
M_4	611.1	648.8	603.0	665.9
M_5	618.1	631.1	599.2	611.0
M_6	627.7	650.0	590.6	642.1
M_7	589.7	597.2	582.6	589.1
M_8	620.2	631.4	601.8	626.7
Time total (*B*)	4,952.9	5,081.6	4,811.5	5,010.8

$$\text{Total } SS = \sum X^2 - C.F.$$

$$= \left[(104.5)^2 + (106.5)^2 + \cdots + (106.4)^2\right]$$

$$- 2,053,606.803$$

$$= 8,555.477$$

$$\text{Replication } SS = \frac{\sum R^2}{abs} - C.F.$$

$$= \frac{(6,284.0)^2 + (6,776.7)^2 + (6,796.1)^2}{(8)(4)(2)}$$

$$- 2,053,606.803$$

$$= 2,632.167$$

$$A \text{ (management level) } SS = \frac{\sum A^2}{rbs} - C.F.$$

$$= \frac{(2,606.1)^2 + (2,465.6)^2 + \cdots + (2,480.1)^2}{(3)(4)(2)}$$

$$- 2,053,606.803$$

$$= 1,482.419$$

$$\text{Error}(a) \, SS = \frac{\sum (RA)^2}{bs} - C.F. - \text{Replication } SS - A \, SS$$

$$= \frac{(834.4)^2 + (873.0)^2 + \cdots + (806.9)^2}{(4)(2)}$$

$$- 2,053,606.803 - 2,632.167 - 1,482.419$$

$$= 1,324.296$$

$$B \text{ (time of application) } SS = \frac{\sum B^2}{ras} - C.F.$$

$$= \frac{(4,952.9)^2 + (5,081.6)^2 + (4,811.5)^2 + (5,010.8)^2}{(3)(8)(2)}$$

$$- 2,053,606.803$$

$$= 820.819$$

Table 6.10 The Replication × Management Level × Time of Nitrogen Application Table of Totals Computed from Data in Table 6.7

Management Level	Time of Application	Plant Height Total (RAB)		
		Rep. I	Rep. II	Rep. III
M_1	T_1	211.0	221.3	215.9
	T_2	221.9	218.3	227.1
	T_3	192.7	213.8	228.2
	T_4	208.8	219.6	227.5
M_2	T_1	184.3	222.9	214.3
	T_2	200.5	226.0	205.6
	T_3	191.2	208.3	211.3
	T_4	185.7	222.3	193.2
M_3	T_1	192.3	218.5	205.6
	T_2	183.4	215.4	224.9
	T_3	172.1	206.4	210.3
	T_4	190.2	214.1	214.6
M_4	T_1	189.1	215.1	206.9
	T_2	204.0	227.0	217.8
	T_3	181.2	213.7	208.1
	T_4	207.5	221.3	237.1
M_5	T_1	208.7	207.3	202.1
	T_2	193.1	207.4	230.6
	T_3	200.0	200.5	198.7
	T_4	189.5	206.3	215.2
M_6	T_1	202.5	210.8	214.4
	T_2	202.3	213.4	234.3
	T_3	187.9	192.2	210.5
	T_4	205.1	222.2	214.8
M_7	T_1	175.0	207.0	207.7
	T_2	201.1	193.8	202.3
	T_3	185.7	194.3	202.6
	T_4	184.2	197.3	207.6
M_8	T_1	210.5	214.7	195.0
	T_2	205.2	211.8	214.4
	T_3	204.9	210.7	186.2
	T_4	212.4	203.0	211.3

$$A \times B\ SS = \frac{\sum(AB)^2}{rs} - C.F. - B\ SS - A\ SS$$

$$= \frac{(648.2)^2 + (667.3)^2 + \cdots + (626.7)^2}{(3)(2)}$$

$$- 2{,}053{,}606.803 - 820.819 - 1{,}482.419$$

$$= 475.305$$

$$\text{Error}(b)\ SS = \frac{\sum(RAB)^2}{s} - C.F. - \text{Replication } SS - A\ SS$$

$$- \text{Error}(a)\ SS - B\ SS - A \times B\ SS$$

$$= \frac{(211.0)^2 + (221.3)^2 + \cdots + (211.3)^2}{2}$$

$$-2{,}053{,}606.803 - 2{,}632.167 - 1{,}482.419$$

$$-1{,}324.296 - 820.819 - 475.305$$

$$= 1{,}653.081$$

Sampling error $SS = \text{Total } SS - (\text{sum of all other } SS)$

$$= 8{,}555.477 - (2{,}632.167 + 1{,}482.419 + 1{,}324.296$$

$$+ 820.819 + 475.305 + 1{,}653.081)$$

$$= 167.390$$

☐ STEP 4. For each source of variation, compute the mean square by dividing the SS by its corresponding $d.f.$:

$$\text{Replication } MS = \frac{\text{Replication } SS}{r - 1}$$

$$= \frac{2{,}632.167}{2} = 1{,}316.084$$

$$A\ MS = \frac{A\ SS}{a - 1}$$

$$= \frac{1{,}482.419}{7} = 211.774$$

$$\text{Error}(a)\ MS = \frac{\text{Error}(a)\ SS}{(r - 1)(a - 1)}$$

$$= \frac{1{,}324.296}{(2)(7)} = 94.593$$

$$B\ MS = \frac{B\ SS}{b - 1}$$

$$= \frac{820.819}{3} = 273.606$$

$$A \times B \; MS = \frac{A \times B \; SS}{(a - 1)(b - 1)}$$

$$= \frac{475.305}{(7)(3)} = 22.634$$

$$\text{Error}(b) \; MS = \frac{\text{Error}(b) \; SS}{a(r - 1)(b - 1)}$$

$$= \frac{1,653.081}{(8)(2)(3)} = 34.439$$

$$\text{Sampling error } MS = \frac{\text{Sampling error } SS}{abr(s - 1)}$$

$$= \frac{167.390}{(8)(4)(3)(1)} = 1.744$$

☐ STEP 5. To test the significance of each of the three effects, namely, A, B, and $A \times B$, follow the procedures outlined in Chapter 3, Section 3.4.2, steps 6 to 9.

☐ STEP 6. Enter all values obtained in steps 2 to 5 in the analysis of variance outline of step 1. The final results are shown in Table 6.11. The results indicate that only the main effect of the time of nitrogen application is significant.

☐ STEP 7. For pair comparison, compute the standard error of the mean difference following the appropriate formula given in Chapter 5, Table 5.5, but with one modification—multiply each divisor by s, the sample size in each plot. For example, to compare two subplot treatments at the same main-plot treatment, the standard error of the mean difference is:

$$s_{\bar{d}} = \sqrt{\frac{2E_b}{rs}}$$

where E_b is the error(b) MS from the analysis of variance. And, to compare two subplot treatments (averaged over all main-plot treatments) the standard error of the mean difference is:

$$s_{\bar{d}} = \sqrt{\frac{2E_b}{ras}}$$

For our example, the standard error of the mean difference between any two

Table 6.11 Analysis of Variance (Split-Plot Design with Data from Plot Sampling) of Data in Table 6.7[a]

Source of Variation	Degree of Freedom	Sum of Squares	Mean Square	Computed F^b	Tabular F 5%	Tabular F 1%
Replication	2	2,632.167	1,316.084			
Management level (A)	7	1,482.419	211.774	2.24^{ns}	2.77	4.28
Error(a)	14	1,324.296	94.593			
Time of application (B)	3	820.819	273.606	7.94**	2.80	4.22
A × B	21	475.305	22.634	<1	—	—
Error(b)	48	1,653.081	34.439			
Sampling error	96	167.390	1.744			
Total	191	8,555.477				

[a] $cv(a) = 9.4\%$, $cv(b) = 5.7\%$.
[b] ** = significant at 1% level, ns = not significant.

times of nitrogen application at the same management level is computed as:

$$s_{\bar{d}} = \sqrt{\frac{2(34.439)}{(3)(2)}}$$

$$= 3.388$$

And the standard error of the mean difference between any two times of nitrogen application averaged over all management levels is computed as:

$$s_{\bar{d}} = \sqrt{\frac{2(34.439)}{(3)(8)(2)}}$$

$$= 1.198$$

☐ STEP 8.   Compute the estimates of two variance components: the experimental error associated with the smallest experimental unit [i.e., error(b) in this case] and the sampling error as:

$$s_E^2 = \frac{MS_2 - MS_1}{s}$$

$$s_S^2 = MS_1$$

where MS_1 is the sampling error MS, MS_2 is the experimental error MS, and s is the number of sampling units per plot. For our example, the two variance estimates are computed as:

$$s_E^2 = \frac{34.439 - 1.744}{2} = 16.348$$

$$s_S^2 = 1.744$$

with the corresponding cv values of

$$cv(E) = \frac{\sqrt{16.348}}{103.4} \times 100 = 3.9\%$$

$$cv(S) = \frac{\sqrt{1.744}}{103.4} \times 100 = 1.3\%$$

The results indicate a relatively small sampling error compared to the experimental error.

For further details in the use of the variance estimates for developing sampling techniques, see Chapter 15, Section 15.2.

6.2 MEASUREMENT OVER TIME

When a character in an experiment is measured over time, the researcher is usually interested in examining the rate of change from one time period to another. For example, when a researcher measures weight of dry matter of rice plants at different growth stages, interest is usually on the effects of treatment on the growth pattern (or the rate of change over time) based on weight of dry matter (Figure 6.1) rather than on the effects of treatment on weight of dry matter at the individual growth stage. In other words, it is important to determine the interaction effect between treatment and stage of observation, but that cannot be done if the analysis of variance is obtained separately for each stage of observation. Hence, the common approach is to combine data from all stages of observation and obtain a single analysis of variance.

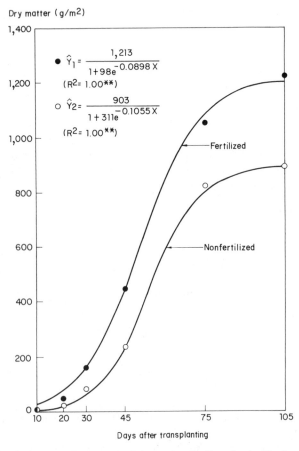

Figure 6.1 Growth response (dry matter) of rice variety IR48 under fertilized and nonfertilized conditions.

Table 6.12 Format of the Pooled Analysis of Variance for Measurements over Time, from a RCB Design

Source of Variation	Degree of Freedom[a]	Sum of Squares	Mean Square	Computed F	Tabular F 5%	1%
Replication	$r - 1$					
Treatment (T)	$t - 1$					
Error(a)	$(r - 1)(t - 1)$					
Time of observation (P)	$p - 1$					
$T \times P$	$(t - 1)(p - 1)$					
Error(b)	$t(r - 1)(p - 1)$					
Total	$rtp - 1$					

[a] r = number of replications, t = number of treatments, and p = number of times of observation.

Such an analysis of variance is accomplished by considering *time of observation* as an additional factor in the experiment and treating it as if it were a *subplot* or the smallest experimental unit. Thus, the format of the pooled analysis of variance for measurements over time based on a RCB design, shown in Table 6.12, is similar to that for the standard split-plot design with treatments as main-plot and times of observation as subplot treatments. The format of the pooled analysis of variance for measurements over time based on a split-plot design, shown in Table 6.13, is similar to that for the standard split-split-plot design, and so on.

Table 6.13 Format of the Pooled Analysis of Variance for Measurements over Time, from a Split-Plot Design

Source of Variation	Degree of Freedom[a]	Sum of Squares	Mean Square	Computed F	Tabular F 5%	1%
Replication	$r - 1$					
Main-plot treatment (A)	$a - 1$					
Error(a)	$(r - 1)(a - 1)$					
Subplot treatment (B)	$b - 1$					
$A \times B$	$(a - 1)(b - 1)$					
Error(b)	$a(r - 1)(b - 1)$					
Time of observation (C)	$p - 1$					
$A \times C$	$(a - 1)(p - 1)$					
$B \times C$	$(b - 1)(p - 1)$					
$A \times B \times C$	$(a - 1)(b - 1)(p - 1)$					
Error(c)	$ab(r - 1)(p - 1)$					
Total	$rabp - 1$					

[a] r = number of replications, a = number of main-plot treatments, b = number of subplot treatments, and p = number of times of observation.

6.2.1 RCB Design

Our computations here use data from a RCB experiment to test nitrogen fertilizer efficiency in medium-deepwater rice plots. Eight fertilizer treatments were tested in four replications. Data on nitrogen content of the soil, collected at three growth stages of the rice crop, are shown in Table 6.14.

Let t denote the number of treatments, r the number of replications, and p the number of times data were collected from each plot. The step-by-step procedures for data analysis are:

☐ STEP 1. Compute an analysis of variance for each of the p stages of observation, following the procedure for standard analysis of variance based on the experimental design used. For our example, the design is RCB. Following the procedure described in Chapter 2, Section 2.2.3, the $p = 3$ analyses of variance are computed and the results are shown in Table 6.15.

☐ STEP 2. Test the homogeneity of the p error variances, following the procedure of Chapter 11, Section 11.2. For our example, the chi-square test for homogeneity of variance is applied to the three error mean squares:

- Compute the χ^2 value as:

$$\chi^2 = \frac{(2.3026)(f)\left(p \log s_p^2 - \sum \log s_i^2\right)}{1 + \dfrac{(p+1)}{3pf}}$$

$$= \frac{(2.3026)(21)\left[3 \log 0.03507 - (-4.38948)\right]}{1 + \dfrac{(3+1)}{3(3)(21)}}$$

$$= 1.15$$

- Compare the computed χ^2 value to the tabular χ^2 value, with $(p - 1) = 2$ degrees of freedom. Because the computed χ^2 value is smaller than the corresponding tabular χ^2 value at the 5% level of 5.99, heterogeneity of variance is not indicated.

☐ STEP 3. Based on the result of the test for homogeneity of variance of step 2, apply the appropriate analysis of variance:

- If heterogeneity of variance is indicated, choose an appropriate data transformation (see Chapter 7, Section 7.2.2.1) that can stabilize the error variances and compute the pooled analysis of variance based on the transformed data. For our example, because heterogeneity of error variance is not indicated, no data transformation is needed. (For an example

Table 6.14 Data on Nitrogen Content of the Soil, Subjected to Eight Fertilizer Treatments in a RCB Design with Four Replications, Collected at Three Growth Stages[a] of the Rice Crop

| Treatment Number | Soil Nitrogen Content, % | | | | | | | | | | | |
| | Rep. I | | | Rep. II | | | Rep. III | | | Rep. IV | | |
	P_1	P_2	P_3	P_1	P_2	P_3	P_1	P_2	P_3	P_1	P_2	P_3
1	3.26	1.88	1.40	2.98	1.74	1.24	2.78	1.76	1.44	2.77	2.00	1.25
2	3.84	2.36	1.53	3.74	2.14	1.21	3.09	1.75	1.28	3.36	1.57	1.17
3	3.50	2.20	1.33	3.49	2.28	1.54	3.03	2.48	1.46	3.36	2.47	1.41
4	3.43	2.32	1.61	3.45	2.33	1.33	2.81	2.16	1.40	3.32	1.99	1.12
5	3.43	1.98	1.11	3.24	1.70	1.25	3.45	1.78	1.39	3.09	1.74	1.20
6	3.68	2.01	1.26	3.24	2.33	1.44	2.84	2.22	1.12	2.91	2.00	1.24
7	2.97	2.66	1.87	2.90	2.74	1.81	2.92	2.67	1.31	2.42	2.98	1.56
8	3.11	2.53	1.76	3.04	2.22	1.28	3.20	2.61	1.23	2.81	2.22	1.29

[a]At 15 days after transplanting (P_1), at 40 days after transplanting (P_2), and at panicle initiation stage (P_3).

on pooled analysis of variance based on transformed data, see Section 6.2.2).

- If heterogeneity of error variance is not indicated, compute a pooled analysis of variance based on original data from all p stages of observation. For our example, because heterogeneity of error variance is not indicated, the pooled analysis of variance is computed.

The general format of the pooled analysis of variance for measurements over time from a RCB design is outlined in Table 6.12. The required computational procedure follows that for a standard analysis of variance of a split-plot design described in Chapter 3, Section 3.4.2, with the t treatments treated as the a main-plot treatments and the p times of observation as the b subplot treatments.

Table 6.15 Three Individual Analyses of Variance (RCB Design), One for Each Stage of Observation, of Data in Table 6.14

| Source of Variation | Degree of Freedom | Mean Square[a] | | |
		P_1	P_2	P_3
Replication	3	0.302412	0.019654	0.061458
Treatment	7	0.211634**	0.415005**	0.063686[ns]
Error	21	0.039843	0.039356	0.026011
Total	31			

[a]** = F test significant at 1% level, [ns] = F test not significant.

The result of the pooled analysis of variance, for our example, is shown in Table 6.16. The results show a highly significant interaction between treatment and growth stage, indicating that the treatment effects varied significantly among the different growth stages.

☐ STEP 4.   For pair comparison, follow the standard procedure for a split-plot design (see Chapter 5, Section 5.1.1.4) based on the result of the pooled analysis of variance obtained in step 3.

For our example, because the interaction between treatment and growth stage is highly significant (Table 6.16), pair comparison should be made only between treatments at the same growth stage. The standard error of the difference between two treatments at the same growth stage is computed as:

$$s_{\bar{d}} = \sqrt{\frac{2[(p-1)E_b + E_a]}{rp}}$$

$$= \sqrt{\frac{2[(2)(0.036349) + 0.036642]}{(4)(3)}}$$

$$= 0.13$$

where E_a is the error(a) MS and E_b is the error(b) MS in the pooled analysis of variance.

The result of the Duncan's multiple range test (see Chapter 5, Section 5.1.2) for comparing treatment means at each growth stage is shown in Table 6.17.

Table 6.16   Pooled Analysis of Variance for Measurements over Time (RCB Design), from Data in Table 6.14[a]

Source of Variation	Degree of Freedom	Sum of Squares	Mean Square	Computed F[b]	Tabular F 5%	Tabular F 1%
Replication	3	0.845742	0.281914			
Treatment (T)	7	1.265833	0.180833	4.94**	2.49	3.65
Error(a)	21	0.769492	0.036642			
Growth stage (P)	2	52.042858	26.021429	715.88**	3.19	5.08
$T \times P$	14	3.566442	0.254746	7.01**	1.90	2.48
Error(b)	48	1.744767	0.036349			
Total	95	60.235134				

[a]$cv(a) = 8.5\%$, $cv(b) = 8.5\%$.
[b]** = significant at 1% level.

Table 6.17 Duncan's Multiple Range Test (DMRT) for Comparing Eight Fertilizer Treatment Means at Each Growth Stage, Computed from Data in Tables 6.14 and 6.16

Treatment Number	Mean Nitrogen Content, %[a]		
	P_1	P_2	P_3
1	2.95 de	1.85 de	1.33 b
2	3.51 a	1.96 cde	1.30 b
3	3.35 ab	2.36 b	1.44 ab
4	3.25 abcd	2.20 bc	1.37 ab
5	3.30 abc	1.80 e	1.24 b
6	3.17 bcd	2.14 bcd	1.27 b
7	2.80 e	2.76 a	1.64 a
8	3.04 cde	2.40 b	1.39 ab

[a]Average of four replications. In a column, means followed by a common letter are not significantly different at the 5% level.

☐ STEP 5. If the interaction between treatment and time of observation is significant, apply an appropriate mean-comparison method to examine the nature of the interaction. The choice of the methods to be used depends on whether the time of observation is a quantitative factor.

If the time of observation is quantitative in nature (such as the chronological age of the crop expressed in terms of the number of days after planting, the age of the confined insects, etc.) and there are at least three stages involved, either one, or both, of the following procedures can be applied:

A. Partition the interaction SS, based on an appropriate choice of trend comparison on the time of observation (see Chapter 5, Section 5.2.3).

B. Apply an appropriate regression technique of Chapter 9 to estimate a functional relationship between the response (Y) and the time of observation (X) separately for each treatment (see, for example, Figure 6.1) and compare these regressions across treatments.

If the time of observation is not quantitative, or if there are less than three stages of observation involved, partition the interaction SS, based on an appropriate set of between-group comparisons either on the treatments or on the growth stages, or both (see Chapter 5, Section 5.2.1).

For our example, because the interaction between treatment and time of observation is significant, the interaction SS should be properly examined. Because the time of observation is represented by a discrete set of crop growth stages (i.e., with one stage of observation representing chronological

Table 6.18   Partitioning of the Treatment × Growth Stage Interaction *SS* in the Analysis of Variance Shown in Table 6.16

Source of Variation	Degree of Freedom	Sum of Squares	Mean Square	Computed F^a	Tabular F 5%	1%
$T \times P$	14	3.566442	0.254746	7.01**	1.90	2.48
(T_7 vs. T_1 to T_6 and T_8) $\times P$	(2)	(2.136675)	(1.068338)	29.39**	3.19	5.08
(T_8 vs. T_1 to T_6) $\times P$	(2)	(0.536011)	(0.268006)	7.37**	3.19	5.08
(T_2, T_5 vs. T_1, T_3, T_4, T_6) $\times P$	(2)	(0.642872)	(0.321436)	8.84**	3.19	5.08
(T_2 vs. T_5) $\times P$	(2)	(0.021700)	(0.010850)	< 1	—	—
(T_1, T_3, T_4, T_6) $\times P$	(6)	(0.229184)	(0.038197)	1.05^{ns}	2.30	3.20
Error(b)	48	1.744767	0.036349			

a** = significant at 1% level, ^{ns} = not significant.

age and another representing physiological age of the crop), the between-group comparisons on treatments or on growth stages should be made. We choose to partition the interaction *SS* based on between-group comparisons on treatments. The results are shown in Table 6.18. Results indicate that the eight fertilizer treatments can be classified, based on similarity of the changes in nitrogen content over time, into four groups: the first group consists of T_7, the second consists of T_8, the third consists of T_2 and T_5, and the fourth consists of T_1, T_3, T_4, and T_6.

6.2.2   Split-Plot Design

We show the computations with data from a split-plot experiment, which sought the optimum time of herbicide application (main-plot treatment) in relation to the application of a protectant (subplot treatment) in wet-seeded rice. Data on plant height, measured at three growth stages of the rice crop, are shown in Table 6.19.

Let A denote the main-plot factor, B the subplot factor, a the levels of factor A, b the levels of factor B, r the number of replications, and p the number of times that data were collected from each plot. The step-by-step procedures for data analysis are:

□ STEP 1.   Compute a standard analysis of variance for each of the p stages of observation, based on the experimental design used. For our example, following the procedure for standard analysis of variance of a split-plot design described in Chapter 3, Section 3.4.2, the $p = 3$ analyses of variance, one for each stage of observation, are shown in Table 6.20.

□ STEP 2.   Test the homogeneity of the p error(b) *MS*, following the procedure of Chapter 11, Section 11.2. For our example, the chi-square test for

Table 6.19 Height of Rice Plants, Subjected to Different Herbicide and Protectant Treatments in a Split-Plot Design, Measured at Three Growth Stages[a]

| | Treatment | | Plant Height, cm | | | | | | | | | | | |
| | Time of Herbicide Application[b] | Protectant Application | Rep I | | | Rep II | | | Rep III | | | Rep IV | | |
Number			P_1	P_2	P_3	P_1	P_2	P_3	P_1	P_2	P_3	P_1	P_2	P_3
1	6 DBS (A_1)	Yes (B_1)	15.8	29.3	52.1	16.0	29.2	52.7	15.3	29.7	58.0	15.0	30.1	63.2
2		No (B_2)	15.3	30.0	52.5	16.5	28.9	50.9	16.3	30.8	55.0	15.9	32.0	62.1
3	3 DBS (A_2)	Yes	14.9	32.3	65.2	14.8	30.3	54.7	14.8	28.5	53.8	14.7	29.4	58.5
4		No	14.8	31.8	56.0	14.2	31.9	57.1	14.4	27.2	48.9	13.6	29.1	57.3
5	0 DAS (A_3)	Yes	13.0	29.8	60.1	13.2	29.5	57.6	13.3	29.9	56.8	15.2	28.4	54.9
6		No	12.8	30.3	56.4	16.5	32.5	49.8	13.8	29.0	57.5	15.1	28.9	55.2
7	3 DAS (A_4)	Yes	14.8	29.1	53.4	14.7	29.6	54.4	15.2	30.5	64.0	14.2	29.4	61.7
8		No	13.1	31.2	58.1	14.3	29.9	52.2	12.1	30.2	65.2	13.6	30.5	59.0
9	6 DAS (A_5)	Yes	12.0	28.2	57.1	14.8	30.3	58.8	13.9	27.2	51.3	13.2	31.2	61.5
10		No	12.0	30.1	58.8	11.2	28.1	57.8	12.6	29.6	56.2	13.2	28.8	50.9
11	10 DAS (A_6)	Yes	15.0	29.9	54.9	13.2	30.0	61.0	13.8	30.7	59.9	14.5	29.7	59.9
12		No	14.7	28.9	53.9	13.0	28.4	58.3	13.6	32.6	60.7	13.6	28.8	60.3
13	No application (A_7)	Yes	15.1	29.8	59.5	13.9	30.6	60.5	15.5	28.9	60.7	13.3	29.7	58.7
14		No	15.3	30.2	57.2	13.7	33.9	58.8	15.2	29.3	59.9	14.2	29.6	50.9

[a] P_1, P_2, and P_3 refer to 14, 25, and 50 days after transplanting, respectively.

[b] DBS = days before seeding; DAS = days after seeding.

263

Table 6.20   Three Individual Analyses of Variance (Split-Plot Design), One for Each Stage of Observation, for Data in Table 6.19

Source of Variation	Degree of Freedom	Mean Square[a]		
		P_1	P_2	P_3
Replication	3	0.027798	1.388750	32.538333
Herbicide (A)	6	6.067857**	0.906429[ns]	16.715357[ns]
Error(a)	18	1.030714	2.807222	23.831944
Protectant (B)	1	1.290179[ns]	2.361607[ns]	14.000000[ns]
$A \times B$	6	1.478095[ns]	0.666190[ns]	4.873333[ns]
Error(b)	21	0.584583	1.255536	4.875238
Total	55			

[a]P_1, P_2, and P_3 refer to 14, 25, and 50 days after transplanting, respectively; ** $= F$ test significant at 1% level, [ns] $= F$ test not significant.

homogeneity of variance is applied to the three error(b) MS, as:

$$\chi^2 = \frac{(2.3026)(21)\left[3(\log 2.238452) - 0.553671\right]}{\left[1 + \dfrac{(3+1)}{(3)(3)(21)}\right]}$$

$$= 23.49$$

The result of the χ^2 test indicates a highly significant difference between error variances across the three growth stages. For data such as plant height, where the values are expected to differ greatly from one growth stage to another (as is the case with this example), the presence of heterogeneity of error variances is not unexpected.

☐ STEP 3.   Based on the result of the test for homogeneity of variance of step 2, apply the appropriate analysis of variance as follows:

·   If heterogeneity of variance is not indicated, compute a pooled analysis of variance using the original data from all p stages of observation.

·   If the heterogeneity of variance is indicated, choose an appropriate data transformation (see Chapter 7, Section 7.2.2.1) that can stabilize the error variances and compute a pooled analysis of variance based on the transformed data. For our example, the logarithmic transformation is applied. The form of the pooled analysis of variance for measurements over time from a split-plot design is shown in Table 6.13. The required computational procedure follows that for a standard analysis of variance of a split-split-plot design described in Chapter 4, Section 4.3.2, with the p times of observation treated as the c sub-subplot treatments. The final result of the pooled analysis of variance based on transformed data is shown in Table 6.21. The results show a highly significant interaction between herbicide treatment and growth stage, indicating that the effects

Table 6.21 Pooled Analysis of Variance for Measurements over Time (Split-Plot Design) Based on Transformed Data (Logarithmic Transformation) for Data in Table 6.19

Source of Variation	Degree of Freedom	Sum of Squares	Mean Square	Computed F^a	Tabular F 5%	1%
Replication	3	0.001082	0.000361			
Herbicide (A)	6	0.013471	0.002245	2.44^{ns}	2.66	4.01
Error(a)	18	0.016576	0.000921			
Protectant (B)	1	0.000700	0.000700	1.72^{ns}	4.32	8.02
$A \times B$	6	0.002675	0.000446	1.09^{ns}	2.57	3.81
Error(b)	21	0.008576	0.000408			
Growth stage (P)	2	10.325742	5.162871	$7,637.38^{**}$	3.11	4.87
$A \times P$	12	0.028467	0.002372	3.51^{**}	1.87	2.40
$B \times P$	2	0.002042	0.001021	1.51^{ns}	3.11	4.87
$A \times B \times P$	12	0.008564	0.000714	1.06^{ns}	1.87	2.40
Error(c)	84	0.056799	0.000676			
Total	167	10.464694				

$^a** $ = significant at 1% level, ^{ns} = not significant.

of herbicide treatment differed significantly between the three growth stages.

☐ STEP 4. Make pair comparison between treatment means by applying the standard procedure for a split-split-plot design (see Chapter 5, Section 5.1.1.6 or Chapter 7, Section 7.2.2.1) based on the result of the pooled analysis of variance obtained in step 3. For our example, because the pooled analysis of variance was based on transformed data, the procedure of Chapter 7, Section 7.2.2.1 should be followed. Because the only significant interaction effect is that between herbicide treatment and growth stage and the effect of the protectant is not significant, the appropriate pair comparison is one between herbicide means at each growth stage. For a pair comparison between two main-plot treatment means (averaged over all subplot treatments) at the same sub-subplot treatment, the standard error of the mean difference is computed, based on the formula in Table 5.10 of Chapter 5 as:

$$s_{\bar{d}} = \sqrt{\frac{2[(p-1)E_c + E_a]}{rbp}}$$

$$= \sqrt{\frac{2[(2)(0.000676) + 0.000921]}{(4)(2)(3)}}$$

$$= 0.01376$$

Table 6.22 Duncan's Multiple Range Test (DMRT) on Mean Plant Height of Seven Herbicide Treatments, at Each Growth Stage, Computed from Data in Tables 6.19 and 6.21

Herbicide Treatment	Mean Plant Height, cm[a]		
	P_1	P_2	P_3
A_1	15.8 a	30.0 a	55.8 a
A_2	14.5 b	30.1 a	56.4 a
A_3	14.1 b	29.8 a	56.0 a
A_4	14.0 b	30.0 a	58.5 a
A_5	12.9 c	29.2 a	57.8 a
A_6	13.9 b	29.9 a	58.6 a
A_7	14.5 b	30.2 a	59.5 a
Av.	14.2	29.9	57.5

[a]Average of two protectant treatments and four replications; P_1, P_2, and P_3 refer to 14, 24, and 50 days after transplanting, respectively. In each column, means followed by a common letter are not significantly different at the 5% level.

The result of the DMRT for comparing herbicide treatment means (averaged over the two protectant treatments), separately at each growth stage, is shown in Table 6.22.

☐ STEP 5.   Follow the procedure outlined in step 5 of Section 6.2.1. For our example, only the interaction between herbicide and growth stage is significant. Hence, only the SS of this interaction needs to be partitioned. Based on the results of the pair comparison between treatment means at each growth stage obtained in step 4 (Table 6.22), there was no significant difference between herbicide treatments at any of the two later growth stages. At the first growth stage, only the fifth herbicide treatment (i.e., application of herbicide at 6 DAS or A_5) and the first (i.e., application of herbicide at 6 DBS or A_1) gave results distinctly different from the rest of the herbicide treatments, with the tallest plants exhibited by A_1 and the shortest by A_5. Thus, an appropriate partitioning of the herbicide × growth stage interaction SS is that shown in Table 6.23. The result of the SS partitioning confirms the observation made.

6.3   MEASUREMENT OVER TIME WITH PLOT SAMPLING

When a character is measured at several stages of observation and plot sampling is done at each stage, the resulting set of data is referred to as

Table 6.23 Partitioning of the Herbicide × Growth Stage Interaction *SS* in the Analysis of Variance of Table 6.21

Source of Variation	Degree of Freedom	Sum of Squares	Mean Square	Computed F^a	Tabular F	
					5%	1%
$A \times P$	12	0.028467	0.002372	3.51**	1.87	2.40
$(A_5$ vs. A_1 to $A_4, A_6, A_7) \times (P_1$ vs. $P_2, P_3)$	(1)	(0.010174)	(0.010174)	15.05**	3.96	6.95
$(A_1$ vs. $A_2, A_3, A_4, A_6, A_7) \times (P_1$ vs. $P_2, P_3)$	(1)	(0.012590)	(0.012590)	18.62**	3.96	6.95
$(A_2, A_3, A_4, A_6, A_7) \times (P_1$ vs. $P_2, P_3)$	(4)	(0.002430)	(0.000608)	< 1	—	—
$(A_5$ vs. A_1 to $A_4, A_6, A_7) \times (P_2$ vs. $P_3)$	(1)	(0.000727)	(0.000727)	1.08^{ns}	3.96	6.95
$(A_1$ vs. $A_2, A_3, A_4, A_6, A_7) \times (P_2$ vs. $P_3)$	(1)	(0.000874)	(0.000874)	1.29^{ns}	3.96	6.95
$(A_2, A_3, A_4, A_6, A_7) \times (P_2$ vs. $P_3)$	(4)	(0.001672)	(0.000418)	< 1	—	—
Error(c)	84	0.056799	0.000676			

^{a}** = significant at 1% level, ^{ns} = not significant.

measurement over time with plot sampling. An illustration of this type of data is shown in Table 6.24. Data on tiller count, obtained from a RCB experiment, with nine fertilizer treatments and four replications, were measured at two growth stages of the rice crop. At each stage, measurement was from four 2×2-hill sampling units per plot.

We use the data in Table 6.24 to illustrate the procedure for analyzing data based on measurement over time with plot sampling. Let t denote the number of treatments, r the number of replications, s the sample size per plot, and p the number of times that data were collected from each plot.

□ STEP 1.   At each stage of observation, compute an analysis of variance of data from plot sampling, according to the basic design involved (see Section 6.1). For our example, because the basic design is RCB, each of the $p = 2$ analyses of variance is computed following the procedure of Section 6.1.1. The results are shown in Table 6.25.

□ STEP 2.   Test the homogeneity of the p experimental error variances from the p analyses of variance of step 1. For our example, because $p = 2$, the F test for homogeneity of variance is applied (instead of the chi-square test) as:

$$F = \frac{297.84}{172.21} = 1.73$$

Because the computed F value is smaller than the corresponding tabular F value of 1.98, with $f_1 = f_2 = 24$ degrees of freedom and at the 5% level of significance, the F test is not significant and the heterogeneity of the experimental error variances over the two growth stages is not indicated.

□ STEP 3.   Compute a pooled analysis of variance using the plot data (i.e., mean over all s sampling units per plot). The pooled analysis of variance should be made based on transformed data (see Sections 6.2.1 and 6.2.2, step 3) if the test for homogeneity of experimental error variances of step 2 is significant; and on the original data, otherwise. For our example, because the heterogeneity of the experimental error variances is not indicated, the pooled analysis of variance is made, using the plot data shown in Table 6.26 and following the procedure of Section 6.2.1. The result of the pooled analysis of variance is shown in Table 6.27. The results show a highly significant interaction between treatment and stage of observation, indicating that the treatment effect is not the same at both stages of observation.

Appropriate procedures for mean comparisons are described in Section 6.2.1.

□ STEP 4.   Test the homogeneity of p sampling error variances, from individual analyses of variance of step 1. If heterogeneity is not indicated by the test, compute the pooled sampling error variances as the arithmetic mean of

Table 6.24 Data on Tiller Count, Tested in a RCB Trial with Nine Fertilizer Treatments, Measured from Four 2 × 2-hill Sampling Units per Plot (S_1, S_2, S_3, and S_4) and at Two Growth States (P_1 and P_2)[a] of the Rice Crop

Tillers, no./4 hills

Treatment Number	Rep. I P_1 $S_1 S_2 S_3 S_4$	Rep. I P_2 $S_1 S_2 S_3 S_4$	Rep. II P_1 $S_1 S_2 S_3 S_4$	Rep. II P_2 $S_1 S_2 S_3 S_4$	Rep. III P_1 $S_1 S_2 S_3 S_4$	Rep. III P_2 $S_1 S_2 S_3 S_4$	Rep. IV P_1 $S_1 S_2 S_3 S_4$	Rep. IV P_2 $S_1 S_2 S_3 S_4$
1	26 25 25 28	30 23 27 22	20 29 26 26	22 26 32 25	20 29 40 31	34 26 30 24	23 34 25 34	40 42 37 25
2	63 71 70 73	48 46 33 42	64 73 103 52	57 60 38 50	68 61 81 67	67 64 63 58	63 41 69 89	40 37 36 60
3	44 44 39 71	52 47 61 46	43 50 50 59	49 41 43 70	59 57 52 62	52 48 54 56	67 58 67 68	50 61 58 74
4	61 70 98 99	45 51 73 55	66 56 71 107	65 62 79 54	95 108 71 82	75 56 75 75	66 71 53 70	58 41 47 58
5	63 65 65 91	52 62 56 52	66 81 78 78	50 72 51 51	72 54 74 52	56 39 49 59	61 59 67 85	53 53 40 72
6	69 97 85 99	62 63 56 43	58 59 57 87	52 48 54 56	67 67 71 83	74 58 48 51	38 60 67 49	63 59 46 52
7	44 66 49 62	58 46 63 55	70 81 73 60	47 50 70 53	61 69 74 68	75 48 73 52	60 60 69 74	66 75 72 74
8	57 68 51 59	63 56 59 49	64 67 85 83	47 53 60 68	58 83 83 78	47 58 65 78	87 68 80 61	63 76 30 68
9	79 43 84 90	70 72 72 49	77 47 54 61	55 44 42 52	40 41 50 66	69 55 56 59	38 48 66 71	53 52 44 49

[a]P_1 = 30 days after transplanting, P_2 = crop maturity.

Table 6.25 Two Individual Analyses of Variance (RCB) of Data from Plot Sampling, One for Each Stage of Observation, for Data in Table 6.24

Source of Variation	Degree of Freedom	Mean Square[a] P_1	Mean Square[a] P_2
Replication	3	106.60	208.41
Treatment	8	3,451.50**	1,576.19**
Experimental error	24	297.84	172.21
Sampling error	108	149.43	76.09
Total	143		

[a]P_1 = 30 days after transplanting and P_2 = crop maturity; ** = F test significant at 1% level.

Table 6.26 Plot Means (Average of Four Sampling Units per Plot) Computed from Data in Table 6.24

Treatment Number	Tillers, no./4 hills Rep. I P_1	Rep. I P_2	Rep. II P_1	Rep. II P_2	Rep. III P_1	Rep. III P_2	Rep. IV P_1	Rep. IV P_2
1	26.0	25.5	25.2	26.2	30.0	28.5	29.0	36.2
2	69.2	42.2	73.0	51.2	69.2	63.0	65.5	43.2
3	49.5	51.5	50.5	50.8	57.5	52.5	65.0	60.8
4	82.0	56.0	75.0	65.0	89.0	70.2	65.0	51.0
5	71.0	55.5	75.8	56.0	63.0	50.8	68.0	54.5
6	87.5	56.0	65.2	52.5	72.0	57.8	53.5	55.0
7	55.2	55.5	71.0	55.0	68.0	62.0	65.8	72.0
8	58.8	56.8	74.8	57.0	75.5	62.0	74.0	70.2
9	74.0	65.8	59.8	48.2	49.2	59.8	55.8	49.5

Table 6.27 Pooled Analysis of Variance of Data from Plot Sampling (RCB Design) in Table 6.24

Source of Variation	Degree of Freedom[a]	Sum of Squares	Mean Square	Computed F[b]	Tabular F 5%	Tabular F 1%
Replication	$r - 1 = 3$	86.36	28.79			
Treatment (T)	$t - 1 = 8$	9,151.68	1,143.96	13.20**	2.36	3.36
Error(a)	$(r - 1)(t - 1) = 24$	2,079.61	86.65			
Growth stage (P)	$p - 1 = 1$	1,538.28	1,538.28	46.47**	4.21	7.68
$T \times P$	$(t - 1)(p - 1) = 8$	913.01	114.13	3.45**	2.30	3.26
Error(b)	$t(r - 1)(p - 1) = 27$	893.81	33.10			
Total	$rtp - 1 = 71$	14,662.75				

[a]t = number of treatments, r = number of replications, and p = number of stages of observation.
[b]** = significant at 1% level.

270

the sampling error MS. For our example, the F test for homogeneity of variance is applied to the two sampling error MS:

$$F = \frac{149.43}{76.09} = 1.96$$

The computed F value is larger than the corresponding tabular F value of 1.56, with $f_1 = f_2 = 108$ degrees of freedom and at the 1% level of significance. Hence, the F test is highly significant indicating that the sampling error variances at the two growth stages differ significantly. The sampling error variance is significantly higher at P_1 than at P_2. Thus, no pooled sampling error variance is computed.

Note that information on the pooled sampling error variance, or on the individual sampling error variances, is useful in the development of sampling technique (see Chapter 15, Section 15.2).

CHAPTER 7

Problem Data

Analysis of variance, which we discuss in Chapters 2 through 6, is valid for use only if the basic research data satisfy certain conditions. Some of those conditions are implied, others are specified. In field experiments, for example, it is implied that all plots are grown successfully and all necessary data are taken and recorded. In addition, it is specified that the data satisfy all the mathematical assumptions underlying the analysis of variance.

We use the term *problem data* for any set of data that does not satisfy the implied or the stated conditions for a valid analysis of variance. In this chapter, we examine two groups of problem data that are commonly encountered in agricultural research:

· Missing data.
· Data that violate some assumptions of the analysis of variance.

For each group, we discuss the common causes of the problem data's occurrence and the corresponding remedial measures.

7.1 MISSING DATA

A missing data situation occurs whenever a valid observation is not available for any one of the experimental units. Occurrence of missing data results in two major difficulties—loss of information and nonapplicability of the standard analysis of variance. We examine some of the more common causes of data loss in agricultural research, the corresponding guidelines for declaring such data as missing, and the procedure for analyzing data with one or more missing observations.

7.1.1 Common Causes of Missing Data

Even though data gathering in field experiments is usually done with extreme care, numerous factors beyond the researcher's control can contribute to missing data.

7.1.1.1 Improper Treatment. Improper treatment is declared when an experiment has one or more experimental plots that do not receive the intended treatment. Nonapplication, application of an incorrect dose, and wrong timing of application are common cases of improper treatment. Any observation made on a plot where treatment has not been properly applied should be considered invalid. There is, however, an exception when improper treatment occurs in all replications of a treatment. If the researcher wishes to retain the modified treatment, all measurements can be considered valid if the treatment and the experimental objectives are properly revised.

7.1.1.2 Destruction of Experimental Plants. Most field experiments aim for a perfect stand in all experimental plots but that is not always achieved. Poor germination, physical damage during crop culture, and pest damage are common causes of the destruction of experimental plants. When the percentage of destroyed plants in a plot is small, as is usually the case, proper thinning (Chapter 14, Section 14.3) or correction for missing plants (Chapter 13, Section 13.3.3) will usually result in a valid observation and avoidance of a case of missing data. However, in rare instances, the percentage of destroyed plants in a plot may be so high that no valid observation can be made for the particular plot. When that happens, missing data must be declared.

It is extremely important, however, to carefully examine a stand-deficient plot before declaring missing data. The destruction of the experimental plants must not be the result of the treatment effect. If a plot has no surviving plants because it has been grazed by stray cattle or vandalized by thieves, each of which is clearly not treatment related, missing data should be appropriately declared. But, for example, if a control plot (i.e., nontreated plot) in an insecticide trial is totally damaged by the insects being controlled, the destruction is a logical consequence of treatment. Thus, the corresponding plot data should be entered (i.e., zero yield if all plants in the plot are destroyed, or the actual low yield value if some plants survive) instead of treating it as missing data.

An incorrect declaration of missing data can easily lead to an incorrect conclusion. The usual result of an incorrect declaration of missing data on crop yield is the inflation of the associated treatment mean. For example, for a treatment with all plants in one plot destroyed by stray cattle and, therefore, declared missing, the computation of its mean is based on the average over the remaining $(r - 1)$ replications, where r is the total number of replications. If, on the other hand, the cause of the plot destruction is treatment related and plot yield is, therefore, recorded as zero, then the treatment mean is computed as the mean of r instead of $(r - 1)$ replications.

In most instances, the distinction between a treatment-related cause and a nontreatment-related cause of plot destruction is not clear cut. We give two examples to illustrate the difficulties commonly encountered and provide guidelines for arriving at a correct decision.

Example 1. If the plants in a plot of a rice variety trial are destroyed by brown planthoppers, should the researcher consider its yield value to be zero or should he treat it as missing data? What if the destruction were caused by rats or by drought?

To answer any of the foregoing questions, the researcher must examine the relationship between the objective of the experiment and the cause of plot destruction. Obviously, the objective of the trial is to evaluate the relative performance of the test varieties. In such a trial the superiority of one variety over another is usually defined in terms of a prescribed set of criteria, which depends on the specific test conditions. Thus, if one of the criteria for superiority is resistance to brown planthoppers, the destruction of the rice plants by brown planthoppers is definitely treatment related.

On the other hand, if the trial's objective is to estimate yield potential under complete pest protection and the brown planthopper infestation in the particular plot was solely due to the researcher's failure to implement proper control, the plot destruction should be considered as nontreatment related and missing data declared.

In the same manner, plots destroyed by rats or drought are usually classified as missing data, unless the trial is designed to evaluate varietal resistance to rats or to drought.

Example 2. When all plants in a plot are destroyed, what values should be given to yield components and other plant characters measured at harvest? Should their values be automatically classified as missing or should they be taken as zero? For example, when all rice plants in a plot are destroyed, what value should be entered for 100-grain weight? What about plant height, panicle number, panicle length, or percent unfilled grains?

To answer these questions, a researcher must first determine whether the value of yield in the affected plot is considered as zero or as missing data. If yield of the affected plot is treated as missing data, all plant characters measured at harvest and all yield components of that plot should also be considered as missing data. If, however, yield is considered to be zero (i.e., if the destruction is considered to be treatment related) the following guidelines apply:

- For characters whose measurement depends on the existence of some yield, such as 100-grain weight and panicle length, they should be treated as missing data.
- For characters that can be measured even if no yield is available, such as plant height, panicle number, and percent unfilled grains, the decision should be based on how the data are to be used. If the data are used to assist in explaining the yield differences among treatments, its actual values should be taken. For example, information on short stunted plants or on the 100% unfilled grains would be useful in explaining the cause of the zero yield obtained.

7.1.1.3 Loss of Harvested Samples. Many plant characters cannot be conveniently recorded, either in the field or immediately after harvest.

- Harvested samples may require additional processing before the required data can be measured. For example, grain yield of rice can be measured only after drying, threshing, and cleaning are completed.
- Some characters may involve long sampling and measurement processes or may require specialized and elaborate measuring devices. Leaf area, 100-grain weight, and protein content are generally measured in a laboratory instead of in the field.

For such data, field samples (leaves in the case of leaf area; matured grains in the case of yield, 100-grain weight, or protein content) are usually removed from each plot and processed in a laboratory before the required data are recorded. It is not uncommon for some portion of the samples to be lost between the time of harvesting and the actual data recording. Because no measurement of such characters is possible, missing data should be declared.

7.1.1.4 Illogical Data. In contrast to the cases of missing data where the problem is recognized before data are recorded, illogical data are usually recognized after the data have been recorded and transcribed.

Data may be considered illogical if their values are too extreme to be considered within the logical range of the normal behavior of the experimental materials. However, only illogical data resulting from some kind of error can be considered as missing. Common errors resulting in illogical data are misread observation, incorrect transcription, and improper application of the sampling technique or the measuring instrument.

If illogical data are detected early enough, their causes, or the specific types of error committed, can usually be traced and the data corrected, or adjusted, accordingly. For example, a misread or incorrectly recorded observation in the measurement of plant height, if detected immediately, can be corrected by remeasuring the sample plants. For characters in which the samples used for determination are not destroyed immediately after measurement, such as seed weight and protein content, a remeasurement is generally possible. Thus, it is a good practice for the researcher to examine all data sets immediately after data collection so that subsequent correction of suspicious or illogical data is possible.

We emphasize at this point that data that a researcher suspects to be illogical should not be treated as missing simply because they do not conform to the researcher's preconceived ideas or hypotheses. An observation considered to be illogical by virtue of the fact that it falls outside the researcher's expected range of values can be judged missing only if it can be shown to be caused by an error, as previously discussed. An observation must not be rejected and treated as missing data without proper justification.

7.1.2 Missing Data Formula Technique

When an experiment has one or more observations missing, the standard computational procedures of the analysis of variance for the various designs (as described in Chapters 2 through 4), except CRD, no longer apply. In such cases, either the *missing data formula technique* or the *analysis of covariance technique* should be applied. We describe the missing data formula technique here. The analysis of covariance technique is explained in Chapter 10.

In the missing data formula technique, an estimate of a single missing observation is provided through an appropriate formula according to the experimental design used. This estimate is used to replace the missing data and the augmented data set is then subjected, with some slight modifications, to the standard analysis of variance.

We emphasize here that an estimate of the missing data obtained through the missing data formula technique does not supply any additional information to the incomplete set of data—once the data is lost, no amount of statistical manipulation can retrieve it. What the procedure attempts to do is to allow the researcher to compute the analysis of variance in the usual manner (i.e., as if the data were complete) without resorting to the more complex procedures needed for incomplete data sets.

The missing data formula technique is described for five experimental designs: randomized complete block, latin square, split-plot, strip-plot, and split-split-plot. For each design, the formula for estimating the missing data and the modifications needed in the analysis of variance and in pair comparisons* of treatment means are given. The iterative procedure for cases with more than one missing observation is also discussed.

7.1.2.1 Randomized Complete Block Design.   The missing data in a randomized complete block design is estimated as:

$$X = \frac{rB_0 + tT_0 - G_0}{(r-1)(t-1)}$$

where

$X =$ estimate of the missing data
$t =$ number of treatments
$r =$ number of replications
$B_o =$ total of observed values of the replication that contains the missing data
$T_o =$ total of observed values of the treatment that contains the missing data
$G_o =$ grand total of all observed values

---

*Procedures for all other types of mean comparison discussed in Chapter 5 can be directly applied to the augmented data set without modification.

The missing data is replaced by the computed value of X and the usual computational procedures for the analysis of variance (Chapter 2, Section 2.2.3) are applied to the augmented data set with some modifications.

The procedures are illustrated with data of Table 2.5 of Chapter 2, with the value of the fourth treatment (100 kg seed/ha) in replication II (i.e., yield of 4,831 kg/ha) assumed to be missing, as shown in Table 7.1. The procedures for the computation of the analysis of variance and pair comparisons of treatment means are:

□ STEP 1.   Estimate the missing data, using the preceding formula and the values of totals in Table 7.1 as:

$$X = \frac{4(26,453) + 6(14,560) - 114,199}{(4 - 1)(6 - 1)}$$

$$= 5,265 \text{ kg/ha}$$

□ STEP 2.   Replace the missing data of Table 7.1 by its estimated value computed in step 1, as shown in Table 7.2; and do analysis of variance of the augmented data set based on the standard procedure of Chapter 2, Section 2.2.3.

□ STEP 3.   Make the following modifications to the analysis of variance obtained in step 2:

• Subtract one from both the total and error $d.f.$ For our example, the total $d.f.$ of 23 becomes 22 and the error $d.f.$ of 15 becomes 14.

Table 7.1   Data from a RCB Design, with One Missing Observation

Treatment, kg seed/ha	Grain Yield, kg/ha				Treatment Total
	Rep. I	Rep. II	Rep. III	Rep. IV	
25	5,113	5,398	5,307	4,678	20,496
50	5,346	5,952	4,719	4,264	20,281
75	5,272	5,713	5,483	4,749	21,217
100	5,164	m^a	4,986	4,410	$(14,560 = T_o)$
125	4,804	4,848	4,432	4,748	18,832
150	5,254	4,542	4,919	4,098	18,813
Rep. total	30,953	$(26,453 = B_o)$	29,846	26,947	
Grand total					$(114,199 = G_o)$

$^a m$ = missing data.

Table 7.2   Data in Table 7.1 with the Missing Data Replaced by the Value Estimated from the Missing Data Formula Technique

Treatment, kg seed/ha	Grain Yield, kg/ha				Treatment Total
	Rep. I	Rep. II	Rep. III	Rep. IV	
25	5,113	5,398	5,307	4,678	20,496
50	5,346	5,952	4,719	4,264	20,281
75	5,272	5,713	5,483	4,749	21,217
100	5,164	5,265[a]	4,986	4,410	19,825
125	4,804	4,848	4,432	4,748	18,832
150	5,254	4,542	4,919	4,098	18,813
Rep. total	30,953	31,718	29,846	26,947	
Grand total					119,464

[a]Estimate of the missing data from the missing data formula technique.

- Compute the correction factor for bias B as:

$$B = \frac{\left[B_o - (t-1)X\right]^2}{t(t-1)}$$

$$= \frac{\left[26,453 - (6-1)(5,265)\right]^2}{6(6-1)}$$

$$= 546$$

And subtract the computed B value of 546 from the treatment sum of squares and the total sum of squares. For our example, the total SS and the treatment SS, computed in step 2 from the augmented data of Table 7.2, are 4,869,966 and 1,140,501, respectively. Subtracting the B value of 546 from these SS values, we obtain the adjusted treatment SS and the adjusted total SS as:

$$\text{Adjusted treatment } SS = 1,140,501 - 546$$

$$= 1,139,955$$

$$\text{Adjusted total } SS = 4,869,966 - 546$$

$$= 4,869,420$$

The resulting analysis of variance is shown in Table 7.3.

☐ STEP 4.   For pair comparisons of treatment means where one of the treatments has missing data, compute the standard error of the mean

Table 7.3 Analysis of Variance (RCB Design) of Data in Table 7.2 with One Missing Value Estimated by the Missing Data Formula Technique

Source of Variation	Degree of Freedom	Sum of Squares	Mean Square	Computed F^a	Tabular F 5%	Tabular F 1%
Replication	3	2,188,739	729,580			
Treatment	5	1,139,955	227,991	2.07^{ns}	2.96	4.69
Error	14	1,540,726	110,052			
Total	22	4,869,420				

^{a} ^{ns} = not significant.

difference $s_{\bar{d}}$ as:

$$s_{\bar{d}} = \sqrt{s^2 \left[\frac{2}{r} + \frac{t}{r(r-1)(t-1)} \right]}$$

where s^2 is the error mean square from the analysis of variance of step 3, r is the number of replications, and t is the number of treatments.

For example, to compare the mean of the fourth treatment (the treatment with missing data) with any one of the other treatments, $s_{\bar{d}}$ is computed as:

$$s_{\bar{d}} = \sqrt{110,052 \left[\frac{2}{4} + \frac{6}{(4)(3)(5)} \right]}$$

$$= 257 \text{ kg/ha}$$

This computed $s_{\bar{d}}$ is appropriate for use either in the computation of the LSD values (Chapter 5, Section 5.1.1) or the DMRT values (Chapter 5, Section 5.1.2). For illustration, the computation of the LSD values is shown even though the F test in the analysis of variance is not significant. Using t_α as the tabular t value at the α level of significance, obtained from Appendix C with 14 $d.f.$, the LSD values for comparing the fourth treatment and any other treatment is computed as:

$$\text{LSD}_\alpha = (t_\alpha)(s_{\bar{d}})$$

$$\text{LSD}_{.05} = (2.145)(257) = 551 \text{ kg/ha}$$

$$\text{LSD}_{.01} = (2.977)(257) = 765 \text{ kg/ha}$$

7.1.2.2 Latin Square Design.   The missing data in a Latin square design is estimated as:

$$X = \frac{t(R_o + C_o + T_o) - 2G_o}{(t-1)(t-2)}$$

where

> t = number of treatments
> R_o = total of observed values of the row that contains the missing data
> C_o = total of observed values of the column that contains the missing data
> T_o = total of observed values of the treatment that contains the missing data
> G_o = grand total of all observed values

For illustration, we use data from a Latin square design shown in Table 2.7 of Chapter 2. We assume that the yield value in the fourth row and the third column (i.e., 1.655) is missing. The procedures involved are:

☐ STEP 1.   Compute the estimate of the missing data, using the foregoing formula:

$$X = \frac{[4(3.515 + 4.490 + 4.200) - 2(19.710)]}{(3)(2)}$$

$$= 1.567 \ \text{t/ha}$$

☐ STEP 2.   Enter the estimated value obtained in step 1 in the table with all other observed values, and perform the usual analysis of variance on the augmented data set, with the following modifications:

- Subtract one from both the total and error $d.f.$ For our example, the total $d.f.$ of 15 becomes 14 and the error $d.f.$ of 6 becomes 5.
- Compute the correction factor for bias B as:

$$B = \frac{[G_o - R_o - C_o - (t-1)T_o]^2}{[(t-1)(t-2)]^2}$$

$$= \frac{[19.710 - 3.515 - 4.490 - (4-1)(4.200)]^2}{[(4-1)(4-2)]^2}$$

$$= 0.022251$$

And subtract this computed B value from the treatment SS and the total SS.

The final analysis of variance is shown in Table 7.4.

□ STEP 3   For pair comparisons of treatment means where one of the treatments has missing data, compute the standard error of the mean difference as:

$$s_{\bar{d}} = \sqrt{s^2\left[\frac{2}{t} + \frac{1}{(t-1)(t-2)}\right]}$$

where s^2 is the error mean square from the analysis of variance. For our example, to compare the mean of treatment A (the treatment with missing data) with any one of the other treatments, $s_{\bar{d}}$ is computed as:

$$s_{\bar{d}} = \sqrt{0.025332\left[\frac{2}{4} + \frac{1}{(4-1)(4-2)}\right]}$$

$$= 0.12995$$

7.1.2.3 Split-Plot Design.   The missing data in a split-plot design is estimated as:

$$X = \frac{rM_o + bT_o - P_o}{(b-1)(r-1)}$$

Table 7.4   Analysis of Variance (Latin Square Design) of Data in Table 2.7 (Chapter 2), with One Value[a] Assumed Missing and Estimated by the Missing Data Formula Technique

Source of Variation	Degree of Freedom	Sum of Squares	Mean Square	Computed F[b]	Tabular F 5%	Tabular F 1%
Row	3	0.039142	0.013047	< 1	—	—
Column	3	0.793429	0.264476	10.44*	5.41	12.06
Treatment	3	0.383438	0.127813	5.05[ns]	5.41	12.06
Error[c]	5	0.126658	0.025332			
Total	14	1.342667				

[a] Yield value of 1.655 t/ha in the fourth row and the third column is assumed missing.
[b] * = significant at 5% level, [ns] = not significant.
[c] Although error d.f. is inadequate for valid test of significance (see Chapter 2, Section 2.1.2.1, step 6), for illustration purposes, such deficiency is ignored.

where

> b = level of subplot factor
> r = number of replications
> M_o = total of observed values of the specific main plot
> that contains the missing data
> T_o = total of observed values of the treatment combina-
> tion that contains the missing data
> P_o = total of observed values of the main-plot treatment
> that contains the missing data

Note that the foregoing missing data formula for a split-plot design is the same as that for the randomized complete block design (Section 7.1.2.1) with main plot replacing replication. For illustration, we assume that the yield of treatment N_2V_1 in replication II of Table 3.7 of Chapter 3 (i.e., 6,420) is missing. The procedures of the missing data formula technique follow:

☐ STEP 1. Compute the estimate of the missing data, using the foregoing formula. For our example, the values of the parameters needed for estimating the missing data are:

$b = 4$, the level of the subplot factor (i.e., variety)

$r = 3$, the number of replications

$M_o = 17,595$, the observed total of the N_2 main plot
  in replication II (6,127 + 5,724 + 5,744)

$T_o = 12,780$, the observed total of N_2V_1 (6,076 + 6,704)

$P_o = 63,975$, the observed total of N_2 (6,076 + 6,704 + 6,008 + 6,127

$+ \cdots + 4,146$)

Thus, the estimate of the missing data is computed as:

$$X = \frac{3(17,595) + 4(12,780) - 63,975}{3(2)}$$

$$= 6,655 \text{ kg/ha}$$

☐ STEP 2. Enter the estimate of the missing data, computed in step 1, in the table with the other observed values and construct the analysis of variance on the augmented data set in the usual manner, with one subtracted from both the total $d.f.$ and the error(b) $d.f.$ For our example, the total $d.f.$ and the error(b) $d.f.$ become 70 and 35, and the final analysis of variance is shown in Table 7.5.

☐ STEP 3. For pair comparisons of treatment means where one of the treatments has missing data, compute the standard error of the mean difference $s_{\bar{d}}$ following the appropriate formula given in Table 7.6. For our example, to compare the mean of N_2 and the mean of any other nitrogen level, under V_1, the $s_{\bar{d}}$ is computed as:

$$s_{\bar{d}} = \sqrt{\frac{2\left\{E_a + E_b\left[(b-1) + \dfrac{b^2}{2(r-1)(b-1)}\right]\right\}}{rb}}$$

$$= \sqrt{\frac{2\left\{141{,}148 + 358{,}779\left[(4-1) + \dfrac{16}{2(2)(3)}\right]\right\}}{(3)(4)}}$$

$$= 531.64 \text{ kg/ha}$$

To compare the mean of V_1 and the mean of any other variety under the

Table 7.5 Analysis of Variance (Split-Plot Design) of Data in Table 3.7 (Chapter 3) with One Value[a] Assumed Missing and Estimated by the Missing Data Formula Technique

Source of Variation	Degree of Freedom	Sum of Squares	Mean Square	Computed F^b	Tabular F 5%	Tabular F 1%
Replication	2	1,164,605	582,302			
Nitrogen (N)	5	30,615,088	6,123,018	43.38**	3.33	5.64
Error(a)	10	1,411,480	141,148			
Variety (V)	3	90,395,489	30,131,830	83.98**	2.87	4.40
$N \times V$	15	69,100,768	4,606,718	12.84**	1.96	2.60
Error(b)	35	12,557,261	358,779			
Total	70	205,244,690				

[a]Yield value of 6,420 kg/ha of treatment N_2V_1 in replication II is assumed missing.
[b]** = significant at 1% level.

Table 7.6   Standard Error of the Mean Difference ($s_{\bar{d}}$) in a Split-Plot Design with Missing Data

Type of Pair Comparison		$s_{\bar{d}}$ [a]
Number	Between	
1	Two main-plot means (averaged over all subplot treatments)	$\sqrt{\dfrac{2(E_a + fE_b)}{rb}}$
2	Two subplot means (averaged over all main-plot treatments)	$\sqrt{\dfrac{2E_b\left(1 + \dfrac{fb}{a}\right)}{ra}}$
3	Two subplot means at the same main-plot treatment	$\sqrt{\dfrac{2E_b\left(1 + \dfrac{fb}{a}\right)}{r}}$
4	Two main-plot means at the same or different subplot treatments	$\sqrt{\dfrac{2\{E_a + E_b[(b-1) + fb^2]\}}{rb}}$

[a] For one missing observation, $f = 1/[2(r-1)(b-1)]$ and, for more than one missing observation, $f = k/[2(r-d)(b-k+c-1)]$ (see Section 7.1.2.6). $E_a = \text{Error}(a)\ MS$, $E_b = \text{Error}(b)\ MS$, $r = $ number of replications, $a = $ number of main-plot treatments, and $b = $ number of subplot treatments.

same nitrogen level, on the other hand, the $s_{\bar{d}}$ is computed as:

$$s_{\bar{d}} = \sqrt{\frac{2E_b\left[1 + \dfrac{b}{2a(r-1)(b-1)}\right]}{r}}$$

$$= \sqrt{\frac{2(358{,}779)\left[1 + \dfrac{4}{2(6)(2)(3)}\right]}{3}}$$

$$= 502.47 \text{ kg/ha}$$

7.1.2.4 Strip-Plot Design.   The missing data in a strip-plot design is estimated as:

$$X = \frac{a(bT_o - P_o) + r(aH_o + bV_o - B_o) - bS_o + G_o}{(a-1)(b-1)(r-1)}$$

whcrc

a = level of horizontal factor
b — level of vertical factor
r = number of replications
T_o = total of observed values of the treatment that contains the missing data
P_o = total of observed values of the specific level of the horizontal factor that contains the missing data
H_o = total of observed values of the horizontal strip that contains the missing data
V_o = total of observed values of the vertical strip that contains the missing data
B_o = total of observed values of the replication that contains the missing data
S_o = total of observed values of the specific level of the vertical factor that contains the missing data
G_o = total of all observed values

For illustration, assume that the yield of treatment $V_6 N_2$ in replication III of Table 3.11 of Chapter 3 (i.e., 4,425) is missing. The procedures are:

☐ STEP 1. Compute the estimate of the missing data, using the foregoing formula. For our example, the values of the parameters needed for estimating the missing data are:

a = 6, the level of horizontal factor (i.e., variety).
b = 3, the level of vertical factor (i.e., nitrogen)
r = 3, the number of replications
T_o = 6,718, the observed total of treatment $V_6 N_2$ (3,896 + 2,822)
P_o = 23,816, the observed total of V_6 (2,572 + 3,724 + 3,326 + ⋯ + 3,214)
H_o = 6,540, the observed total of V_6 in replication III (3,326 + 3,214)
V_o = 29,912, the observed total of N_2 in replication III (4,889 + 7,177 + 7,019 + 4,816 + 6,011)
B_o = 96,094, the observed total of replication III (4,384 + 4,889 + 8,582 + ⋯ + 3,326 + 3,214)
S_o = 94,183, the observed total of N_2 (4,076 + 6,431 + 4,889 + 5,630 + ⋯ + 2,822)
G_o = 281,232, the observed total of all values

The estimate of the missing data is then computed as:

$$X = \{6[3(6,718) - 23,816] + 3[6(6,540) + 3(29,912) - 96,094]$$

$$- 3(94,183) + 281,232\}/(5)(2)(2)$$

$$= 3,768 \text{ kg/ha}$$

☐ STEP 2. Enter the estimate of the missing data, obtained in step 1, in the table with the other observed values and compute the analysis of variance based on the augmented data set in the usual manner, but with one subtracted from both the total $d.f.$ and the error(c) $d.f.$. The final analysis of variance is shown in Table 7.8.

7.1.2.5 Split-Split-Plot Design. The missing data in a split-split-plot design is estimated as:

$$X = \frac{rM_o + cT_o - P_o}{(c-1)(r-1)}$$

where

c = level of the sub-subplot factor
r = number of replications
M_o = total of observed values of the specific subplot
that contains the missing data
T_o = total of observed values of the treatment that
contains the missing data
P_o = total of observed values of all subplots containing
the same set of treatments as that of the missing data

Table 7.7 Analysis of Variance (Strip-Plot Design) of Data in Table 3.11 (Chapter 3) with One Value[a] Assumed Missing and Estimated by the Missing Data Formula Technique

Source of Variation	Degree of Freedom	Sum of Squares	Mean Square	Computed F^b	Tabular F 5%	1%
Replication	2	8,850,049	4,425,024			
Variety (V)	5	59,967,970	11,993,594	8.15**	3.33	5.64
Error(a)	10	14,709,970	1,470,997			
Nitrogen (N)	2	50,444,651	25,222,326	c	—	—
Error(b)	4	3,072,363	768,091			
$V \times N$	10	23,447,863	2,344,786	5.52**	2.38	3.43
Error(c)	19	8,072,974	424,893			
Total	52	168,565,841				

[a]Yield value of 4,425 kg/ha of treatment $V_6 N_2$ in replication III is assumed missing.
[b]** = significant at 1% level.
[c]Error(b) $d.f.$ is not adequate for valid test of significance.

For illustration, we assume that the yield of treatment $N_4 M_2 V_1$ in replication III of Table 4.4 of Chapter 4 (i.e., 5.345) is missing. The procedures are.

⊔ STEP 1.   Compute the values of the parameters needed for estimating the missing data, using the foregoing formula:

$c = 3$, the level of the sub-subplot factor (i.e., variety)
$r = 3$, the number of replications
$M_o = 13.942$, the observed total of subplot $N_4 M_2$ in replication III (6.164 + 7.778)
$T_o = 10.997$, the observed total of treatment $N_4 M_2 V_1$ (5.255 + 5.742)
$P_o = 57.883$, the observed total of all subplots containing $N_4 M_2$ (5.255 + 5.742 + 6.992 + $\cdots$ + 7.778)

Then compute the estimate of the missing data as:

$$X = \frac{3(13.942) + 3(10.997) - 57.883}{(2)(2)} = 4.234 \text{ t/ha}$$

□ STEP 2.   Enter the estimate of the missing data, obtained in step 1, in the table with the other observed values and compute the analysis of variance based on the augmented data set in the usual manner, but with one subtracted from both the total $d.f.$ and the error(c) $d.f.$. The final analysis of variance is shown in Table 7.8.

7.1.2.6   More Than One Missing Observation.

The missing data formula technique, which is discussed for the various designs in Sections 7.1.2.1 through 7.1.2.5, is not directly applicable to the case of more than one missing observation in any design, except for a split-plot or a split-split-plot design where the two or more missing data satisfy the following conditions:

1.   For a split-plot design, no two missing data share the same treatment combination or the same main-plot treatment. For example, the two missing data could be $a_0 b_1$ of replication I and $a_1 b_1$ of replication III.

2.   For a split-split-plot design, no two missing data share the same treatment combination or the same subplot × main-plot treatment combination. For example, the two missing data could be $a_0 b_1 c_1$ of replication I and $a_1 b_2 c_3$ of replication III.

When the missing data formula technique is not applicable, the iterative procedure should be applied. Because the basic principle of the iterative procedure is similar for all designs, we illustrate it only once. RCB data in Table 2.5 of Chapter 2 is used, by assuming that the values of the fourth

Table 7.8 Analysis of Variance (Split-Split-Plot Design) of Data in Table 4.4 (Chapter 4) with One Value[a] Assumed Missing and Estimated by the Missing Data Formula Technique

Source of Variation	Degree of Freedom	Sum of Squares	Mean Square	Computed F[b]	Tabular F 5%	Tabular F 1%
Replication	2	0.925	0.4627			
Nitrogen (N)	4	60.826	15.2065	23.78**	3.84	7.01
Error(a)	8	5.116	0.6395			
Management (M)	2	43.106	21.5530	72.11**	3.49	5.85
$N \times M$	8	0.829	0.1036	< 1	—	—
Error(b)	20	5.978	0.2989			
Variety (V)	2	209.204	104.6020	211.49**	3.15	4.99
$N \times V$	8	14.362	1.7952	3.63**	2.10	2.83
$M \times V$	4	3.897	0.9742	1.97ns	2.52	3.66
$N \times M \times V$	16	4.026	0.2516	< 1	—	—
Error(c)	59	29.183	0.4946			
Total	133	377.453				

[a]Yield value of 5.345 t/ha of treatment $N_4 M_2 V_1$ in replication III is assumed missing.
[b]** = significant at 1% level, ns = not significant.

treatment in replication II (i.e., 4,831) and of the first treatment in replication I (i.e., 5,113) are both missing. The step-by-step procedure for estimating the two missing values and for obtaining the analysis of variance is:

☐ STEP 1. Assign *initial values* to all missing data except one. Although any value can be used as the initial value, unusually large or unusually small values can result in a lengthy computation. The most commonly used initial value for each missing observation is the average of its marginal means:

$$\overline{X}_{ij} = \frac{\overline{t}_i + \overline{b}_j}{2}$$

where $\overline{X}_{ij}$ is the initial value of the ith treatment and the jth replication, $\overline{t}_i$ is the mean of all observed values of the ith treatment, and $\overline{b}_j$ is the mean of all observed values of the jth replication.

For our example, because there are two missing observations, only one initial value needs to be assigned. Assuming that we wish to assign the initial value to the first treatment in replication I using the average of marginal means, the computation would be:

A. Compute $\bar{t}_1$ and $\bar{b}_1$ using values from Table 2.5 as:

$$\bar{t}_1 = \frac{5,398 + 5,307 + 4,678}{3}$$

$$= 5,128 \text{ kg/ha}$$

$$\bar{b}_1 = \frac{5,346 + 5,272 + 5,164 + 4,804 + 5,254}{5}$$

$$= 5,168 \text{ kg/ha}$$

B. Compute the marginal mean $\overline{X}_{11}$ as:

$$\overline{X}_{11} = \frac{\bar{t}_1 + \bar{b}_1}{2}$$

$$= \frac{5,128 + 5,168}{2} = 5,148 \text{ kg/ha}$$

The initial value assigned to the first treatment in replication I is therefore $X_{11} = 5,148$ kg/ha. Note that the two subscripts of X are the treatment number (i) and the replication number (j).

☐ STEP 2. Enter all initial values assigned in step 1, in the table of observed values and estimate the one remaining missing observation by using the appropriate missing data formula technique of Sections 7.1.2.1, 7.1.2.2, 7.1.2.3, 7.1.2.4, or 7.1.2.5, according to the design used.

For our example, the only initial value assigned in step 1 for the first treatment in replication I is entered in the table of observed values, as shown in Table 7.9; and the value of the fourth treatment of replication II is estimated, using the missing data formula technique for the randomized complete block design of Section 7.1.2.1, as follows:

$$X_{42} = \frac{4(26,453) + 6(14,560) - 114,234}{(3)(5)}$$

$$= 5,263 \text{ kg/ha}$$

☐ STEP 3. Enter the estimate of the missing data obtained in step 2, in the table consisting of all observed values and the initial value (or values) assigned in step 1. Then,

· Remove one initial value. The order in which the initial values are removed is not important at this stage but the order used here must be followed in the succeeding steps.

- Treat the removed value as the missing data, and estimate it following the same missing data formula technique used in step 2.

Repeat the foregoing procedure for the third missing observation, then for the fourth missing observation, and so on, until all missing data have been estimated once through the missing data formula technique. This then completes the first cycle of iteration.

For our example, enter the estimated value of $X_{42} = 5,263$ kg/ha, obtained in step 2, in Table 7.9. Remove the initial value given to the first treatment of replication I and apply the missing data formula technique to reestimate the value of X_{11} as:

$$X'_{11} = \frac{4(25,840) + 6(15,383) - 114,349}{(3)(5)}$$

$$= 5,421 \text{ kg/ha}$$

Then, with the value of X'_{11} replacing X_{11}, reestimate X_{42}, using the same missing data formula as:

$$X'_{42} = \frac{4(26,453) + 6(14,560) - 114,507}{15}$$

$$= 5,244 \text{ kg/ha}$$

□ STEP 4. Repeat step 3 for the second cycle of iteration, following the same order of missing data previously used. Compare the new set of estimates to that obtained in the first cycle. If the differences are satisfactorily small, the

Table 7.9   Data from a RCB Design, with Two Missing Values, One of Which is Assigned an Initial Value, as the First Step in Estimating Missing Data Through the Iterative Procedure

Treatment, kg seed/ha	Yield, kg/ha				Treatment Total
	Rep. I	Rep. II	Rep. III	Rep. IV	
25	$(5,148)^a$	5,398	5,307	4,678	(20,531)
50	5,346	5,952	4,719	4,264	20,281
75	5,272	5,713	5,483	4,749	21,217
100	5,164	m^b	4,986	4,410	(14,560)
125	4,804	4,848	4,432	4,748	18,832
150	5,254	4,542	4,919	4,098	18,813
Rep. total	(30,988)	(26,453)	29,846	26,947	
Grand total					(114,234)

[a] The initial value assigned to one of the two missing data.
[b] m = missing data.

new set of estimates can be accepted and the iteration process terminated. Otherwise, the third cycle of iteration should be initiated and the process should be continued until the difference between the last two sets of estimate (i.e., from the last two cycles of iteration) is satisfactorily small. Note that this difference becomes smaller as more cycles are applied. Thus, the decision as to when to stop the iteration process is usually a balance between the degree of precision desired and the computational resources available.

For our example, repeat step 3 and start the second cycle of iteration, following the same order of missing data previously used. For that, the two missing values are reestimated as:

$$X_{11}'' = \frac{4(25,840) + 6(15,383) - 114,330}{15}$$

$$= 5,422 \text{ kg/ha}$$

$$X_{42}'' = \frac{4(26,453) + 6(14,560) - 114,508}{15}$$

$$= 5,244 \text{ kg/ha}$$

Because the new estimates from the second cycle are close to those from the first cycle, the process can be terminated.

The estimates of the two missing values, namely the fourth treatment of replication II and the first treatment of replication I, are, therefore, 5,244 kg/ha and 5,422 kg/ha.

Note that it is usually not necessary to carry out the process until the values of the two successive cycles are almost exactly the same, as in this example. The process could be terminated as soon as the difference between the estimates in successive cycles is, in the researcher's judgment, sufficiently small.

☐ STEP 5. Use the set of estimates from the last cycle of iteration together with all other observed data to compute the analysis of variance in the usual manner, but with m subtracted from both the total *d.f.* and the error *d.f.*, where m is the total number of missing values.

For our example, the estimates of the two missing values are entered in the table with the other observed values and the analysis of variance is constructed on the augmented data set in the usual manner, with two (number of missing data) subtracted from both the total *d.f.* and the error *d.f.* The completed analysis of variance is shown in Table 7.10.

☐ STEP 6. For pair comparisons of treatment means where only one of the treatments has missing data, the procedures for the case of one missing observation, described for the various designs in Sections 7.1.2.1 through

Table 7.10   Analysis of Variance of Data in Table 7.9 with Two Missing Data Estimated through the Iterative Procedure

Source of Variation	Degree of Freedom	Sum of Squares	Mean Square	Computed F^a	Tabular F 5%	1%
Replication	3	2,300,267	766,756			
Treatment	5	1,252,358	250,472	2.20^{ns}	3.02	4.86
Error	13	1,481,374	113,952			
Total	21	5,033,999				

[a] ns = not significant.

7.1.2.3, can be applied. For pair comparisons where both treatments have missing data, or where one of the two treatments involves more than one missing observation, the $s_{\bar{d}}$ should be computed as:

- For RCB design,

$$s_{\bar{d}} = \sqrt{s^2\left(\frac{1}{r'_A} + \frac{1}{r'_B}\right)}$$

where r'_A and r'_B are the *effective numbers of replications* assigned to treatments A and B. For a treatment to be compared, say treatment A, a replication is counted as 1 if it has data for both treatments, as $\frac{1}{2}$ if it has data for treatment A but not for treatment B, and as 0 if it does not have data for treatment A.

For our example (see Table 7.9), if the first treatment is to be compared with the fourth treatment (i.e., both treatments having one missing observation each), the computation would be:

(i) Assign an effective number of replications to each of the two treatments. For example, for the first treatment, replication I is counted as zero because the first treatment is missing in replication I, replication II is counted as $\frac{1}{2}$ because the first treatment has data but the fourth treatment does not, and replications III and IV are counted as 1 each because both treatments are present in both replications. Thus, the effective number of replications for the first treatment is computed as $0 + \frac{1}{2} + 1 + 1 = 2.5$. In the same manner, the effective number of replications for the fourth treatment is computed as 2.5.

(ii) Compute the $s_{\bar{d}}$ as:

$$s_{\bar{d}} = \sqrt{s^2\left(\frac{1}{2.5} + \frac{1}{2.5}\right)}$$

$$= \sqrt{(113,952)(0.8)} = 301.930$$

- For split-plot design, the formulas given in Table 7.6 can be used with the parameter f computed as:

$$f = \frac{k}{2(r - d)(b - k + c - 1)}$$

where b = level of the subplot factor; and k, c, and d, which refer only to missing observations for the two treatments being compared, are defined as:

k = number of missing data
c = number of replications that contain at least one missing observation
d = number of missing observations in the treatment combination that has the largest number of missing data

For illustration, we use the same set of data used in Section 7.1.2.3. Assume that, in addition to the missing data of N_2V_1 in replication II, the yield of treatment N_2V_4 of replication I is also missing. Following the procedure described in steps 1 to 5, the estimates of the two missing data are 6,621 kg/ha for N_2V_1 in replication II and 4,749 kg/ha for N_2V_4 in replication I.

The $s_{\bar{d}}$ value for comparing any two subplot treatment means under each main-plot treatment, when either one or both means involve a missing observation, is computed as:

$$s_{\bar{d}} = \sqrt{\frac{2E_b\left[1 + \dfrac{bk}{2a(r - d)(b - k + c - 1)}\right]}{r}}$$

The $s_{\bar{d}}$ value for comparing any two main-plot treatment means at same or different subplot treatments, when either one or both means involve a missing observation, is computed as:

$$s_{\bar{d}} = \sqrt{\frac{2\left\{E_a + E_b\left[(b - 1) + \dfrac{b^2(k)}{2(r - d)(b - k + c - 1)}\right]\right\}}{rb}}$$

For our example, if we wish to compare the mean of N_2V_1 and the mean of N_2V_4, the $s_{\bar{d}}$ value is:

$$s_{\bar{d}} = \sqrt{\frac{2(368,742)\left[1 + \dfrac{4(2)}{2(6)(3-1)(4-2+2-1)}\right]}{3}}$$

$$= 522.6 \text{ kg/ha}$$

To compare N_2V_1 mean with any other treatment of the same nitrogen level, except for N_2V_4, the $s_{\bar{d}}$ value is:

$$s_{\bar{d}} = \sqrt{\frac{2(368,742)\left[1 + \dfrac{4(1)}{2(6)(3-1)(4-1+1-1)}\right]}{3}}$$

$$= 509.4 \text{ kg/ha}$$

And, to compare N_2V_1 mean with any other treatment involving different nitrogen level, say N_1V_2 mean, the $s_{\bar{d}}$ value is:

$$s_{\bar{d}} = \sqrt{\frac{2\left\{139,612 + 368,742\left[(4-1) + \dfrac{16(1)}{2(3-1)(4-1+1-1)}\right]\right\}}{3(4)}}$$

$$= 538.1 \text{ kg/ha}$$

7.2 DATA THAT VIOLATE SOME ASSUMPTIONS OF THE ANALYSIS OF VARIANCE

The usual interpretation of the analysis of variance is valid only when certain mathematical assumptions concerning the data are met. These assumptions are:

Additive Effects. Treatment effects and environmental effects are additive.
Independence of Errors. Experimental errors are independent.
Homogeneity of Variance. Experimental errors have common variance.
Normal Distribution. Experimental errors are normally distributed.

Failure to meet one or more of these assumptions affects both the level of significance and the sensitivity of the F test in the analysis of variance. Thus, any drastic departure from one or more of the assumptions must be corrected before the analysis of variance is applied.

7.2.1 Common Violations in Agricultural Experiments

The assumptions underlying the analysis of variance are reasonably satisfied for most experimental data in agricultural research, but there are certain types of experiment that are notorious for frequent violations of these assumptions. We describe some of the assumptions that are usually violated and give examples of experiments wherein these violations can be expected.

7.2.1.1 Nonadditive Effects. The effects of two factors, say treatment and replication, are said to be additive if the effect of one factor remains constant over all levels of the other factor. In other words, if the treatment effect remains constant for all replications and the replication effect remains constant for all treatments, then the effects of treatment and replication are additive. A hypothetical set of data from a RCB design, having two treatments and two replications, with additive effects, is illustrated in Table 7.11. Here, the treatment effect is equal to 20 for both replications and the replication effect is equal to 60 for both treatments.

A common departure from the assumption of additivity in agricultural experiments is one where the effects are multiplicative instead of additive. Two factors are said to have multiplicative effects if their effects are additive only when expressed in terms of percentages.

Table 7.12 illustrates a hypothetical set of data with multiplicative effects. In this case, the effect of treatment is 30 in replication I and 20 in replication II while the effect of replication is 60 for treatment A and 50 for treatment B. That is, the treatment effect is not constant over replications and the replication effect is not constant over treatments. However, when both treatment effect and block effect are expressed in terms of percentages, an entirely different pattern emerges. The replication effect is 50% for both treatments and the treatment effect is 20% in both replications.

Table 7.11 A Hypothetical Set of Data with Additive Effects of Treatment and Replication

Treatment	Replication		Replication Effect $(I - II)$
	I	II	
A	180	120	60
B	160	100	60
Treatment effect $(A - B)$	20	20	

Table 7.12   A Hypothetical Set of Data with Multiplicative Effects of Treatment and Replication

Treatment	Replication		Replication Effect	
	I	II	I − II	100(I − II)/II
A	180	120	60	50
B	150	100	50	50
Treatment effect $A - B$	30	20		
$100(A - B)/B$	20	20		

The multiplicative effect is commonly encountered in experiments designed to evaluate the incidence of insects or diseases. This happens because the changes in insect and disease incidence usually follow a pattern that is in multiples of the initial incidence.

7.2.1.2   Nonindependence of Errors.   The assumption of independence of experimental errors requires that the error of an observation is not related to, or dependent upon, that of another. This assumption is usually assured with the use of proper randomization (i.e., treatments are assigned at random to experimental units). However, in a *systematic design*, where the experimental units are arranged systematically instead of at random, the assumption of independence of errors is usually violated. This is so because the design requires that certain pairs of treatments be placed in adjacent plots whereas others are always placed some distance apart. Figure 7.1 illustrates a layout of a systematic design where certain pairs of treatments, say, *A* and *B*, are adjacent to each other in all replications, and other pairs of treatments, say, *B* and *E*, are always separated by two intervening plots. Because experimental errors of adjacent plots tend to be more alike than those farther apart, the error for plots of treatments *A* and *B* can be expected to be more related compared to that between plots of treatments *B* and *E*.

Replication I	A	B	C	D	E	F
Replication II	F	A	B	C	D	E
Replication III	E	F	A	B	C	D
Replication IV	D	E	F	A	B	C

Figure 7.1   Layout of a systematic design involving six treatments (*A*, *B*, *C*, *D*, *E*, and *F*) and four replications: a source of nonindependence of errors.

Because proper randomization usually ensures the independence of experimental errors, the simplest way to detect nonindependence of errors is to check the experimental layout. If there is a systematic pattern in the arrangement of treatments from one replication to another, nonindependence of errors may be expected.

7.2.1.3 Variance Heterogeneity and Nonnormality.

7.2.1.3 Variance Heterogeneity and Nonnormality.   Heterogeneity of variance can be classified into two types:

- Where the variance is functionally related to the mean
- Where there is no functional relationship between the variance and the mean

The first type of variance heterogeneity is usually associated with data whose distribution is not normal. For example, count data, such as the number of infested plants per plot or the number of lesions per leaf, usually follow the *poisson* distribution wherein the variance is equal to the mean; that is, $s^2 = \overline{X}$. Another example is the *binomial* distribution, which is expected in data such as percent survival of insects or percent plants infected with a disease. Such data describe the proportion of occurrences in which each occurrence can only be one of the two possible outcomes (e.g., alive or dead and infested or not infested). For this latter type of data, the variance and the mean are related as $s^2 = \overline{X}(1 - \overline{X})$.

The second type of variance heterogeneity (where variance and mean are not related) usually occurs in experiments where, due to the nature of the treatments tested, some treatments have errors that are substantially higher (or lower) than others. Examples of this type of variance heterogeneity are:

- In variety trials where various types of breeding material are being compared, the size of the variance between plots of a particular variety will depend on the degree of genetic homogeneity of the material being tested. The variance of the F_2 generation, for example, can be expected to be higher than that of the F_1 generation because genetic variability in F_2 is much higher than that in F_1. The variances of varieties that are highly tolerant of, or highly susceptible to, the stress being tested are expected to be smaller than those of varieties with moderate degrees of tolerance.
- In testing yield response to chemical treatment, such as fertilizer, insecticide, or herbicide, the nonuniform application of chemical treatments may result to a higher variability in the treated plots than that in the nontreated plots.

7.2.2 Remedial Measures for Handling Variance Heterogeneity

The most common symptom of experimental data that violate one or more of the assumptions of the analysis of variance is variance heterogeneity. In this section we focus on the two remedial measures for handling variance heterogeneity. These are:

* The method of data transformation for variances that are functionally related to the mean
* The method of error partitioning for variances that are not functionally related to the mean

For data with heterogeneous variances, a correct diagnosis of the specific type of variance heterogeneity present in the data must be made before an appropriate remedial measure can be selected. We present the following simplified procedure for detecting the presence of variance heterogeneity and for diagnosing the type of variance heterogeneity:

☐ STEP 1.   For each treatment, compute the variance and the mean across replications (the range can be used in place of the variance if ease of computation is required).

☐ STEP 2.   Plot a scatter diagram between the mean value and the variance (or the range). The number of points in the scatter diagram equals the number of treatments.

☐ STEP 3.   Visually examine the scatter diagram to identify the pattern of relationship, if any, between the mean and the variance. Figure 7.2 illustrates three possible outcomes of such an examination:

* Homogeneous variance (Figure 7.2*a*)
* Heterogeneous variance when the variance is functionally related to the mean (Figure 7.2*b*)
* Heterogeneous variance when there is no functional relationship between the variance and the mean (Figure 7.2*c*)

7.2.2.1 Data Transformation.   Data transformation is the most appropriate remedial measure for variance heterogeneity where the variance and the mean are functionally related. With this technique, the original data are

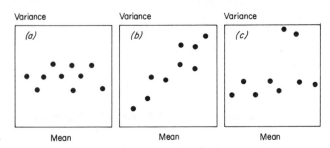

Figure 7.2   Illustration of different types of variance heterogeneity: (*a*) homogeneous variance, (*b*) heterogeneous variance where variance is proportional to the mean, and (*c*) heterogeneous variance without any functional relationship between variance and mean.

Table 7.13 Application of the Logarithmic Transformation to Data in Table 7.12, to Convert the Multiplicative Effect between Replication and Treatment to an Additive Effect

Treatment	Replication		Replication Effect
	I	II	
A	2.255	2.079	0.176
B	2.176	2.000	0.176
Treatment effect	0.079	0.079	

converted into a new scale resulting in a new data set that is expected to satisfy the condition of homogeneity of variance. Because a common transformation scale is applied to all observations, the comparative values between treatments are not altered and comparisons between them remain valid.

The appropriate data transformation to be used depends on the specific type of relationship between the variance and the mean. We explain three of the most commonly used transformations for data in agricultural research.

7.2.2.1.1 Logarithmic Transformation. The logarithmic transformation is most appropriate for data where the standard deviation is proportional to the mean or where the effects are multiplicative. These conditions are generally found in data that are whole numbers and cover a wide range of values. Data on the number of insects per plot or the number of egg masses per plant (or per unit area) are typical examples.

To transform a data set into the logarithmic scale, simply take the logarithm of each and every component of the data set. For example, to convert Table 7.12 to a logarithmic scale, each of the numbers in the four data cells are converted into its logarithm. The results are shown in Table 7.13. Note that although the effects of treatment and replication are multiplicative in the original data of Table 7.12, the effects become additive with the transformed data of Table 7.13. In Table 7.13 the treatment effect is 0.079 in both replications and the replication effect is 0.176 for both treatments. This illustrates the effectiveness of logarithmic transformation in converting multiplicative effect to additive effect.

If the data set involves small values (e.g., less than 10), $\log(X + 1)$ should be used instead of $\log X$, where X is the original data. To illustrate the procedure for applying the logarithmic transformation, we use data on the number of living larvae on rice plants treated with various rates of an insecticide from a RCB experiment with four replications (Table 7.14). The step-by-step procedures are:

☐ STEP 1. Verify the functional relationship between the mean and the variance using the scatter-diagram procedure described earlier. For our example, we use range instead of variance. The result (Figure 7.3*a*) shows a

Table 7.14 Number of Living Larvae Recovered Following Different Insecticide Treatments

Treatment		Larvae, no.				Treatment	Treatment
Number	Description[a]	Rep. I	Rep. II	Rep. III	Rep. IV	Total	Mean
1	Diazinon (4)	9	12	0	1	22	5.50
2	Diazinon (3)	4	8	5	1	18	4.50
3	Diazinon (2)	6	15	6	2	29	7.25
4	Diazinon (1)	9	6	4	5	24	6.00
5	Diazinon (2) + MLVC (2)	27	17	10	10	64	16.00
6	Diazinon (2) + MLVC + SLVC (2)	35	28	2	15	80	20.00
7	Diazinon (1) at 15% DH infestation	1	0	0	0	1	0.25
8	Diazinon (1) at 20% DH infestation	10	0	2	1	13	3.25
9	Control	4	10	15	5	34	8.50
	Total	105	96	44	40	285	

[a]Number in parentheses refers to number of times the chemicals were applied.

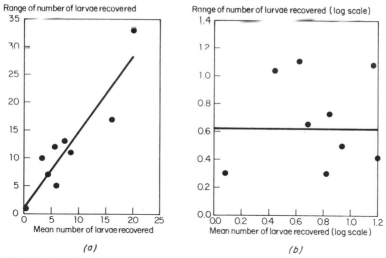

Figure 7.3 Relationships between the range and the treatment mean of data given in Table 7.14: (*a*) before logarithmic transformation and (*b*) after logarithmic transformation.

linear relationship between the range and the mean (i.e., the range increases proportionally with the mean), suggesting the use of logarithmic transformation.

☐ STEP 2. Because some of the values in Table 7.14 are less than 10, $\log(X + 1)$ is applied instead of $\log X$. Compute $\log(X + 1)$, where X is the original data in Table 7.14, for all values. The transformed data are shown in Table 7.15.

☐ STEP 3. Verify the success of the logarithmic transformation in achieving the desired homogeneity of variance, by applying step 1 to the transformed

Table 7.15   Data in Table 7.14 Transformed to Logarithmic Scale, $\log(X + 1)$

Treatment Number	\multicolumn{5}{c}{Larvae Number in Log Scale}		Mean $(\bar{X})$	Antilog of $\bar{X}$
	Rep. I	Rep. II	Rep. III	Rep. IV	Total		
1	1.0000	1.1139	0.0000	0.3010	2.4149	0.6037	4.02
2	0.6990	0.9542	0.7782	0.3010	2.7324	0.6831	4.82
3	0.8451	1.2041	0.8451	0.4771	3.3714	0.8428	6.96
4	1.0000	0.8451	0.6990	0.7782	3.3223	0.8306	6.77
5	1.4472	1.2553	1.0414	1.0414	4.7853	1.1963	15.71
6	1.5563	1.4624	0.4771	1.2041	4.6999	1.1750	14.96
7	0.3010	0.0000	0.0000	0.0000	0.3010	0.0752	1.19
8	1.0414	0.0000	0.4771	0.3010	1.8195	0.4549	2.85
9	0.6990	1.0414	1.2041	0.7782	3.7227	0.9307	8.53
Total	8.5890	7.8764	5.5220	5.1820	27.1694		

Table 7.16   Analysis of Variance of Data in Table 7.15

Source of Variation	Degree of Freedom	Sum of Squares	Mean Square	Computed F^a
Replication	3	0.95666	0.31889	
Treatment	8	3.98235	0.49779	5.70**
Error	24	2.09615	0.08734	
Total	35	7.03516		

[a]** = significant at 1% level.

data in Table 7.15. The result (Figure 7.2b) based on transformed data shows no apparent relationship between the range and the mean, indicating the success of the logarithmic transformation.

☐ STEP 4. Construct the analysis of variance, in the usual manner (see Chapter 2, Section 2.2.3), on the transformed data in Table 7.15. Results are shown in Table 7.16.

☐ STEP 5. Perform the desired mean comparisons by applying the standard procedures described in Chapter 5 directly to the transformed data. The test for pair comparison, either LSD test or DMRT, must be made based on the transformed means even though the final presentation of the treatment means is in the original scale. For example, only apply the LSD values in the transformed scale and only to the transformed means.

For our example, the computation involved in the application of the LSD test and the DMRT to the pair comparisons of treatment means is:

- *Least Significant difference (LSD) Test.* If the researcher wishes to determine the effectivity of each insecticide treatment, a valid comparison between each treatment and the control treatment can be made using the LSD test. The LSD values are computed using results of the analysis of variance in Table 7.16 and the formula in Chapter 5, Section 5.1.1.1.1, as:

$$LSD_{.05} = 2.064\sqrt{\frac{2(0.08734)}{4}} = 0.4313$$

$$LSD_{.01} = 2.797\sqrt{\frac{2(0.08734)}{4}} = 0.5845$$

As indicated before, these LSD values are in the transformed scale and must be applied to the transformed means in Table 7.15. For example, to compare the control treatment (treatment 9) with each of the other treatments, the transformed mean of the control should be subtracted from the transformed mean of each of the other treatments, and each of these differences is then compared to the LSD values just computed. For example, the mean difference between treatment 1 and the control is computed as 0.9307 − 0.6037 = 0.3270, which is smaller than the computed LSD$_{05}$ value of 0.4111. Hence, treatment 1 is not significantly different from the control treatment.

- *Duncan's Multiple Range Test (DMRT)*. For comparisons of all possible pairs of means, first apply the standard DMRT procedure (see Chapter 5, Section 5.1.2) to the transformed means. For our example, the results are shown under Preliminary Step in Table 7.17.

Next, compute the treatment means based on the original scale, either by computing the treatment means from original data in Table 7.14 or converting the transformed means in Table 7.15 to the original scale (i.e., by computing the antilog of each transformed mean and reducing by 1). Theoretically, the latter procedure of converting the transformed means is more appropriate but, in practice, computing treatment means from original data is more frequently used because of its simplicity.

Even though the procedure of computing treatment means from original data is adequate in most cases, care should be taken to ensure that the order of ranking of the means in the original scale and in the transformed scale are not so different as to affect the order of mean comparison. In some rare cases, this condition is not satisfied, as illustrated by Table 7.18, in which the original mean of treatment 5 is higher than both treatments 4 and 6 but is the lowest on the basis of the transformed means. In such cases, the procedure of converting the transformed means is preferred.

Table 7.17 Application of Duncan's Multiple Range Test (DMRT) for Comparing the Treatment Means in Table 7.14, through DMRT Comparisons of the Transformed Means in Table 7.15

| Treatment Number | Preliminary Step | | Final Presentation | |
	Transformed Mean	DMRT[a]	Original Mean	DMRT[a]
1	0.6037	b	5.50	b
2	0.6831	b	4.50	b
3	0.8428	bc	7.25	bc
4	0.8306	bc	6.00	bc
5	1.1963	c	16.00	c
6	1.1750	c	20.00	c
7	0.0752	a	0.25	a
8	0.4549	ab	3.25	ab
9	0.9307	bc	8.50	bc

[a] Means followed by a common letter are not significantly different at the 5% level.

Table 7.18   Illustration of the Difference in the Ranking Order of Treatment Means Based on Original Scale and Transformed Scale[a]

Treatment Number	Green leafhopper, no./cage		Transformed Scale, logarithmic		Antilog of Transformed Mean
	Mean	Rank	Mean	Rank	
1	73.0	4	1.75872	4	57.4
2	22.8	10	1.34197	7	22.0
3	112.3	1	2.01669	1	103.9
4	25.5	9	1.26622	9	18.5
5	36.0	5	1.01940	10	10.5
6	35.0	6	1.47483	5	29.8
7	101.0	2	1.92451	2	84.0
8	94.3	3	1.91985	3	83.1
9	27.8	7	1.37783	6	23.9
10	26.0	8	1.30138	8	20.0

[a]Average of four replications (original replication data are not shown).

For our example, the first procedure of computing treatment means from the original data in Table 7.14 is used and the results are presented in Table 7.17 under the first column of final presentation. The DMRT results applied earlier to the transformed means are then directly transferred to the original means (see last column of Table 7.17).

7.2.2.1.2 Square-Root Transformation.   Square-root transformation is appropriate for data consisting of small whole numbers, for example, data obtained in counting rare events, such as the number of infested plants in a plot, the number of insects caught in traps, or the number of weeds per plot. For these data, the variance tends to be proportional to the mean.

The square-root transformation is also appropriate for percentage data where the range is between 0 and 30% or between 70 and 100%. For other ranges of percentage data, see discussion on the use of the arc sine transformation in the next section.

If most of the values in the data set are small (e.g., less than 10), especially with zeroes present, $(X + 0.5)^{1/2}$ should be used instead of $X^{1/2}$, where X is the original data.

For illustration, we use data on percentage of white heads from a rice variety trial of the 14 entries in a randomized complete block design with three replications (Table 7.19). The range of data is from 0 to 26.39%. Because many of the values are less than 10, data are transformed into $(X + 0.5)^{1/2}$, as shown in Table 7.20. Analysis of variance is then constructed on the transformed data in Table 7.20. The result is in Table 7.21.

Table 7.19   Percentage of White Heads of 14 Rice Varieties Tested in a RCB Design with Three Replications

Variety	White Heads, %			Total	Mean
	Rep. I	Rep. II	Rep. III		
IR5	1.39	0.92	2.63	4.94	1.65
IR20-1	8.43	4.38	6.94	19.75	6.58
C4-63G	7.58	3.79	1.91	13.28	4.43
C168-134	8.95	12.81	3.22	24.98	8.33
BPI-76	4.16	17.39	8.06	29.61	9.87
MRC 407-1	4.68	1.32	2.09	8.09	2.70
PARC 2-2	2.37	5.32	4.86	12.55	4.18
TN1	0.95	0.70	0.98	2.63	0.88
Rexoro	26.09	25.36	15.69	67.14	22.38
Luma-1	26.39	22.29	1.98	50.66	16.89
IR127-80-1	21.99	12.88	5.15	40.02	13.34
IR1108-3-5	3.58	2.62	2.91	9.11	3.04
IR1561-228-3-3	0.19	0.00	0.61	0.80	0.27
IR2061-464-2	0.00	3.64	4.44	8.08	2.69
Rep. total	116.75	113.42	61.47		
Grand total				291.64	

Table 7.20   Transformation of Data in Table 7.19 Using Square-Root Transformation, $(X + 0.5)^{1/2}$

Variety	Transformed Data			Total	Mean
	Rep. I	Rep. II	Rep. III		
IR5	1.37	1.19	1.77	4.33	1.44
IR20-1	2.99	2.21	2.73	7.93	2.64
C4-C3G	2.84	2.07	1.55	6.46	2.15
C168-134	3.07	3.65	1.93	8.65	2.88
BPI-76	2.16	4.23	2.93	9.32	3.11
MRC 407-1	2.28	1.35	1.61	5.24	1.75
PARC 2-2	1.69	2.41	2.32	6.42	2.14
TN1	1.20	1.10	1.22	3.52	1.17
Rexoro	5.16	5.09	4.02	14.27	4.76
Luma-1	5.19	4.77	1.57	11.53	3.84
IR127-80-1	4.74	3.66	2.38	10.78	3.59
IR1108-3-5	2.02	1.77	1.85	5.64	1.88
IR1561-228-3-3	0.83	0.71	1.05	2.59	0.86
IR2061-464-2	0.71	2.03	2.22	4.96	1.65
Rep. total	36.25	36.24	29.15		
Grand total				101.64	

Table 7.21 Analysis of Variance of Data in Table 7.20

Source of Variation	Degree of Freedom	Sum of Squares	Mean Square	Computed F^a
Replication	2	2.3971	1.1986	
Variety	13	48.0366	3.6951	5.88**
Error	26	16.3275	0.6280	
Total	41	66.7612		

[a]** = significant at 1% level.

For comparisons of all possible pairs of means, the DMRT (see Chapter 5, Section 5.1.2) is first applied to the transformed means and then transferred to the original means, following similar procedure described earlier for the logarithmic transformation in Section 7.2.2.1.1. The result is shown in Table 7.22.

7.2.2.1.3 Arc Sine Transformation. An arc sine or angular transformation is appropriate for data on proportions, data obtained from a count, and data expressed as decimal fractions or percentages. Note that percentage data that

Table 7.22 Application of Duncan's Multiple Range Test (DMRT) for Comparing the Treatment Means in Table 7.19, through DMRT Comparisons of the Transformed Means in Table 7.20

Variety	Transformed Mean[a]	Original Mean[a]
IR5	1.44 abc	1.65 abc
IR20-1	2.64 b–f	6.58 b–f
C4-63G	2.15 a–e	4.43 a–e
C168-134	2.88 c–f	8.33 c–f
BPI-76	3.11 def	9.87 def
MRC 407-1	1.75 a–d	2.70 a–d
PARC 2-2	2.14 a–e	4.18 a–e
TN1	1.17 ab	0.88 ab
Rexoro	4.76 g	22.38 g
Luma-1	3.84 fg	16.89 fg
IR127-80-1	3.59 efg	13.34 efg
IR1108-3-5	1.88 a–d	3.04 a–d
IR1561-228-3-3	0.86 a	0.27 a
IR2061-464-2	1.65 a–d	2.69 a–d

[a]Means followed by a common letter are not significantly different at the 5% level.

are derived from count data, such as percentage barren tillers (which is derived from the ratio of the number of nonbearing tillers to the total number of tillers), should be clearly distinguished from other types of percentage data, such as percentage protein or percentage carbohydrates, which are not derived from count data.

The mechanics of data transformation are greatly facilitated by using a table of the arc sine transformation (Appendix J). The value of 0% should be substituted by $(1/4n)$ and the value of 100% by $(100 - 1/4n)$, where n is the number of units upon which the percentage data was based (i.e., the denominator used in computing the percentage).

Not all percentage data need to be transformed and, even if they do, arc sine transformation is not the only transformation possible. As we indicated before, the square-root transformation is occasionally used for percentage data. The following rules may be useful in choosing the proper transformation scale for percentage data derived from count data.

RULE 1. For percentage data lying within the range of 30 to 70%, no transformation is needed.

RULE 2. For percentage data lying within the range of either 0 to 30% or 70 to 100%, but not both, the square-root transformation should be used.

RULE 3. For percentage data that do not follow the ranges specified in either rule 1 or rule 2, the arc sine transformation should be used.

We illustrate the application of arc sine transformation with data on percentage of insect survival in a rice variety trial with 12 varieties in a completely randomized design with three replications (Table 7.23). For each plant, 75 insects were caged and the number of surviving insects determined.

Table 7.23 Percentage Survival of Zigzag Leafhoppers on 12 Rice Varieties Tested in a CRD Experiment with Three Replications

Variety	Survival, %			Total	Mean
	Rep. I	Rep. II	Rep. III		
ASD 7	44.00	25.33	48.00	117.33	39.11
Mudgo	21.33	49.33	80.00	150.66	50.22
Ptb 21	0.00	0.00	0.00	0.00	0.00
D 204-1	25.33	26.66	49.33	101.32	33.77
Su-Yai 20	24.00	26.66	54.66	105.32	35.11
Balamawee	0.00	0.00	20.00	20.00	6.67
DNJ 24	32.00	29.33	28.00	89.33	29.78
Ptb 27	0.00	0.00	0.00	0.00	0.00
Rathu Heenati	17.33	33.33	10.66	61.32	20.44
Taichung (N)1	93.33	100.00	100.00	293.33	97.78
DS 1	13.33	36.00	33.33	82.66	27.55
BJ 1	46.66	46.66	16.00	109.32	36.44

Table 7.24   Transformation of Data in Table 7.23 Using Arc Sine Transformation

Variety	Rep. I	Rep. II	Rep. III	Total	Mean
		Survival (arc sine scale)			
ASD 7	41.55	30.22	43.85	115.62	38.54
Mudgo	27.51	44.62	63.43	135.56	45.19
Ptb 21	0.33	0.33	0.33	0.99	0.33
D 204-1	30.22	31.09	44.62	105.93	35.31
Su-Yai 20	29.33	31.09	47.67	108.09	36.03
Balamawee	0.33	0.33	26.57	27.23	9.08
DNJ 24	34.45	32.79	31.95	99.19	33.06
Ptb 27	0.33	0.33	0.33	0.99	0.33
Rathu Heenati	24.60	35.26	19.06	78.92	26.31
Taichung (N)1	75.03	89.67	89.67	254.37	84.79
DS 1	21.41	36.87	35.26	93.54	31.18
BJ 1	43.08	43.08	23.58	109.74	36.58

Based on rule 3, the arc sine transformation should be used because the percentage data ranged from 0 to 100%. Before transformation, all zero values are replaced by $[1/4(75)]$ and all 100 values by $\{100 - [1/4(75)]\}$. The transformed data are shown in Table 7.24 and its analysis of variance (following procedure of Chapter 2, Section 2.1.2) is in Table 7.25. The DMRT (see Chapter 5, Section 5.1.2) was first applied to the transformed means and then transferred to the original means, following a procedure similar to that earlier described for the logarithmic transformation. The result is shown in Table 7.26.

7.2.2.2   Error Partitioning.   Heterogeneity of variance, in which no functional relationship between the variance and the mean exists, is almost always due to the presence of one or more treatments whose associated errors are different from that of the others. These unusually large or unusually small errors are generally due to two major causes:

- They involve treatments which, by their own nature, exhibit either large or small variances (see Section 7.2.1.3 for examples).

Table 7.25   Analysis of Variance of Data in Table 7.24

Source of Variation	Degree of Freedom	Sum of Squares	Mean Square	Computed F^a
Variety	11	16,838.6368	1,530.7852	16.50**
Error	24	2,225.9723	92.7488	
Total	35	19,064.6091		

[a]** = significant at 1% level.

Table 7.26   Application of Duncan's Multiple Range Test (DMRT) for Comparing the Treatment Means in Table 7.23, through DMRT Comparison of the Transformed Means in Table 7.24

Variety	Transformed Mean[a]	Original Mean[a]
ASD 7	38.54 bc	39.11 bc
Mudgo	45.19 c	50.22 c
Ptb 21	0.33 a	0.00 a
D 204-1	35.31 bc	33.77 bc
Su-Yai 20	36.03 bc	35.11 bc
Balamawee	9.08 a	6.67 a
DNJ 24	33.06 bc	29.78 bc
Ptb 27	0.33 a	0.00 a
Rathu Heenati	26.31 b	20.44 b
Taichung (N)1	84.79 d	97.78 d
DS 1	31.18 bc	27.55 bc
BJ 1	36.58 bc	36.44 bc

[a] Means followed by a common letter are not significantly different at the 5% level.

- They involve gross errors; that is, some unusually large or unusually small values may have been mistakenly recorded in some plots resulting in unusually large error variances for the treatments involved.

Error partitioning is a commonly used procedure to handle data that have heterogeneous variances that are not functionally related to the mean. Error partitioning should not be used, however, when variance heterogeneity is due to gross errors. In other words, error partitioning should be applied only after the presence of gross errors has been thoroughly examined and eliminated.

We describe the step-by-step procedures for detecting gross errors and for applying the error partitioning method. We use yield data from a variety trial with 35 entries, consisting of 15 hybrids, 17 parents, and 3 checks, tested in a RCB design with three replications (Table 7.27).

☐ STEP 1.   Detect gross errors, as follows:

- Identify treatments that have extremely large differences between observations of different replications. For each of these treatments, identify the specific plot whose value is greatly different from the rest (i.e., plots with unusually large or unusually small values).
- For each plot in question, check the records or the daily logbook to see if any special observations or remarks were noted by the researcher to explain the extreme value.
- On the field layout, mark all plots having extreme values by putting a plus sign on the plot with unusually high value and a minus sign on the

Table 7.27   Yields of 35 Entries Tested in a RCB Design with Three Replications

Entry Number[a]	Yield, t/ha			Range
	Rep. I	Rep. II	Rep. III	
1	8.171	7.951	8.074	0.220
2	7.049	7.792	7.626	0.743
3	8.067	8.597	6.772	1.825
4	7.855	7.601	7.273	0.582
5	8.815	8.259	7.621	1.194
6	7.211	8.115	8.488	1.277
7	6.557	8.388	6.895	1.831
8	7.999	8.701	8.253	0.702
9	9.310	8.310	9.194	1.000
10	7.372	8.198	8.246	0.874
11	7.142	6.980	8.653	1.673
12	8.265	9.097	8.514	0.832
13	7.413	8.807	10.128	2.715
14	7.130	7.990	8.088	0.958
15	7.089	8.543	7.893	1.454
16	5.832	5.671	6.042	0.371
17	7.619	5.580	8.488	2.908
18	8.427	8.327	7.065	1.362
19	7.311	6.984	7.240	0.327
20	6.010	7.124	6.536	1.114
21	6.514	7.366	7.240	0.852
22	7.832	7.251	7.116	0.716
23	7.914	7.994	7.519	0.475
24	7.448	7.808	7.327	0.481
25	7.014	8.799	7.301	1.785
26	6.375	7.716	6.590	1.341
27	7.042	6.531	6.699	0.511
28	5.998	6.888	6.926	0.928
29	7.175	7.756	7.528	0.581
30	7.425	7.531	7.091	0.440
31	7.453	7.568	7.607	0.154
32	7.073	8.244	6.839	1.405
33	7.235	7.362	7.445	0.210
34	6.984	7.723	7.735	0.751
35	7.185	6.958	7.417	0.459

[a]Entries 1 to 15 are hybrids, 16 to 32 are parents, and 33 to 35 are checks.

plot with unusually low value. Examine the proximity of the plots with pluses and minuses to pinpoint possible causes that are related to plot location in the experimental area.

- For plots whose causes of extreme values were identified as gross errors, retrieve the correct values if possible. If retrieval is not possible, the suspected data should be rejected and missing data is declared (see Section 7.1.1). For plots whose causes of extreme values cannot be determined, the suspected data should be retained.

For our example, based on the range values shown in Table 7.27, two entries (13 and 17) are identified to have extremely large range values. The range value of entry 13 is 2.715 t/ha (10.128 − 7.413) and that of entry 17 is 2.908 t/ha (8.488 − 5.580).

By a thorough check (following the procedures just described), it was found that the extremely low value of 5.580 t/ha for entry 17 was the result of a transcription error and that the correct value should have been 7.580 t/ha. On the other hand, the cause for the extremely high value of 10.128 t/ha for entry 13 could not be determined; hence, the value of 10.128 t/ha is retained.

☐ STEP 2. Construct the standard analysis of variance on the revised data, following the standard procedures of Chapters 2 to 4. For our example, the revised data set is the same as that shown in Table 7.27 except that the value of entry 17 in replication II of 5.580 t/ha is replaced by 7.580 t/ha. The standard analysis of variance for a RCB design with $t = 35$ treatments and $r = 3$ replications is computed from the revised set of data, following the procedure of Chapter 2, Section 2.2.3. The result is shown in Table 7.28.

☐ STEP 3. Classify the treatments into s groups, each group containing treatments with homogeneous variance. For our example, a logical way of grouping is to classify the entries into three groups, namely, hybrids, parents, and checks. Thus, there are $s = 3$ groups with 15 entries in the first group, 17 in the second group, and 3 in the third group.

Table 7.28 The Analysis of Variance (RCB Design) for the Revised[a] Data in Table 7.27

Source of Variation	Degree of Freedom	Sum of Squares	Mean Square	Computed F	Tabular F 5%	1%
Replication	2	3.306077	1.653038			
Entry	34	40.019870	1.177055	4.00**	1.59	1.94
Error	68	20.014056	0.294324			
Total	104	63.340003				

[a] The value of 5.580 of entry number 17 in replication II is replaced by 7.580.

☐ STEP 4. Partition the treatment sum of squares and the error sum of squares in the analysis of variance, following procedures of Chapter 5, Section 5.2, into the following components:

- Components of treatment SS:

$$C_0: \text{Between groups}$$

$$C_1: \text{Between treatments within group 1}$$

$$C_2: \text{Between treatments within group 2}$$

$$\vdots$$

$$C_s: \text{Between treatments within group } s$$

- Components of error SS:

$$C_0 \times \text{replication}$$

$$C_1 \times \text{replication}$$

$$C_2 \times \text{replication}$$

$$\vdots$$

$$C_s \times \text{replication}$$

For our example, the treatment SS and the error SS are partitioned into four components, as shown in Table 7.29.

☐ STEP 5. Exclude the first component of error (i.e., $C_0 \times$ replication) and any component having less than 6 $d.f.$ Apply a test for homogeneity of variance of Chapter 11, Section 11.2, to the rest of the components of error. Those not found to be significantly different are then pooled. If any pooling or removal of error components is performed, the corresponding pooling and removal of the treatment components should also be done.

The pooled sum of squares over k components, each with f_i $d.f.$, has $f = \sum_{i=1}^{k} f_i$ degrees of freedom and is computed as:

$$SS_p = \sum_{i=1}^{k} SS_i$$

where SS_i is the ith component of the error sum of squares. Thus, the

Table 7.29 Partitioning of Treatment and Error Sums of Squares, of the Analysis of Variance in Table 7.28, Based on Homogeneous Groupings of Treatments

Source of Variation	Degree of Freedom	Sum of Squares	Mean Square	Computed F^a	Tabular F 5%	Tabular F 1%
Replication	2	3.306077	1.653038			
Entry	34	40.019870	1.177055	4.00**	1.55	1.94
Between groups	(2)	15.109932	7.554966	b	—	—
Entries within group 1 (hybrid)	(14)	9.911929	0.707995	1.63ns	2.06	2.80
Entries within group 2 (parent)	(16)	14.867981	0.929249	4.38**	1.97	2.62
Entries within group 3 (check)	(2)	0.130028	0.065014	c	—	—
Error	68	20.014056	0.294324			
Replication × Between groups	(4)	0.830352	0.207588			
Replication × Hybrid	(28)	12.127555	0.433127			
Replication × Parent	(32)	6.795953	0.212374			
Replication × Check	(4)	0.260196	0.065049			

[a] ** = significant at 1% level, ns = not significant.

[b] Replication × between groups $d.f.$ is not adequate for valid test of significance.

[c] Replication × check $d.f.$ is not adequate for valid test of significance.

313

pooled mean square is computed as:

$$MS_p = \frac{SS_p}{f}$$

For our example, because the fourth error component has only 4 *d.f.*, it is excluded. The F value for testing the homogeneity of variance between the second and third components of the error SS is computed as:

$$F = \frac{0.433127}{0.212374} = 2.04$$

The computed F value is significant at the 5% level. Thus, the two error components, one concerning hybrids and the other concerning parents, are not homogeneous. Hence, the two error variances are not pooled.

☐ STEP 6.   Test each of the components (pooled or nonpooled) of the treatment SS against its corresponding error component. For our example, the hybrid MS is tested against the replication × hybrid MS and the parent MS is tested against the replication × parent MS:

$$F \text{ (hybrid)} = \frac{0.707995}{0.433127} = 1.63^{\text{ns}}$$

$$F \text{ (parent)} = \frac{0.929249}{0.212374} = 4.38^{**}$$

Note that if the first and fourth error components had had sufficient *d.f.*, the between groups component and the between checks component would have been also tested.

☐ STEP 7.   For mean comparison involving treatments from the same group (pooled or nonpooled), the standard test procedures described in Chapter 5 can be applied directly, using the appropriate error terms. For pair comparisons of treatments coming from different groups, the standard error of the mean difference, where one treatment comes from the ith group and the other treatment comes from the jth group, is computed as:

$$s_{\bar{d}} = \sqrt{\frac{1}{r}\left(s_i^2 + s_j^2\right)}$$

where r is the number of replications; and s_i^2 and s_j^2 are the components of the error mean squares corresponding to the ith group and the jth group, respectively. For the computation of the LSD values (see Chapter 5, Section 5.1.1), the tabular t values are obtained as follows:

· If the error degrees of freedom are the same for the two groups, the tabular t value is obtained directly from Appendix C with n equals the common error *df*.

- If the error degrees of freedom in the two groups differ, with n_i $d.f.$ for the ith group and n_j $d.f.$ for the jth group, the tabular t value is computed as:

$$t = \frac{s_i^2 t_i + s_j^2 t_j}{s_i^2 + s_j^2}$$

where t_i and t_j are the tabular t values obtained from Appendix C with n_i and n_j degrees of freedom, respectively.

For our example, the standard errors for the three types of pair comparison are computed as:

- Between two hybrid means:

$$s_{\bar{d}} = \sqrt{\tfrac{2}{3}(0.433127)}$$

$$= 0.54$$

- Between two parent means:

$$s_{\bar{d}} = \sqrt{\tfrac{2}{3}(0.212374)}$$

$$= 0.38$$

- Between one hybrid mean and one parent mean:

$$s_{\bar{d}} = \sqrt{\tfrac{1}{3}(0.433127 + 0.212374)}$$

$$= 0.46$$

CHAPTER 8

Analysis of Data From a Series of Experiments

Crop performance depends on the genotype, the environment in which the crops are grown, and the interaction between the genotype and the environment. Genotype and some factors of the environment, such as fertilizer rate, plant population, and pest control, can be controlled by the researcher. But other factors of the environment, such as sunshine, rainfall, and some soil properties, are generally fixed and difficult to modify for a given site and planting season. Thus, a researcher with a one-time experiment at a single site can vary and evaluate only the controllable factors but not the environmental factors that are beyond his control.

The effect of the uncontrollable environmental factors on crop performance is as important as, if not more important than, that of the controllable factors; and the evaluation and quantification of their effects are just as essential. Because the uncontrollable factors are expected to change with season and site, and because these changes are measurable, their effects on treatment performance can be evaluated. In crop research, the most commonly used way to evaluate the effects of the uncontrollable environmental factors on crop response is to repeat the experiment at several sites, or over several crop seasons, or both.

Experiments that are conducted at several sites and repeated over several seasons can be classified into four groups according to their objectives. These are:

- Preliminary evaluation experiments, which are designed to identify—from a large number of new technologies—a few technologies that give a consistently superior performance in the area where they are developed.
- Technology adaptation experiments, which are designed to determine the range of geographical adaptability of the few superior technologies identified in one or more preliminary evaluation experiments.
- Long-term experiments, which are designed to characterize a new technology with respect to its long-term effect on productivity.

- Response prediction experiments, which are designed to identify a functional relationship between crop response and some environmental factors and predict productivity over a wide range of environments.

The primary focus of this chapter is to describe these four types of experiment and the corresponding data analyses that are critical in answering the prescribed objectives. Our descriptions assume that appropriate analyses for each individual trial (see Chapters 2 to 7) have been completed and we discuss only the combined analysis over several trials. We emphasize the use of specific statistical techniques that are suited to the specific objectives of the experiment.

8.1 PRELIMINARY EVALUATION EXPERIMENT

The development of an improved crop production technology usually involves a series of elimination processes starting with a large number of new technologies and ending with a few superior ones that are identified and to be recommended for commercial use. Much of the elimination processes is done at a single site where the new technologies are developed and assembled.

The preliminary evaluation experiment is a part of the elimination process at a single site where trials are replicated and repeated over several planting seasons. Its primary objective is to identify a technology (or technologies) that is consistently superior in at least one site where consistent superiority is defined as top ranking performance for at least two planting times—either over seasons in the same year or over years of the same season, or both.

8.1.1 Analysis Over Seasons

For a given crop at a specific site, planting is usually not staggered uniformly over a 12-month period but is distinctly bunched in some well-defined periods that are consistently repeated over years. A good example is maize grown in temperate climates where only one crop is generally planted at the start of summer. Even in the humid tropics, where two to three crops of maize can be grown within a 12-month period, the planting seasons usually remain distinct with respect to planting date and expected environmental features. Consequently, the planting season within a year is considered a *fixed variable*: a superior technology can be separately identified to fit a specific season. In fact, the primary objective of a combined analysis over seasons is to examine the interaction between season and treatment and to determine the necessity of a separate technology recommendation for each planting season.

We give the procedures for analyzing data of experiments over crop seasons using a fertilizer trial with five nitrogen rates tested on rice for two seasons, using a RCB design with three replications. Grain yield data are shown in Table 8.1 and the individual analyses of variance are shown in Table 8.2. The

Table 8.1 Grain Yield of Rice Tested with Five Rates of Nitrogen in Two Crop Seasons

Nitrogen Rate, kg/ha	Grain Yield, t/ha			Total	Mean
	Rep. I	Rep. II	Rep. III		
Dry Season					
0 (N_0)	4.891	2.577	4.541	12.009	4.003
60 (N_1)	6.009	6.625	5.672	18.306	6.102
90 (N_2)	6.712	6.693	6.799	20.204	6.735
120 (N_3)	6.458	6.675	6.636	19.769	6.590
150 (N_4)	5.683	6.868	5.692	18.243	6.081
Total				88.531	
Wet Season					
0	4.999	3.503	5.356	13.858	4.619
60	6.351	6.316	6.582	19.249	6.416
90	6.071	5.969	5.893	17.933	5.978
120	4.818	4.024	5.813	14.655	4.885
150	3.436	4.047	3.740	11.223	3.741
Total				76.918	

Table 8.2 Individual Analyses of Variance (RCB Design) for a Rice Fertilizer Trial with Five Treatments and Three Replications, by Crop Season

Source of Variation	Degree of Freedom	Sum of Squares	Mean Squares	Computed F^a
Dry Season				
Replication	2	0.018627	0.009314	
Nitrogen	4	14.533384	3.633346	6.43*
Error	8	4.522162	0.565270	
Wet Season				
Replication	2	1.242944	0.621472	
Nitrogen	4	13.869888	3.467472	10.91**
Error	8	2.541472	0.317684	

[a]** = significant at 1% level, * = significant at 5% level.

step-by-step procedures for the analysis of data combined over seasons are:

☐ STEP 1.   Construct the outline of the combined analysis of variance over crop seasons, based on the basic experimental design used. For our example, the outline of the combined analysis of variance over seasons based on RCB design is shown in Table 8.3.

☐ STEP 2.   Compute the various SS in the combined analysis of variance of step 1:

• Compute the replications within season SS as the sum over s replication SS, and the pooled error SS as the sum over s error SS, from the individual analyses of variance. That is,

$$\text{Reps. within season } SS = \sum_{i=1}^{s} (\text{Rep. } SS)_i$$

$$\text{Pooled error } SS = \sum_{i=1}^{s} (\text{Error } SS)_i$$

where $(\text{Rep. } SS)_i$ and $(\text{Error } SS)_i$ are the replication SS and error SS from the analysis of variance of the ith season.

For our example, using the SS values in Table 8.2, the foregoing two SS are computed as:

$$\text{Reps. within season } SS = (\text{Rep. } SS)_D + (\text{Rep. } SS)_W$$

$$= 0.018627 + 1.242944$$

$$= 1.261571$$

Table 8.3   Outline of the Combined Analysis of Variance Over s Crop Seasons,[a] Based on RCB Design with t Treatments and r Replications

Source of Variation	Degree of Freedom	Mean Square	Computed F
Season (S)	$s - 1$	$S\ MS$	$\dfrac{S\ MS}{R\ MS}$
Reps. within season	$s(r - 1)$	$R\ MS$	
Treatment (T)	$t - 1$	$T\ MS$	$\dfrac{T\ MS}{E\ MS}$
$S \times T$	$(s - 1)(t - 1)$	$S \times T\ MS$	$\dfrac{S \times T\ MS}{E\ MS}$
Pooled error	$s(r - 1)(t - 1)$	$E\ MS$	
Total	$srt - 1$		

[a]Crop season is considered as a fixed variable.

$$\text{Pooled error } SS = (\text{Error } SS)_D + (\text{Error } SS)_W$$

$$= 4.522162 + 2.541472$$

$$= 7.063634$$

where subscripts D and W indicate dry season and wet season.

- Compute the other SS values, either from total or mean values:

(i) *Based on Totals.* The computations of the various SS follow the standard procedure as:

$$C.F. = \frac{\left(\sum\limits_{i=1}^{s} G_i \right)^2}{srt}$$

$$S\ SS = \sum_{i=1}^{s} \frac{G_i^2}{rt} - C.F.$$

$$T\ SS = \sum_{j=1}^{t} \frac{T_j^2}{sr} - C.F.$$

$$S \times T\ SS = \sum_{i=1}^{s} \sum_{j=1}^{t} \frac{(ST)_{ij}^2}{r} - C.F. - S\ SS - T\ SS$$

where G_i is the grand total of the ith season, $(ST)_{ij}$ is the total of the jth treatment in the ith season, and T_j is the total of the jth treatment over all s seasons.

(ii) *Based on Means.* The computation of the various SS is:

$$C.F. = srt(\bar{G})^2$$

$$S\ SS = rt\left(\sum_{i=1}^{s} \bar{S}_i^2 \right) - C.F.$$

$$T\ SS = sr\left(\sum_{j=1}^{t} \bar{T}_j^2 \right) - C.F.$$

$$S \times T\ SS = r\left[\sum_{i=1}^{s} \sum_{j=1}^{t} (\overline{ST})_{ij}^2 \right] - C.F. - S\ SS - T\ SS$$

where $\bar{G}$ is the grand mean, $(\overline{ST})_{ij}$ is the mean of the jth treatment in the ith season, $\bar{S}_i$ is the mean of the ith season, and $\bar{T}_j$ is the mean of the jth treatment.

Note that the mean values used in the computation of *SS* should have the same number of significant digits as is used in the totals. This warning in necessary because mean values used for presentation purposes are usually rounded off to minimum significant digits.

For our example, computation is made using the total values in Table 8.1, as follows:

$$C.F. = \frac{(88.531 + 76.918)^2}{30}$$

$$= 912.44572$$

$$S\ SS = \frac{(88.531)^2 + (76.918)^2}{3(5)} - 912.44572$$

$$= 4.495392$$

$$T\ SS = \left\{ \left[(25.867)^2 + (37.555)^2 + (38.137)^2 \right. \right.$$

$$\left. \left. + (34.424)^2 + (29.466)^2 \right] / 6 \right\} - 912.44572$$

$$= 18.748849$$

$$S \times T\ SS = \frac{(12.009)^2 + (18.306)^2 + \cdots + (11.223)^2}{3}$$

$$- 912.44572 - 4.495392 - 18.748849$$

$$= 9.654423$$

☐ STEP 3. For each source of variation, compute the mean square by dividing each *SS* by its *d.f.* For our example, the results of the computation are shown in Table 8.4.

Table 8.4 Combined Analysis of Variance over Two Crop Seasons, Computed from Data in Tables 8.1 and 8.2

Source of Variation	Degree of Freedom	Sum of Squares	Mean Square	Computed F^a
Season (S)	1	4.495392	4.495392	[b]
Reps. within season	4	1.261571	0.315393	
Nitrogen (N)	4	18.748849	4.687212	10.62**
S × N	4	9.654423	2.413606	5.47**
Pooled error	16	7.063634	0.441477	
Total	29			

[a]** = significant at 1% level.
[b]Reps. within season *d.f.* is not adequate for valid test of significance.

☐ STEP 4. Test the homogeneity of the s error MS from the individual analyses of variance, through the application of the F test or the chi-square test (Chapter 11, Section 11.2). For most purposes, if the highest error MS is not three-fold larger than the smallest error MS, the error variances can be considered homogeneous.

For our example, because there are only two error MS, the F test is applied as:

$$F = \frac{\text{Larger error } MS}{\text{Smaller error } MS}$$

$$= \frac{0.565270}{0.317684} = 1.78$$

The computed F value is smaller than 3.44, the corresponding tabular F value with $f_1 = f_2 = 8$ degrees of freedom at the 5% level of significance. Thus, the hypothesis of homogeneous error variances over seasons cannot be rejected.

☐ STEP 5. If homogeneity of error variances cannot be established in step 4, proceed to step 7. Otherwise, compute the F values for testing the various effects, as indicated in Table 8.3, as:

$$F(S) = \frac{S\ MS}{\text{Reps. within season } MS}$$

$$F(T) = \frac{T\ MS}{\text{Pooled error } MS}$$

$$F(S \times T) = \frac{S \times T\ MS}{\text{Pooled error } MS}$$

For our example, the homogeneity test of step 4 is not significant. Hence, the F values are computed, using the foregoing formulas, as:

$$F(N) = \frac{4.687212}{0.441477} = 10.62$$

$$F(S \times N) = \frac{2.413606}{0.441477} = 5.47$$

Note that the F value for the season effect is not computed due to the inadequate $d.f.$ for the replications within season MS.

The corresponding tabular F values for each of the two computed F values, with $f_1 = 4$ and $f_2 = 16$ $d.f.$, are 3.01 at the 5% level of significance and 4.77 at the 1% level. Thus, both the treatment main effect and its interaction with crop season are highly significant; there is a significant yield response to treatment (nitrogen application) but the response differed between the two crop seasons.

· □ STEP 6. If the season × treatment interaction ($S \times T$) is significant, parti tion the interaction SS into a set of orthogonal contrasts (see Chapter 5, Section 5.2.4) that is most likely to provide information on the nature of the interaction; why the relative performance of the treatments differed over seasons.

For our example, with nitrogen rates as treatments, the $S \times T$ interaction corresponds to $S \times N$ which is highly significant (Table 8.4). The most natural set of contrasts that can explain the nature of such an interaction is one involving the orthogonal polynomials on nitrogen. That is, the $S \times N$ SS should be partitioned into $S \times N_{linear}$, $S \times N_{quadratic}$, and so on. Because the nitrogen rates tested in this experiment have unequal intervals, the orthogonal polynomial coefficients are derived following the procedures given in Chapter 5, Section 5.2.3.2. The derived coefficients for the linear and quadratic polynomials are shown below, and the results of the partitioned SS are shown in Table 8.5.

Nitrogen Rate, kg/ha	Orthogonal Polynomial Coefficient	
	Linear	Quadratic
0	−14	22
60	−4	−21
90	1	−21
120	6	−5
150	11	25

Table 8.5 Combined Analysis of Variance in Table 8.4, with Partitioning of SS

Source of Variation	Degree of Freedom	Sum of Squares	Mean Square	Computed F^a
Season (S)	1	4.495392	4.495392	b
Reps. within season	4	1.261571	0.315393	
Nitrogen (N)	4	18.748849	4.687212	10.62**
N_L	(1)	1.435356	1.435356	3.25^{ns}
N_Q	(1)	17.185048	17.185048	38.93**
$N_{Res.}$	(2)	0.128445	0.064222	< 1
$S \times N$	4	9.654423	2.413606	5.47**
$S \times N_L$	(1)	8.807778	8.807778	19.95**
$S \times N_Q$	(1)	0.547297	0.547297	1.24^{ns}
$S \times N_{Res.}$	(2)	0.299348	0.149674	< 1
Pooled error	16	7.063634	0.441477	

^{a}** = significant at 1% level, ^{ns} = not significant.
^{b}Reps. within season $d.f.$ is not adequate for valid test of significance.

The analysis indicates that only the linear part of the response function varied significantly with season. Because only the quadratic component of the N SS is significant, the quadratic function is fitted to the treatment mean values, separately for each season, following the regression technique of Chapter 9, Section 9.2.2. The results are:

Dry season: $\hat{Y}_1 = 3.983 + 0.0523N - 0.000255N^2$ ($R^2 = 1.00**$)

Wet season: $\hat{Y}_2 = 4.675 + 0.0477N - 0.000366N^2$ ($R^2 = .97*$)

The two regression equations are represented graphically in Figure 8.1. Visual examination of Figure 8.1, together with the large and significant R^2 value of each regression, indicates that the quadratic response fitted the data reasonably well. The rate of yield increase with increase in the rate of nitrogen application (i.e., the linear component of the function) is higher in the dry season than in the wet season.

With the estimated nitrogen response functions, the two types of optimum nitrogen rate can be computed as:

- The nitrogen rate that maximizes yield:

$$N_y = \frac{-b}{2c}$$

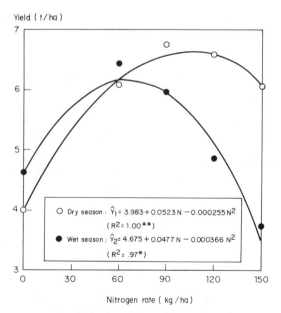

Yield (t / ha)

O Dry season : $\hat{Y}_1 = 3.983 + 0.0523N - 0.000255N^2$
($R^2 = 1.00**$)

● Wet season : $\hat{Y}_2 = 4.675 + 0.0477N - 0.000366N^2$
($R^2 = .97*$)

Nitrogen rate (kg / ha)

Figure 8.1   Nitrogen response curves estimated from data in Table 8.1, separately for dry and wet season.

- The nitrogen rate that maximizes profit:

$$N_p = \frac{1}{2c}\left(\frac{P_f}{P_y} - b\right)$$

where b and c are the estimates of the regression coefficients in $\hat{Y} = a + bN + cN^2$, and P_f and P_y are the prices of nitrogen and rice, respectively.

For our example, the N_y values for both regressions are estimated as:

$$\text{Dry season: } N_y = \frac{-0.0523}{(2)(-0.000255)} = 102.5 \text{ kg N/ha}$$

$$\text{Wet season: } N_y = \frac{-0.0477}{(2)(-0.000366)} = 65.2 \text{ kg N/ha}$$

To compute the N_p values, we assume that the ratio of the price of nitrogen (kg/ha) over the price of rice (t/ha) is $= 0.005$. With this price ratio, the N_p values are estimated as:

$$\text{Dry season: } N_p = \frac{0.005 - 0.0523}{2(-0.000255)} = 92.7 \text{ kg N/ha}$$

$$\text{Wet season: } N_p = \frac{0.005 - 0.0477}{2(-0.000366)} = 58.3 \text{ kg N/ha}$$

The experimental results seem to indicate the need to have different nitrogen recommendations for dry and wet seasons.

☐ STEP 7. If the test for homogeneity of variances in step 4 is significant, follow the partitioning procedure of the $S \times T$ interaction SS outlined in step 6, but instead of using the pooled error MS as the error term for the F test, the pooled error SS must first be partitioned into components corresponding to those of the $S \times T$ SS. The F value is then computed for each component of the $S \times T$ interaction using the corresponding component in the pooled error as its error term.

For our example, this analysis is not needed because the homogeneity test of step 4 is not significant. However, for illustration purposes, the analysis will be performed by using the same partitioning of the $S \times N$ SS given in step 6. The corresponding components of the pooled error SS are computed using the formula:

$$(\text{Reps. within } S) \times N_L = \sum_{i=1}^{s}\left[\sum_{k=1}^{r}(N_L\,SS)_{ki} - (N_L\,SS)_i\right]$$

$$(\text{Reps. within } S) \times N_Q = \sum_{i=1}^{s}\left[\sum_{k=1}^{r}(N_Q\,SS)_{ki} - (N_Q\,SS)_i\right]$$

where $(N_L\ SS)_{ki}$ is the linear component of the $N\ SS$ computed from the data of the kth replication in the ith season, $(N_L\ SS)_i$ is the corresponding component computed from the totals over all replications, and $(N_Q\ SS)_{ki}$ and $(N_Q\ SS)_i$ are similarly defined but for the quadratic component.

The computations of these sums of squares, done separately for each crop season, are shown in Table 8.6. For example, the linear component of the N SS for replication I of the dry season is computed as:

$$(N_L\ SS)_{11} = \frac{[-14(4.891) - 4(6.009) + (6.712) + 6(6.458) + 11(5.683)]^2}{(-14)^2 + (-4)^2 + (1)^2 + (6)^2 + (11)^2}$$

$$= \frac{(15.463)^2}{370} = 0.646228$$

And, the linear component of the N SS for the dry season is computed as:

$$(N_L\ SS)_1 = \frac{[-14(12.009) - 4(18.306) + (20.204) + 6(19.769) + 11(18.243)]^2}{3[(-14)^2 + (-4)^2 + (1)^2 + (6)^2 + (11)^2]}$$

$$= \frac{(98.141)^2}{(3)(370)} = 8.677167$$

Table 8.6   Computation of the Linear and Quadratic Components of the Pooled Error SS in Table 8.5, Data are from Table 8.1

Replication Number	Treatment Total					Sum of Squares[a]	
	N_0	N_1	N_2	N_3	N_4	Linear	Quadratic
			Dry Season				
I	4.891	6.009	6.712	6.458	5.683	0.646228	1.227907
II	2.577	6.625	6.693	6.675	6.868	9.636871	3.555132
III	4.541	5.672	6.799	6.636	5.692	1.425382	1.386474
Sum	12.009	18.306	20.204	19.769	18.243	8.677167	5.799363
			Wet Season				
I	4.999	6.351	6.071	4.818	3.436	1.382265	3.935604
II	3.503	6.316	5.969	4.024	4.047	0.000284	4.946835
III	5.356	6.582	5.893	5.813	3.740	1.017294	3.151471
Sum	13.858	19.249	17.933	14.655	11.223	1.565967	11.932982

[a] The orthogonal polynomial coefficients are $(-14, -4, 1, 6, 11)$ for linear and $(22, -21, -21, -5, 25)$ for quadratic components.

Finally, the components of the pooled error SS are computed as:

$$(\text{Reps. within } S) \times N_L = [(0.640228 + 9.636871 + 1.425382) - 8.677167]$$
$$+ [(1.382265 + 0.000284 + 1.017294) - 1.565967]$$
$$= 3.031314 + 0.833876 = 3.865190$$

$$(\text{Reps. within } S) \times N_Q = [(1.227907 + 3.555132 + 1.386474) - 5.799363]$$
$$+ [(3.935604 + 4.946835 + 3.151471) - 11.932982]$$
$$= 0.370150 + 0.100928 = 0.471078$$

$$(\text{Reps. within } S) \times N_{\text{Res.}} = \text{Pooled error } SS - (\text{Reps. within } S) \times N_L SS$$
$$- (\text{Reps. within } S) \times N_Q SS$$
$$= 7.063634 - 3.865190 - 0.471078 = 2.727366$$

These results are summarized as:

Source of Variation	Degree of Freedom	Sum of Squares	Mean Square
Pooled error	(16)	(7.063634)	(0.441477)
(Reps. within S) $\times N_L$	4	3.865190	0.966298
(Reps. within S) $\times N_Q$	4	0.471078	0.117770
(Reps. within S) $\times N_{\text{Res.}}$	8	2.727366	0.340921

The three components of the pooled error are then used to test the significance of the corresponding components of the $S \times N$ interaction in Table 8.5 as:

$$F(S \times N_L) = \frac{S \times N_L\ MS}{(\text{Reps. within } S) \times N_L\ MS}$$
$$= \frac{8.807778}{0.966298} = 9.11$$

$$F(S \times N_Q) = \frac{S \times N_Q\ MS}{(\text{Reps. within } S) \times N_Q\ MS}$$
$$= \frac{0.547297}{0.117770} = 4.65*$$

$$F(S \times N_{\text{Res.}}) = \frac{S \times N_{\text{Res.}}\ MS}{(\text{Reps. within } S) \times N_{\text{Res.}}\ MS}$$
$$= \frac{0.149674}{0.340921} < 1*$$

*Although degrees of freedom of the (Reps. within S) $\times N_L$ MS and (Reps. within S) $\times N_Q$ MS are not adequate for valid test of significance (see Chapter 2, Section 2.1.2.1, step 6), for illustration purposes, such deficiency has been ignored.

8.1.2   Analysis Over Years

In contrast to the planting seasons within a year, which are characterized by some distinct and predictable environmental features (see Section 8.1.1), the variability of the environment over years is usually unpredictable. For example, it is not reasonable to expect that odd-numbered years are drier than even-numbered years or that sunshine is increasing over years. Because of the absence of any predictable pattern, years are generally considered as a *random variable*: a superior technology must show consistent superiority over several years for at least one planting season. Thus, the primary objective of a combined analysis of data over years is to identify technologies whose average effect over years is stable and high. The interaction between treatment and year has no agronomic meaning and, therefore, is much less important than the interaction between treatment and season.

We illustrate the procedure for combining data over years with a trial, with seven rice varieties tested in two years using a RCB design with three replications. Grain yield data are shown in Table 8.7 and individual analyses of variance in Table 8.8. Variety means are presented for each year in Table 8.9.

The step-by-step procedures for the analysis of data combined over years are:

☐ STEP 1.   Construct an outline of the combined analysis of variance over years, based on the basic design used. For our example, the outline of the combined analysis of variance over years based on RCB design is shown in Table 8.10. Note that the primary difference between this combined analysis of variance over years and that over seasons (Section 8.1.1, Table 8.3) is in

Table 8.7   Grain Yield (t / ha) of Seven Rice Varieties Tested in RCB Design with Three Replications, in the Same Season for Two Consecutive Years

Variety Number	Year 1				Year 2[a]			
	Rep. I	Rep. II	Rep. III	Total	Rep. I	Rep. II	Rep. III	Total
1	3.036	4.177	3.884	11.097	1.981	3.198	3.726	8.905
2	1.369	1.554	1.899	4.822	3.751	2.391	3.714	9.856
3	5.311	5.091	4.839	15.241	3.868	3.134	3.487	10.489
4	2.559	3.980	3.853	10.392	2.729	2.786	2.598	8.113
5	1.291	1.705	2.130	5.126	3.222	3.554	2.452	9.228
6	3.452	3.548	4.640	11.640	4.250	4.134	3.339	11.723
7	1.812	2.914	0.958	5.684	3.336	4.073	2.885	10.294
Total				64.002				68.608

[a]Plot layout was rerandomized in the second year.

Table 8.8 Individual Analyses of Variance, by Year, of a Variety Trial in RCB Design with Seven Varieties and Three Replications, Computed from Data in Table 8.7

Source of Variation	Degree of Freedom	Sum of Squares	Mean Square	Computed F^a	Tabular F 5%	1%
		Year 1				
Replication	2	1.38549	0.69274			
Variety	6	31.85691	5.30948	16.11**	3.00	4.82
Error	12	3.95588	0.32966			
		Year 2				
Replication	2	0.09698	0.04849			
Variety	6	2.79800	0.46633	1.16^{ns}	3.00	4.82
Error	12	4.84346	0.40362			

^{a}** = significant at 1% level, ^{ns} = not significant.

the F test for testing the significance of the treatment effect. When data are combined over seasons (i.e., fixed variable) the error term is the pooled error MS; when combined over years (i.e., random variable) the error term is the year $\times$ treatment interaction MS.

☐ STEP 2. Compute the various SS and MS needed for the combined analysis of variance outlined in step 1, following the procedures described in steps 2 to 3 of Section 8.1.1, by replacing season by year. For our example, the

Table 8.9 Mean Yields of Seven Rice Varieties Tested in Two Consecutive Years, Computed from Data in Table 8.7

Variety Number	Mean Yield, t/ha			
	Year 1	Year 2	Av.	Difference
1	3.699	2.968	3.334	−0.731
2	1.607	3.285	2.446	1.678
3	5.080	3.496	4.288	−1.584
4	3.464	2.704	3.084	−0.760
5	1.709	3.076	2.392	1.367
6	3.880	3.908	3.894	0.028
7	1.895	3.431	2.663	1.536
Av.	3.048	3.267	3.158	0.219

Table 8.10   An Outline of the Combined Analysis of Variance over s Years,[a] Based on RCB Design with t Treatments and r Replications

Source of Variation	Degree of Freedom	Mean Square	Computed F
Year (Y)	$s - 1$	$Y\,MS$	$\dfrac{Y\,MS}{R\,MS}$
Reps. within year	$s(r - 1)$	$R\,MS$	
Treatment (T)	$t - 1$	$T\,MS$	$\dfrac{T\,MS}{Y \times T\,MS}$
$Y \times T$	$(s - 1)(t - 1)$	$Y \times T\,MS$	$\dfrac{Y \times T\,MS}{E\,MS}$
Pooled error	$s(r - 1)(t - 1)$	$E\,MS$	
Total	$srt - 1$		

[a] Year is considered as a random variable.

computations are:

$$\text{Reps. within } Y\,SS = 1.38549 + 0.09698 = 1.48247$$

$$\text{Pooled error } SS = 3.95588 + 4.84346 = 8.79934$$

$$C.F. = \frac{(64.002 + 68.608)^2}{42} = 418.70029$$

$$Y\,SS = \frac{(64.002)^2 + (68.608)^2}{3(7)} - 418.70029$$

$$= 0.50512$$

$$T\,SS = \frac{\left[(20,002)^2 + (14,678)^2 + \cdots + (15.978)^2\right]}{(3)(2)} - 418.70029$$

$$= 19.15891$$

$$Y \times T\,SS = \frac{\left[(11.097)^2 + (4.822)^2 + \cdots + (10.294)^2\right]}{3}$$

$$- 418.70029 - 0.50512 - 19.15891$$

$$= 15.49600$$

The *MS* value, for each *SS* just computed, is computed in the usual manner. The results are shown in Table 8.11.

☐ STEP 3.   Test the homogeneity of the *s* error *MS*, from the individual analyses of variance, through the application of the *F* test or the chi-square test (Chapter 11, Section 11.2).

Table 8.11 Combined Analysis of Variance of RCB Experiments over Two Years, Computed from Data in Tables 8.7 and 8.8

Source of Variation	Degree of Freedom	Sum of Squares	Mean Square	Computed F^a	Tabular F 5%	1%
Year (Y)	1	0.50512	0.50512	b	—	—
Reps. within year	4	1.48247	0.37062			
Variety (V)	6	19.15891	3.19315	1.24^{ns}	4.28	8.47
$Y \times V$	6	15.49600	2.58267	7.04^{**}	2.51	3.67
Pooled error	24	8.79934	0.36664			
Total	41	45.44184				

$^a** =$ significant at 1% level, $^{ns} =$ not significant.
^{b} Reps. within year $d.f.$ is not adequate for valid test of significance.

For our example, the F test is applied as:

$$F = \frac{\text{Larger error } MS}{\text{Smaller error } MS}$$

$$= \frac{0.40362}{0.32966} = 1.22$$

The computed F value is smaller than the corresponding tabular F value of 2.69, with $f_1 = f_2 = 12$ degree of freedom at the 5% level of significance. Thus, the hypothesis of homogeneous error variances is not rejected.

☐ STEP 4. If homogeneity of error variances cannot be established in step 3, proceed to step 5. Otherwise, compute the F values for testing significance of the various effects, as indicated in Table 8.10.

For our example, because the F test for homogeneity of error variances in step 4 gave a nonsignificant result, the computation of the F values follows the formulas in Table 8.10. Because the degree of freedom for the replications within year SS is only 4, no F value is computed to test the main effect of year (Chapter 2, Section 2.1.2). The other two F values are computed as:

$$F(V) = \frac{V \ MS}{Y \times V \ MS}$$

$$= \frac{3.19315}{2.58267} = 1.24$$

$$F(Y \times V) = \frac{Y \times V \ MS}{\text{Pooled error } MS}$$

$$= \frac{2.58267}{0.36664} = 7.04$$

Comparison of these computed F values to the corresponding tabular F values, in the usual manner, shows $F(V)$ to be nonsignificant and $F(Y \times V)$ to be significant at the 1% level. Thus, the interaction effect between variety and year is highly significant but the average varietal effect is not.

☐ STEP 5. If the test for homogeneity of variance in step 4 is significant, partition the pooled error SS based on the set of desired contrasts on $Y \times T$ SS, following the procedure outlined and illustrated in step 6 of Section 8.1.1.

For our example, because the homogeneity of error variances is established, no partitioning of the pooled error SS is necessary. That is, the $Y \times V$ effect, as well as any of its components, can be tested using the pooled error MS as the error term.

☐ STEP 6. If the treatment × year interaction is significant, examine its size relative to the average effect of the treatments. If the interaction effect is small relative to the average effect, the ranking of the treatments over years is expected to be stable (e.g., treatment A performs better than treatment B in all years although in some years the difference is slightly larger than in the others) and the interaction can be ignored.

On the other hand, if the interaction effect is relatively large and the ranking of treatments changes over years (e.g., when treatment A performs better than treatment B in some years and no better or even poorer in others) then an examination of the nature of interaction would be useful.

For our example, the interaction is so large that the average effect of variety is not significant. Upon examination of the mean difference between years for each of the seven varieties (Table 8.9, last column) three groups of varieties can be identified. The first group consists of varieties 2, 5, and 7, which had higher yields in year 2 than in year 1; the second consists of varieties 1, 3, and 4, which performed better in year 1 than in year 2; and the third consists of variety 6, which gave similar yields in both years.

Even though it is clear that a consistently superior variety cannot be identified in this trial, emphasis in future trials should probably be given to variety 6, which exhibited a high degree of consistency and to variety 3 which, although giving the highest mean yield (4.288 t/ha) over years, performed extremely well only in the first year (5.08 t/ha).

8.2 TECHNOLOGY ADAPTATION EXPERIMENT: ANALYSIS OVER SITES

Technology adaptation experiments are designed to estimate the range of adaptability of new production technologies, where adaptability of a technology at a given site is defined in terms of its superiority over other technologies tested simultaneously at that site. The primary objective of such a trial is to recommend one or more new practices that are an improvement upon, or can be substituted for, the currently used farmers' practices. Thus, a technology adaptation experiment has three primary features:

1.   The primary objective is the identification of the range of adaptability of a technology. A particular technology is said to be adapted to a particular site if it is among the top performers at that site. Furthermore, its range of adaptability includes areas represented by the test sites in which the technology has shown superior performance.

2.   The primary basis for selecting test sites is representation of a geographical area. The specific sites for the technology adaptation experiments are purposely selected to represent the geographical area, or a range of environments, in which the range of adaptability of technology is to be identified. Such areas are not selected *at random*. In most cases, these test sites are research stations in different geographical area. However, when such research stations are not available, farmers' fields are sometimes used as test sites (see Chapter 16 for discussion on technology-generation experiments in farmers' fields).

3.   The treatments consist mainly of promising technologies. Only those technologies that have shown excellent promise in at least one environment (e.g., selection from a preliminary evaluation experiment) are tested. In addition, at least one of the treatments tested is usually a *control*, which represents either a no-technology treatment (such as no fertilizer application or no insect control) or a currently used technology (such as local variety).

Two common examples of technology adaptation experiments are crop variety trials or a series of fertilizer trials at different research stations in a region or country. For the variety trials, a few promising newly developed varieties of a given crop are tested, at several test sites and for several crop seasons, together with the most widely grown variety in a particular area. The results of such trials are used as the primary basis for identifying the *best* varieties as well as the range of adaptability of each of these varieties. For fertilizer trials, on the other hand, several fertilizer rates may be tested at different test sites and for several crop seasons—in order to identify groups of sites having similar fertilizer responses.

Because technology adaptation experiments are generally at a large number of sites, the size of each trial is usually small and its design simple. If factorial experiments are used, the number of factors generally does not exceed two. Thus, the two most commonly used designs are a randomized complete block and a split-plot design.

Technology adaptation experiments at a series of sites generally have the same set of treatments and use the same experimental design, a situation that greatly simplifies the required analysis. Data from a series of experiments at several sites are generally analyzed together at the end of each crop season to examine the treatment × site interaction effect and the average effects of the treatments over homogeneous sites. These effects are the primary basis for identifying the best performers, and their range of adaptability, among the different technologies tested.

Table 8.12 Individual Analysis of Variance (RCB), One for Each of the Nine Sites (L_1 to L_9), of Data from a Variety Trial with Seven Rice Varieties

Source of Variation	Degree of Freedom	Sum of Squares[a]								
		L_1	L_2	L_3	L_4	L_5	L_6	L_7	L_8	L_9
Replication	2	0.01581	1.36015	0.26245	0.13993	0.03304	0.09698	4.33975	0.70832	0.08317
Variety	6	6.75774*	5.91469ns	7.44175ns	8.60013**	8.34556**	2.79800ns	7.76741ns	3.07978ns	3.49240**
Error	12	3.25070	4.25357	20.58217	1.33921	0.51767	4.84346	5.33916	4.15965	0.52724
cv (%)		10.9	16.2	17.0	6.5	14.7	19.4	9.0	14.5	4.4

[a]** = F test significant at 1% level, * = F test significant at 5% level, ns = F test not significant.

8.2.1 Variety Trial in Randomized Complete Block Design

The procedures for combining data from a variety adaptation trial at m sites are shown below for a trial involving seven rice varieties tested in a RCB design at nine sites. The results of the individual analysis of variance at each site, based on grain yield data, are shown in Table 8.12. The variety mean yields at each site are shown in Table 8.13. The step-by-step procedures are:

☐ STEP 1. Apply the chi-square test for homogeneity of variances (Chapter 11, Section 11.2) to the m error variances from the individual analyses of variance. If the test is significant (which is generally expected), all sites whose coefficients of variation are extremely large (i.e., $cv > 20\%$) can be excluded from the combined analysis.

For our example, the chi-square test applied to the nine error mean squares in Table 8.12 results in a χ^2 value of 59.65, which is significant at the 1% level of significance. However, because no site has a cv value that is larger than 20%, all nine sites are included in the combined analysis.

☐ STEP 2. Construct an outline of the combined analysis of variance over m sites and compute the various SS and MS, following the procedure outlined for combining data over years in Section 8.1.2 with year (Y) replaced by site (L).

The computations of SS in Section 8.1.2 are based on the total values. For this example, we illustrate the computations using the mean values in Table 8.13 and the sum of squares in Table 8.12 as:

$$\text{Reps. within site } SS = 0.01581 + 1.36015 + \cdots + 0.08317$$

$$= 7.03960$$

$$\text{Pooled error } SS = 3.25070 + 4.25357 + \cdots + 0.52724$$

$$= 44.81283$$

Table 8.13 Mean Yields of Seven Rice Varieties (V_1 to V_7) Tested at Nine Sites (L_1 to L_9)[a]

Variety	Mean Yield, t/ha									
	L_1	L_2	L_3	L_4	L_5	L_6	L_7	L_8	L_9	Av.
V_1	4.835	4.288	7.882	5.219	2.052	2.968	8.066	4.568	5.122	5.000
V_2	4.412	3.694	8.110	5.036	1.042	3.285	6.584	3.889	4.445	4.500
V_3	4.888	3.963	7.495	5.054	1.342	3.496	8.387	4.136	4.945	4.856
V_4	3.717	2.675	6.568	3.725	2.643	2.704	7.182	4.136	5.200	4.283
V_5	5.635	3.351	7.475	5.858	0.772	3.076	7.679	3.765	4.159	4.641
V_6	4.808	4.288	7.672	5.403	1.042	3.908	6.796	4.506	5.012	4.826
V_7	5.271	3.495	8.652	5.645	1.000	3.431	7.261	3.395	4.224	4.708
Av.	4.795	3.679	7.693	5.134	1.413	3.267	7.422	4.056	4.730	4.688

[a] Individual analysis of variance is in Table 8.12.

$$C.F. = (3)(7)(9)(4.688)^2 = 4{,}153.71802$$

$$L\ SS = (3)(7)\big[(4.795)^2 + \cdots + (4.729)^2\big] - 4{,}153.71802$$

$$= 647.87421$$

$$V\ SS = (3)(9)\big[(5.000)^2 + \cdots + (4.708)^2\big] - 4{,}153.71802$$

$$= 8.85162$$

$$V \times L\ SS = (3)\big[(4.835)^2 + (4.288)^2 + \cdots + (4.224)^2\big]$$

$$- 4{,}153.71802 - 647.87421 - 8.85162$$

$$= 45.63398$$

The *MS* values are then computed in the usual manner. The results are shown in Table 8.14.

☐ STEP 3. Identify the range of adaptability of the treatments through suitable partitionings of the treatment × site interaction *SS*. This can be done by one or a combination of the following approaches:

- *Homogeneous Site Approach.* Classify sites into homogeneous groups so that the interaction between treatment and site within each group is not significant. A separate set of varieties can then be identified as adapted varieties for recommendation to each set of homogeneous sites.

- *Homogeneous Treatment Approach.* Identify one or more treatments whose contribution to the interaction between treatment and site is high. Treatments whose performance fluctuates widely over sites are the major contributors to the treatment × site interaction and are expected to have a narrow range of adaptability.

To decide which approach to take, examine the data by (1) plotting mean values of each variety in the *Y* axis against the site code (i.e., L_1, L_2, L_3,...)

Table 8.14 Preliminary Combined Analysis of Variance Over Sites, Computed from Data in Tables 8.12 and 8.13

Source of Variation	Degree of Freedom	Sum of Squares	Mean Square
Site (*L*)	8	647.87421	80.98428
Reps. within site	18	7.03960	0.39109
Variety (*V*)	6	8.85162	1.47527
V × *L*	48	45.63398	0.95071
Pooled Error	108	44.81283	0.41493
Total	188	754.21224	

in the X axis, (2) plotting mean values of each site in the Y axis against the variety code in the X axis, or (3) plotting both. If graph (1) reveals one or more varieties that are clearly different from the others, the homogeneous treatment approach is suggested. If, however, visual observation of graph (2) indicates a clear-cut grouping of sites with similar varietal responses, the homogeneous site approach is suggested.

For our example, graph (1) is shown in Figure 8.2. It shows V_4 to give a trend that is distinctly different from the rest. Thus, the homogeneous treatment approach is suggested, and $V\ SS$ and $V \times L\ SS$ should be

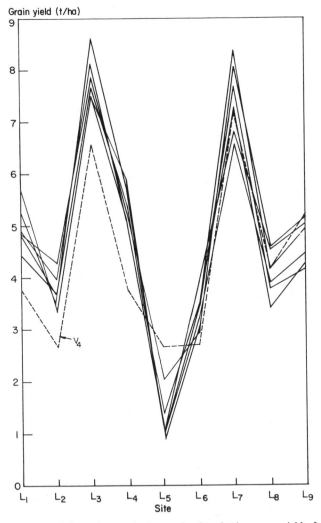

Figure 8.2 Examination of the variety × site interaction by plotting mean yield of each of the seven varieties against sites, from data in Table 8.13.

Table 8.15   Partitioning of *SS* in the Preliminary Combined Analysis of Variance in Table 8.14, to Cope with Heterogeneity of Error Variances over Sites

Source of Variation	Degree of Freedom	Sum of Squares	Mean Square	Computed F^a	Tabular F 5%	1%
Site (L)	8	647.87421	80.98428	207.01**	2.51	3.71
Reps. within site	18	7.03960	0.39109			
Variety (V)	6	8.85162	1.47527	3.56**	2.18	2.96
E_1: V_4 vs. others	(1)	5.15950	5.15950	12.43**	3.93	6.88
E_2: Between others	(5)	3.69212	0.73842	1.78^{ns}	2.30	3.19
$V \times L$	48	45.63398	0.95071	2.29**	1.48	1.76
$E_1 \times L$	(8)	21.22237	2.65280	6.39**	2.02	2.68
$E_2 \times L$	(40)	24.41161	0.61029	1.47^{ns}	1.50	1.78
Pooled error	108	44.81283	0.41493			
(Reps. within site) $\times E_1$	(18)	5.05481	0.28082			
(Reps. within site) $\times E_2$	(90)	39.75802	0.44176			

^{a}** = significant at 1% level, ^{ns} = not significant.

partitioned into the following components:

$$E_1: \ V_4 \text{ vs. other varieties}$$

$$E_2: \ \text{Between other varieties}$$

The results of this partitioning are shown in Table 8.15.

□ STEP 4.   The specific procedures for computing F values to test the significance of V *SS*, $V \times L$ *SS*, and their components depend on the significance of the test for homogeneity of error variances over sites computed in step 1. Procedures discussed in steps 5 to 7 of Section 8.1.1 should be followed.

For our example, the test for homogeneity of error variances in step 1 was significant. Hence, the partitioning of the pooled error *SS* corresponding to the desired set of contrasts on varieties derived in step 3, is performed. The results are shown in Table 8.15.

The F test for the homogeneity of the two components of the pooled error *SS* gives the computed F value of 1.57, which is not significant. Thus, the pooled error *MS* can be used as the denominator of the F values for testing the significance of either the V effect, the $V \times L$ interaction effect, or their components. For example,

$$F(V) = \frac{V \ MS}{\text{Pooled error } MS}$$

$$= \frac{1.47527}{0.41493} = 3.56$$

which is significant at the 1% level. Results for all tests, shown in Table 8.15, indicate that, except for V_4, there was no significant difference between the test varieties and this was consistently so at all sites. This means that, except for V_4, all varieties are equally adapted to the nine sites. The performance of V_4 (Table 8.13), however, fluctuated substantially over sites, indicating its narrower range of adaptability. Furthermore, because its average yield is also significantly lower than the others, its potential for recommendation, even for a few selected sites where its yields are relatively high, is doubtful.

8.2.2 Fertilizer Trial in Split-Plot Design

We illustrate the step-by-step procedures, for combining data from a series of split-plot experiments at m sites, with a factorial experiment involving two rice varieties and six rates of nitrogen tested at three sites. Each site had three replications. Nitrogen rate was the main plot and variety was the subplot factor. Grain yield data are shown in Table 8.16 and the individual analyses of variance, one for each site, are shown in Table 8.17.

Table 8.16 Grain Yield of Two Rice Varieties (V_1 and V_2) Tested with Six Rates of Nitrogen (N_1 to N_6) at Three Sites (L_1 to L_3) in a Split-Plot Design with Three Replications

| | | Grain Yield, kg/ha | | | | | | | |
| | | Rep. I | | Rep. II | | Rep. III | | Total | |
Site	Nitrogen	V_1	V_2	V_1	V_2	V_1	V_2	V_1	V_2
L_1	N_1	1,979	5,301	1,511	1,883	3,664	3,571	7,154	10,755
	N_2	4,572	5,655	4,340	5,100	4,132	5,385	13,044	16,140
	N_3	5,630	6,339	6,780	6,622	4,933	6,332	17,343	19,293
	N_4	7,153	8,108	6,504	8,583	6,326	7,637	19,983	24,328
	N_5	7,223	7,530	7,107	7,097	6,051	6,667	20,381	21,294
	N_6	7,239	7,853	6,829	7,105	5,874	7,443	19,942	22,401
	Total							(97,847)	(114,211)
L_2	N_1	3,617	3,447	3,580	3,560	3,939	3,516	11,136	10,523
	N_2	6,065	5,905	5,463	5,969	5,435	6,026	16,963	17,900
	N_3	6,092	5,322	6,571	5,883	6,084	6,489	18,747	17,694
	N_4	5,916	6,513	6,982	6,556	7,145	7,853	20,043	20,922
	N_5	7,191	8,153	6,109	7,208	7,967	6,685	21,267	22,046
	N_6	5,805	7,290	6,890	6,564	7,113	7,401	19,808	21,255
	Total							(107,964)	(110,340)
L_3	N_1	4,320	4,891	4,068	2,577	3,856	4,541	12,244	12,009
	N_2	5,862	6,009	4,626	6,625	4,913	5,672	15,401	18,306
	N_3	5,136	6,712	5,836	6,693	4,898	6,799	15,870	20,204
	N_4	6,336	6,458	5,456	6,675	5,663	6,636	17,455	19,769
	N_5	5,571	5,683	5,854	6,868	5,533	5,692	16,958	18,243
	N_6	6,765	6,335	5,263	6,064	3,910	5,949	15,938	18,348
	Total							(93,866)	(106,879)

Table 8.17 Individual Analyses of Variance for Data in Table 8.16 (Split-Plot design), One for Each of the Three Sites (L_1 to L_3)

Source of Variation	Degree of Freedom	L_1			L_2			L_3		
		SS	MS	F^a	SS	MS	F^a	SS	MS	F^a
Replication	2	1,984,474	992,237		1,040,419	520,210		1,520,023	760,012	
Nitrogen rate (N)	5	85,511,456	17,102,291	25.21**	51,671,994	10,334,399	37.35**	18,257,311	3,651,462	8.50**
Error(a)	10	6,782,715	678,272		2,766,813	276,681		4,294,864	429,486	
Variety (V)	1	7,438,348	7,438,348	17.93**	156,816	156,816	< 1	4,703,837	4,703,837	11.79**
$N \times V$	5	1,247,351	249,470	< 1	815,823	163,165	< 1	1,978,118	395,624	< 1
Error(b)	12	4,979,194	414,933		3,667,250	305,604		4,788,886	399,074	
$cv(a)\%$			14.0			8.7			11.8	
$cv(b)\%$			10.9			9.1			11.3	

^{a}** = significant at 1% level.

☐ STEP 1. Apply the procedure described in step 1 of Section 8.2.1 to both error(a) MS and error(b) MS. For our example, the χ^2 value for testing the homogeneity of the three error(a) MS of Table 8.17 is 1.90; that for the corresponding error(b) MS is 0.31; both are nonsignificant.

☐ STEP 2. Construct an outline of the combined analysis of variance over m sites, based on a split-plot design with a main-plot treatments, b subplot treatments, and r replications, as shown in Table 8.18.

☐ STEP 3. Compute the replications within site SS, the pooled error(a) SS, and the pooled error(b) SS as:

$$\text{Reps. within site } SS = \sum_{i=1}^{m} (\text{Rep. } SS)_i$$

$$\text{Pooled error}(a) \ SS = \sum_{i=1}^{m} [\text{Error}(a) \ SS]_i$$

$$\text{Pooled error}(b) \ SS = \sum_{i=1}^{m} [\text{Error}(b) \ SS]_i$$

where subscript i refers to the ith site.

Table 8.18   Outline of the Combined Analysis of Variance Over Sites, Based on Split-Plot Design

Source of Variation	Degree of Freedom[a]	Mean Square	Computed F
Sites (L)	$m-1$	$L\ MS$	$L\ MS/R\ MS$
Reps. within site	$m(r-1)$	$R\ MS$	
Main-plot factor (A)	$a-1$	$A\ MS$	$A\ MS/E_a\ MS$
$L \times A$	$(m-1)(a-1)$	$L \times A\ MS$	$L \times A\ MS/E_a\ MS$
Pooled error(a)	$m(r-1)(a-1)$	$E_a\ MS$	
Subplot factor (B)	$b-1$	$B\ MS$	$B\ MS/E_b\ MS$
$L \times B$	$(m-1)(b-1)$	$L \times B\ MS$	$L \times B\ MS/E_b\ MS$
$A \times B$	$(a-1)(b-1)$	$A \times B\ MS$	$A \times B\ MS/E_b\ MS$
$L \times A \times B$	$(m-1)(a-1)(b-1)$	$L \times A \times B\ MS$	$L \times A \times B\ MS/E_b\ MS$
Pooled error(b)	$ma(r-1)(b-1)$	$E_b\ MS$	
Total	$mrab-1$		

[a]a = number of main-plot treatments, b = number of subplot treatments, r = number of replications, and m = number of sites.

For our example, the three SS values are computed, using data in Table 8.17, as:

$$\text{Reps. within site } SS = 1{,}984{,}474 + 1{,}040{,}419 + 1{,}520{,}023$$

$$= 4{,}544{,}916$$

$$\text{Pooled error}(a)\ SS = 6{,}782{,}715 + 2{,}766{,}813 + 4{,}294{,}864$$

$$= 13{,}844{,}392$$

$$\text{Pooled error}(b)\ SS = 4{,}979{,}194 + 3{,}667{,}250 + 4{,}788{,}886$$

$$= 13{,}435{,}330$$

☐ STEP 4.   Compute the remaining SS needed in Table 8.18, either from mean values or from total values, as:

A.   *Based on totals.* The computation of the various SS follows the standard procedure:

$$C.F. = \frac{\left(\sum\limits_{i=1}^{m} G_i\right)^2}{mabr}$$

$$L\ SS = \frac{\sum\limits_{i=1}^{m} G_i^2}{abr} - C.F.$$

$$A\ SS = \frac{\sum\limits_{j=1}^{a} A_j^2}{mbr} - C.F.$$

$$L \times A\ SS = \frac{\sum\limits_{i=1}^{m}\sum\limits_{j=1}^{a} (LA)_{ij}^2}{br} - C.F. - L\ SS - A\ SS$$

$$B\ SS = \frac{\sum\limits_{k=1}^{b} B_k^2}{mar} - C.F.$$

$$L \times B\ SS = \frac{\sum\limits_{i=1}^{m}\sum\limits_{k=1}^{b} (LB)_{ik}^2}{ar} - C.F. - L\ SS - B\ SS$$

$$A \times B \ SS = \frac{\displaystyle\sum_{j=1}^{a} \sum_{1}^{b} (AB)_{jk}^2}{mr} \quad C.F. - A \ SS - B \ SS$$

$$L \times A \times B \ SS = \frac{\displaystyle\sum_{i=1}^{m} \sum_{j=1}^{a} \sum_{k=1}^{b} (LAB)_{ijk}^2}{r} - C.F. - L \ SS - A \ SS$$

$$- B \ SS - L \times A \ SS - L \times B \ SS - A \times B \ SS$$

where G_i is the grand total of the ith site, A_j is the total of the jth level of factor A, $(LA)_{ij}$ is the total of the jth level of factor A at the ith site, B_k is the total of the kth level of factor B, $(LB)_{ik}$ is the total of the kth level of factor B at the ith site, $(AB)_{jk}$ is the total of the treatment involving the jth level of factor A and the kth level of factor B, and $(LAB)_{ijk}$ is the total of the treatment involving the jth level of factor A and the kth level of factor B at the ith site.

B. *Based on means.* The computation of the various SS is:

$$C.F. = mabr(\bar{G})^2$$

$$L \ SS = abr\left(\sum_{i=1}^{m} \bar{L}_i^2\right) - C.F.$$

$$A \ SS = mbr\left(\sum_{j=1}^{a} \bar{A}_j^2\right) - C.F.$$

$$L \times A \ SS = br\left[\sum_{i=1}^{m} \sum_{j=1}^{a} (\overline{LA})_{ij}^2\right] - C.F. - L \ SS - A \ SS$$

$$B \ SS = mar\left(\sum_{k=1}^{b} \bar{B}_k^2\right) - C.F.$$

$$L \times B \ SS = ar\left[\sum_{i=1}^{m} \sum_{k=1}^{b} (\overline{LB})_{ik}^2\right] - C.F. - L \ SS - B \ SS$$

$$A \times B \ SS = mr\left[\sum_{j=1}^{a} \sum_{k=1}^{b} (\overline{AB})_{jk}^2\right] - C.F. - A \ SS - B \ SS$$

$$L \times A \times B \ SS = r\left[\sum_{i=1}^{m} \sum_{j=1}^{a} \sum_{k=1}^{b} (\overline{LAB})_{ijk}^2\right] - C.F. - L \ SS - A \ SS$$

$$- B \ SS - L \times A \ SS - L \times B \ SS - A \times B \ SS$$

Table 8.19 The Site × Nitrogen ($L \times N$) Totals, Computed from Data in Table 8.16

| Nitrogen | Yield Total | | | Nitrogen Total |
	L_1	L_2	L_3	
N_1	17,909	21,659	24,253	63,821
N_2	29,184	34,863	33,707	97,754
N_3	36,636	36,441	36,074	109,151
N_4	44,311	40,965	37,224	122,500
N_5	41,675	43,313	35,201	120,189
N_6	42,343	41,063	34,286	117,692
Total	212,058	218,304	200,745	631,107

where $\bar{\bar{G}}$ is the grand mean, $\bar{L}_i$ is the mean of the ith site, $\bar{A}_j$ is the mean of the jth level of A, $(\overline{LA})_{ij}$ is the mean of the jth level of A at the ith site, $\bar{B}_k$ is the mean of the kth level of factor B, $(\overline{LB})_{ik}$ is the mean of the kth level of factor B at the ith site, $(\overline{AB})_{jk}$ is the mean of the treatment involving the jth level of A and the kth level of B, and $(\overline{LAB})_{ijk}$ is the mean of the treatment involving the jth level of A at the kth level of B at the ith site.

For our example, the computation is made based on the total values in Table 8.16. First, the $L \times N$ and the $N \times V$ tables of totals are computed, as shown in Tables 8.19 and 8.20. Then, the various SS are computed in the standard manner as:

$$C.F. = \frac{(212,058 + 218,304 + 200,745)^2}{(3)(6)(2)(3)}$$

$$= \frac{(631,107)^2}{108} = 3,687,926,346$$

Table 8.20 The Nitrogen × Variety ($N \times V$) Totals, Computed from Data in Table 8.16

| Nitrogen | Yield Total | |
	V_1	V_2
N_1	30,534	33,287
N_2	45,408	52,346
N_3	51,960	57,191
N_4	57,481	65,019
N_5	58,606	61,583
N_6	55,688	62,004
Variety total	299,677	331,430

$$L\ SS = \frac{(212{,}058)^2 + (218{,}304)^2 + (200{,}745)^2}{(6)(2)(3)}$$

$$- 3{,}687{,}926{,}346$$

$$= 4{,}401{,}065$$

$$N\ SS = \frac{(63{,}821)^2 + (97{,}754)^2 + \cdots + (117{,}692)^2}{(3)(2)(3)}$$

$$- 3{,}687{,}926{,}346$$

$$= 136{,}849{,}095$$

$$L \times N\ SS = \frac{(17{,}909)^2 + (21{,}659)^2 + \cdots + (34{,}286)^2}{(2)(3)}$$

$$- 3{,}687{,}926{,}346 - 4{,}401{,}065 - 136{,}849{,}095$$

$$= 18{,}591{,}664$$

$$V\ SS = \frac{(299{,}677)^2 + (331{,}430)^2}{(3)(6)(3)} - 3{,}687{,}926{,}346$$

$$= 9{,}335{,}676$$

$$L \times V\ SS = \frac{(97{,}847)^2 + (114{,}211)^2 + \cdots + (106{,}879)^2}{(6)(3)}$$

$$- 3{,}687{,}926{,}346 - 4{,}401{,}065 - 9{,}335{,}676$$

$$= 2{,}963{,}325$$

$$N \times V\ SS = \frac{(30{,}534)^2 + (33{,}287)^2 + \cdots + (62{,}004)^2}{(3)(3)}$$

$$- 3{,}687{,}926{,}346 - 136{,}849{,}095 - 9{,}335{,}676$$

$$= 1{,}145{,}104$$

$$L \times N \times V\ SS = \frac{(7{,}154)^2 + (10{,}755)^2 + \cdots + (18{,}348)^2}{3}$$

$$- 3{,}687{,}926{,}346 - 4{,}401{,}065 - 136{,}849{,}095$$

$$- 9{,}335{,}676 - 18{,}591{,}664 - 2{,}963{,}325$$

$$- 1{,}145{,}104$$

$$= 2{,}896{,}188$$

□ STEP 5.   For each source of variation, compute the mean square by dividing the SS by its $d.f.$ For our example, the results are shown in Table 8.21.

□ STEP 6.   Compute the F value to test the significance of the location effect as:

$$F(L) = \frac{L\ MS}{\text{Reps. within site } MS}$$

$$= \frac{2{,}200{,}532}{757{,}486} = 2.91$$

Because the computed F value is less than the corresponding tabular F value of 5.14, with $f_1 = 2$ and $f_2 = 6$ degrees of freedom at the 5% level of significance, the site effect is not significant.

□ STEP 7.   If homogeneity of error(a) MS over sites is established (step 1), compute the F values to test the significance of the A effect and the $L \times A$ effect as:

$$F(A) = \frac{A\ MS}{\text{Pooled error}(a)\ MS}$$

$$F(L \times A) = \frac{L \times A\ MS}{\text{Pooled error}(a)\ MS}$$

Table 8.21   Combined Analysis of Variance over Sites (Split-Plot Design) for Data in Tables 8.16 and 8.17

Source of Variation	Degree of Freedom	Sum of Squares	Mean Square	Computed F^a	Tabular F 5%	1%
Site (L)	2	4,401,065	2,200,532	2.91^{ns}	5.14	10.92
Reps. within site	6	4,544,916	757,486			
Nitrogen rate (N)	5	136,849,095	27,369,819	59.31**	2.53	3.70
$L \times N$	10	18,591,664	1,859,166	4.03**	2.16	2.98
Pooled error(a)	30	13,844,392	461,480			
Variety (V)	1	9,335,676	9,335,676	25.01**	4.11	7.39
$L \times V$	2	2,963,325	1,481,662	3.97*	3.26	5.25
$N \times V$	5	1,145,104	229,021	< 1	—	—
$L \times N \times V$	10	2,896,188	289,619	< 1	—	—
Pooled error(b)	36	13,435,330	373,204			
Total	107					

^{a}** = significant at 1% level, * = significant at 5% level, ^{ns} = not significant.

Otherwise, partition the A SS, $L \times A$ SS, and the pooled error(a) SS, following the procedures outlined in Section 8.1.1.

For our example, because the homogeneity of the three error(a) MS was established, the two F values are computed based on these formulas as:

$$F(N) = \frac{27,369,819}{461,480} = 59.31$$

$$F(L \times N) = \frac{1,859,166}{461,480} = 4.03$$

The corresponding tabular F values for $F(N)$ are 2.53 at the 5% level of significance and 3.70 at the 1% level; those for $F(L \times N)$ are 2.16 and 2.98, respectively. Hence, both the nitrogen effect and its interaction with site are highly significant. The experimental evidence clearly indicates that, although the average nitrogen response is significant, the response varied between test sites.

☐ STEP 8.   If homogeneity of error(b) MS over sites is established (step 1), compute the F values for all effects involving the subplot factor B as:

$$F(B) = \frac{B\ MS}{\text{Pooled error}(b)\ MS}$$

$$F(L \times B) = \frac{L \times B\ MS}{\text{Pooled error}(b)\ MS}$$

$$F(A \times B) = \frac{A \times B\ MS}{\text{Pooled error}(b)\ MS}$$

$$F(L \times A \times B) = \frac{L \times A \times B\ MS}{\text{Pooled error}(b)\ MS}$$

Otherwise, partition all effects involving factor B and the pooled error SS, following the procedures outlined in Section 8.1.1.

For our example, because homogeneity of error(b) MS over the three sites was established, the F values are computed based on these formulas as:

$$F(V) = \frac{9,335,676}{373,204} = 25.01$$

$$F(L \times V) = \frac{1,481,662}{373,204} = 3.97$$

$$F(N \times V) = \frac{229,021}{373,204} < 1$$

$$F(L \times N \times V) = \frac{289,619}{373,204} < 1$$

Table 8.22 Partitioning of the Site × Nitrogen ($L \times N$) Interaction *SS* of the Combined Analysis of Variance in Table 8.11

Source of Variation	Degree of Freedom	Sum of Squares	Mean Square	Computed F^a
$L \times N$	10	18,591,664	1,859,166	4.03**
$[L_3 \text{ vs. } (L_1, L_2)] \times N_L$	(1)	13,233,863	13,233,863	28.68**
$[L_3 \text{ vs. } (L_1, L_2)] \times N_Q$	(1)	273,658	273,658	< 1
$[L_3 \text{ vs. } (L_1, L_2)] \times N_{\text{Res.}}$	(3)	470,259	156,753	< 1
$(L_1 \text{ vs. } L_2) \times N_L$	(1)	1,945,357	1,945,357	4.22*
$(L_1 \text{ vs. } L_2) \times N_Q$	(1)	365,600	365,600	< 1
$(L_1 \text{ vs. } L_2) \times N_{\text{Res.}}$	(3)	2,302,927	767,642	1.66[ns]
Pooled error(a)	30	13,844,392	461,480	

[a]** = significant at 1% level, * = significant at 5% level, [ns] = not significant.

□ STEP 9.   Enter all computations obtained from steps 3 to 8 in the outline of the combined analysis of variance obtained in step 2. For our example, the final results are shown in Table 8.21. These can be summarized as:

- The experimental evidence failed to indicate any significant interaction between nitrogen and variety, and this result was consistent at all sites tested (i.e., $L \times N \times V$ is not significant).
- The main effects of both factors (nitrogen and variety) are highly significant but both the nitrogen response and the varietal difference varied between the sites tested (i.e., both $L \times N$ and $L \times V$ are significant).
- There was no significant difference between the site means.

□ STEP 10.   Make appropriate examination of all interaction effects involving sites that are significant. For our example, there are two such interactions, namely, the $L \times N$ and the $L \times V$ interaction effects:

- For the $L \times N$ interaction, partition the $L \times N$ *SS* and fit the regression equation describing yield response to nitrogen at each site, following the procedure of step 7 in Section 8.1.1. The results of the partitioning of the $L \times N$ *SS*, shown in Table 8.22, indicate that only the linear portion of the nitrogen response curves varied between sites. The estimated quadratic regressions, one for each site, are graphically represented in Figure 8.3.
- For the $L \times V$ interaction, because variety is a discrete factor and because there are only two varieties, the table of varietal mean difference by site is constructed as shown in Table 8.23. Results indicate that V_2 gave higher yields than V_1 at all sites but significantly so only at sites L_1 and L_3.

□ STEP 11.   Evaluate the results so far obtained to summarize the findings and draw up conclusions and, if possible, recommendations. For our example,

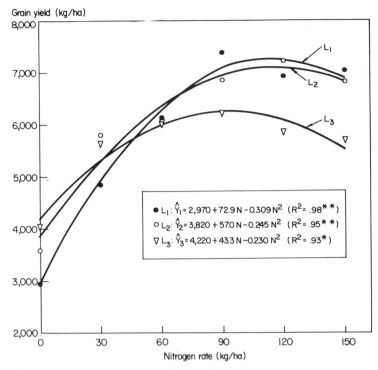

Figure 8.3 Estimates of rice yield response to nitrogen at three locations (L_1, L_2, and L_3) computed from data in Table 8.19.

the findings can be summarized as follows:

- In terms of variety, because V_2 was either better than or equal to V_1 in all test sites, V_2 is a potential variety for recommendation in areas represented by the three test sites.
- In terms of nitrogen response, the linear portion of the response curves differed between sites. The estimated quadratic response equations for

Table 8.23 Varietal Difference at Each Test Site, Computed from Data in Table 8.16

Site	Mean Yield, kg/ha		
	V_1	V_2	Difference[a]
L_1	5,436	6,345	909**
L_2	5,998	6,130	132[ns]
L_3	5,215	5,938	723**
Av.	5,550	6,138	

[a]** = significant at 1% level, [ns] = not significant.

the three sites are given in Figure 8.3. The nitrogen rates that maximize yield are estimated as 118 kg N/ha at L_1, 116 kg N/ha at L_2, and 94 kg N/ha at L_3.

8.3   LONG-TERM EXPERIMENTS

The use of new crop production practices, such as fertilization, insect and disease control, and cropping pattern, will generally change the important physical and biological factors of the environment. This process of change can take many years. Consequently, the productivity of some types of technology must be evaluated by a series of trials repeated over time. Such trials are generally referred to as long-term experiments. Their distinguishing features are:

- Change over time is the primary performance index. Even though the average performance over time remains as an important measure of productivity, the change over time, in either crop or environmental traits, or both, is the more critical parameter. Obviously, an increasing rather than a decreasing productivity trend is an important feature of a desirable technology. In addition, although the initial productivity may be high, the buildup of pests or the depletion of soil nutrients resulting from continuous use of the technology can be serious enough to cause the technology to be abandoned.
- The experimental field, the treatments, and the plot layout remain constant over time. Randomization of treatments to plots is done only once during the first crop season and the layout in all subsequent cropping seasons exactly follows the initial layout.

Some examples of long-term experiments are:

- Long-term fertility trials, which are designed to evaluate changes in soil properties and nutrients as a consequence of the application of some soil amendments over time.
- Maximum yield trials, which are designed to measure crop yields and change over time, in both physical and biological environments under intensive cropping and best management.
- Weed control trials, which are designed to measure the change in weed population over time following different types of weed control measures.

The first step in the analysis of a long-term experiment is to identify one or more characters to be used as an index of crop productivity. Because random fluctuation can mask changes over long time periods, a good index is one that is least sensitive to, or is least affected by, the random changes over time with respect to environmental factors such as climate and pest incidences. Crop yield and soil properties are commonly used indices. Although crop yield is an excellent index of productivity, it is greatly influenced by the changes in both the climate and biological environments over time. Soil properties, on the other hand, are less affected by environments but may not be as closely related to productivity as yield. Thus, more than one index is generally analyzed in a long-term experiment. We illustrate the analysis of long-term experiments with a rice fertilizer trial conducted consecutively for 13 years with two crop seasons

Table 8.24   Grain Yield of Rice Tested with Five Fertilizer Treatments[a] in a Series of Long-Term Fertility Trials, for 13 Years with Two Crop Seasons per Year

		Mean Yield, t/ha[b]				
Year	Season	F_1	F_2	F_3	F_4	F_5
1	Dry	3.98	6.67	7.05	6.00	6.99
	Wet	2.84	4.84	5.15	5.11	5.59
2	Dry	3.10	5.69	5.80	5.58	5.83
	Wet	3.48	4.40	4.55	4.36	5.08
3	Dry	2.69	5.09	5.75	5.03	6.27
	Wet	3.17	5.04	5.29	5.26	5.47
4	Dry	3.70	5.13	6.50	5.55	7.15
	Wet	3.30	4.33	4.57	4.78	5.27
5	Dry	3.46	5.48	6.13	5.44	7.01
	Wet	2.63	3.42	3.96	3.48	4.63
6	Dry	2.66	4.14	5.29	4.35	5.83
	Wet	2.60	3.12	3.65	3.15	4.28
7	Dry	2.99	4.55	4.99	4.97	5.71
	Wet	2.26	2.83	2.92	2.78	3.56
8	Dry	2.91	3.85	4.08	4.37	5.41
	Wet	2.34	3.11	3.47	2.99	3.63
9	Dry	3.71	4.46	5.31	4.75	6.75
	Wet	3.19	3.44	4.15	4.00	4.88
10	Dry	2.11	3.24	3.34	3.49	5.16
	Wet	3.41	3.67	4.46	3.92	5.00
11	Dry	2.56	2.87	4.12	3.18	3.77
	Wet	1.99	2.08	2.33	2.09	2.80
12	Dry	3.20	4.70	5.71	5.48	7.44
	Wet	3.24	4.06	5.02	4.15	5.68
13	Dry	2.22	3.40	6.10	3.49	6.61
	Wet	2.50	3.28	4.10	3.42	4.31
Av.		2.93	4.11	4.76	4.28	5.39

[a]Combinations of N − P − K are: $F_1 = 0 - 0 - 0$, $F_2 = N - 0 - 0$, $F_3 = N - P - 0$, $F_4 = N - 0 - K$, and $F_5 = N - P - K$.
[b]Average of three replications.

Table 8.25   Outline of a Combined Analysis of Variance over Time of Data from a Series of Long-Term Fertility Trials; RCB with t Treatments, r Replications, and c Crops

Source of Variation	Degree of Freedom	Mean Square	Computed F
Replication	$r - 1$	$R\ MS$	$R\ MS/E_a\ MS$
Treatment (T)	$t - 1$	$T\ MS$	$T\ MS/E_a\ MS$
Error(a)	$(r - 1)(t - 1)$	$E_a\ MS$	
Crop (C)	$c - 1$	$C\ MS$	$C\ MS/E_b\ MS$
$C \times T$	$(c - 1)(t - 1)$	$C \times T\ MS$	$C \times T\ MS/E_b\ MS$
Error(b)	$t(r - 1)(c - 1)$	$E_b\ MS$	
Total	$crt - 1$		

per year. A RCB design with three replications was used to test five N-P-K fertilizer treatments: 0-0-0, N-0-0, N-P-0, N-0-K, and N-P-K, where 0 represents no fertilizer application. The mean yield data for each treatment in each of the 26 crops are shown in Table 8.24. The step-by-step procedures in the analysis of crop yield data over 26 crops are:

☐ STEP 1.   Outline the combined analysis of variance over c crops, based on the basic design used. For our example, the outline, based on RCB design with t treatments and r replications, is shown in Table 8.25.

☐ STEP 2.   Compute the SS and MS for each source of variation in the combined analysis of variance outlined in step 1. For our example, the computations follow the procedures for the standard split-plot design (Chapter 3, Section 3.4) with fertilizer as main-plot and crop as subplot factors. The results are shown in Table 8.26. Note that there was a highly

Table 8.26   Combined Analysis of Variance over Time of Data from a Series of Long-Term Fertility Trials, Involving Five Fertilizer Treatments and Three Replications, over 26 Crops

Source of Variation	Degree of Freedom	Sum of Squares	Mean Square	Computed F^a
Replication	2	3.43030	1.71515	
Treatment (T)	4	257.65765	64.41441	172.42**
Error(a)	8	2.98876	0.37360	
Crop (C)	25	321.16860	12.84674	108.87**
$C \times T$	100	69.38389	0.69384	5.88**
Error(b)	250	29.49909	0.11800	
Total	389	684.12829		

^{a}** = significant at 1% level.

significant interaction between treatment and crop, indicating that treatment responses varied among crops.

☐ STEP 3. Examine the nature of the treatment × crop interaction (if found significant in step 2) through proper partitioning of sums of squares, based either on appropriate contrasts of treatment *SS*, or crop *SS*, or both.

- *On Treatment SS*: In our example, the four treatments, besides control, represent the 2 × 2 factorial treatments of the two factors *P* and *K*. Thus, the most appropriate set of contrasts would be the following four single *d.f.* contrasts:

Description	Contrast
Control vs. treated (T_0)	$4 F_1 - F_2 - F_3 - F_4 - F_5$
P	$-F_2 + F_3 - F_4 + F_5$
K	$-F_2 - F_3 + F_4 + F_5$
P × *K*	$F_2 - F_3 - F_4 + F_5$

The results of the partitioned treatment *SS* and partitioned treatment × crop *SS*, based on the foregoing set of single *d.f.* contrasts, are given in Table 8.27. All components of both treatment *SS* and interaction *SS*, except for the *C* × *P* × *K* interaction, were highly significant.

- *On Crop SS*: In our example, the 26 crops represent 13 years and two crop seasons per year. Hence, a standard factorial partitioning of crop *SS* into year, season, and year × season would be most appropriate. The

Table 8.27 Partitioning of the Treatment *SS* and the Crop × Treatment *SS* in the Combined Analysis of Variance in Table 8.26

Source of Variation	Degree of Freedom	Sum of Squares	Mean Square	Computed F^a
Treatment (*T*)	4			
T_0	(1)	180.63827	180.63827	483.51**
P	(1)	60.52265	60.52265	162.00**
K	(1)	12.26956	12.26956	32.84**
P × *K*	(1)	4.22717	4.22717	11.31**
Error(*a*)	8	2.98876	0.37360	
C × *T*	100			
C × T_0	(25)	33.11770	1.32471	11.23**
C × *P*	(25)	23.56402	0.94256	7.99**
C × *K*	(25)	9.24122	0.36965	3.13**
C × *P* × *K*	(25)	3.46095	0.13844	1.17ns
Error(*b*)	250	29.49909	0.11800	

^{a}** = significant at 1% level, ns = not significant.

Table 8.28 Partitioning of the Crop *SS* and the Crop × Treatment *SS*, in the Combined Analysis of Variance in Tables 8.26 and 8.27

Source of Variation	Degree of Freedom	Sum of Squares	Mean Square	Computed F^a
Year (Y)	12	193.27160	16.10597	136.49**
Season (S)	1	89.97732	89.97732	762.52**
$Y \times S$	12	37.91968	3.15997	26.78**
$C \times T_0$	25			
$\quad Y \times T_0$	(12)	16.30327	1.35861	11.51**
$\quad S \times T_0$	(1)	14.81539	14.81539	125.55**
$\quad Y \times S \times T_0$	(12)	1.99904	0.16659	1.41^{ns}
$C \times P$	25			
$\quad Y \times P$	(12)	12.79748	1.06646	9.04**
$\quad S \times P$	(1)	4.48712	4.48712	38.03**
$\quad Y \times S \times P$	(12)	6.27942	0.52328	4.43**
$C \times K$	25			
$\quad Y \times K$	(12)	4.99842	0.41654	3.53**
$\quad S \times K$	(1)	0.39645	0.39645	3.36^{ns}
$\quad Y \times S \times K$	(12)	3.84635	0.32053	2.72**
Error(b)	250	29.49909	0.11800	

^{a}** = significant at 1% level, ^{ns} = not significant.

results of the partitioned crop *SS* and partitioned crop × treatment *SS*, based on the factorial components of crop and the results in Table 8.27, are shown in Table 8.28. All individual components of treatment interact significantly with both year and season.

□ STEP 4.   Evaluate the trend over time with respect to the specific treatment response. For our example, because the treatment × season × year interaction is significant, time should be represented by year, and the study of trend should be done separately for each season. For the treatment response, because there was no significant interaction between crop and $P \times K$, the trends of P effect and of K effect over years are examined, separately for each season.

In a long-term experiment, time is a quantitative factor. Hence, a common practice is to partition time into components associated with orthogonal polynomials. In most cases, a simple linear trend—either positive or negative—should be sufficient to describe a long term trend. However, in cases where the performance index fluctuates substantially over time, as is the case with our crop yield data, the trend may not be easily described. This is clearly demonstrated by the example we used where variation in yield response, to either P or K, is greatly dominated by the random fluctuations (Figure 8.4). Thus, no partitioning of the sums of squares based on year is made.

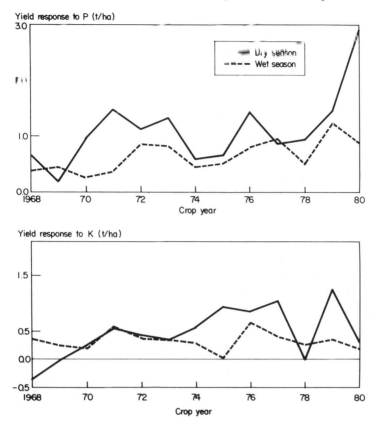

Figure 8.4 Changes in the yield response to phosphorus (*P*) and potassium (*K*) with successive croppings, 1968–1980, dry and wet season.

8.4   RESPONSE PREDICTION EXPERIMENT

Response prediction experiments are designed to characterize the behavior of treatment effects over a wide range of environments. The primary objective is to identify a functional relationship between crop performance and some important environmental factors that can adequately predict crop productivity within a fairly wide range of environments. The important features of a response prediction experiment are:

1.   Functional relationship between crop performance and environment is the primary focus. Management practices, which can be controlled by the researcher, and the soil and climate, which are difficult to modify, constitute the environmental factors. Their effects on crop yield are to be quantified and put together in a functional relationship that can predict crop performance over a wide range of environments. Consequently, the choice of season and test site, as well as the treatments to be tested, should be made to facilitate the development of the functional relationship.

2. Variability among selected environmental factors is the primary basis for selecting test sites. In contrast to technology adaptation experiments, where the test sites are selected to represent the environment of a specified geographical area, those for response prediction experiments are selected to represent a wide range of variability for some preselected environmental factors. For example, if water availability and soil texture are identified as the two most important determinants of crop performance, the test sites to be used should have adequate variability with regards to these two variables. Test sites used should include those that represent all existing combinations of the two factors, e.g., low, medium, and high water availability on one hand; and fine, medium, and coarse soil textures on the other hand.

3. Variability between selected management practices is the primary basis for selecting the treatments. The controllable factors that are suspected to have the largest impact on crop performance are used as the primary variables in specifying the treatments to be tested. For example, if nitrogen fertilization and planting density are considered important in influencing crop yields, then the treatments to be tested should reflect variability in these two controllable factors, that is, several rates of nitrogen fertilizer and several distances of planting included as treatments.

We stress that there is a clear distinction between a response prediction experiment and the other types of experiment described in earlier sections. For the others, identification of superior technology is the primary objective; for a response prediction experiment, the primary focus is on the identification of a functional relationship between crop performance and environment. Thus, procedures for data analysis described earlier are not applicable to the response prediction experiment. The analytical procedures for the development of a functional relationship involving many environmental factors are much more complex than those for the identification of a superior technology.

The common tools for the development of a functional relationship are complex regression analysis and mathematical modeling, which are beyond the coverage of this book. Analysis of a series of experiments for the development of a functional relationship between crop performance and the environment should be attempted only with the aid of a competent statistician.

CHAPTER 9

Regression And Correlation Analysis

Three groups of variables are normally recorded in crop experiments. These are:

1. Treatments, such as fertilizer rates, varieties, and weed control methods, which are generated from one or more management practices and are the primary focus of the experiment.
2. Environmental factors, such as rainfall and solar radiation, which represent the portion of the environment that is not within the researcher's control.
3. Responses, which represent the biological and physical features of the experimental units that are expected to be affected by the treatments being tested.

Response to treatments can be exhibited either by the crop, in terms of changes in such biological features as grain yield and plant height (to be called *crop response*), or by the surrounding environment in terms of changes in such features as insect incidence in an entomological trial and soil nutrient in a fertility trial (to be called *noncrop response*).

Because agricultural research focuses primarily on the behavior of biological organisms in a specified environment, the associations among treatments, environmental factors, and responses that are usually evaluated in crop research are:

1. Association between Response Variables. Crop performance is a product of several crop and noncrop characters. Each, in turn, is affected by the treatments. All these characters are usually measured simultaneously, and their association with each other can provide useful information about how the treatments influenced crop response. For example, in a trial to determine the effect of plant density on rice yield, the association between yield and its components, such as number of tillers or panicle weight, is a good indicator of the indirect effect of treatments; grain yield is increased as a result of increased tiller numbers, or larger panicle size, or a combination of the two.

Another example is in a varietal improvement program designed to produce rice varieties with both high yield and high protein content. A positive association between the two characters would indicate that varieties with both high yield and high protein content are easy to find, whereas a negative association would indicate the low frequency of desirable varieties.

2. Association between Response and Treatment. When the treatments are quantitative, such as kilograms of nitrogen applied per hectare and numbers of plants per m^2, it is possible to describe the association between treatment and response. By characterizing such an association, the relationship between treatment and response is specified not only for the treatment levels actually tested but for all other intermediate points within the range of the treatments tested. For example, in a fertilizer trial designed to evaluate crop yield at 0, 30, 60, and 90 kg N/ha, the relationship between yield and nitrogen rate specifies the yields that can be obtained not only for the four nitrogen rates actually tested but also for all other rates of application between zero and 90 kg N/ha.

3. Association between Response and Environment. For a new crop management practice to be acceptable, its superiority must hold over diverse environments. Thus, agricultural experiments are usually repeated in different areas or in different crop seasons and years. In such experiments, association between the environmental factors (sunshine, rainfall, temperature, soil nutrients) and the crop response is important.

In characterizing the association between characters, there is a need for statistical procedures that can simultaneously handle several variables. If two plant characters are measured to represent crop response, the analysis of variance and mean comparison procedures (Chapters 2 to 5) can evaluate only one character at a time, even though response in one character may affect the other, or treatment effects may simultaneously influence both characters. Regression and correlation analysis allows a researcher to examine any one or a combination of the three types of association described earlier provided that the variables concerned are expressed quantitatively.

Regression analysis describes the effect of one or more variables (designated as *independent variables*) on a single variable (designated as the *dependent variable*) by expressing the latter as a function of the former. For this analysis, it is important to clearly distinguish between the dependent and independent variable, a distinction that is not always obvious. For instance, in experiments on yield response to nitrogen, yield is obviously the dependent variable and nitrogen rate is the independent variable. On the other hand, in the example on grain yield and protein content, identification of variables is not obvious. Generally, however, the character of major importance, say grain yield, becomes the dependent variable and the factors or characters that influence grain yield become the independent variables.

Correlation analysis, on the other hand, provides a measure of the degree of association between the variables or the goodness of fit of a prescribed relationship to the data at hand.

Regression and correlation procedures can be classified according to the number of variables involved and the form of the functional relationship between the dependent variable and the independent variables. The procedure is termed *simple* if only two variables (one dependent and one independent variable) are involved and *multiple*, otherwise. The procedure is termed *linear* if the form of the underlying relationship is linear and *nonlinear*, otherwise. Thus, regression and correlation analysis can be classified into four types:

1. simple linear regression and correlation
2. multiple linear regression and correlation
3. simple nonlinear regression and correlation
4. multiple nonlinear regression and correlation

We describe:

- The statistical procedure for applying each of the four types of regression and correlation analysis, with emphasis on simple linear regression and correlation because of its simplicity and wide usage in agricultural research.
- The statistical procedures for selecting the best functional form to describe the relationship between the dependent variable and the independent variables of interest.
- The common misuses of regression and correlation analysis in agricultural research and the guidelines for avoiding them.

9.1 LINEAR RELATIONSHIP

The relationship between any two variables is linear if the change is constant throughout the whole range under consideration. The graphical representation of a linear relationship is a straight line, as illustrated in Figure 9.1*a*. Here, Y constantly increases two units for each unit change in X throughout the whole range of X values from 0 to 5: Y increases from 1 to 3 as X changes from 0 to 1, and Y increases from 3 to 5 as X changes from 1 to 2, and so on.

The functional form of the linear relationship between a dependent variable Y and an independent variable X is represented by the equation:

$$Y = \alpha + \beta X$$

where α is the intercept of the line on the Y axis and β, the linear regression coefficient, is the slope of the line or the amount of change in Y for each unit

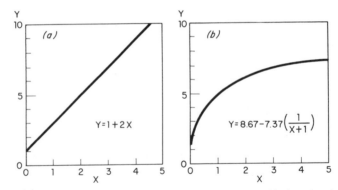

Figure 9.1 Illustration of a linear (*a*), and a nonlinear (*b*), relationship between the dependent variable *Y* and the independent variable *X*.

change in *X*. For example, for the linear relationship of Figure 9.1*a*, with the intercept α of 1 and the linear regression coefficient β of 2, the relationship is expressed as:

$$Y = 1 + 2X \quad \text{for } 0 \le X \le 5$$

With the two parameters of the linear relationship (i.e., α and β) specified, the value of the dependent variable *Y*, corresponding to a given value of the independent variable *X* within the range of *X* values considered, can be immediately determined by replacing *X* in the equation with the desired value and computing for *Y*. For example, for $X = 1.5$, *Y* is computed as $1 + 2(1.5) = 4$, and for $X = 3$, *Y* is computed as $1 + 2(3) = 7$.

When there is more than one independent variable, say *k* independent variables $(X_1, X_2, \ldots, X_k)$, the simple linear functional form of the equation $Y = \alpha + \beta X$ can be extended to the multiple linear functional form of

$$Y = \alpha + \beta_1 X_1 + \beta_2 X_2 + \cdots + \beta_k X_k$$

where α is the intercept (i.e., the value of *Y* when all *X*'s are zeroes) and β_i ($i = 1, \ldots, k$), the *partial regression coefficient* associated with the independent variable X_i, represents the amount of change in *Y* for each unit change in X_i. Thus, in the multiple linear functional form with *k* independent variables, there are $(k + 1)$ parameters (i.e., $\alpha, \beta_1, \beta_2, \ldots, \beta_k$) that need to be estimated.

The presence of β_i (i.e., when the value of β_i is not zero) indicates the dependence of *Y* on X_i. In other words, if $\beta_i = 0$, then *Y* does not depend on X_i (i.e., there is no association between *Y* and X_i in the manner prescribed). Thus, the test of significance of each β_i to determine whether or not $\beta_i = 0$ is an essential part of the regression analysis.

In some situations, the researcher may also wish to test the significance of the intercept α (i.e., to check whether $\alpha = \alpha_0$, where α_0 is the value specified by the researcher). For example, if the researcher wishes to determine whether $Y = 0$ when the value of X in the equation $Y = \alpha + \beta X$ is zero, which is equivalent to checking whether the line passes through the origin, then he must test whether or not $\alpha = 0$.

9.1.1 Simple Linear Regression and Correlation

For the simple linear regression analysis to be applicable, the following conditions must hold:

- There is only one independent variable X affecting the dependent variable Y.
- The relationship betwen Y and X is known, or can be assumed, to be linear.

Although these two conditions may seem too restrictive, they are often satisfied for data from controlled experiments.

Most controlled experiments are designed to keep the many factors that can simultaneously influence the dependent variable constant, and to vary only the factor (treatment) being investigated. In a nitrogen fertilizer trial, for example, all other factors that can affect yield, such as phosphorus application, potassium application, plant population, variety, and weed control, are carefully controlled throughout the experiment. Only nitrogen rate is varied. Consequently, the assumption that the rate of nitrogen application is the major determinant of the variation in the yield data is satisfied.

In contrast, if data on grain yield and the corresponding rate of nitrogen application were collected from an experiment where other production factors were allowed to vary, the assumption of one independent variable would not be satisfied and, consequently, the use of a simple regression analysis would be inappropriate.

Although the assumption of a linear relationship between any two characters in biological materials seldom holds, it is usually adequate within a relatively small range in the values of the independent variable. An example is the commonly observed behavior of plant growth over time, as illustrated in Figure 9.1*b*. Typically, growth rate is rapid when the plant is young and declines considerably as the plant becomes older. Thus, the relationship between plant growth (as measured by weight or height) and plant age is not linear over the whole life cycle. However, within some limited region—for example, within the range of X values from 0 to 1 or from 1 to 2—the relationship could be adequately described by a straight line. Because the range of a crop response is generally limited by the range of the treatments under test, and because this range is fairly narrow in most controlled experiments, the assumption of linearity is generally satisfied.

9.1.1.1 Simple Linear Regression Analysis. The simple linear regression analysis deals with the estimation and tests of significance concerning the two parameters α and β in the equation $Y = \alpha + \beta X$. It should be noted that because the simple linear regression analysis is performed under the assumption that there is a linear relationship between X and Y, it does not provide any test as to whether the *best* functional relationship between X and Y is indeed linear.

The data required for the application of the simple linear regression analysis are the n pairs (with $n > 2$) of Y and X values. For example, in the study of nitrogen response using data from a fertilizer trial involving t nitrogen rates, the n pairs of Y and X values would be the t pairs of mean yield (Y) and nitrogen rate (X).

We illustrate the procedure for the simple linear regression analysis with the rice yield data from a trial with four levels of nitrogen, as shown in Table 9.1. The primary objective of the analysis is to estimate a linear response in rice yield to the rate of nitrogen applied, and to test whether this linear response is significant. The step-by-step procedures are:

☐ STEP 1.   Compute the means $\bar{X}$ and $\bar{Y}$, the corrected sums of squares Σx^2 and Σy^2, and the corrected sum of cross products Σxy, of variables X and Y as:

$$\bar{X} = \frac{\Sigma X}{n}$$

$$\bar{Y} = \frac{\Sigma Y}{n}$$

$$\Sigma x^2 = \sum_{i=1}^{n} (X_i - \bar{X})^2$$

$$\Sigma y^2 = \sum_{i=1}^{n} (Y_i - \bar{Y})^2$$

$$\Sigma xy = \sum_{i=1}^{n} (X_i - \bar{X})(Y_i - \bar{Y})$$

where (X_i, Y_i) represents the ith pair of the X and Y values.

For our example, $n = 4$ pairs of values of rice yield (Y) and nitrogen rate (X). Their means, corrected sums of squares, and corrected sum of cross products are computed as shown in Table 9.1.

Table 9.1   Computation of a Simple Linear Regression Equation between Grain Yield and Nitrogen Rate Using Data from a Fertilizer Experiment in Rico

Nitrogen Rate, kg/ha (X)	Grain Yield, kg/ha (Y)	Deviation from Means		Square of Deviate		Product of Deviates
		x	y	x^2	y^2	$(x)(y)$
0	4,230	−75	−1,640.75	5,625	2,692,061	123,056
50	5,442	−25	−428.75	625	183,827	10,719
100	6,661	25	790.25	625	624,495	19,756
150	7,150	75	1,279.25	5,625	1,636,481	95,944
			Sum			
300	23,483	0	0.00	12,500	5,136,864	249,475
			Mean			
75	5,870.75					

☐ STEP 2.   Compute the estimates of the regression parameters α and β as:

$$a = \bar{Y} - b\bar{X}$$

$$b = \frac{\sum xy}{\sum x^2}$$

where a is the estimate of α; and b, the estimate of β.

For our example, the estimates of the two regression parameters are:

$$b = \frac{249,475}{12,500} = 19.96$$

$$a = 5,870.75 - (19.96)(75) = 4,374$$

Thus, the estimated linear regression is

$$\hat{Y} = a + bX$$

$$= 4,374 + 19.96X \quad \text{for } 0 \leq X \leq 150$$

☐ STEP 3.   Plot the observed points and draw a graphical representation of the estimated regression equation of step 2:

- Plot the n observed points. For our example, the four observed points (i.e., the X and Y values in Table 9.1) are plotted in Figure 9.2.
- Using the estimated linear regression of step 2, compute the $\hat{Y}$ values, one corresponding to the smallest X value (i.e., X_{min}) and the other corre-

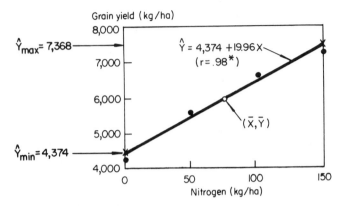

Figure 9.2 The estimated linear regression between grain yield (Y) and nitrogen rate (X), computed from data in Table 9.1.

sponding to the largest X value (i.e., X_{max}):

$$\hat{Y}_{min} = a + b(X_{min})$$

$$\hat{Y}_{max} = a + b(X_{max})$$

For our example, with $X_{min} = 0$ kg N/ha and $X_{max} = 150$ kg N/ha, the corresponding $\hat{Y}_{min}$ and $\hat{Y}_{max}$ values are computed as:

$$\hat{Y}_{min} = 4{,}374 + 19.96(0) = 4{,}374 \text{ kg/ha}$$

$$\hat{Y}_{max} = 4{,}374 + 19.96(150) = 7{,}368 \text{ kg/ha}$$

- Plot the two points ($X_{min}, \hat{Y}_{min}$) and ($X_{max}, \hat{Y}_{max}$) on the (X, Y) plane and draw the line between the two points, as shown in Figure 9.3. The following features of a graphical representation of a linear regression, such as that of Figure 9.3, should be noted:

(i)   The line must be drawn within the range of values of X_{min} and X_{max}. It is not valid to extrapolate the line outside this range.
(ii) The line must pass through the point ($\bar{X}, \bar{Y}$), where $\bar{X}$ and $\bar{Y}$ are the means of variables X and Y, respectively.
(iii) The slope of the line is b.
(iv) The line, if extended, must intersect the Y axis at the Y value of a.

For our example, the two points $(0, 4374)$ and $(150, 7368)$ are plotted and the line drawn between them, as shown in Figure 9.2.

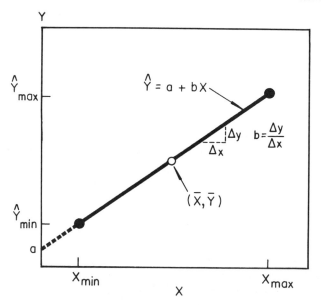

Figure 9.3 Graphical representation of an estimated regression line: $\hat{Y} = a + bX$.

☐ STEP 4. Test the significance of β:

• Compute the residual mean square as:

$$s^2_{y \cdot x} = \frac{\sum y^2 - \dfrac{\left(\sum xy\right)^2}{\sum x^2}}{n - 2}$$

where the values of $\sum y^2$, $\sum xy$, and $\sum x^2$ are those computed in step 1, as recorded in Table 9.1.

• Compute the t_b value as:

$$t_b = \frac{b}{\sqrt{\dfrac{s^2_{y \cdot x}}{\sum x^2}}}$$

• Compare the computed t_b value to the tabular t values of Appendix C, with $(n - 2)$ degrees of freedom. β is judged to be significantly different from zero if the absolute value of the computed t_b value is greater than the tabular t value at the prescribed level of significance.

For our example, the residual mean square and the t_b value are computed as:

$$s_{y \cdot x}^2 = \frac{5{,}136{,}864 - \dfrac{(249{,}475)^2}{12{,}500}}{4 - 2} = 78{,}921$$

$$t_b = \frac{19.96}{\sqrt{\dfrac{78{,}921}{12{,}500}}} = 7.94$$

The tabular t values at the 5% and 1% levels of significance, with $(n - 2) = 2$ degrees of freedom, are 4.303 and 9.925, respectively. Because the computed t_b value is greater than the tabular t value at the 5% level of significance but smaller than the tabular t value at the 1% level, the linear response of rice yield to changes in the rate of nitrogen application, within the range of 0 to 150 kg N/ha, is significant at the 5% level of significance.

☐ STEP 5.   Construct the $(100 - \alpha)\%$ confidence interval for β, as:

$$C.I. = b \pm t_\alpha \sqrt{\frac{s_{y \cdot x}^2}{\sum x^2}}$$

where t_α is the tabular t value, from Appendix C, with $(n - 2)$ degrees of freedom and at α level of significance. For our example, the 95% confidence interval for β is computed as:

$$C.I.(95\%) = b \pm t_{.05} \sqrt{\frac{s_{y \cdot x}^2}{\sum x^2}}$$

$$= 19.96 \pm 4.303 \sqrt{\frac{78{,}921}{12{,}500}}$$

$$= 19.96 \pm 10.81$$

$$= (9.15, 30.77)$$

Thus, the increase in grain yield for every 1 kg/ha increase in the rate of nitrogen applied, within the range of 0 to 150 kg N/ha, is expected to fall between 9.15 kg/ha and 30.77 kg/ha, 95% of the time.

☐ STEP 6.   Test the hypothesis that $\alpha = \alpha_0$:

• Compute the t_a value as:

$$t_a = \frac{a - \alpha_0}{\sqrt{s_{y \cdot x}^2 \left(\dfrac{1}{n} + \dfrac{\bar{X}^2}{\sum x^2} \right)}}$$

• Compare the computed t_a value to the tabular t value, from Appendix C, with $(n - 2)$ degrees of freedom and at a prescribed level of significance. Reject the hypothesis that $\alpha = \alpha_0$ if the absolute value of the computed t_a value is greater than the corresponding tabular t value.

For our example, although there is probably no need to make the test of significance on α, we illustrate the test procedure by testing whether α (i.e., yield at 0 kg N/ha) is significantly different from 4,000 kg/ha. The t_a value is computed as:

$$t_a = \frac{4{,}374 - 4{,}000}{\sqrt{78{,}921 \left[\dfrac{1}{4} + \dfrac{(75)^2}{12{,}500} \right]}} = 1.59$$

Because the t_a value is smaller than the tabular t value with $(n - 2) = 2$ degrees of freedom at the 5% level of significance of 4.303, the α value is not significantly different from 4,000 kg/ha.

9.1.1.2 Simple Linear Correlation Analysis.

The simple linear correlation analysis deals with the estimation and test of significance of the simple linear correlation coefficient r, which is a measure of the degree of linear association between two variables X and Y. Computation of the simple linear correlation coefficient is based on the amount of variability in one variable that can be explained by a linear function of the other variable. The result is the same whether Y is expressed as a linear function of X, or X is expressed as a linear function of Y. Thus, in the computation of the simple linear correlation coefficient, there is no need to specify which variable is the cause and which is the consequence, or to distinctly differentiate between the dependent and the independent variable, as is required in the regression analysis.

The value of r lies within the range of -1 and $+1$, with the extreme value indicating perfect linear association and the midvalue of zero indicating no linear association between the two variables. An intermediate value of r indicates the portion of variation in one variable that can be accounted for by the linear function of the other variable. For example, with an r value of .8, the implication is that 64% $[(100)(r^2) = (100)(.8)^2 = 64]$ of the variation in the variable Y can be explained by the linear function of the variable X. The minus

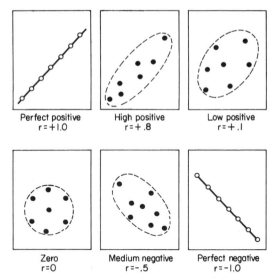

Figure 9.4   Graphical representations of various values of simple correlation coefficient *r*.

or plus sign attached to the *r* value indicates the direction of change in one variable relative to the change in the other. That is, the value of *r* is negative when a positive change in one variable is associated with a negative change in another, and positive when the two variables change in the same direction. Figure 9.4 illustrates graphically the various degrees of association between two variables as reflected in the *r* values.

Even though the zero *r* value indicates the absence of a linear relationship between two variables, it does not indicate the absence of *any* relationship between them. It is possible for the two variables to have a nonlinear relationship, such as the quadratic form of Figure 9.5, with an *r* value of zero. This is why we prefer to use the word *linear*, as in simple linear correlation coefficient, instead of the more conventional names of simple correlation coefficient or merely correlation coefficient. The word linear emphasizes the underlying assumption of linearity in the computation of *r*.

The procedures for the estimation and test of significance of a simple linear correlation coefficient between two variables *X* and *Y* are:

☐ STEP 1. Compute the means $\overline{X}$ and $\overline{Y}$, the corrected sums of squares $\sum x^2$ and $\sum y^2$, and the corrected sum of cross products $\sum xy$, of the two variables, following the procedure in step 1 of Section 9.1.1.1.

☐ STEP 2. Compute the simple linear correlation coefficient as:

$$r = \frac{\sum xy}{\sqrt{\left(\sum x^2\right)\left(\sum y^2\right)}}$$

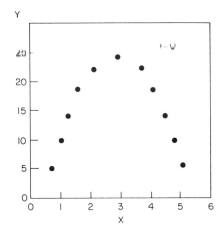

Figure 9.5 Illustration of a quadratic relationship, between two variables Y and X, that results in a simple linear correlation coefficient r being 0.

☐ STEP 3.   Test the significance of the simple linear correlation coefficient by comparing the computed r value of step 2 to the tabular r value of Appendix H with $(n - 2)$ degrees of freedom. The simple linear correlation coefficient is declared significant at the α level of significance if the absolute value of the computed r value is greater than the corresponding tabular r value at the α level of significance.

In agricultural research, there are two common applications of the simple linear correlation analysis:

- It is used to measure the degree of association between two variables with a well-defined cause and effect relationship that can be defined by the linear regression equation $Y = \alpha + \beta X$.
- It is used to measure the degree of linear association between two variables in which there is no clear-cut cause and effect relationship.

We illustrate the linear correlation procedure, with two examples. Each example represents one of the two types of application.

Example 1.   We illustrate the association between response and treatment with the data used to illustrate the simple linear regression analysis in Section 9.1.1.1. Because the data was obtained from an experiment in which all other environmental factors except the treatments were kept constant, it is logical to assume that the treatments are the primary cause of variation in the crop response. Thus, we apply the simple linear correlation analysis to determine the strength of the linear relationship between crop response (represented by grain yield) as the dependent variable and treatment as the independent variable. The step-by-step procedures are:

☐ STEP 1. Compute the means, corrected sums of squares, and corrected sum of cross products of the two variables (nitrogen rate and yield), as shown in Table 9.1.

☐ STEP 2. Compute the simple linear correlation coefficient r, as:

$$r = \frac{\sum xy}{\sqrt{(\sum x^2)(\sum y^2)}}$$

$$= \frac{249,475}{\sqrt{(12,500)(5,136,864)}} = .985$$

☐ STEP 3. Compare the absolute value of the computed r value to the tabular r values with $(n - 2) = 2$ degrees of freedom, which are .950 at the 5% level of significance and .990 at the 1% level. Because the computed r value is greater than the tabular r value at the 5% level but smaller than the tabular r value at the 1% level, the simple linear correlation coefficient is declared significant at the 5% level of significance. The computed r value of .985 indicates that $(100)(.985)^2 = 97\%$ of the variation in the mean yield is accounted for by the linear function of the rate of nitrogen applied. The relatively high r value obtained is also indicative of the *closeness* between the estimated regression line and the observed points, as shown in Figure 9.2. Within the range of 0 to 150 kg N/ha, the linear relationship between mean yield and rate of nitrogen applied seems to fit the data adequately.

We add a note of caution here concerning the magnitude of the computed r value and its corresponding degree of freedom. It is clear that the tabular r values in Appendix H decrease sharply with the increase in the degree of freedom, which is a function of n (i.e., the number of pairs of observations used in the computation of the r value). Thus, the smaller n is, the larger the computed r value must be to be declared significant. In our example with $n = 4$, the seemingly high value of the computed r of .985 is still not significant at the 1% level. On the other hand, with $n = 9$, a computed r value of .8 would have been declared significant at the 1% level. Thus, the practical importance of the significance and the size of the r value must be judged in relation to the sample size n. It is, therefore, a good practice to always specify n in the presentation of the regression and correlation result (for more discussion, see Section 9.4).

Example 2. To illustrate the association between two responses, we use data on soluble protein nitrogen (variable X_1) and total chlorophyll (variable X_2) in the leaves obtained from seven samples of the rice variety IR8 (Table 9.2). In

Table 9.2 Computation of a Simple Linear Correlation Coefficient between Soluble Protein Nitrogen (X_1) and Total Chlorophyll (X_2) in the Leaves of Rice Variety IR8

Sample Number	Soluble Protein N, mg/leaf (X_1)	Total Chlorophyll, mg/leaf (X_2)	Deviate x_1	x_2	Square of Deviate x_1^2	x_2^2	Product of Deviates $(x_1)(x_2)$
1	0.60	0.44	−0.37	−0.38	0.1369	0.1444	0.1406
2	1.12	0.96	0.15	0.14	0.0225	0.0196	0.0210
3	2.10	1.90	1.13	1.08	1.2769	1.1664	1.2204
4	1.16	1.51	0.19	0.69	0.0361	0.4761	0.1311
5	0.70	0.46	−0.27	−0.36	0.0729	0.1296	0.0972
6	0.80	0.44	−0.17	−0.38	0.0289	0.1444	0.0646
7	0.32	0.04	−0.65	−0.78	0.4225	0.6084	0.5070
Total	6.80	5.75	0.01[a]	0.01[a]	1.9967	2.6889	2.1819
Mean	0.97	0.82					

[a]The nonzero values are only due to rounding.

this case, it is not clear whether there is a cause and effect relationship between the two variables and, even if there were one, it would be difficult to specify which is the cause and which is the effect. Hence, the simple linear correlation analysis is applied to measure the degree of linear association between the two variables without specifying the causal relationship. The step-by-step procedures are:

☐ STEP 1. Compute the means, corrected sums of squares, and corrected sum of cross products, following the procedure in step 1 of Section 9.1.1.1. Results are shown in Table 9.2.

☐ STEP 2. Compute the simple linear correlation coefficient r:

$$r = \frac{2.1819}{\sqrt{(1.9967)(2.6889)}} = .942$$

☐ STEP 3. Compare the absolute value of the computed r value to the tabular r values from Appendix H, with $(n - 2) = 5$ degrees of freedom, which are .754 at the 5% level of significance and .874 at the 1% level. Because the computed r value exceeds both tabular r values, we conclude that the simple linear correlation coefficient is significantly different from zero at the 1% probability level. This significant, high r value indicates that there is strong evidence that the soluble protein nitrogen and the total chlorophyll in the leaves of IR8 are highly associated with one another in a linear way: leaves with high soluble protein nitrogen also have a high total chlorophyll, and vice versa.

9.1.1.3 Homogeneity of Regression Coefficients.

In a single-factor experiment where only one factor is allowed to vary, the association between response and treatment is clearly defined. On the other hand, in a factorial experiment, where more than one factor varies, the linear relationship between response and a given factor may have to be examined over different levels of the other factors. For example, with data from a two-factor experiment involving four rice varieties and five levels of nitrogen fertilization, the linear regression of yield on nitrogen level may have to be examined separately for each variety. Or, with data from a three-factor experiment involving three varieties, four plant densities, and five levels of nitrogen fertilization; twelve separate regressions between yield and nitrogen levels may need to be estimated for each of the $3 \times 4 = 12$ treatment combinations of variety and plant density. Similarly, if the researcher is interested in examining the relationship between yield and plant density, he can estimate a regression equation for each of the $3 \times 5 = 15$ treatment combinations of variety and nitrogen.

In the same manner, for experiments in a series of environments (i.e., at different sites or seasons or years), the regression analysis may need to be applied separately for each experiment.

When several linear regressions are estimated, it is usually important to determine whether the various regression coefficients or the slopes of the various regression lines differ from each other. For example, in a two-factor experiment involving variety and rate of nitrogen, it would be important to know whether the rate of change in yield for every incremental change in nitrogen fertilization varies from one variety to another. Such a question is answered by comparing the regression coefficients of the different varieties. This is referred to as *testing the homogeneity of regression coefficients*.

The concept of homogeneity of regression coefficients is closely related to the concept of interaction between factors, which is discussed in Chapter 3, Section 3.1, and Chapter 4, Section 4.1. Regression lines with equal slopes are parallel to one another, which also means that there is no interaction between the factors involved. In other words, the response to the levels of factor A remains the same over the levels of factor B.

Note also that homogeneity of regression coefficients does not imply equivalence of the regression lines. For two or more regression lines to coincide (one on top of another) the regression coefficients β and the intercepts α must be homogeneous. In agricultural research where regression analysis is usually applied to data from controlled experiments, researchers are generally more interested in comparing the rates of change (β) than the intercepts (α). However, if a researcher wishes to determine whether a single regression line can be used to represent several regression lines with homogeneous regression coefficients, the appropriate comparison of treatment means (at the X level of zero) can be made following the procedures outlined in Chapter 5. If the difference between these means is not significant, then a single regression line can be used.

We present procedures for testing the homogeneity of regression coefficients for two cases, one where only two regression coefficients are involved and another where there are three or more regression coefficients. In doing this, we concentrate on simplicity of the procedure for two regression coefficients, which is most commonly used in agricultural research.

9.1.1.3.1 Two Regression Coefficients. The procedure for testing the hypothesis that $\beta_1 = \beta_2$ in two regression lines, represented by $Y_1 = \alpha_1 + \beta_1 X$ and $Y_2 = \alpha_2 + \beta_2 X$, is illustrated using data of grain yield (Y) and tiller number (X) shown in Table 9.3. The objective is to determine whether the regression coefficients in the linear relationships between grain yield and tiller number are the same for the two varieties. The step-by-step procedures are:

☐ STEP 1. Apply the simple linear regression procedure of Section 9.1.1.1 to each of the two sets of data, one for each variety, to obtain the estimates of

Table 9.3 Computation of Two Simple Linear Regression Coefficients between Grain Yield (Y) and Tiller Number (X), One for Each of the Two Rice Varieties Milfor 6(2) and Taichung Native 1

Milfor 6(2)		Taichung Native 1	
Grain Yield, kg/ha	Tillers, no./m²	Grain Yield, kg/ha	Tillers, no./m²
4,862	160	5,380	293
5,244	175	5,510	325
5,128	192	6,000	332
5,052	195	5,840	342
5,298	238	6,416	342
5,410	240	6,666	378
5,234	252	7,016	380
5,608	282	6,994	410

$$\bar{X}_1 = 217 \qquad\qquad \bar{X}_2 = 350$$
$$\bar{Y}_1 = 5,230 \qquad\qquad \bar{Y}_2 = 6,228$$
$$\sum x_1^2 = 12,542 \qquad\qquad \sum x_2^2 = 9,610$$
$$\sum y_1^2 = 357,630 \qquad\qquad \sum y_2^2 = 2,872,044$$
$$\sum x_1 y_1 = 57,131 \qquad\qquad \sum x_2 y_2 = 153,854$$
$$b_1 = \frac{57,131}{12,542} = 4.56 \qquad\qquad b_2 = \frac{153,854}{9,610} = 16.01$$
$$a_1 = 5,230 - (4.56)(217) \qquad a_2 = 6,228 - (16.01)(350)$$
$$= 4,240 \qquad\qquad\qquad = 624$$

the two linear regressions:

$$\hat{Y}_1 = a_1 + b_1 X$$

$$\hat{Y}_2 = a_2 + b_2 X$$

For our example, the computations are shown in Table 9.3 and the two estimated regression lines are:

$$\hat{Y}_1 = 4{,}240 + 4.56X \quad \text{for variety Milfor 6(2)}$$

$$\hat{Y}_2 = 624 + 16.01X \quad \text{for variety Taichung Native 1}$$

These two estimated linear regressions are represented graphically in Figure 9.6.

☐ STEP 2.   Compute the residual mean square (Section 9.1.1.1, step 4) for each set of data. For our example, the two values of the residual mean square are:

$$s_{y \cdot x}^2(1) = \frac{357{,}630 - \dfrac{(57{,}131)^2}{12{,}542}}{8 - 2}$$

$$= 16{,}231.39$$

$$s_{y \cdot x}^2(2) = \frac{2{,}872{,}044 - \dfrac{(153{,}854)^2}{9{,}610}}{8 - 2}$$

$$= 68{,}145.85$$

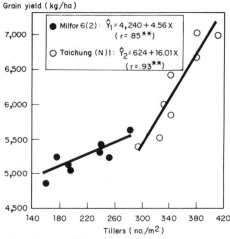

Figure 9.6 Estimated linear relationships between grain yield (Y) and tiller number (X) of two rice varieties: Milfor 6(2) and Taichung (N)1.

☐ STEP 3. Compute the pooled residual mean square as:

$$s_p^2 = \frac{(n_1 - 2)s_{y \cdot x}^2(1) + (n_2 - 2)s_{y \cdot x}^2(2)}{n_1 + n_2 - 4}$$

where $s_{y \cdot x}^2(1)$ and $s_{y \cdot x}^2(2)$ are the values of the residual mean square computed in step 2, and n_1 and n_2 are the numbers of paired observations for the first and second set of data. If $n_1 = n_2$, the s_p^2 is simply the standard arithmetic mean of the two residual mean squares. For our example, because $n_1 = n_2 = 8$, the pooled residual mean square is computed as:

$$s_p^2 = \frac{16{,}231.39 + 68{,}145.85}{2}$$

$$= 42{,}188.62$$

☐ STEP 4. Compute the t value as:

$$t = \frac{b_1 - b_2}{\sqrt{s_p^2 \left(\dfrac{1}{\sum x_1^2} + \dfrac{1}{\sum x_2^2} \right)}}$$

where b_1 and b_2 are the estimated regression coefficients, and $\sum x_1^2$ and $\sum x_2^2$ are the values of the corrected sum of squares, for the first and second set of data.

For our example, the t value is computed as:

$$t = \frac{4.56 - 16.01}{\sqrt{42{,}188.62 \left(\dfrac{1}{12{,}542} + \dfrac{1}{9{,}610} \right)}}$$

$$= -4.11$$

☐ STEP 5. Compare the computed t value with the tabular t values, from Appendix C, with $(n_1 + n_2 - 4)$ degrees of freedom. Reject the hypothesis that the two regression coefficients are the same ($\beta_1 = \beta_2$) if the absolute value of the computed t value is greater than the corresponding tabular t value at the prescribed level of significance.

For our example, the tabular t values obtained from Appendix C with $(n_1 + n_2 - 4) = 12$ degrees of freedom are 2.179 for the 5% level of significance and 3.055 for the 1% level. Because the absolute value of the computed t value is larger than the tabular t value at the 1% level of significance, the hypothesis that the regression coefficients are the same for the two varieties is rejected. We therefore conclude that the two regression coefficients are not homogeneous. The rate of increase in grain yield due to an incremental change in tiller number is significantly faster for variety Taichung Native 1 than for variety Milfor 6(2).

9.1.1.3.2   Three or More Regression Coefficients.   For the k linear regression lines, $\hat{Y}_i = \alpha_i + \beta_i X$ ($i = 1,\ldots,k$), the procedure for testing the hypothesis that $\beta_1 = \beta_2 = \cdots = \beta_k$ is illustrated using data of grain yield (Y) and tiller number (X) of three rice varieties shown in Table 9.4. The step-by-step procedures are:

☐ STEP 1.   Apply the simple linear regression procedure of Section 9.1.1.1 to each of the k sets of data, with seven pairs of observations per set, to obtain k estimated linear regressions:

$$\hat{Y}_i = a_i + b_i X \quad \text{for } i = 1,\ldots,k$$

For our example, the estimated linear regressions between grain yield (Y) and tiller number (X) for $k = 3$ varieties are computed as:

$$\hat{Y}_1 = 1{,}436 + 536.70X \qquad \text{for IR1514A-E666}$$

$$\hat{Y}_2 = -3{,}414 + 860.70X \quad \text{for IR8}$$

$$\hat{Y}_3 = 8{,}771 - 942.38X \qquad \text{for Peta}$$

These three estimated linear regressions are represented graphically in Figure 9.7.

☐ STEP 2.   Compute the value of the residual sum of squares for each of the k sets of data as:

$$A_i = \sum y^2 - \frac{\left(\sum xy\right)^2}{\sum x^2} \quad \text{for } i = 1,\ldots,k$$

For our example, the three values of the residual sum of squares for IR1514A-E666, IR8, and Peta are:

$$A_1 = 4{,}068{,}187 - \frac{(4{,}519)^2}{8.42}$$

$$= 1{,}642{,}847$$

Table 9.4 Computation for Testing Homogeneity of Three Regression Coefficients, in the Linear Relationships between Grain Yield (*Y*) and Tiller Number (*X*), of Rice Varieties IR1514A-E666, IR8, and Peta

IR1514A-E666		IR8		Peta	
Grain Yield, kg/ha	Tillers, no./hill	Grain Yield, kg/ha	Tillers, no./hill	Grain Yield, kg/ha	Tillers, no./hill
5,932	7.98	4,876	9.98	1,528	7.90
4,050	5.72	3,267	7.67	2,858	6.39
4,164	4.95	2,051	6.67	3,857	5.78
4,862	7.82	4,322	8.33	3,796	5.55
5,596	6.67	4,557	9.16	1,507	6.50
5,570	6.55	4,832	9.67	2,078	7.67
4,002	5.28	2,322	6.75	1,638	7.03
$\bar{X}_1 = 6.42$		$\bar{X}_2 = 8.32$		$\bar{X}_3 = 6.69$	
$\bar{Y}_1 = 4,882$		$\bar{Y}_2 = 3,747$		$\bar{Y}_3 = 2,466$	
$\sum x_1^2 = 8.42$		$\sum x_2^2 = 10.89$		$\sum x_3^2 = 4.79$	
$\sum y_1^2 = 4,068,187$		$\sum y_2^2 = 8,576,031$		$\sum y_3^2 = 6,493,098$	
$\sum x_1 y_1 = 4,519$		$\sum x_2 y_2 = 9,373$		$\sum x_3 y_3 = -4,514$	

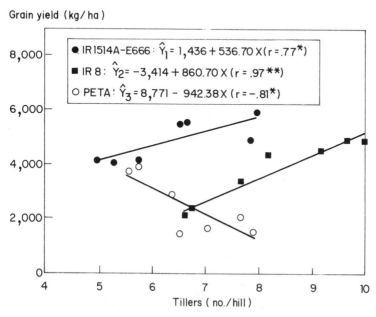

Figure 9.7 Estimated linear relationships between grain yield (*Y*) and tiller number (*X*) of three rice varieties: IR1514A-E666, IR8, and Peta.

$$A_2 = 8,576,031 - \frac{(9,373)^2}{10.89}$$

$$= 508,710$$

$$A_3 = 6,493,098 - \frac{(-4,514)^2}{4.79}$$

$$= 2,239,195$$

☐ STEP 3. Compute the sum of the k values of the residual sum of squares computed in step 2 as:

$$B = \sum_{i=1}^{k} A_i$$

For our example, the sum of the $k = 3$ values of the residual sum of squares is:

$$B = 1,642,847 + 508,710 + 2,239,195 = 4,390,752$$

☐ STEP 4. Compute the sum of the k corrected sums of squares for each of the two variables, and the sum of the k corrected sums of cross products between the two variables as:

$$C = \sum x_1^2 + \sum x_2^2 + \cdots + \sum x_k^2$$

$$D = \sum y_1^2 + \sum y_2^2 + \cdots + \sum y_k^2$$

$$E = \sum x_1 y_1 + \sum x_2 y_2 + \cdots + \sum x_k y_k$$

For our example, the values of C, D, and E are:

$$C = 8.42 + 10.89 + 4.79 = 24.10$$

$$D = 4,068,187 + 8,576,031 + 6,493,098$$

$$= 19,137,316$$

$$E = 4,519 + 9,373 + (-4,514) = 9,378$$

☐ STEP 5. Compute the F value as:

$$F = \frac{[D - (E^2/C) - B]/(k-1)}{B / \left(\sum_{i=1}^{k} n_i - 2k \right)}$$

where n_i ($i = 1,\ldots,k$) is the number of paired observations in the ith set of data. For our example, the F value is:

$$F = \frac{\{19,137,316 - [(9,378)^2/(24.10)] - 4,390,752\}/(3 - 1)}{4,390,752/[(7 + 7 + 7) - 2(3)]}$$

$$= 18.96$$

□ STEP 6. Compare the computed F value to the tabular F values (Appendix E) with $f_1 = (k - 1)$ and $f_2 = (\Sigma_{i=1}^{k} n_i - 2k)$ degrees of freedom. Reject the hypothesis of homogeneity of k regression coefficients ($\beta_1 = \beta_2 = \cdots = \beta_k$) if the computed F value is greater than the corresponding tabular F value at the prescribed level of significance.

For our example, the tabular F values with $f_1 = 2$ and $f_2 = 15$ degrees of freedom are 3.68 at the 5% level of significance and 6.36 at the 1% level. Because the computed F value is greater than the tabular F value at the 1% level of significance, the hypothesis of homogeneity between the three regression coefficients is rejected.

9.1.1.4 Homogeneity of Correlation Coefficients.

To determine whether the degree of the linear association between two variables remains the same at different levels of a third or fourth variable, homogeneity of the simple linear correlation coefficients can be tested. For example, the simple linear correlation coefficients between grain yield and ear weight in maize grown with different rates of nitrogen fertilizer maybe tested for homogeneity to determine whether the linear association between grain yield and ear weight is affected by the nitrogen level.

We illustrate the procedure for testing the homogeneity of k simple linear correlation coefficients with data of soluble protein nitrogen (X_1) and total chlorophyll (X_2) in the leaves of rice varieties IR8 and IR22. Data for IR8 is in Table 9.2; data for IR22 is in Table 9.5. The objective is to determine whether the degrees of the linear association between soluble protein nitrogen and total chlorophyll in the leaves are the same for the two rice varieties. The step-by-step procedures are:

□ STEP 1. Compute the simple linear correlation coefficient, following the procedure of Section 9.1.1.2, for each of the k sets of data. Denote these coefficients as $r_1, r_2,\ldots,r_k$. For our example, the $k = 2$ simple linear correlation coefficients are computed as $r_1 = .942$ for IR8 (Table 9.2) and $r_2 = .944$ for IR22 (Table 9.5).

□ STEP 2. For each computed r value, compute the corresponding z value as:

$$z = 0.5 \ln \frac{1 + r}{1 - r}$$

Table 9.5 Computation of a Simple Linear Correlation Coefficient between Soluble Protein Nitrogen (X_1) and Total Chlorophyll (X_2) in the Leaves of Rice Variety IR22

Sample Number	Soluble Protein N, mg/leaf (X_1)	Total Chlorophyll, mg/leaf (X_2)	Deviate x_1	Deviate x_2	Square of Deviate x_1^2	Square of Deviate x_2^2	Product of Deviates $(x_1)(x_2)$
1	0.84	0.55	-0.52	-0.70	0.2704	0.4900	0.3640
2	1.24	1.24	-0.12	-0.01	0.0144	0.0001	0.0012
3	2.10	1.56	0.74	0.31	0.5476	0.0961	0.2294
4	2.64	2.52	1.28	1.27	1.6384	1.6129	1.6256
5	1.31	1.64	-0.05	0.39	0.0025	0.1521	-0.0195
6	1.22	1.17	-0.14	-0.08	0.0196	0.0064	0.0112
7	0.19	0.04	-1.17	-1.21	1.3689	1.4641	1.4157
Total	9.54	8.72	0.02^a	-0.03^a	3.8618	3.8217	3.6276
Mean	1.36	1.25					

^{a} The nonzero values are only due to rounding.

where ln refers to the natural logarithm (i.e., log base e). For our example, the two z values are computed as:

$$z_1 = 0.5 \ln \frac{(1 + 0.942)}{(1 - 0.942)}$$

$$= 1.756$$

$$z_2 = 0.5 \ln \frac{(1 + 0.944)}{(1 - 0.944)}$$

$$= 1.774$$

□ STEP 3. Compute the weighted mean of the z values as:

$$\bar{z}_w = \frac{\sum\limits_{i=1}^{k} (n_i - 3) z_i}{\sum\limits_{i=1}^{k} (n_i - 3)}$$

where n_i ($i = 1, \ldots, k$) is the number of paired observations in the ith set of data used in the computation of the r_i value in step 1. For our example, the

$\bar{z}_w$ value is computed as:

$$\bar{z}_w = \frac{(7-3)(1.756)+(7-3)(1.774)}{(7-3)+(7-3)} = 1.765$$

Note that if n_i is the same for all sets of data (as is the case with this example), the $\bar{z}_w$ value is simply the standard arithmetic mean of the z_i values.

☐ STEP 4. Compute the χ^2 value (chi-square value) as:

$$\chi^2 = \sum_{i=1}^{k} (n_i - 3)(z_i - \bar{z}_w)^2$$

For our example, the χ^2 value is computed as:

$$\chi^2 = (7-3)(1.756 - 1.765)^2 + (7-3)(1.774 - 1.765)^2 = 0.001$$

☐ STEP 5. Compare the computed χ^2 value to the tabular χ^2 values, from Appendix D, with $(k-1)$ degrees of freedom. Reject the hypothesis of homogeneity of the k simple linear correlation coefficients if the computed χ^2 value is greater than the corresponding tabular χ^2 value at the prescribed level of significance.

For our example, the tabular χ^2 values from Appendix D, with $(k-1)$ = 1 degree of freedom, are 3.84 at the 5% level of significance and 6.63 at the 1% level. Because the computed χ^2 value is smaller than the tabular χ^2 value at the 5% level of significance, the test is not significant and the hypothesis of homogeneity cannot be rejected.

☐ STEP 6. If the χ^2 test is not significant, obtain the value of the pooled simple linear correlation coefficient from Appendix I, based on the $\bar{z}_w$ value computed in step 3. If the $\bar{z}_w$ value is not available in Appendix I, the pooled simple linear correlation coefficient can be computed as:

$$r_p = \frac{e^{2\bar{z}_w} - 1}{e^{2\bar{z}_w} + 1}$$

For our example, the pooled simple linear correlation coefficient is computed as:

$$r_p = \frac{e^{2(1.765)} - 1}{e^{2(1.765)} + 1} = .943$$

Thus, the simple linear correlations between soluble protein nitrogen and total chlorophyll in the leaves of IR8 and IR22 can be measured with a single coefficient of .943.

9.1.2 Multiple Linear Regression and Correlation

The simple linear regression and correlation analysis, as described in Section 9.1.1, has one major limitation. That is, that it is applicable only to cases with one independent variable. However, with the increasingly accepted perception of the interdependence between factors of production and with the increasing availability of experimental procedures that can simultaneously evaluate several factors, researchers are increasing the use of factorial experiments. Thus, there is a corresponding increase in need for use of regression procedures that can simultaneously handle several independent variables.

Regression analysis involving more than one independent variable is called *multiple regression analysis*. When all independent variables are assumed to affect the dependent variable in a linear fashion and independently of one another, the procedure is called *multiple linear regression analysis*. A multiple linear regression is said to be operating if the relationship of the dependent variable Y to the k independent variables $X_1, X_2, \ldots, X_k$ can be expressed as

$$Y = \alpha + \beta_1 X_1 + \beta_2 X_2 + \cdots + \beta_k X_k$$

The data required for the application of the multiple linear regression analysis involving k independent variables are the $(n)(k + 1)$ observations described here:

Observation Number	Observation Value					
	Y	X_1	X_2	X_3	$\ldots$	X_k
1	Y_1	X_{11}	X_{21}	X_{31}	$\ldots$	X_{k1}
2	Y_2	X_{12}	X_{22}	X_{32}	$\ldots$	X_{k2}
3	Y_3	X_{13}	X_{23}	X_{33}	$\ldots$	X_{k3}
.	.	.	.	.	$\ldots$	.
.	.	.	.	.	$\ldots$	.
.	.	.	.	.	$\ldots$	.
n	Y_n	X_{1n}	X_{2n}	X_{3n}	$\ldots$	X_{kn}

The $(k + 1)$ variables $Y, X_1, X_2, \ldots, X_k$ must be measured simultaneously for each of the n units of observation (i.e., experimental unit or sampling unit). In addition, there must be enough observations to make n greater than $(k + 1)$.

The multiple linear regression procedure involves the estimation and test of significance of the $(k + 1)$ parameters of the multiple linear regression equation. We illustrate the procedure for a case where $k = 2$, using the data on grain yield (Y), plant height (X_1), and tiller number (X_2) in Table 9.6. With $k = 2$, the multiple linear regression equation is expressed as:

$$Y = \alpha + \beta_1 X_1 + \beta_2 X_2$$

The step-by-step procedures are:

☐ STEP 1. Compute the mean and the corrected sum of squares for each of the $(k + 1)$ variables Y, X_1, X_2,..., X_k, and the corrected sums of cross products for all possible pair-combinations of the $(k + 1)$ variables, following the procedure described in step 1 of Section 9.1.1.1. A summary of the parameters to be computed, together with the variables involved, is shown here.

| Variable | Mean | \multicolumn{5}{c}{Corrected Sum of Squares and Cross Products} |
		X_1	X_2	...	X_k	Y
X_1	$\overline{X}_1$	$\sum x_1^2$	$\sum x_1 x_2$	...	$\sum x_1 x_k$	$\sum x_1 y$
X_2	$\overline{X}_2$		$\sum x_2^2$	...	$\sum x_2 x_k$	$\sum x_2 y$
.	.		...			
.	.		...			
.	.			...		
X_k	$\overline{X}_k$				$\sum x_k^2$	$\sum x_k y$
Y	$\overline{Y}$					$\sum y^2$

For our example, the results of the computation are shown in Table 9.6.

Table 9.6 Computation of a Multiple Linear Regression Equation Relating Plant Height (X_1) and Tiller Number (X_2) to Yield (Y), over Eight Rice Varieties

Variety Number	Grain Yield, kg/ha (Y)	Plant Height, cm (X_1)	Tiller, no./hill (X_2)
1	5,755	110.5	14.5
2	5,939	105.4	16.0
3	6,010	118.1	14.6
4	6,545	104.5	18.2
5	6,730	93.6	15.4
6	6,750	84.1	17.6
7	6,899	77.8	17.9
8	7,862	75.6	19.4
Mean	6,561	96.2	16.7

$\sum x_1^2 = 1{,}753.72$ $\sum x_2^2 = 23.22$ $\sum x_1 y = -65{,}194$

$\sum x_2 y = 7{,}210$ $\sum x_1 x_2 = -156.65$ $\sum y^2 = 3{,}211{,}504$

□ STEP 2.   Solve for $b_1, b_2, \ldots, b_k$ from the following k simultaneous equations, which are generally referred to as the *normal equations*:

$$b_1 \sum x_1^2 + b_2 \sum x_1 x_2 + \cdots + b_k \sum x_1 x_k = \sum x_1 y$$

$$b_1 \sum x_1 x_2 + b_2 \sum x_2^2 + \cdots + b_k \sum x_2 x_k = \sum x_2 y$$

$$\begin{array}{cccc} \cdot & \cdot & \cdot & \cdot \\ \cdot & \cdot & \cdot & \cdot \\ \cdot & \cdot & \cdot & \cdot \end{array}$$

$$b_1 \sum x_1 x_k + b_2 \sum x_2 x_k + \cdots + b_k \sum x_k^2 = \sum x_k y$$

where $b_1, b_2, \ldots, b_k$ are the estimates of $\beta_1, \beta_2, \ldots, \beta_k$ of the multiple linear regression equation, and the values of the sum of squares and sum of cross products of the $(k + 1)$ variables are those computed in step 1.

There are many standardized procedures for solving the k simultaneous equations for the k unknowns, either manually or with the aid of computers (Simmons, 1948; Anderson and Bancroft, 1952; Nie et al., 1975; and Barr et al., 1979).*

For our example, with $k = 2$, the normal equations are:

$$b_1 \sum x_1^2 + b_2 \sum x_1 x_2 = \sum x_1 y$$

$$b_1 \sum x_1 x_2 + b_2 \sum x_2^2 = \sum x_2 y$$

and the solutions for b_1 and b_2 are:

$$b_1 = \frac{\left(\sum x_2^2\right)\left(\sum x_1 y\right) - \left(\sum x_1 x_2\right)\left(\sum x_2 y\right)}{\left(\sum x_1^2\right)\left(\sum x_2^2\right) - \left(\sum x_1 x_2\right)^2}$$

$$= \frac{(23.22)(-65{,}194) - (-156.65)(7{,}210)}{(1{,}753.72)(23.22) - (-156.65)^2}$$

$$= -23.75$$

*H. A. Simmons. *College Algebra.* New York: Macmillan, 1948. pp. 123–147.
R. L. Anderson and T. A. Bancroft. *Statistical Theory in Research.* USA: McGraw-Hill, 1952. pp. 192–200.
N. H. Nie et al. *Statistical Package for the Social Sciences,* 2nd ed. USA: McGraw-Hill, 1975. pp. 320–360.
A. J. Barr et al. *Statistical Analysis System User's Guide.* USA: SAS Institute, 1979. pp. 237–263.

$$b_2 = \frac{\left(\sum x_1^2\right)\left(\sum x_2 y\right) - \left(\sum x_1 x_2\right)\left(\sum x_1 y\right)}{\left(\sum x_1^2\right)\left(\sum x_2^2\right) - \left(\sum x_1 x_2\right)^2}$$

$$= \frac{(1{,}753.72)(7{,}210) - (-156.65)(-65{,}194)}{(1{,}753.72)(23.22) - (-156.65)^2}$$

$$= 150.27$$

☐ STEP 3. Compute the estimate of the intercept α as:

$$a = \overline{Y} - b_1 \overline{X}_1 - b_2 \overline{X}_2 - \cdots - b_k \overline{X}_k$$

where $\overline{Y}, \overline{X}_1, \overline{X}_2, \ldots, \overline{X}_k$ are the means of the $(k + 1)$ variables computed in step 1.

Thus, the estimated multiple linear regression is:

$$\hat{Y} = a + b_1 X_1 + b_2 X_2 + \cdots + b_k X_k$$

For our example, the estimate of the intercept α is computed as:

$$a = \overline{Y} - b_1 \overline{X}_1 - b_2 \overline{X}_2$$

$$= 6{,}561 - (-23.75)(96.2) - (150.27)(16.7)$$

$$= 6{,}336$$

And the estimated multiple linear regression equation relating plant height (X_1) and tiller number (X_2) to yield (Y) is:

$$\hat{Y} = 6{,}336 - 23.75 X_1 + 150.27 X_2$$

☐ STEP 4. Compute:

- The sum of squares due to regression, as:

$$SSR = \sum_{i=1}^{k} (b_i)\left(\sum x_i y\right)$$

- The residual sum of squares, as:

$$SSE = \sum y^2 - SSR$$

- The coefficient of determination, as:

$$R^2 = \frac{SSR}{\sum y^2}$$

The coefficient of determination R^2 measures the contribution of the linear function of k independent variables to the variation in Y. It is usually expressed in percentage. Its square root (i.e., R) is referred to as the *multiple correlation coefficient*.

For our example, the values of SSR, R^2, and SSE are computed as:

$$SSR = b_1 \sum x_1 y + b_2 \sum x_2 y$$

$$= (-23.75)(-65,194) + (150.27)(7,210)$$

$$= 2,631,804$$

$$SSE = \sum y^2 - SSR = 3,211,504 - 2,631,804$$

$$= 579,700$$

$$R^2 = \frac{SSR}{\sum y^2} = \frac{2,631,804}{3,211,504} = .82$$

Thus, 82% of the total variation in the yields of eight rice varieties can be accounted for by a linear function, involving plant height and tiller number, as expressed in step 3.

☐ STEP 5.   Test the significance of R^2:

• Compute the F value as:

$$F = \frac{SSR/k}{SSE/(n - k - 1)}$$

For our example, the F value is:

$$F = \frac{2,631,804/2}{579,700/(8 - 2 - 1)}$$

$$= 11.35$$

• Compare the computed F value to the tabular F values (Appendix E) with $f_1 = k$ and $f_2 = (n - k - 1)$ degrees of freedom. The coefficient of determination R^2 is said to be significant (significantly different from zero) if the computed F value is greater than the corresponding tabular F value at the prescribed level of significance.

For our example, the tabular F values (Appendix E) with $f_1 = 2$ and $f_2 = 5$ degrees of freedom are 5.79 at the 5% level of significance and 13.27 at the 1% level. Because the computed F value is larger than the corresponding tabular F value at the 5% level of significance, but smaller than the tabular F value at the 1% level, the estimated multiple linear regression $\hat{Y} = 6,336 - 23.75X_1 + 150.27X_2$ is significant at the 5% level of significance. Thus, the combined linear effects of plant height and tiller number contribute significantly to the variation in yield.

We make a point here concerning the significance of the F test (which indicates the significance of R^2) and the size of the R^2 value. Although the significance of the linear regression implies that some portion of the variability in Y is indeed explained by the linear function of the independent variables, the size of the R^2 value provides information on the size of that portion. Obviously, the larger the R^2 value is, the more important the regression equation is in characterizing Y. On the other hand, if the value of R^2 is low, even if the F test is significant, the estimated regression equation may not be meaningful. For example, an R^2 value of .26, even if significant, indicates that only 26% of the total variation in the dependent variable Y is explained by the linear function of the independent variables considered. In other words, 74% of the variation in Y cannot be accounted for by the regression. With such low level of influence, the estimated regression equation would not be useful in estimating, much less predicting, the values of Y.

The following are important pointers for applying the multiple linear regression procedure just described:

- The procedure is applicable only if two conditions are satisfied. First, the effect of each and all k independent variables $X_1, X_2, \ldots, X_k$ on the dependent variable Y must be linear. That is, the amount of change in Y per unit change in each X_i is constant throughout the range of X_i values under study. Second, the effect of each X_i on Y is independent of the other X. That is, the amount of change in Y per unit change in each X_i is the same regardless of the values of the other X. Whenever any one or both of the foregoing conditions need to be relaxed, use of the procedures of Section 9.2 should be considered.

- The procedure for estimating and testing the significance of the $(k + 1)$ parameters in a multiple linear regression involving k independent variables is lengthy and time consuming, and becomes increasingly so as k becomes larger. This is where the use of a computer should be seriously considered. Numerous computer programs and statistical computer packages that can perform the multiple linear regression analysis are available at most computer centers and for most computers. Some of the more commonly used packages are the Barr et al., 1972; Nie et al., 1975; and Dixon, 1975.*

---

*A. J. Barr et al., *SAS*, pp. 237–263.

N. H. Nie et al., *SPSS*, pp. 320–360.

W. J. Dixon, Ed. *BMDP Biomedical Computer Programs*. Berkeley: University of California Press, 1975.

9.2   NONLINEAR RELATIONSHIP

The functional relationship between two variables is nonlinear if the rate of change in Y associated with a unit change in X is not constant over a specified range of X values. A nonlinear relationship among variables is common in biological organisms, especially if the range of values is wide. Two typical examples are:

- The response of rice yield to nitrogen fertilization, which is usually rapid at low levels, slower at the intermediate levels, and could become negative at high levels of nitrogen (Figure 9.8).
- The pattern of plant growth over time, which usually starts slowly, increases to a fast rate at intermediate growth stages, and slows toward the end of the life cycle (Figure 9.1b).

When the relationship among the variables under consideration is not linear, the regression procedures outlined in Section 9.1 are inadequate and a researcher must turn to nonlinear regression analysis procedures.

9.2.1   Simple Nonlinear Regression

There are numerous functional forms that can describe a nonlinear relationship between two variables, and choice of the appropriate regression and correlation technique depends on the functional form involved. We focus primarily on one technique: that involving the linearization of the nonlinear form, either through transformation of variables or through creation of new variables. We focus on a single technique, for two important reasons:

1.   The technique is simple because, after linearization, the regression procedures of Section 9.1 are directly applicable.

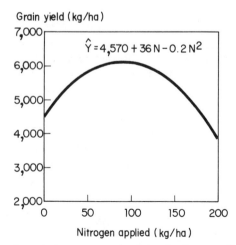

Grain yield (kg/ha)

$$\hat{Y} = 4{,}570 + 36\,N - 0.2\,N^2$$

Nitrogen applied (kg/ha)

Figure 9.8   An estimated nitrogen response of rice showing a nonlinear relationship.

2. The technique has wide applicability because most of the nonlinear relationships found in agricultural research can be linearized through variable transformation or variable creation.

9.2.1.1   Transformation of Variable.   A nonlinear relationship between two variables can be linearized by transforming one or both of the variables. Below are examples of some nonlinear forms, commonly encountered in agricultural research, that can be linearized through variable transformation.

Example 1.   The nonlinear form

$$Y = \alpha e^{\beta X}$$

can be linearized by transforming the dependent variable Y to $\ln Y$, where $\ln$ denotes the natural logarithm (log base e). Thus, the linearized form is:

$$Y' = \alpha' + \beta X$$

where $Y' = \ln Y$ and $\alpha' = \ln \alpha$.

Example 2.   The nonlinear form

$$Y = \alpha \beta^{X}$$

can be linearized by transforming the dependent variable Y to $\log Y$, where $\log$ denotes the logarithm base 10. Thus, the linearized form is:

$$Y' = \alpha' + \beta' X$$

where $Y' = \log Y$, $\alpha' = \log \alpha$, and $\beta' = \log \beta$.

Example 3.   The nonlinear form

$$\frac{1}{Y} = \alpha + \beta X$$

can be linearized by transforming the dependent variable Y to $1/Y$. Thus, the linearized form is:

$$Y' = \alpha + \beta X$$

where $Y' = 1/Y$.

Example 4.   The nonlinear form

$$Y = \alpha + \frac{\beta}{X}$$

can be linearized by transforming the independent variable X to $1/X$. Thus, the linearized form is:

$$Y = \alpha + \beta X'$$

where $X' = 1/X$.

Example 5. The nonlinear form

$$Y = \left(\alpha + \frac{\beta}{X} \right)^{-1}$$

can be linearized by transforming the dependent variable Y to $1/Y$ and the independent variable X to $1/X$. Thus, the linearized form is:

$$Y' = \alpha + \beta X'$$

where $Y' = 1/Y$ and $X' = 1/X$.

After linearization through variable transformation, the simple linear regression and correlation procedures described in Section 9.1.1 can be directly applied. We illustrate the procedure for applying the transformation technique to a simple nonlinear regression form with data on light transmission ratio (Y) and leaf-area index (X), in Table 9.7, which are to be fitted to the nonlinear

Table 9.7   Computation of a Nonlinear Regression Equation between Light Transmission Ratio (Y) and Leaf-Area Index (X) of Rice Variety IR8, by the Variable-Transformation Method

Observation Number	Light Transmission Ratio (Y)	Leaf-Area Index (X)	$Y' = \ln Y$
1	75.0	0.50	4.31749
2	72.0	0.60	4.27667
3	42.0	1.80	3.73767
4	29.0	2.50	3.36730
5	27.0	2.80	3.29584
6	10.0	5.45	2.30259
7	9.0	5.60	2.19722
8	5.0	7.20	1.60944
9	2.0	8.75	0.69315
10	2.0	9.60	0.69315
11	1.0	10.40	0.00000
12	0.9	12.00	−0.10536
Mean	22.9	5.60	2.19876
$\sum y'^2 = 28.77936$	$\sum x^2 = 175.40500$		$\sum xy' = -70.76124$

form of example 1; that is,

$$Y = \alpha e^{\beta X}$$

The step-by-step procedures are:

☐ STEP 1.   Linearize the prescribed nonlinear functional form through a proper transformation of one or both variables. For our example, the linearized form of the foregoing function is:

$$Y' = \alpha' + \beta X$$

where $Y' = \ln Y$ and $\alpha' = \ln \alpha$.

☐ STEP 2.   Compute the transformed values for all n units of observation of each variable that is transformed in step 1. For our example, values of $Y' = \ln Y$ are computed for the $n = 12$ pairs of (X, Y) values, as shown in Table 9.7.

☐ STEP 3.   Apply the simple linear regression technique of Section 9.1.1.1 to the data derived in step 2, based on the linearized form derived in step 1. For our example, the estimates of the two parameters α' and β are computed as:

- Compute the means $\bar{Y}'$ and $\bar{X}$, sums of squares $\Sigma y'^2$ and Σx^2, and sum of cross products $\Sigma xy'$, of the two variables Y' and X, as shown in Table 9.7.
- Compute the estimates of α' and β, following the formulas in Section 9.1.1.1, step 2, as:

$$b = \frac{\Sigma xy'}{\Sigma x^2}$$

$$= \frac{-70.76124}{175.40500} = -0.40342$$

$$a' = \bar{Y}' - b\bar{X}$$

$$= 2.19876 - (-0.40342)(5.60)$$

$$= 4.45791$$

☐ STEP 4.   Using estimates of the regression parameters of the linearized form obtained in step 3, derive an appropriate estimate of the original regression based on the specific transformation used in step 1.

For our example, conversion of the estimated regression parameters is needed. That is, in order to derive the estimate of the original nonlinear regression, its regression parameter α needs to be computed as the antilog of α':

$$a = \text{antilog of } a' = \text{antilog of } 4.45791$$

$$= 86.31$$

And, using the estimate of β computed in step 3, the required estimate of the regression equation is obtained as:

$$\hat{Y} = 86.31e^{-0.40342X}$$

The graphical representation of the estimated nonlinear regression is shown in Figure 9.9.

9.2.1.2 *Creation of New Variable.* Some nonlinear relationship between two variables can be linearized through the creation of one or more variables such that they can account for the nonlinear component of the original

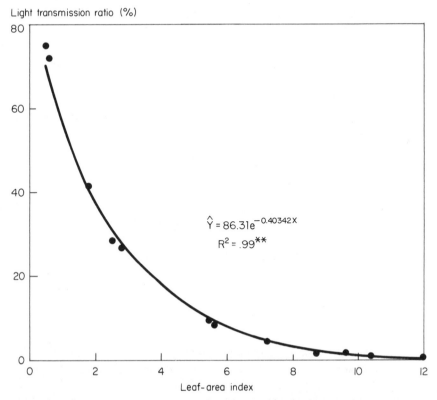

Figure 9.9 An estimated nonlinear relationship between light transmission ratio (Y) and leaf-area index (X) of rice variety IR8.

function. In agricultural research, this technique is most commonly applied to the kth degree polynomial:

$$Y = \alpha + \beta_1 X + \beta_2 X^2 + \cdots + \beta_k X^k$$

Such an equation can be linearized by creating k new variables: $Z_1, Z_2, \ldots, Z_k$, to form a multiple linear equation of the form:

$$Y = \alpha + \beta_1 Z_1 + \beta_2 Z_2 + \cdots + \beta_k Z_k$$

where $Z_1 = X, Z_2 = X^2, \ldots,$ and $Z_k = X^k$.

With the linearized form resulting from the creation of new variables, the procedure for multiple linear regression and correlation analysis of Section 9.1.2 can be directly applied.

We illustrate this technique using the data of Table 9.8 where the relationship between yield (Y) and nitrogen rate (X) is assumed to be quadratic (second-degree polynomial):

$$Y = \alpha + \beta_1 X + \beta_2 X^2$$

The step-by-step procedures for fitting the second-degree polynomial using the creation-of-new-variable technique are:

☐ STEP 1. Linearize the prescribed nonlinear functional form through the creation of an appropriate set of new variables.

Table 9.8 **Computation of an Estimated Quadratic Regression Equation $\hat{Y} = a + b_1 X + b_2 X^2$, or its Linearized form: $\hat{Y} = a + b_1 Z_1 + b_2 Z_2$, to Describe the Yield Response of Rice Variety IR661-1-170 to Nitrogen Fertilizer**

Pair Number	Grain Yield, kg/ha (Y)	Nitrogen Rate, kg/ha $(Z_1 = X)$	$Z_2 = X^2$
1	4,878	0	0
2	5,506	30	900
3	6,083	60	3,600
4	6,291	90	8,100
5	6,361	120	14,400
Mean	5,824	60	5,400
$\sum y^2 = 1,569,579$		$\sum z_1^2 = 9,000$	$\sum z_2^2 = 140,940,000$
$\sum z_1 y = 112,530$		$\sum z_2 y = 12,167,100$	$\sum z_1 z_2 = 1,080,000$

For our example, the linearized form of the second-degree polynomial is:

$$Y = \alpha + \beta_1 Z_1 + \beta_2 Z_2$$

where the two newly created variables Z_1 and Z_2 are defined as $Z_1 = X$ and $Z_2 = X^2$.

☐ STEP 2.   Compute the values of each newly created variable for all n units of observation. For our example, only the values of the variable Z_2 need to be computed because the values of the variable Z_1 are the same as those of the original variable X. The Z_2 values are computed by squaring the corresponding values of the original variable X, as shown in Table 9.8.

☐ STEP 3.   Apply the appropriate multiple linear regression technique of Section 9.1.2 to the linearized form derived in step 1, using the data derived in step 2. For our example, because the linearized form consists of two independent variables Z_1 and Z_2, the multiple linear regression procedure for two independent variables, as described in Section 9.1.2.2, is applied as:

- Compute the means, sums of squares, and sum of cross products for the three variables Y, Z_1, and Z_2, as shown in Table 9.8.
- Compute the estimates of the three parameters: α, β_1, and β_2, following the formulas in step 2 of Section 9.1.2, as:

$$b_1 = \frac{\left(\sum z_2^2\right)\left(\sum z_1 y\right) - \left(\sum z_1 z_2\right)\left(\sum z_2 y\right)}{\left(\sum z_1^2\right)\left(\sum z_2^2\right) - \left(\sum z_1 z_2\right)^2}$$

$$= \frac{(140{,}940{,}000)(112{,}530) - (1{,}080{,}000)(12{,}167{,}100)}{(9{,}000)(140{,}940{,}000) - (1{,}080{,}000)^2}$$

$$= 26.65$$

$$b_2 = \frac{\left(\sum z_1^2\right)\left(\sum z_2 y\right) - \left(\sum z_1 z_2\right)\left(\sum z_1 y\right)}{\left(\sum z_1^2\right)\left(\sum z_2^2\right) - \left(\sum z_1 z_2\right)^2}$$

$$= \frac{(9{,}000)(12{,}167{,}100) - (1{,}080{,}000)(112{,}530)}{(9{,}000)(140{,}940{,}000) - (1{,}080{,}000)^2}$$

$$= -0.118$$

$$a = \bar{Y} - b_1 \bar{Z}_1 - b_2 \bar{Z}_2$$

$$= 5{,}824 - (26.65)(60) - (-0.118)(5{,}400)$$

$$= 4{,}862$$

Thus, the second-degree polynomial regression equation describing the yield response of rice selection IR661-1-170 to the nitrogen rates applied, within the range of 0 to 120 kg N/ha, is estimated as:

$$\hat{Y} = 4{,}862 + 26.65X - 0.118X^2 \quad \text{for } 0 \le X \le 120$$

- Compute the coefficient of determination, as:

$$R^2 = \frac{b_1 \sum z_1 y + b_2 \sum z_2 y}{\sum y^2}$$

$$= \frac{(26.65)(112{,}530) + (-0.118)(12{,}167{,}100)}{1{,}569{,}579}$$

$$= \frac{1{,}563{,}207}{1{,}569{,}579} = .996$$

- Compute the F value, as:

$$F = \frac{(n - k - 1)\left(b_1 \sum z_1 y + b_2 \sum z_2 y\right)}{k\left(\sum y^2 - b_1 \sum z_1 y - b_2 \sum z_2 y\right)}$$

$$= \frac{(5 - 2 - 1)(1{,}563{,}207)}{(2)(1{,}569{,}579 - 1{,}563{,}207)}$$

$$= 245.32$$

Because the computed F value exceeds the tabular F value with $f_1 = f_2 = 2$ degrees of freedom at the 1% level of significance of 99.0, the estimated quadratic regression equation is significant at the 1% level. Results indicate that the yield response of the rice variety IR661-1-170 to nitrogen fertilization can be adequately described by the quadratic equation. The computed R^2 value of .996 indicates that 99.6% of the total variation in the mean yields was explained by the quadratic regression equation estimated.

9.2.2 Multiple Nonlinear Regression

When the relationship between the dependent variable Y and the k independent variables $X_1, X_2, \ldots, X_k$, where $k > 1$, does not follow the multiple linear relationship, a *multiple nonlinear relationship* exists. The occurrence of a multiple nonlinear relationship may be the result of any of the following:

- At least one of the independent variables exhibits a nonlinear relationship with the dependent variable Y. For example, with two independent variables X_1 and X_2, a multiple nonlinear relationship exists if either or both of the two variables exhibit a nonlinear relationship with the dependent variable. If both X_1 and X_2 are related to Y in a quadratic manner, for instance, the corresponding multiple nonlinear regression equation representing their relationship to Y would be:

$$Y = \alpha + \beta_1 X_1 + \beta_2 X_1^2 + \beta_3 X_2 + \beta_4 X_2^2$$

- At least two independent variables interact with each other. For example, with two independent variables X_1 and X_2, each of which separately affects Y in a linear fashion, the multiple regression equation may be nonlinear if the effect of factor X_1 on Y varies with the level of factor X_2, and vice versa. In such a case, the multiple nonlinear regression equation may be represented by:

$$Y = \alpha + \beta_1 X_1 + \beta_2 X_2 + \beta_3 X_1 X_2$$

where the last quantity in the equation represents the interaction term.

- Both of the foregoing cases occur simultaneously. That is, at least one of the independent variables has a nonlinear relationship with the dependent variable and at least two independent variables interact with each other. Putting the two equations together, we may have:

$$Y = \alpha + \beta_1 X_1 + \beta_2 X_1^2 + \beta_3 X_2 + \beta_4 X_2^2 + \beta_5 X_1 X_2$$

or, more generally:

$$Y = \alpha + \beta_1 X_1 + \beta_2 X_1^2 + \beta_3 X_2 + \beta_4 X_2^2 + \beta_5 X_1 X_2$$
$$+ \beta_6 X_1^2 X_2 + \beta_7 X_1 X_2^2 + \beta_8 X_1^2 X_2^2$$

The analytical technique we discuss in this section is essentially an extension of the linearization technique for the simple nonlinear regression discussed in Section 9.2.1. The multiple nonlinear form is first linearized so that the multiple linear regression analysis discussed in Section 9.1.2 can be directly applied. Three examples of the multiple nonlinear form and their corresponding linearized forms are:

Example 1. Linearization of equation:

$$Y = \alpha + \beta_1 X_1 + \beta_2 X_1^2 + \beta_3 X_2 + \beta_4 X_2^2$$

through the creation of two new variables: $Z_1 = X_1^2$ and $Z_2 = X_2^2$. Thus, the linearized form is:

$$Y = \alpha + \beta_1 X_1 + \beta_2 Z_1 + \beta_3 X_2 + \beta_4 Z_2$$

Example 2. Linearization of the Cobb-Douglas equation:

$$Y = \alpha \beta_1^{X_1} \beta_2^{X_2} \beta_3^{X_3} \cdots \beta_k^{X_k}$$

through the transformation of the dependent variable Y to $\log Y$ (see Section 9.2.1.1, example 2). Thus, the linearized form is:

$$Y' = \alpha' + \beta_1' X_1 + \beta_2' X_2 + \cdots + \beta_k' X_k$$

where $Y' = \log Y$, $\alpha' = \log \alpha$, and $\beta_i' = \log \beta_i$ $(i = 1, \ldots, k)$.

Example 3. Linearization of a nonlinear form, caused by the presence of one or more interaction terms, through the creation of one new variable for every interaction term. For example, with the following equation having one interaction term:

$$Y = \alpha + \beta_1 X_1 + \beta_2 X_2 + \beta_3 X_1 X_2$$

only one new variable $Z_1 = X_1 X_2$ needs to be created. The linearized form is then:

$$Y = \alpha + \beta_1 X_1 + \beta_2 X_2 + \beta_3 Z_1$$

Once the linearized form is derived, the procedure for the multiple linear regression analysis of Section 9.1.2 can be directly applied.

9.3 SEARCHING FOR THE BEST REGRESSION

There are essentially two ways in which the relationship between the dependent variable Y and the k independent variables $X_1, X_2, \ldots, X_k$ may be specified.

First, based on accepted biological concepts, secondary data, or past experiences, the researcher postulates, even before the data is gathered, one or more functional forms (shapes of curves) that should adequately describe the relationship among the variables of interest. For example, biological concepts and past experiences in rice strongly suggest that the response of grain yield to nitrogen application follows the quadratic function (i.e., the second-degree polynomial) as shown in Figure 9.8; or the relationship between light transmission ratio and leaf-area index follows an exponential function of the type shown in Figure 9.9. The specification of the functional form based on previous experience and accepted biological concepts is preferred because the regression parameters can usually be easily associated to some known biological phenomena and its biological implication is easily identified.

Second, based on data gathered in the experiment itself, the researcher identifies one or more functional forms that are most likely to best fit the current data. This approach is frequently used in practice because the relationship among variables is rarely known before the experiment is started. In fact, a common objective of many biological experiments is to identify the form of the relationship itself.

In practice, both procedures are generally used jointly. The task of specifying the functional relationship between a dependent variable and more than one independent variable is a two-stage process. The first stage involves the specification of an appropriate functional form between the dependent variable and each of the independent variables; the second stage involves the specification of the terms representing interaction effects between the independent variables.

Consider, for example, the case where the researcher wishes to specify the relationship between two independent variables, nitrogen and phosphorus, and the dependent variable, crop yield. The first stage in the process consists of specifying the relationships between yield and nitrogen and yield and phosphorous separately; and the second stage consists of identifying the type of interaction, if any, between nitrogen and phosphorous.

We discuss four procedures that are commonly used for the specification of an appropriate functional form between a dependent variable and one or more independent variables:

- The scatter diagram technique is used primarily for the case of one independent variable (i.e., simple regression).
- The analysis of variance technique is used primarily for the specification of interaction terms in a multiple regression.
- The test of significance technique is used primarily for the elimination of *unnecessary* regression terms in a specified multiple regression.
- The step-wise regression technique is used primarily for identifying the sequence of importance in which each regression term should be included in the multiple regression equation based on their relative contributions.

9.3.1 The Scatter Diagram Technique

The scatter diagram is the simplest and most commonly used procedure for examining the relationship between two variables. The steps involved are:

☐ STEP 1. For any pair of variables, say X and Y, that are suspected to be associated with each other, plot all pairs of values as points in an $X - Y$ plane, such as those shown in Figure 9.10.

☐ STEP 2. Examine the scatter diagram, constructed in step 1, for the presence of outliers (i.e., points that are apart from the bulk of the other data points). Outliers usually exert a considerable influence on the kind of functional form that may be perceived from the data and should, therefore,

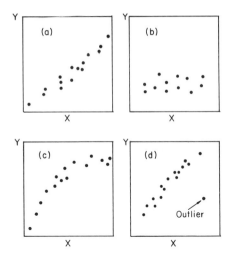

Figure 9.10 Scatter diagrams showing various types and degrees of relationship between two variables Y and X: (a) linear relationship, (b) no relationship, (c) nonlinear relationship, and (d) the presence of an outlier.

be examined carefully to ensure that their presence is not due to measurement or transcription errors.

The procedures outlined in Chapter 7, Section 7.2.2.2, for the verification of the validity of suspiciously large (or small) data should be used. It is usually tempting to discard outright an outlier, especially if by so doing a good fit to the hypothesized relationship can be ensured. As discussed in Chapter 7, Section 7.2.2.2, however, the decision to exclude an outlier should be made with extreme caution. If the discarded outlier represents a legitimate biological event rather than an error, its elimination will obviously result in incorrect interpretation of data. More importantly, the researcher would have missed an opportunity to explore probable causes of unusual events, a rare opportunity that could lead to an important discovery.

☐ STEP 3. Examine the scatter diagram to identify the functional relationship, if any, between the two variables. For example, a set of points arranged in a narrow band stretching from two opposite corners of the $X - Y$ plane, as shown in Figure 9.10a, indicates a strong linear relationship, whereas an evenly scattered set of points whose boundaries approach that of a square, rectangle, or circle, as shown in Figure 9.10b, suggests the lack of any relationship.

☐ STEP 4. If no clear relationship is indicated in step 3, confirm the absence of any linear relationship by computing a simple linear correlation coefficient r and testing its significance following the procedure outlined in Section 9.1.1.2. The lack of any obvious relationship through visual observation of step 3 plus a nonsignificant r is usually a good indication of the absence of any relationship between the two variables. When such is the case, the procedure can be terminated at this point.

If the presence of a relationship is indicated in step 3, identify the specific functional form that can best fit the observed relationship. A useful practice here is to first check whether the linear approximation may be adequate. If not, other possible nonlinear functional forms can be examined and the specific form that can best describe the relationship between the two variables chosen. Because there are innumerable forms of nonlinear relationship, this task is not easy for a nonstatistician. Fortunately, most of the relationships between characters in biological materials can be adequately described by few nonlinear forms. The most common ones are the high-degree polynomials, the sigmoid, the logarithmic, and the exponential curves. These are illustrated in Figure 9.11.

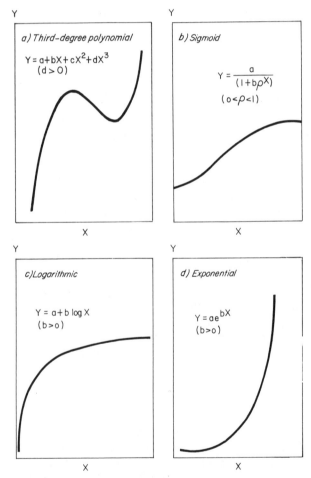

Figure 9.11 Examples of four functional forms found useful in describing relationships between two variables Y and X.

9.3.2   The Analysis of Variance Technique

The analysis of variance technique is most suited for data from a factorial experiment in which the independent variables of the regression equation are the factors being tested. An example is the regression equation

$$Y = \alpha + \beta_1 X_1 + \beta_2 X_1^2 + \beta_3 X_2 + \beta_4 X_2^2$$

which could have been derived from a two-factor experiment involving, say, the rate of nitrogen application as one factor (X_1) and the rate of phosphorus application as the other factor (X_2).

The analysis of variance procedure involves an appropriate partitioning of the sums of squares, in the analysis of variance, that are associated with the main effects and the interaction effects, into single *d.f.* contrasts. The partitioning is done so that the functional form that best describes the relationship between the variable of interest (Y) and the factors being tested ($X_1, X_2, \ldots, X_k$) can be indicated. We illustrate the procedure with data on nitrogen uptake of the rice plants (Y) from a two-factor experiment involving duration of water stress (X_1) and level of nitrogen application (X_2) in Table 9.9.

The experiment was a greenhouse trial in a split-plot design with four water-stress days as main-plot treatments and four nitrogen rates as subplot

Table 9.9   Nitrogen Uptake of the Rice Plants, Grown with Four Degrees of Water Stress and Four Rates of Nitrogen Application

Treatment Number	Water Stress, days	Nitrogen Rate, kg/ha	Nitrogen Uptake, g/pot				
			Rep. I	Rep. II	Rep. III	Rep. IV	Mean
1	0 (W_1)	0 (N_1)	0.250	0.321	0.373	0.327	0.318
2		90 (N_2)	0.503	0.493	0.534	0.537	0.517
3		180 (N_3)	0.595	0.836	0.739	0.974	0.786
4		270 (N_4)	1.089	1.297	1.007	0.677	1.018
5	10 (W_2)	0	0.254	0.373	0.349	0.367	0.336
6		90	0.506	0.613	0.588	0.625	0.583
7		180	0.692	0.754	0.548	0.713	0.677
8		270	1.033	0.757	1.034	0.831	0.914
9	20 (W_3)	0	0.248	0.234	0.267	0.305	0.264
10		90	0.428	0.397	0.493	0.587	0.476
11		180	0.484	0.453	0.457	0.372	0.442
12		270	0.507	0.498	0.477	0.619	0.525
13	40 (W_4)	0	0.099	0.103	0.093	0.084	0.095
14		90	0.154	0.142	0.133	0.129	0.140
15		180	0.111	0.102	0.098	0.152	0.116
16		270	0.089	0.142	0.138	0.141	0.128

treatments, in four replications. The step-by-step procedures for the use of the analysis of variance technique to arrive at an appropriate form of the multiple regression equation between the response (i.e., nitrogen uptake) and the two factors tested (i.e., water stress and nitrogen rate) are:

☐ STEP 1.   Compute a standard analysis of variance based on the experimental design used. Follow the procedures in Chapters 2 to 4. Then, make suitable trend comparisons for all quantitative factors, both for the main effects and the respective interaction effects, based on the procedure in Chapter 5, Section 5.2.3.

Table 9.10   Analysis of Variance of a Two-Factor Experiment (Split-Plot Design), with Polynomial Partitioning[a] of All Main-Effect and Interaction SS of Data in Table 9.9

Source of Variation	Degree of Freedom	Sum of Squares	Mean Square	Computed F^b	Tabular F 5%	Tabular F 1%
Replication	3	0.0080792	0.0026931			
Water stress (W)	3	2.9587214	0.9862405	208.09**	3.86	6.99
W_L	(1)	2.6530149	2.6530149	559.77**	5.12	10.56
W_Q	(1)	0.3026375	0.3026375	63.85**	5.12	10.56
W_C	(1)	0.0030690	0.0030690	< 1	—	—
Error(a)	9	0.0426555	0.0047950			
Nitrogen rate (N)	3	1.2872277	0.4290759	40.23**	2.86	4.38
N_L	(1)	1.2606476	1.2606476	118.21**	4.11	7.39
N_Q	(1)	0.0048825	0.0048825	< 1	—	—
N_C	(1)	0.0216976	0.0216976	2.03^{ns}	4.11	7.39
$W \times N$	9	0.6844155	0.0760462	7.13**	2.15	2.94
$W_L \times N_L$	(1)	0.6334966	0.6334966	59.40**	4.11	7.39
$W_L \times N_Q$	(1)	0.0049691	0.0049691	< 1	—	—
$W_L \times N_C$	(1)	0.0049738	0.0049738	< 1	—	—
$W_Q \times N_L$	(1)	0.0009146	0.0009146	< 1	—	—
$W_Q \times N_Q$	(1)	0.0048129	0.0048129	< 1	—	—
$W_Q \times N_C$	(1)	0.0222278	0.0222278	2.08^{ns}	4.11	7.39
$W_C \times N_L$	(1)	0.0087938	0.0087938	< 1	—	—
$W_C \times N_Q$	(1)	0.0042268	0.0042268	< 1	4.11	7.39
$W_C \times N_C$	(1)	0.0000001	0.0000001	< 1	—	—
Error(b)	36	0.3839131	0.0106643			
Total	63	5.3650124				

[a]Subscripts L, Q, and C for each factor refer to linear, quadratic, and cubic, respectively.
[b]** = significant at 1% level, ^{ns} = not significant.

For our example, the experimental design is a split-plot. The standard analysis of variance is, thus, based on the procedure of Chapter 3, Section 3.4.2. Because both factors—water stress and nitrogen rate—are quantitative, trend comparisons are applied for both main effects as well as their interaction. The orthogonal polynomial procedure of Chapter 5, Section 5.2.4, is used. The result of the analysis of variance with polynomial partitioning of the sums of squares (SS) associated with main effects and their interaction is shown in Table 9.10.

Note that because there are four levels for each factor, the largest degree polynomial that can be evaluated is the third degree (cubic). Thus, there are three single $d.f.$ components for each of the two main effects and nine for the interaction between the two factors.

☐ STEP 2. Using crop response as the dependent variable Y and the quantitative factors being tested as the independent variables $X_1, \ldots, X_p$, determine the number of multiple regression equations to be fitted following these guidelines:

A. When all test factors are quantitative, only one equation needs to be fitted. The data to be used are the treatment means, averaged over all replications.

B. When one or more test factors are discrete (i.e., not quantitative), the following alternatives should be considered:

· Estimate a single regression equation by including the discrete factors in the multiple regression equation as *dummy variables*. Appropriate discrete codes (such as -1 and $+1$ or 0, 1, 2, etc.) are assigned to each dummy variable. Because of the practical difficulty in (1) assigning suitable codes, (2) specifying appropriate interaction terms involving both discrete and quantitative factors, and (3) interpreting the results; user's experience and availability of proper advise are needed in adopting this approach. This procedure is not discussed in this book. Interested researchers should see Draper and Smith, 1966; Johnston, 1963; Teh-wei Hu, 1973; A. Koutsoyiannis, 1977; and S. R. Searle, 1971.*

· Estimate m regression equations, where m is the product of the levels of all discrete factors. However, if one of the discrete factors, say factor A, gave neither significant main effect nor interaction with any of the quantitative factors, the level of factor A, that is, a, is not

---

*N. R. Draper and H. Smith. *Applied Regression Analysis*. New York: Wiley, 1966. pp. 134–141.
J. Johnston. *Econometric Methods*. New York: McGraw-Hill, 1963. pp. 176–186.
Teh-wei Hu. *Econometrics, An Introductory Analysis*. Baltimore: University Park Press, 1973. pp. 65–70.
A. Koutsoyiannis. *Theory of Econometrics*. London: MacMillan, 1977. pp. 281–284.
S. R. Searle. *Linear Models*. USA: Wiley, 1971. pp. 135–163.

included in the computation of m. That is, there will be $m' = m/a$ regression equations instead.

For our example, there are two factors and both are quantitative. Hence, only one equation needs to be fitted. The equation has the nitrogen uptake as the dependent variable Y, water stress as the first independent variable X_1, and nitrogen rate as the second independent variable X_2.

Note that if our example were a three-factor experiment with variety as the third factor (i.e., discrete factor), there would be one multiple regression equation corresponding to each variety being tested. However, if the main effect of variety and its interaction with either water stress or nitrogen rate, or both, is not significant, then only one regression needs to be fitted, using the means of all treatment combinations between water stress and nitrogen rate averaged over the three varieties and four replications.

☐ STEP 3.   Specify the form of each multiple regression equation based on the results of the partitioning of SS performed in step 1. All significant components associated with each and all quantitative factors, whether main effects or interactions, should be included as regression terms in the equations.

For our example, the results of SS partitioning in step 1 indicate that four single $d.f.$ components are significant. They are:

- The linear component of water stress (W_L)
- The quadratic component of water stress (W_Q)
- The linear component of nitrogen rate (N_L)
- The linear component of the interaction ($W_L \times N_L$)

Hence, the regression terms to be included in the single multiple regression equation are X_1, X_1^2, X_2, and $X_1 X_2$. That is, the form of the multiple regression equation between nitrogen uptake (Y) and the two factors—water stress (X_1) and nitrogen rate (X_2)—can be represented by:

$$Y = \alpha + \beta_1 X_1 + \beta_2 X_1^2 + \beta_3 X_2 + \beta_4 X_1 X_2$$

☐ STEP 4.   Apply an appropriate multiple regression technique (Section 9.1.2 or Section 9.2.2) to estimate the regression parameters of each of the equations specified in step 3.

For our example, the form of the single regression equation specified in step 3 is a multiple nonlinear regression. Following the linearization procedure of Section 9.2.1.2, the linearized form is:

$$Y = \alpha + \beta_1 X_1 + \beta_2 Z_1 + \beta_3 X_2 + \beta_4 Z_2$$

where $Z_1 = X_1^2$, and $Z_2 = X_1 X_2$.

The data to be used are the 16 treatment means in the last column of Table 9.9. The standard multiple linear regression technique of Section 9.1.2

is then applied to obtain the following estimate of the regression equation:

$$\hat{Y} = 0.33300 + 0.00123 X_1 - 0.00016 Z_1 + 0.00254 X_2$$

$$- 0.00007 Z_2$$

That is, the regression equation expressed in terms of the original variables is:

$$\hat{Y} = 0.33300 + 0.00123 X_1 - 0.00016 X_1^2 + 0.00254 X_2$$

$$- 0.00007 X_1 X_2$$

The coefficient of determination R^2 is .97 which is significant at the 1% level. That is 97% of the variation between the 16 treatment means can be explained by the specific function of the two factors as specified above.

9.3.3 The Test of Significance Technique

In this approach, a multiple regression equation, involving as many terms as possible that have some possibilities of contributing to the variation in the dependent variable, is first constructed and the corresponding regression coefficients estimated. Appropriate tests of significance are then applied, either separately to each individual regression term or jointly to a group of regression terms. Based on these tests, regression terms that give significant contribution to the variation in the dependent variable are retained and the terms that give nonsignificant contributions are dropped. In contrast to the analysis of variance technique, which is applicable only to data from controlled experiments, this approach can also be applied to data that cannot be subjected to the analysis of variance technique.

9.3.3.1 Testing of Individual Regression Terms. This technique involves the fitting of data to a multiple regression equation that has as many regression terms as seem logical. The estimated regression coefficient of each term is then tested for its significance, separately and independently, and only those regression terms with significant coefficients are retained. A revised regression equation, with fewer terms, is then re-fitted and used as the chosen regression. We illustrate the procedure for the testing of individual regression terms with the same set of data used in Section 9.3.2. There are two independent variables with data on nitrogen uptake (Y), water stress (X_1), and nitrogen rate (X_2), as shown in Table 9.9. The step-by-step procedures are:

☐ STEP 1. Determine the number of regression equations to be fitted following the procedure outlined in step 2 of Section 9.3.2. For our example, there is only one multiple regression equation. The data to be used are the means (averaged over all replications) of the 16 treatment combinations.

☐ STEP 2. For each regression equation determined in step 1, specify a regression equation that would include as many terms as are suspected to influence the dependent variable. The choice of the number of terms to be included, say k, is crucial; a k that is too large reduces the chance of detecting significance in the individual regression terms, and a k that is too small increases the chance that some important terms are missed. The k value is primarily limited by the number of observations involved (n). Definitely, k must not be larger than $(n - 2)$.

For our example, assuming that prior information indicates that the dependent variable Y may be influenced by the quadratic form of variable X_1, the linear form of variable X_2, the interaction between the linear component of X_1 and the linear component of X_2, and the interaction between the quadratic component of X_1 and the linear component of X_2; the following multiple regression equation may be initially prescribed:

$$Y = \alpha + \beta_1 X_1 + \beta_2 X_1^2 + \beta_3 X_2 + \beta_4 X_1 X_2 + \beta_5 X_1^2 X_2$$

Here, there are five regression terms involving the two independent variables X_1 and X_2. Thus $k = 5$.

☐ STEP 3. Apply an appropriate multiple regression technique (Section 9.1.2 or Section 9.2.2) to fit the prescribed regression equation or equations specified in step 1.

For our example, the equation in step 2 should first be linearized into:

$$Y = \alpha + \beta_1 Z_1 + \beta_2 Z_2 + \beta_3 Z_3 + \beta_4 Z_4 + \beta_5 Z_5$$

where

$$Z_1 = X_1, Z_2 = X_1^2, Z_3 = X_2, Z_4 = X_1 X_2, \text{ and } Z_5 = X_1^2 X_2.$$

Then the standard multiple linear regression technique of Section 9.1.2 can be applied to obtain an estimate of the linearized equation as:

$$\hat{Y} = 0.30809 + 0.00639 Z_1 - 0.00029 Z_2 + 0.00273 Z_3$$

$$- 0.00010 Z_4 + 0.000001 Z_5$$

That is,

$$\hat{Y} = 0.30809 + 0.00639 X_1 - 0.00029 X_1^2 + 0.00273 X_2$$

$$- 0.00010 X_1 X_2 + 0.000001 X_1^2 X_2$$

The coefficient of determination R^2 of the estimated regression is .97, which is significant at the 1% level.

☐ STEP 4. Apply the test of significance to each regression coefficient obtained in step 3, by computing the corresponding t value as:

- Compute the standard error of each of the k estimated regression coefficients as:

$$s(b_i) = \sqrt{\frac{(c_{ii})(SSE)}{(n - k - 1)}}$$

where b_i is the estimate of the ith regression coefficient, SSE is the residual sum of squares defined in Section 9.1.2, step 4, and c_{ii} is the ith diagonal element of the C-matrix which is the inverse of the following matrix*:

$$A = \begin{bmatrix} \sum x_1^2 & \sum x_1 x_2 & \cdots & \sum x_1 x_k \\ \sum x_1 x_2 & \sum x_2^2 & \cdots & \sum x_2 x_k \\ \cdot & \cdot & \cdots & \cdot \\ \cdot & \cdot & \cdots & \cdot \\ \cdot & \cdot & \cdots & \cdot \\ \sum x_1 x_k & \sum x_2 x_k & & \sum x_k^2 \end{bmatrix}$$

where $\sum x_i^2$ and $\sum x_i x_j$ are the sum of squares and sum of cross products defined in Section 9.1.2.

For our example, the C-matrix is

$$C = \begin{bmatrix} 0.0103568182 & -0.0002306818 & 0.0002863636 & -0.0000493182 & 0.0000010985 \\ & 0.0000055682 & -0.0000053030 & 0.0000010985 & -0.0000000265 \\ & & 0.0000226712 & -0.0000021212 & 0.0000000393 \\ & & & 0.0000003653 & -0.0000000081 \\ & & & & 0.0000000002 \end{bmatrix}$$

and the value of SSE is computed as:

$$SSE = \sum y^2 - \sum_{i=1}^{k} (b_i)\left(\sum x_i y\right)$$

$$= 0.034276$$

---

*The definition of a matrix, its notation, and the derivation of its inverse are beyond the scope of this book. Interested readers should refer to:

S. Brandt. *Statistical and Computational Methods in Data Analysis*. Amsterdam: North Holland, 1970. pp. 248–263.

B. Ostle and R. W. Mensing. *Statistics in Research*, 3rd ed., USA: Iowa State University Press, 1975. pp. 193–202.

R. K. Singh and B. D. Chaudhary. *Biometrical Methods in Quantitative Genetic Analysis*. Indiana: Kalyani, 1977. pp. 19–38.

R. G. D. Steel and J. H. Torrie. *Principles and Procedures of Statistics*. USA: McGraw-Hill, 1960. pp. 290–297.

M. S. Younger. *Handbook for Linear Regression*. Massachusetts: Duxbury Press, 1979. pp. 298–307.

The standard error of each estimated regression coefficient is then computed as:

$$s(b_1) = \sqrt{(0.0103568182)\frac{(0.034276)}{10}} = 0.00595811$$

$$s(b_2) = \sqrt{(0.0000055682)\frac{(0.034276)}{10}} = 0.00013815$$

$$s(b_3) = \sqrt{(0.0000226712)\frac{(0.034276)}{10}} = 0.00027876$$

$$s(b_4) = \sqrt{(0.0000003653)\frac{(0.034276)}{10}} = 0.00003539$$

$$s(b_5) = \sqrt{(0.0000000002)\frac{(0.034276)}{10}} = 0.00000083$$

- Compute the t value, for each estimated regression coefficient b_i, as:

$$t_i = \frac{b_i}{s(b_i)}$$

where b_i is the regression coefficient computed in step 3 and $s(b_i)$ is the corresponding standard error computed above.

For our example, the five t values are:

$$t_1 = \frac{0.00639}{0.00595811} = 1.07$$

$$t_2 = \frac{-0.00029}{0.00013815} = -2.10$$

$$t_3 = \frac{0.00273}{0.00027876} = 9.79$$

$$t_4 = \frac{-0.00010}{0.00003539} = -2.83$$

$$t_5 = \frac{0.000001}{0.00000083} = 1.20$$

☐ STEP 5. Compare each computed t value to the tabular t value, from Appendix C, with $(n - k - 1)$ degrees of freedom. Reject the hypothesis that $\beta_i = 0$, at α level of significance, if the absolute value of the computed t value is greater than the corresponding tabular t value at α level of significance.

For our example, the tabular t values with $(n - k - 1) = 10$ degrees of freedom are 2.228 at the 5% level of significance and 3.169 at the 1% level. Results show that only the absolute value of the computed t_3 value is greater than the tabular t value at the 1% level, and the absolute value of the computed t_4 value is greater than the tabular t value at the 5% level. This indicates that, of the five regression terms originally suspected to influence the dependent variable, only two, namely, X_2 and $X_1 X_2$, are significant.

□ STEP 6.   Drop all regression terms whose regression coefficients are shown to be not significant and refit the data to the regression equation with $(k - m)$ regression terms, where m is the total number of the regression terms dropped. For our example, because only two regression coefficients (β_3 and β_4) are significant, all three other regression terms are dropped from the original regression equation and the revised equation is:

$$Y = \alpha + \gamma_1 X_2 + \gamma_2 X_1 X_2$$

which is to be refitted, following steps 3 to 5, and using the same set of data. The estimate of the revised regression equation is obtained as:

$$\hat{Y} = 0.27015 + 0.00299 X_2 - 0.00009 X_1 X_2$$

with the coefficient of determination R^2 of .93, which is significant at the 1% level.

9.3.3.2   Joint Testing of Regression Terms.   Instead of testing each regression term separately, as is the case with the procedure of Section 9.3.3.1, the joint testing technique applies the test of significance jointly to a group consisting of q regression terms, where $q \geq 2$. Depending on the result of the test of significance, the whole group of q regression terms is either retained, or dropped, as a unit.

We illustrate the procedure for joint testing with the same set of data used in Sections 9.3.2 and 9.3.3.1 (Table 9.9). Assume that we wish to test the significance of the joint contribution of the terms involving X_1^2 and $X_1^2 X_2$ of the equation:

$$Y = \alpha + \beta_1 X_1 + \beta_2 X_1^2 + \beta_3 X_2 + \beta_4 X_1 X_2 + \beta_5 X_1^2 X_2$$

The step-by-step procedures are:

□ STEP 1.   Follow steps 1 and 2 of Section 9.3.3.1. For our example, the single multiple regression equation to be fitted has $k = 5$ regression terms (excluding intercept).

□ STEP 2.   Specify the q (where $q < k$) regression terms whose joint contribution to the regression equation prescribed in step 1 is to be tested. The set of q regression terms is usually chosen so that the terms within the set are closely related to each other.

For our example, the set of $q = 2$ regression terms, one involving X_1^2 and another involving $X_1^2 X_2$, is chosen. Note that both terms are closely related because both have the quadratic component of X_1.

□ STEP 3.   Apply the appropriate regression analysis to the prescribed regression equation based on k regression terms. Designate the sum of squares due to this regression as $SSR(k)$ and the residual sum of squares as $SSE(k)$.

For our example, the multiple regression analysis is already applied to the equation in Section 9.3.3.1, and $SSR(k)$ and $SSE(k)$ are computed as:

$$SSR(k) = SSR(5) = 1.197854$$

$$SSE(k) = SSE(5) = 0.034276$$

□ STEP 4.   Apply the appropriate regression analysis to the regression equation based on $(k - q)$ regression terms. Designate the sum of squares due to this regression as $SSR(k - q)$. For our example, the regression equation involving $(k - q) = 5 - 2 = 3$ regression terms is:

$$Y = \alpha + \beta_1 X_1 + \beta_2 X_2 + \beta_3 X_1 X_2$$

Following the procedure of the multiple regression analysis applied to the regression of this form discussed in Section 9.1.2, the value of the sum of squares due to this regression (with $k - q = 3$ regression terms) is:

$$SSR(k - q) = SSR(3) = 1.180536$$

□ STEP 5.   Compute the difference between the two sums of squares due to regression, computed in steps 3 and 4, as:

$$SSR_q = SSR(k) - SSR(k - q)$$

where SSR_q represents the joint contribution of the q regression terms to the variation in the dependent variable.

For our example, the value of SSR_q is computed as:

$$SSR_2 = SSR(5) - SSR(3)$$

$$= 1.197854 - 1.180536 = 0.017318$$

□ STEP 6.   Compute the F value as:

$$F = \frac{SSR_q/q}{SSE(k)/(n - k - 1)}$$

$$= \frac{(0.017318)/2}{(0.034276)/10} = 2.53$$

☐ STEP 7.   Compare the computed F value with the tabular F value (Appendix E) at a specified level of significance with $f_1 = q$ and $f_2 = (n - k - 1)$ degrees of freedom. If the computed F value is larger than the corresponding tabular F value, the joint contribution of the group of q regression terms to the variation in the dependent variable is significant. In such a case, the use of the regression equation having $(k - q)$ regression terms is not appropriate and the regression equation having k regression terms should be used.

For our example, the tabular F value with $f_1 = 2$ and $f_2 = 10$ degrees of freedom is 4.10 at the 5% level of significance and 7.56 at the 1% level. Because the computed F value is less than the tabular F value at the 5% level of significance, the joint contribution of the two regression terms, associated with X_1^2 and $X_1^2 X_2$, is not significant. Thus, the multiple regression equation with $5 - 2 = 3$ regression terms, as specified in step 4, is appropriate. The estimate of this regression equation is:

$$\hat{Y} = 0.36514 - 0.00543X_1 + 0.00254X_2 - 0.00007X_1 X_2$$

with the coefficient of determination R^2 of .96, which is significant at the 1% level.

9.3.4   Stepwise Regression Technique

The stepwise regression technique is similar to that of the test of significance technique (Section 9.3.3) in that it aims to include in the regression equation only those terms that contribute significantly to the variation in the dependent variable. However, this objective is achieved in the stepwise regression technique by systematically adding terms, one at a time, to the regression equation, instead of removing terms, singly or jointly, from an initially large equation. We illustrate the procedure with the same set of data used in Sections 9.3.2 and 9.3.3 (Table 9.9). The procedures are:

☐ STEP 1.   Specify a set of f regression terms, which are associated with the independent variables of interest and have possibilities of being included in the regression equation. A large f provides a wider range for screening than a small f and thus increases the chance of including as many regression terms that should be included in the regression equation as possible. The computational procedure is, however, more complicated with a large f but the computational work can be greatly minimized by use of a computer. In general, f is usually larger than k of Section 9.3.3.1 and Section 9.3.3.2.

For our example, we consider a second-degree polynomial relationship for each of the two factors, and all four possible interaction terms between the two factors based on the second-degree polynomial. That is, we consider $f = 8$ regression terms associated with water stress (X_1) and nitrogen rate (X_2):

$$
\begin{aligned}
Z_1 &= X_1 & Z_5 &= X_1 X_2 \\
Z_2 &= X_2 & Z_6 &= X_1 X_2^2 \\
Z_3 &= X_1^2 & Z_7 &= X_1^2 X_2 \\
Z_4 &= X_2^2 & Z_8 &= X_1^2 X_2^2
\end{aligned}
$$

☐ STEP 2.   Treat each of the f regression terms specified in step 1 (i.e., $Z_1, Z_2, \ldots, Z_f$) as a newly created variable and compute a simple correlation coefficient r between the dependent variable Y and each of the Z variables, following the procedures of Section 9.1.1.2. The first Z variable to enter the regression equation is that with the highest r value, which is significant at the α level of significance. The commonly used α value is between 10 and 20%. The larger the α value, the greater the number of variables that have the chance of entering into the regression equation.

For our example, the values of the eight Z variables are first computed from the values of the two original factors X_1 and X_2, as shown in Table 9.11. Then, following procedure of Section 9.1.1.2 and using data of Table 9.11, compute the simple correlation coefficient between the Y variable and each of the Z variables and indicate the significance of each coefficient. The results, with significance based on $\alpha = 20\%$ level of significance, are:

First Z Variable	r^*
Z_1	$-.761^+$
Z_2	$.506^+$
Z_3	$-.759^+$
Z_4	$.476^+$
Z_5	$-.378^+$
Z_6	$-.294$
Z_7	$-.495^+$
Z_8	$-.415^+$

The largest r value is $r = -.761$ (absolute value is considered) exhibited by Z_1, which is significant at the 20% level of significance. Thus, the first Z variable to enter the regression equation is Z_1.

☐ STEP 3.   Fit the following $(f - 1)$ multiple linear regression equations, following the procedure of Section 9.1.2 and using data of Table 9.11:

$$
Y_i = \alpha_i + \beta_{ki} Z_k + \beta_{ii} Z_i \qquad i = 1, \ldots, f \text{ and } i \neq k
$$

$^{*+}$ = significant at the 20% level.

Table 9.11 Computation of Parameters Needed in the Stepwise Regression Technique to Determine an Appropriate Regression Equation between Nitrogen Uptake (Y), Water Stress (X_1), and Nitrogen Rate (X_2); from Data in Table 9.9

Treatment Number	Nitrogen Uptake (Y)	Water Stress ($X_1 = Z_1$)	Nitrogen Rate ($X_2 = Z_2$)	$Z_3 = X_1^2$	$Z_4 = X_2^2$	$Z_5 = X_1X_2$	$Z_6 = X_1X_2^2$	$Z_7 = X_1^2X_2$	$Z_8 = X_1^2X_2^2$
1	0.318	0	0	0	0	0	0	0	0
2	0.517	0	90	0	8,100	0	0	0	0
3	0.786	0	180	0	32,400	0	0	0	0
4	1.018	0	270	0	72,900	0	0	0	0
5	0.336	10	0	100	0	0	0	0	0
6	0.583	10	90	100	8,100	900	81,000	9,000	810,000
7	0.677	10	180	100	32,400	1,800	324,000	18,000	3,240,000
8	0.914	10	270	100	72,900	2,700	729,000	27,000	7,290,000
9	0.264	20	0	400	0	0	0	0	0
10	0.476	20	90	400	8,100	1,800	162,000	36,000	3,240,000
11	0.442	20	180	400	32,400	3,600	648,000	72,000	12,960,000
12	0.525	20	270	400	72,900	5,400	1,458,000	108,000	29,160,000
13	0.095	40	0	1,600	0	0	0	0	0
14	0.140	40	90	1,600	8,100	3,600	324,000	144,000	12,960,000
15	0.116	40	180	1,600	32,400	7,200	1,296,000	288,000	51,840,000
16	0.128	40	270	1,600	72,900	10,800	2,916,000	432,000	116,640,000

$\sum y^2 = 1.2321$

$\sum z_1^2 = 3,500$

$\sum z_2^2 = 162,000$

$\sum z_3^2 = 6,510,000$

$\sum z_4^2 = 12,859,560,000$

$\sum z_5^2 = 148,837,500$

$\sum z_6^2 = 9,564,297,750,000$

$\sum z_7^2 = 229,209,750,000$

$\sum z_8^2 = 14,008,883,175,000,000$

$\sum z_1 y = -49.962$

$\sum z_2 y = 225.945$

$\sum z_3 y = -2,150.675$

$\sum z_4 y = 59,879.250$

$\sum z_5 y = -5,113.238$

$\sum z_6 y = -1,008,277.875$

$\sum z_7 y = -263,233.125$

$\sum z_8 y = -54,506,216.250$

where Z_k is the first Z variable identified in step 2. The Z_i variable, which corresponds to the ith regression equation, with the largest value of the multiple correlation coefficient R and whose coefficient is found to be significant at the α level of significance, is chosen as the second Z variable to enter the regression equation. If none of the Z_i variables gives a significant regression coefficient, the process is terminated and the variable chosen in step 1 is the only one included in the regression equation.

For our example, the multiple correlation coefficient corresponding to each of the $(f - 1) = 7$ regression equations and the t value for testing significance of the regression coefficient corresponding to each of the seven remaining Z variables (Z_2 to Z_8) are shown below. Each regression equation has two independent variables—the first is Z_1 (identified in step 2) and the second is each of the remaining seven Z variables.

Second Z Variable	R	t
Z_2	.914	4.48[+]
Z_3	.768	−0.58
Z_4	.897	3.89[+]
Z_5	.777	0.91
Z_6	.774	0.81
Z_7	.762	0.23
Z_8	.762	0.22

The largest R value is exhibited by Z_2, whose regression coefficient is significant at the 20% level of significance. Thus, Z_2 is the second variable to be entered in the regression equation.

□ STEP 4. Fit the following $(f - 2)$ multiple linear regression equation, following the same procedure used in step 3:

$$Y_i' = \alpha_i' + \beta_{ki}' Z_k + \beta_{mi}' Z_m + \beta_{ii}' Z_i \quad i = 1,\ldots,f \text{ and } i \neq m \text{ or } k$$

where Z_k is the first Z variable identified in step 2, Z_m is the second Z variable identified in step 3, and Z_i refers to each of the $(f - 2)$ remaining Z variables (i.e., except for Z_k and Z_m). And use the procedure in step 3 to identify the third Z variable or to terminate the process.

For our example, the multiple correlation coefficient corresponding to each of the $(f - 2) = 6$ regression equations and the t value for testing significance of the regression coefficient corresponding to each of the six remaining Z variables (Z_3 to Z_8) are shown below. Each regression equation has three independent variables, two of which are Z_1 and Z_2 and the third is each of the remaining six Z variables.

Third Z Variable	R	t
Z_3	.919	-0.90
Z_4	.914	-0.27
Z_5	.979	-5.95^+
Z_6	.973	-5.06^+
Z_7	.974	-5.12^+
Z_8	.970	-4.64^+

The largest R value is exhibited by Z_5, whose regression coefficient is significant at the 20% level of significance. Thus, Z_5 is the third variable to enter the regression equation.

☐ STEP 5. Fit the following $(f-3)$ multiple linear regression equations, following the same procedure used in step 4:

$$Y_i'' = \alpha_i'' + \beta_{ki}'' Z_k + \beta_{mi}'' Z_m + \beta_{ni}'' Z_n + \beta_{ii}'' Z_i$$

$$i = 1,\dots,f \quad \text{and} \quad i \neq m, k \text{ or } n$$

where Z_k, Z_m, and Z_n are the three previously identified Z variables and Z_i refers to each of the $(f-3)$ remaining Z variables (i.e., except for Z_k, Z_m, and Z_n). And use the procedure in step 3 to identify the fourth Z variable or to terminate the process.

For our example, the multiple correlation coefficient corresponding to each of the $(f-3) = 5$ regression equations, each having four independent variables, three of which are Z_1, Z_2, and Z_5 and the fourth is each of the remaining five Z variables (Z_3, Z_4, Z_6, Z_7, and Z_8).

Fourth Z Variable	R	t
Z_3	.984	-1.92^+
Z_4	.979	-0.51
Z_6	.980	-0.65
Z_7	.980	-0.78
Z_8	.980	-0.73

The largest R value is exhibited by Z_3, whose regression coefficient is significant at the 20% level of significance. Thus, Z_3 is the fourth variable to be entered in the regression equation.

□ STEP 6. Fit the following $(f - 4)$ multiple linear regression equations, following the procedure used in step 4:

$$Y_i''' = \alpha_i''' + \beta_{ki}''' Z_k + \beta_{mi}''' Z_m + \beta_{ni}''' Z_n + \beta_{pi}''' Z_p + \beta_{ii}''' Z_i$$

$$i = 1,\ldots,f \text{ and } i \neq m, k, n \text{ or } p$$

where Z_k, Z_m, Z_n, and Z_p are the four previously identified Z variables and Z_i refers to each of the $(f - 4)$ remaining Z variables (i.e., except for Z_k, Z_m, Z_n, and Z_p). And use the procedure in step 3 to identify the fifth Z variable or to terminate the process. For our example, the multiple correlation coefficient corresponding to each of the $(f - 4) = 4$ regression equations, each having five independent variables, four of which are Z_1, Z_2, Z_5, and Z_3, are:

Fifth Z Variable	R	t
Z_4	.985	−0.57
Z_6	.985	−0.72
Z_7	.986	1.12
Z_8	.984	0.20

Because none of the regression coefficients associated with the fifth variable is significant at the 20% level, the process is terminated. Thus, the regression equation that should be used is one that consists of the four Z variables identified in step 5. The estimate of this regression equation is:

$$\hat{Y} = 0.33300 + 0.00123Z_1 + 0.00254Z_2 - 0.00016Z_3$$
$$- 0.00007Z_5$$

That is,

$$\hat{Y} = 0.33300 + 0.00123X_1 + 0.00254X_2 - 0.00016X_1^2$$
$$- 0.00007X_1 X_2$$

9.4 COMMON MISUSES OF CORRELATION AND REGRESSION ANALYSIS IN AGRICULTURAL RESEARCH

The regression and correlation analysis is a powerful tool for analyzing the relationship and association among physical and biological variables. Thus, it is one of the most widely used statistical procedures in agricultural research and its use is consistently increasing. However, the analysis is often misused which frequently causes misleading and incorrect interpretation of results.

The common cause of misuse of regression and correlation analysis in agricultural research is that, unlike the analysis of variance which requires some rigid assumptions about the data to which it is applied, the regression and correlation analysis is applicable to many types of data from almost any source. To illustrate this point, consider the association between grain yield and protein content in rice. The association between these two traits is usually negative when computed from data taken from different varieties. When computed from data covering a wide range of nitrogen fertilization, however, the relationship becomes positive. Although these two results seem conflicting, it is, in reality, consistent with biological expectations. This is so because increased nitrogen fertilization for a single rice variety is expected to increase both grain yield and protein content—and thus have a positive correlation; whereas for different varieties grown at the same level of nitrogen, those having higher grain yield are expected to have lower protein content, and vice versa—and thus have a negative correlation.

The common misuses of the regression and correlation technique in agricultural research can be classified into four categories:

- Improper match between data and objective
- Broad generalization of results
- Use of data from individual replications
- Misinterpretation of the simple linear regression and correlation analysis

9.4.1 Improper Match Between Data and Objective

As indicated previously, the interpretation of the results of a regression and correlation analysis depends greatly on the type of data used. Of particular importance in this regard are the primary sources of variability among the data points. These sources of variability must be selected to properly match the objective of the analysis.

The association between grain yield and protein content illustrates clearly the importance of proper matching of data and objective. In one case, the primary source of variability is variety; in the other case it is the rate of nitrogen fertilizer. Clearly, the negative correlation obtained from data with variety as the primary variable is appropriate for answering the objective of the rice breeders whose aim is to select for high-yield and high-protein varieties, while the positive correlation obtained from data with nitrogen rate as the primary variable is appropriate for answering the objective of the agronomists whose aim is to determine the optimum rate of nitrogen fertilizer. Obviously, a substantial error would have resulted if there were a mismatch between the type of data used and the objective of the analysis.

Although the choice of the appropriate data to match a given objective is fairly obvious in the foregoing example, such a choice is often not simple in practice. Take the case of a researcher who wishes to evaluate the effect of stem borer infestation on grain yield, in a rice variety such as IR8. Such possible sources of data for his study are:

A. Data on grain yield and corresponding stem borer incidence in IR8 obtained from several secondary sources, such as his own experiments, or other researchers' experiments, that were conducted previously for other purposes.

B. Data on grain yield and stem borer incidence measured from sample crop-cuts of IR8 made at different sites in one or more farmers' fields.

C. Data on grain yield and stem borer incidence measured from several subplots that are generated by subdividing an area that is large enough to allow variation in stem borer incidences. The whole area grows IR8 with uniform management and without insecticide application so that variability in insect infestation is induced.

D. Data on grain yield and stem borer incidence collected from each experimental plot of a replicated field trial in which different levels of stem borer control (e.g., different types and rates of insecticide application) constitute the treatments tested.

E. Data on grain yield and stem borer incidence from each experimental unit of an experiment in which the treatments are the varying numbers of stem borers per plant or per unit area. Such an experiment may be a greenhouse trial where one or more experimental pots represent an experimental unit, or a field trial where a unit area (covered with a net to confine the insects) represents an experimental unit.

Although regression and correlation analysis can be applied to any one of these sources of data, the specific manner of application, the degree of accuracy of the estimated regression equation, and the interpretation of the result vary greatly from one data source to another. For example, the degree of accuracy is expected to be lowest with source A and becomes gradually higher with source B, C, D, and E, in that order. On the other hand, the complexity of the analysis and its interpretation is expected to be highest with source A and becomes gradually less with source B, C, D, and E, in that order.

With data from source A, many noncontrolled factors besides stem borer incidence can be expected to contribute to the variation in yield. Different observations could have come from different sites or different crop seasons and, therefore, could have been exposed to different environments (soil and climate), could have been fertilized with different types and rates of fertilizer, and so on. These factors could cause variation in yield that could be larger than that caused by stem borer, which is the factor of major interest. Consequently, the estimate of the relationship between yield and stem borer incidence is expected to be poor. In fact, for such a data source, it may be necessary to use a multiple regression technique that can handle several more variables besides stem borer incidence. For any variable to be eligible for inclusion in the multiple regression analysis, however, it should have been measured uniformly on all experiments comprising the data source. This requirement is difficult to satisfy in practice because data from several nonrelated secondary sources seldom record a uniform set of variables.

With data from source B, some improvements over source A can be expected. For example, seasonal effects are expected to be minimized because the data are collected in the same crop season. Morever, any major difference in management and in soil conditions between farms can be noted during the process of crop cutting. These factors can then be used as additional independent variables in the multiple regression. For this technique to succeed, however, all variable factors that significantly influence yield must be included. Because the number of noncontrolled factors that can potentially affect yield for this source of data is large, the choice of the appropriate variables to measure and to include in the analysis is usually not easy.

With data from source C, variation in the cultural and management practices and in soil and climate is expected to be minimal. However, the major problem of this source of data is that the variability in stem borer incidence, and in turn that for yield, may not be sufficiently large for a meaningful assessment of their relationship.

With data from source D, the above mentioned deficiency of source C is remedied because the treatments are selected to induce different levels of stem borer control and a large variability in the stem borer incidences between observations can be expected.

With data from source E, the problem of controlling incidence of other insects and diseases, which is present in all other data sources (A through D) is expected to be remedied. Although this data source can control many undesirable sources of variation, it has two major problems, namely high cost and high level of artificiality. Yield data may not be realistic in a greenhouse trial; and with the use of screen, or a net in a field trial, the resulting environment may be highly artificial.

We make two major points from the foregoing discussion:

1. The choice of data source is critical in the application of the regression and correlation technique. This is even more critical in the interpretation of results. The important considerations for selecting the data source are the expected accuracy of results and the estimated cost, both in money and time. Often, the correct choice is a compromise between those two considerations. Usually the alternatives with low accuracy are used in preliminary studies because of their low cost and fast turnaround. But when a high degree of accuracy is desired, data from properly controlled experiments may be more appropriate.

2. A critical examination of the data source and identification of possible factors that may affect the result and its interpretation is essential for proper use of the regression and correlation technique.

9.4.2 Broad Generalization of Regression and Correlation Analysis Results

One of the most common misuses associated with a regression and correlation analysis is the tendency to generalize the coverage of the functional relationship, or the association between variables, beyond the scope of the data to which the analysis was applied. This is usually done by extrapolating the result of a regression and correlation analysis outside of the range of X values permitted. For example, the result of a linear relationship between rice yield (Y) and nitrogen rate (X) obtained from data whose X values are in the range of 0 to 90 kg N/ha is valid only within the specified range. It is not valid to extrapolate the results beyond the rate of 90 kg N/ha, because the relationship could easily change beyond that point. Another example is the relationship between yield of IR8 and stem borer incidence as discussed in Section 9.4.1. In this case, it is not valid to extend the result to cover all rice varieties, or even all semidwarf rice varieties such as IR8, because the relationships may vary between varieties with different levels of resistance to this insect or between varieties differing in their yield potential.

Another area that is prone to improper generalization of results is the use of regression and correlation analysis to identify substitute methods for measuring plant characters. Two examples of this type of application are:

- The use of ear weight as a substitute for grain weight in variety trials in maize, because the former is faster and easier to measure and because the two characters are highly correlated.
- The use of leaf length and leaf width in place of leaf area, because the measurement of leaf length and leaf width is not only simple and inexpensive but does not require destruction of plants.

The large size of the simple linear correlation coefficient between the difficult-to-measure character and its potential substitute is generally used as the justification for the use of the substitute method. Obviously, a perfect correlation means no information is lost in the substitution, but a perfect correlation cannot always be expected and the loss of information for a less-than-perfect correlation is proportional to the quantity $(1 - r^2)$.

Even though the rationale for such substitution is valid, mistakes are sometimes committed in interpreting and applying the results. Two common mistakes are:

- The result of a simple linear correlation coefficient obtained from one set of data is overextended to cover all types of data. For example, although the r value between two characters, obtained from data covering a large number of varieties exhibiting a wide range in values of the character under examination, can be adequately extended to cover various variety tests; it may not be applicable for other types of experiment, such as fertilizer trials and insecticide trials. Thus, substituting one character for another could be satisfactory for certain types of trial but not for others. It is, therefore, not correct to assume that a substitute method found satisfactory based on one set of data will be satisfactory for all types of data.

- The simple linear correlation coefficient used to identify a substitute method is usually estimated from data covering an extremely wide range of values of the character under study; and, thus, its high value may no longer hold in situations with a much narrower range. For example, the simple linear correlation coefficient *r* between grain yield and ear weight in maize computed from tests involving only promising maize varieties can be expected to be smaller than the corresponding *r* value computed from a variety test involving varieties randomly chosen from the world germ plasm collection. This is because variability in the latter set of materials is expected to be much higher than that in the first set. Thus, it is extremely important to know the data source used in estimating the correlation coefficient used to select a substitute method of measurement, and to not apply the result beyond the scope of that data.

While it is clear that the extrapolation of a regression or correlation analysis to cover areas beyond the scope of the original data is risky and should be attempted only with the justification of a well-known biological phenomenon, the temptation to do so can be strong in certain cases. To minimize this error, the results of any regression and correlation analysis should always be presented with the following information:

- The range value for each and all independent variables, so that the underlying population in which the results are applicable is properly defined.
- The correlation coefficient (or the coefficient of determination) and its significance, to indicate the amount of variation in the dependent variable that can be explained by the independent variable or variables.
- The sample size (*n*), for proper judgment of the size of the correlation coefficient.
- The source of data, for proper interpretation of the results.
- Graphical presentation of the data points together with the regression and correlation results, to help in the understanding of the underlying basis for the specific relationship derived (the form and the relative sizes of the regression parameters) and the degree of association observed.

9.4.3 Use of Data from Individual Replications

When the regression and correlation analysis is applied to data from replicated trials—for example, to evaluate the relationship between crop response and treatment—there are two choices of data that can be used. One is the data by replication and another is the treatment means averaged over all replications.

As a general rule, although use of the data by replication seems more appealing because the number of data points is larger, use of the treatment means averaged over replications is more appropriate. This is so because variation between experimental units receiving the same treatment (experimental error) need not enter into the evaluation of the association between crop response and treatment. For example, if the researcher wishes to examine yield response to different rates of nitrogen, based on data from a replicated fertilizer trial, the treatment means rather than the replication data should be used. This avoids confounding the residual sum of squares with the experimental error.

The use of replication data is justified only in some rare cases. One of the most common cases is when response to treatments is not the primary objective of the analysis. For example, in evaluating the relationship between grain yield and leaf-area index in rice, several plant spacings may be tested—not to study spacing effects on grain yield and on leaf-area index, but merely to induce variation in both characters so that their association can be examined. In such a case, the replication data may be used instead of the treatment means. However, because the use of replication data involves more than one source of variation, interpretation of results becomes more complex and consultation with a competent statistician is advisable.

9.4.4 Misinterpretation of the Simple Linear Regression and Correlation Analysis

Because of its simplicity, the simple linear regression and correlation analysis is one of the most frequently used techniques in agricultural research. Unfortunately, it is also the most frequently misused. Three common types of misinterpretation of the simple linear regression and correlation analysis are:

- *Spurious Data.* A set of data points is termed spurious when its distribution over the range of the independent variable is uneven. An example is shown by Figure 9.12, where there are two clusters of data points, one at each end of the range of X values. For such type of spurious data, the simple linear regression and correlation coefficients would almost always be significant. However, such a significant result is not a valid indication of a linear relationship, because nothing is known about data points between the two extremes. To avoid such misinterpretation, a scatter diagram of the observed data should be constructed and checked for uneven distribution of data points whenever significant results are obtained, or even before the simple regression and correlation analysis is applied.
- *Cause and Effect.* One of the most common misinterpretations of the simple linear correlation coefficient r is to say that a significant r indicates the presence of a causal relationship between the two variables. Even though the correlation analysis can quantify the degree of association between two characters, it cannot and should not provide *reasons* for such an association.

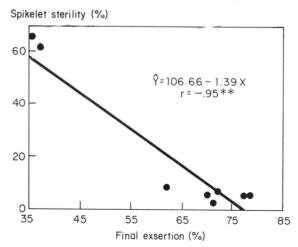

Figure 9.12   A case of spurious data with a highly significant simple linear correlation coefficient *r* between spikelet sterility (*Y*) and panicle exsertion (*X*) of rice variety IR36, but without a clear indication of the type of relationship.

For example, a high positive association between tiller number and plant height observed from a rice fertilizer trial does not indicate that higher tiller number is the effect of higher plant height, or vice versa. In fact, it is the increase in the fertilizer rate that causes both characters to increase simultaneously.

• *Nonsignificant r Value and the Absence of a Functional Relationship.* The simple linear correlation coefficient is designed to detect the presence of a linear association between two variables. It cannot detect any other type of variable association. Thus, a nonsignificant *r* value cannot be taken to imply the absence of any functional relationship between the two variables. As shown in Figure 9.5, two variables may have a strong nonlinear relationship even if their *r* value is low and nonsignificant.

CHAPTER 10
Covariance Analysis

On the premise that the various biophysical features of an experimental plot do not behave independently but are often functionally related to each other, the analysis of covariance simultaneously examines the variances and covariances among selected variables such that the treatment effect on the character of primary interest is more accurately characterized than by use of analysis of variance only.

Analysis of covariance requires measurement of the character of primary interest plus the measurement of one or more variables known as *covariates*. It also requires that the functional relationship of the covariates with the character of primary interest is known beforehand.

Consider the case of a rice variety trial in which weed incidence is used as a covariate. With a known functional relationship between weed incidence and grain yield, the character of primary interest, the covariance analysis can adjust grain yield in each plot to a common level of weed incidence. With this adjustment, the variation in yield due to weed incidence is quantified and effectively separated from that due to varietal difference.

Covariance analysis can be applied to any number of covariates and to any type of functional relationship between variables. In this chapter, we deal primarily with the case of a single covariate whose relationship to the character of primary interest is linear. Although this limited focus greatly simplifies our discussion, we do not expect it to unduly reduce the applicability of the procedures because the condition of a single covariate with a linear relationship to the primary variable is adequate for most agricultural research.

10.1 USES OF COVARIANCE ANALYSIS IN AGRICULTURAL RESEARCH

Three important uses of covariance analysis in agricultural research are:

1. To control experimental error and to adjust treatment means
2. To estimate missing data
3. To aid in the interpretation of experimental results

10.1.1   Error Control and Adjustment of Treatment Means

We have emphasized repeatedly that the size of experimental error is closely related to the variability between experimental units. We have also shown that proper blocking (Chapter 2, Section 2.2.1) can reduce experimental error by maximizing the differences between blocks and thus minimizing differences within blocks. Blocking, however, cannot cope with certain types of variability such as spotty soil heterogeneity and unpredictable insect incidence. In both instances, heterogeneity between experimental plots does not follow a definite pattern, which causes difficulty in getting maximum differences between blocks. Indeed, blocking is ineffective in the case of nonuniform insect incidences because blocking must be done before the incidence occurs. Furthermore, even though it is true that a researcher may have some information on the probable path or direction of insect movement, unless the direction of insect movement coincides with the soil fertility gradient, the choice of whether soil heterogeneity or insect incidence should be the criterion for blocking is difficult. The choice is especially difficult if both sources of variation have about the same importance.

Use of covariance analysis should be considered in experiments in which blocking cannot adequately reduce the experimental error. By measuring an additional variable (i.e., covariate X) that is known to be linearly related to the primary variable Y, the source of variation associated with the covariate can be deducted from experimental error. With that done, the primary variable Y can be adjusted linearly upward or downward, depending on the relative size of its respective covariate. The adjustment accomplishes two important improvements:

1.   The treatment mean is adjusted to a value that it would have had, had there been no differences in the values of the covariate.
2.   The experimental error is reduced and the precision for comparing treatment means is increased.

Although blocking and covariance technique are both used to reduce experimental error, the differences between the two techniques are such that they are usually not interchangeable. The analysis of covariance, for example, can be used only when the covariate representing the heterogeneity between experimental units can be measured quantitatively. However, that is not a necessary condition for blocking. In addition, because blocking is done before the start of the experiment, it can be used only to cope with sources of variation that are known or predictable. Analysis of covariance, on the other hand, can take care of unexpected sources of variation that occur during the experiment. Thus, covariance analysis is useful as a supplementary procedure to take care of sources of variation that cannot be accounted for by blocking.

When covariance analysis is used for error control and adjustment of treatment means, the covariate must not be affected by the treatments being tested. Otherwise, the adjustment removes both the variation due to experimental error and that due to treatment effects. A good example of covariates that are free of treatment effects are those that are measured before the treatments are applied, such as soil analysis and residual effects of treatments applied in past experiments. In other cases, care must be exercised to ensure that the covariates defined are not affected by the treatments being tested (see Section 10.1.1.4).

We describe some specific application of the covariance technique in controlling experimental error and in adjusting treatment means.

10.1.1.1 Soil Heterogeneity.   The covariance technique is effective in controlling experimental error caused by soil heterogeneity when:

- The pattern of soil heterogeneity is spotty or unknown.
- Variability between plots in the same block remains large despite blocking.

Use of covariance technique in such cases involves the measurement, from individual experimental plots, of a covariate that can distinguish differences in the native soil fertility between plots and, at the same time, is linearly related to the character of primary interest. Two types of covariate that are commonly used for controlling experimental error due to soil heterogeneity are uniformity trial data and crop performance data prior to treatment.

10.1.1.1.1 Uniformity Trial Data.   Uniformity trial data are crop performance records of small plots obtained from an experimental field that is managed uniformly (see also Chapter 12, Section 12.2.1).

Based on the premise that a uniform soil cropped uniformly gives uniform crop performance, soil heterogeneity is measured as the differences in crop performance from one plot to another. Thus, in using uniformity trial data as the covariate, the two variables involved are:

- The primary variable Y, recorded from experimental plots after the treatments are applied
- The covariate X, recorded from a uniformity trial in the same area but before the experiment is conducted

Uniformity trial data is an ideal covariate to correct for variability due to soil heterogeneity. It clearly satisfies the requirement that it is not affected by the treatments since measurements are made before the start of the experiment and, thus, before the treatments are applied.

Despite the substantial gains in precision that can be expected from using covariance analysis with uniformity trial data as the covariate, it has not been widely used in agricultural research. There are two important reasons for this:

- A uniformity trial is expensive to conduct. In research stations where experimental fields are limited, the conduct of uniformity trials may take away land that otherwise would have been used for one or more field experiments. In addition, because of the need for small basic units, the uniformity trial involves a more complex data collection scheme than the standard field experiments.

- The data from a uniformity trial is, strictly speaking, valid only in the specific field in which the trial was conducted.

10.1.1.1.2 Data Collected prior to Treatment Implementation. In experiments where there is a time lag between crop establishment and treatment application, some crop characteristics can be measured before treatments are applied. Such data represent the inherent variation between experimental plots. In such cases, one or more plant characters that are closely related to crop growth, such as plant height and tiller number, may be measured for each experimental plot just before the treatments are applied. Because all plots are managed uniformly before treatment, any difference in crop performance between plots at this stage can be attributed primarily to soil heterogeneity.

Crop performance data is clearly easier and cheaper to gather than uniformity trial data because crop performance data is obtained from the crop used for the experiment. However, such data are only available when the treatments are applied late in the life cycle of the experimental plants.

10.1.1.2 Residual Effects of Previous Trials. Some experiments, because of the treatments used, may create residual effects that increase soil heterogeneity in the succeeding crops. In cases where soil heterogeneity caused by residual effects of previous trials is expected to be large, the field should be left in fallow or green manure for a period of time between experiments. This practice, however, is costly and is not commonly used where experimental field areas are limited. When a fallow period or a green manure crop cannot be used, the expected residual effects can be corrected by blocking, or by a covariance technique, or by both. With the covariance technique, the covariate could be plot yield in the previous trial or an index representing the expected residual effects of the previous treatments.

10.1.1.3 Stand Irregularities. Variation in the number of plants per plot often becomes an important source of variation in field experiments. This is especially so in the annual crops where population density is high and accidental loss of one or more plants is quite common. Several methods for stand correction are available (see Chapter 13, Section 13.3.3) but the covariance technique with stand count as the covariate is one of the best alternatives, provided that stand irregularities are not caused by treatments. Such a condition exists when:

- The treatments are applied after the crop is already well established.
- The plants are lost due to mechanical errors in planting or damage during cultivation.

- The plants are damaged in a random fashion from such causes as rat infestation, accidental cattle grazing, or thefts.

Frequently, the cause of stand irregularity is related to the effect of the treatments. In such cases, covariance analysis may still be applied but its purpose is totally different. Such usage of the covariance technique is to aid in the interpretation of results (see Section 10.1.3) rather than to control experimental error and to adjust treatment means.

10.1.1.4 Nonuniformity in Pest Incidence.   The distribution of pest damage is usually spotty and the pattern of occurrence is difficult to predict. Consequently, blocking is usually ineffective and covariance analysis, with pest damage as the covariate, is an ideal alternative. However, to ensure that the difference in pest damage between plots is not due to treatment effects, the following steps may be followed:

☐ STEP 1.   Classify the experiment as either Type A or Type B. Type A experiments are those in which the specific pest incidence (e.g., insect or disease incidence) is a major criterion for answering the experimental objective, such as those where treatments involve control of the insects and diseases or variety trials in which insect and disease resistance is a major criterion. Type B experiments are those that do not belong to Type A.

☐ STEP 2.   For type A experiments, apply the covariance technique only if variation in pest damage occurs before treatment implementation. For example, in a trial testing different insecticides, the covariate could be the insect count made before the insecticides are applied. Or, in a weed control trial, the covariate could be the weed population before treatment. In a variety trial to evaluate insect resistance, however, it is not possible to separate insect infestation from the varietal characteristics and the covariance technique is not applicable for error control.

☐ STEP 3.   For type B experiments, the covariance technique can be applied to adjust for any variability in pest damage either before or after treatment implementation. For example, rat damage is a valid covariate in an insecticide trial or in a variety trial for insect resistance.

10.1.1.5 Nonuniformity in Environmental Stress.   To select varieties that are tolerant of environmental stress, competing genotypes are usually tested at a specified level of stress in a particular environment. Some examples are insect and disease incidence, drought or water logging, salinity, iron toxicity, and low fertility. It is usually difficult to maintain a uniform level of stress for all test varieties. For example, in field screening for varietal resistance to an insect infestation, it is often impossible to ensure a uniform insect exposure for all test varieties. Or in a field test for varietal tolerance for alkalinity, nonuniformity in the level of alkalinity over an experimental field is normal.

To monitor the variation in the stress condition over an experimental area, a commonly used field plot technique is to plant, at regular intervals, a check variety whose reaction to the stress of interest is well established. A susceptible variety is commonly used for this purpose. The reaction of the susceptible check nearest to, or surrounding, each test variety can be used as the covariate to adjust for variability in the stress levels between the test plots.

10.1.1.6 Competition Effects in Greenhouse Trials. Despite the potential for better control of environment in greenhouse trials, variation between experimental units in such trials is generally as large as, and frequently larger than, the variation in field trials. This is mainly because space limitation in the greenhouse usually requires experimental unit sizes that are much smaller than that used in field experiment. With small experimental units, competition effects between adjacent units are large and the standard remedial measures, such as the removal of border plants, are usually not feasible. The covariance technique can control experimental error due to competition effects in greenhouse trials. For example, in a trial where an experimental unit is a pot of plants, the average height of plants in the surrounding pots can be used as the covariate to adjust for varietal competition effects.

For greenhouse trials where there is a time lag between planting and the implementation of treatments, nondestructive measurement of certain plant characters (such as tiller number and plant height) before treatment implementation can also be used as the covariate to correct for variation between experimental plants.

10.1.2   Estimation of Missing Data

The standard analysis of variance procedure is not directly applicable to data with one or more missing observations (see Chapter 7, Section 7.1). The most commonly used procedure for handling missing data is the missing data formula technique, in which the standard analysis of variance is performed after the missing data is estimated.

Covariance analysis offers an alternative to the missing data formula technique. It is applicable to any number of missing data. One covariate is assigned to each missing observation. The technique prescribes an appropriate set of values for each covariate.

10.1.3   Interpretation of Experimental Results

The covariance technique can assist in the interpretation and characterization of the treatment effects on the primary character of interest Y, in much the same way that the regression and correlation analysis is used. By examining the primary character of interest Y together with other characters whose functional relationships to Y are known, the biological processes governing the treatment effects on Y can be characterized more clearly. For example, in a water management trial, with various depths of water applied at different growth stages of the rice plants, the treatments could influence both the grain yield and the weed population. In such an experiment, covariance analysis, with weed population as the covariate, can be used to distinguish between the yield difference caused directly by water management and that caused indirectly by changes in weed population, which is also caused by water management. The manner in which the covariance analysis answers this question is to determine whether the yield differences between treatments, after adjusting for the regression of yield on weeds, remain significant. If the adjustment for the effect of weeds results in a significant reduction in the difference between treatments, then the effect of water management on grain yield is due largely to its effects on weeds.

Another example is the case of rice variety trials in which one of the major evaluation criteria is varietal resistance to insects. With the brown planthopper, for example, covariance analysis on grain yield, using brown planthopper infestation as covariate, can provide information on whether yield differences between the test varieties are due primarily to the difference in their resistance to brown planthopper.

The major difference between the use of covariance analysis for error control (Section 10.1.1) and for assisting in the interpretation of results, is in the type of covariate used. For error control, the covariate should not be influenced by the treatments being tested; but for the interpretation of experimental results, the covariate should be closely associated with the treatment effects. We emphasize that while the computational procedures for both techniques are the same, the use of covariance technique to assist in the interpretation of experimental results requires more skill and experience and, hence, should be attempted only with the help of a competent statistician.

10.2   COMPUTATIONAL PROCEDURES

Covariance analysis is essentially an extension of the analysis of variance and, hence, all the assumptions for a valid analysis of variance apply. In addition, the covariance analysis requires that:

- The relationship between the primary character of interest Y and the covariate X is linear.
- This linear relationship, or more specifically the linear regression coefficient, remains constant over other known sources of variation such as treatments and blocks.

The most important task in the application of covariance analysis is the identification of the covariate, a task influenced greatly by the purpose for which the technique is applied. Once the values of the covariate are assigned, the computational procedures are the same regardless of the type of application. However, because of the unique nature of the covariate assigned in the case of missing data, some of the standard procedures can be simplified. We illustrate computational procedures for error control and for estimation of missing data.

10.2.1 Error Control

The data required for the use of covariance technique for error control are the paired observations (X, Y) measured on each and all experimental units, where X refers to the covariate and Y to the primary character of interest. The covariate should represent the particular source of variation that the analysis aims to control and must satisfy the requirement that the covariate is not affected by treatments. Because the computational procedure varies somewhat with the experimental design used, we give three examples: CRD, RCB design, and split-plot design.

10.2.1.1 Completely Randomized Design. We illustrate the procedure for use of covariance analysis with a CRD by a greenhouse trial on brown planthopper damage to rice plants. The trial tested varying numbers of brown planthopper nymphs per tiller. The main character of interest was number of productive panicles per pot. Tiller count, made before the brown planthopper nymphs were placed on the plant, is used as the covariate. Data on productive panicle (Y) and initial tiller number (X), for each of the 50 experimental pots (10 treatments and 5 replications), are shown in Table 10.1. The computational procedure of the covariance analysis is:

☐ STEP 1. Compute the various sums of squares for each of the two variables following the standard analysis of variance procedure for a CRD, described in Chapter 2, Section 2.1.2.1. For our example, the results are shown in Table 10.2, under column XX for the X variable and column YY for the Y variable.

☐ STEP 2. With r as the number of replications and t as the number of treatments, compute the sum of cross products (SCP) for each source of variation as:

- Compute the correction factor as:

$$C.F. = \frac{G_x G_y}{rt}$$

where G_x is the grand total of the X variable and G_y, the grand total of the Y variable. For our example, using the value of G_x and G_y from Table 10.1, the correction factor is:

$$C.F. = \frac{(443)(425)}{(5)(10)} = 3,765.50$$

Table 10.1   Panicle Number per Hill (*Y*) and Initial Tiller Number per Hill[a] (*X*) in a Greenhouse Study of Brown Planthopper Damage to Rice

Treatment		Rep. I		Rep. II		Rep. III		Rep. IV		Rep. V		Treatment Total	
Number	Nymphs, no./tiller[b]	*X*	*Y*	*X*	*Y*	*X*	*Y*	*X*	*Y*	*X*	*Y*	T_x	T_y
1	0.0	5	5	12	12	11	11	5	8	10	10	43	46
2	0.1	7	7	9	9	14	8	9	8	8	8	47	40
3	0.2	9	9	5	5	12	13	5	7	14	16	45	50
4	0.5	7	6	10	10	6	8	8	8	14	11	45	43
5	1.0	8	8	5	5	13	11	5	5	15	5	46	34
6	2.0	12	11	5	5	9	11	7	8	8	8	41	43
7	5.0	7	7	4	4	11	11	6	5	10	10	38	37
8	10.0	7	8	20	16	6	7	9	9	8	10	50	50
9	25.0	10	4	6	7	12	11	5	5	13	6	46	33
10	50.0	10	10	10	13	7	11	7	8	8	7	42	49
Total		82	75	86	86	101	102	66	71	108	91		
Grand total (*G*)												443	425

[a] Counted four days after transplanting.
[b] Placed on 88-day-old plant.

- Compute the total sum of cross products as:

$$\text{Total } SCP = \sum_{i=1}^{t} \sum_{j=1}^{r} (X_{ij})(Y_{ij}) - C.F.$$

where Y_{ij} is the value of the Y variable for the ith treatment and the jth replication, and X_{ij} is the corresponding value of the X variable.

Table 10.2   Analysis of Covariance of CRD Data in Table 10.1[a]

Source of Variation	Degree of Freedom	Sum of Cross Products				Y Adjusted for X		
		XX	XY	YY	d.f.	SS	MS	F[b]
Treatment	9	20.82	4.90	73.30				
Error	40	515.20	305.60	321.20	39	139.93	3.59	
Total	49	536.02	310.50	394.50	48	214.64		
Treatment adjusted					9	74.71	8.30	2.31*

[a] $cv = 22.3\%$, $R.E. = 223\%$.
[b] * = significant at 5% level.

For our example, the total sum of cross products is computed as:

$$\text{Total } SCP = [(5)(5) + (12)(12) + \cdots + (8)(7)]$$

$$- 3,765.50$$

$$= 310.50$$

- Compute the treatment sum of cross products as:

$$\text{Treatment } SCP = \frac{\Sigma (T_x)(T_y)}{r} - C.F.$$

where T_x and T_y are the treatment totals of the X variable and the Y variable, and the summation is over the t treatments.

For our example, the treatment sum of cross products is computed as:

$$\text{Treatment } SCP = \frac{(43)(46) + (47)(40) + \cdots + (42)(49)}{5}$$

$$- 3,765.50$$

$$= 4.90$$

- Compute the error sum of cross products, by subtraction, as:

$$\text{Error } SCP = \text{Total } SCP - \text{Treatment } SCP$$

$$= 310.50 - 4.90 = 305.60$$

□ STEP 3. For each source of variation, compute the adjusted sum of squares of the Y variable as:

$$\text{Total adjusted } SS \text{ of } Y = \text{Total } SS \text{ of } Y - \frac{(\text{Total } SCP)^2}{\text{Total } SS \text{ of } X}$$

$$= 394.50 - \frac{(310.50)^2}{536.02} = 214.64$$

$$\text{Error adjusted } SS \text{ of } Y = \text{Error } SS \text{ of } Y - \frac{(\text{Error } SCP)^2}{\text{Error } SS \text{ of } X}$$

$$= 321.20 - \frac{(305.60)^2}{515.20} = 139.93$$

Treatment adjusted SS of Y = Total adjusted SS of Y

$- $ Error adjusted SS of Y.

$= 214.64 - 139.93 = 74.71$

☐ STEP 4. For each adjusted SS computed in step 3, compute the corresponding degree of freedom:

Adjusted error $d.f.$ = Error $d.f. - 1$

$= 40 - 1 = 39$

Adjusted total $d.f.$ = Total $d.f. - 1$

$= 49 - 1 = 48$

Adjusted treatment $d.f.$ = Treatment $d.f.$ = 9

☐ STEP 5. Compute the adjusted mean squares of Y for treatment and adjusted error as:

$$\text{Treatment adjusted } MS \text{ of } Y = \frac{\text{Treatment adjusted } SS \text{ of } Y}{\text{Adjusted treatment } d.f.}$$

$$= \frac{74.71}{9} = 8.30$$

$$\text{Error adjusted } MS \text{ of } Y = \frac{\text{Error adjusted } SS \text{ of } Y}{\text{Adjusted error } d.f.}$$

$$= \frac{139.93}{39} = 3.59$$

☐ STEP 6. Compute the F value as:

$$F = \frac{\text{Treatment adjusted } MS \text{ of } Y}{\text{Error adjusted } MS \text{ of } Y}$$

$$= \frac{8.30}{3.59} = 2.31$$

☐ STEP 7. Compare the computed F value to the tabular F values from Appendix E with f_1 = adjusted treatment $d.f.$ and f_2 = adjusted error $d.f.$ For our example, the tabular F values with $f_1 = 9$ and $f_2 = 39$ degrees of freedom are 2.13 at the 5% level of significance and 2.90 at the 1% level. Because the computed F value is greater than the corresponding tabular F

value at the 5% level of significance but smaller than that at the 1% level, it is concluded that there are significant differences between the adjusted treatment means at the 5% probability level.

☐ STEP 8. Compute the relative efficiency of covariance analysis compared to the standard analysis of variance as:

$$R.E. = \frac{(100)(\text{Error } MS \text{ of } Y)}{(\text{Error adjusted } MS \text{ of } Y)\left(1 + \dfrac{\text{Treatment } MS \text{ of } X}{\text{Error } SS \text{ of } X}\right)}$$

$$= \frac{(100)(321.20/40)}{(3.59)\left(1 + \dfrac{20.82/9}{515.20}\right)} = 223\%$$

The result indicates that the use of initial tiller number as the covariate has increased precision in panicle number by 123% over that which would have been obtained had the standard analysis of variance been used.

☐ STEP 9. Compute the coefficient of variation as:

$$cv = \frac{\sqrt{\text{Error adjusted } MS \text{ of } Y}}{\text{Grand mean of } Y} \times 100$$

$$= \frac{\sqrt{3.59}}{8.50} \times 100 = 22.3\%$$

☐ STEP 10. Compute the error regression coefficient as:

$$b_{y \cdot x} = \frac{\text{Error } SCP}{\text{Error } SS \text{ of } X}$$

For our example, the error regression coefficient is computed as:

$$b_{y \cdot x} = \frac{305.60}{515.20} = 0.593$$

☐ STEP 11. Compute the adjusted treatment means as:

$$\overline{Y}_i' = \overline{Y}_i - b_{y \cdot x}(\overline{X}_i - \overline{X})$$

where $\overline{Y}_i'$ and $\overline{Y}_i$ are the adjusted and unadjusted means of the Y variable for the ith treatment, $\overline{X}_i$ is the mean of the X variable for the ith treatment, and $\overline{X}$ is the grand mean of the X variable.

Table 10.3 Computation of the Adjusted Treatment Means for Data on Panicle Number per Hill (Y) and Initial Tiller Number (X) in Table 10.1

Treatment Number	Unadjusted Treatment Mean $\overline{Y}_i$	$\overline{X}_i$	Deviation $(D = \overline{X}_i - \overline{X})$	Adjustment Factor $(C = 0.593\,D)$	Adjusted Treatment Mean $(\overline{Y}_i' = \overline{Y}_i - C)$
1	9.2	8.6	−0.3	−0.18	9.38
2	8.0	9.4	0.5	0.30	7.70
3	10.0	9.0	0.1	0.06	9.94
4	8.6	9.0	0.1	0.06	8.54
5	6.8	9.2	0.3	0.18	6.62
6	8.6	8.2	−0.7	−0.42	9.02
7	7.4	7.6	−1.3	−0.77	8.17
8	10.0	10.0	1.1	0.65	9.35
9	6.6	9.2	0.3	0.18	6.42
10	9.8	8.4	−0.5	−0.30	10.10
Sum	85.0	88.6	−0.4[a]	−0.24[a]	85.24
Av.	8.5 = $\overline{Y}$	8.9 = $\overline{X}$			8.52[b]

[a] Except for rounding error, the value should be zero.
[b] Except for rounding error, this value should equal $\overline{Y}$, the grand mean of the Y variable.

For our example, the computation of the adjusted treatment means for the 10 treatments is shown in Table 10.3. For example, the adjusted mean of treatment 1 is computed as:

$$\overline{Y}_1' = 9.2 - 0.593(8.6 - 8.9) = 9.38$$

☐ STEP 12. Test the significance of the differences between the adjusted treatment means. To make a pair comparison (see Chapter 5), compute the standard error of the mean difference for a pair consisting of the ith and jth treatments as:

$$s_{\overline{d}} = \sqrt{(\text{Error adjusted } MS)\left[\frac{2}{r} + \frac{(\overline{X}_i - \overline{X}_j)^2}{\text{Error } SS \text{ of } X}\right]}$$

where $\overline{X}_i$ and $\overline{X}_j$ are the means of the X variable for the ith and jth treatments, and r is the number of replications which is common to both treatments. In some specific situations, this equation should be modified as:

- When the error $d.f.$ exceeds 20, the following approximate formula may be used:

$$s_{\bar{d}}(\text{approx.}) = \sqrt{\frac{2(\text{Error adjusted } MS)}{r}\left[1 + \frac{\text{Treatment } SS \text{ of } X}{(t-1)(\text{Error } SS \text{ of } X)}\right]}$$

where t is the total number of treatments. With this modification, only one $s_{\bar{d}}$ value is needed for any and all pairs of means being compared.

For our example, the computation of the approximate standard error of the mean difference is:

$$s_{\bar{d}}(\text{approx.}) = \sqrt{\frac{2(3.59)}{5}\left[1 + \frac{20.82}{(10-1)(515.20)}\right]}$$

$$= 1.2010$$

Thus, to compare the adjusted means of any two treatments of Table 10.3, the LSD values may be computed, following procedures of Chapter 5, Section 5.1.1, as:

$$LSD = t_{\alpha}s_{\bar{d}}$$

where t_{α} is the tabular t value, from Appendix C, with the adjusted error $d.f.$ and at α level of significance. For our example, the LSD values at the 5% and 1% levels of significance are computed as:

$$LSD_{.05} = (2.0231)(1.2010) = 2.43$$

$$LSD_{.01} = (2.7086)(1.2010) = 3.25$$

For example, to compare the adjusted means of treatments 1 and 2 in Table 10.3, the mean difference is computed as $9.38 - 7.70 = 1.68$. Because the mean difference is smaller than the computed LSD value at the 5% level of significance, the difference is not significant.

- When the treatments have unequal numbers of replications, replace $2/r$ in both of the preceding equations of $s_{\bar{d}}$ by $[(1/r_i) + (1/r_j)]$, where r_i and r_j are the numbers of replications of the ith and jth treatments.

10.2.1.2 Randomized Complete Block Design. We illustrate the procedure for use of covariance analysis with RCB, by a trial designed to evaluate 15 rice varieties grown in soil with a toxic level of iron. The experiment was in a RCB design with three replications. Guard rows of a susceptible check variety were planted on two sides of each experimental plot. Scores for tolerance for iron toxicity were collected from each experimental plot as well as from guard rows. For each experimental plot, the score of the susceptible check (averaged over two guard rows) constitutes the value of the covariate for that plot. Data on the tolerance score of each test variety (Y variable) and on the score of the corresponding susceptible check (X variable) are shown in Table 10.4. The step-by-step procedures for the computation of the analysis of covariance are:

☐ STEP 1. Compute the various sums of squares for each of the two variables, following the standard analysis of variance procedure for a RCB design as described in Chapter 2, Section 2.2.3. For our example, the results are shown in columns XX and YY in Table 10.5.

☐ STEP 2. With r as the number of replications and t the number of treatments, compute the correction factor, total SCP, and treatment SCP, following formulas in Section 10.2.1.1, step 2. Then, compute the replication SCP and error SCP as:

$$\text{Replication } SCP = \frac{\Sigma B_x B_y}{t} - C.F.$$

$$\text{Error } SCP = \text{Total } SCP - \text{Treatment } SCP - \text{Replication } SCP$$

Table 10.4   Scores for Tolerance for Iron Toxicity (Y) of 15 Rice Varieties and Those of the Corresponding Guard Rows of a Susceptible Check Variety (X) in a RCB Trial

Variety Number	Rep. I		Rep. II		Rep. III		Variety Total	
	X	Y	X	Y	X	Y	T_x	T_y
1	5	2	6	3	6	4	17	9
2	6	4	5	3	5	3	16	10
3	5	4	5	4	5	3	15	11
4	6	3	5	3	5	3	16	9
5	7	7	7	6	6	6	20	19
6	6	4	5	3	5	3	16	10
7	6	3	5	3	6	3	17	9
8	6	6	7	7	6	6	19	19
9	7	4	5	3	5	4	17	11
10	7	7	7	7	5	6	19	20
11	6	5	5	4	5	5	16	14
12	6	5	5	3	5	3	16	11
13	5	4	5	4	6	5	16	13
14	5	5	5	4	5	3	15	12
15	5	4	5	5	6	6	16	15
Total	88	67	82	62	81	63	251	192

Table 10.5 Analysis of Covariance of RCB Data in Table 10.4[a]

Source of variation	Degree of Freedom	Sum of Cross Products			Adjusted for X			
		XX	XY	YY	d.f.	SS	MS	F[b]
Total	44	22.9778	28.0667	82.8000				
Replication	2	1.9111	1.2667	0.9333				
Treatment	14	10.3111	19.4000	68.1333				
Error	28	10.7556	7.4000	13.7333	27	8.6420	0.3201	
Treatment + error	42	21.0667	26.8000	81.8666	41	47.7730		
Treatment adjusted					14	39.1310	2.7951	8.73**

[a]$cv = 13.2\%$, $R.E. = 143\%$.
[b]** = significant at 1% level.

where B_x and B_y are the replication totals of the X variable and the Y variable, and the summation is over the r replications.

For our example, the correction factor and the various SCP are computed as:

$$C.F. = \frac{(251)(192)}{(3)(15)} = 1{,}070.9333$$

$$\text{Total } SCP = [(5)(2) + (6)(3) + \cdots + (6)(6)] - 1{,}070.9333$$

$$= 28.0667$$

$$\text{Treatment } SCP = \frac{[(17)(9) + (16)(10) + \cdots + (16)(15)]}{3}$$

$$- 1{,}070.9333$$

$$= 19.4000$$

$$\text{Replication } SCP = \frac{[(88)(67) + (82)(62) + (81)(63)]}{15} - 1{,}070.9333$$

$$= 1.2667$$

$$\text{Error } SCP = 28.0667 - 19.4000 - 1.2667 = 7.4000$$

□ STEP 3. Compute the error adjusted SS of Y, following formula in Section 10.2.1.1, step 3, as:

$$\text{Error adjusted } SS \text{ of } Y = 13.7333 - \frac{(7.4000)^2}{10.7556} = 8.6420$$

☐ STEP 4.   Compute the (treatment + error) adjusted SS of Y as:

$$\text{(Treatment + error) adjusted } SS \text{ of } Y = A - \frac{C^2}{B}$$

where:

$$A = \text{(Treatment + error)} SS \text{ of } Y$$

$$= \text{Treatment } SS \text{ of } Y + \text{Error } SS \text{ of } Y$$

$$B = \text{(Treatment + error)} SS \text{ of } X$$

$$= \text{Treatment } SS \text{ of } X + \text{Error } SS \text{ of } X$$

$$C = \text{(Treatment + error)} SCP$$

$$= \text{Treatment } SCP + \text{Error } SCP$$

For our example, the computation of the (treatment + error) adjusted SS of Y is as follows:

$$A = 68.1333 + 13.7333 = 81.8666$$

$$B = 10.3111 + 10.7556 = 21.0667$$

$$C = 19.4000 + 7.4000 = 26.8000$$

$$\text{(Treatment + error) adjusted } SS \text{ of } Y = 81.8666 - \frac{(26.8000)^2}{21.0667}$$

$$= 47.7730$$

☐ STEP 5.   Using the results of steps 3 and 4, compute the treatment adjusted SS of Y as:

$$\text{Treatment adjusted } SS \text{ of } Y = \text{(Treatment + error) adjusted } SS \text{ of } Y$$

$$- \text{Error adjusted } SS \text{ of } Y$$

$$= 47.7730 - 8.6420 = 39.1310$$

☐ STEP 6.   For each adjusted SS computed in steps 3 to 5, compute the corresponding degree of freedom as:

$$\text{Adjusted error } d.f. = \text{Error } d.f. - 1$$

$$= 28 - 1 = 27$$

Adjusted(treatment + error) $d.f.$ = Treatment $d.f.$ + Error $d.f.$ − 1

$$= 14 + 28 - 1 = 41$$

Adjusted treatment $d.f.$ = Treatment $d.f.$ = 14

☐ STEP 7.   Compute the treatment adjusted mean square of Y and the error adjusted mean square of Y as:

$$\text{Treatment adjusted } MS \text{ of } Y = \frac{\text{Treatment adjusted } SS \text{ of } Y}{\text{Adjusted treatment } d.f.}$$

$$= \frac{39.1310}{14} = 2.7951$$

$$\text{Error adjusted } MS \text{ of } Y = \frac{\text{Error adjusted } SS \text{ of } Y}{\text{Adjusted error } d.f.}$$

$$= \frac{8.6420}{27} = 0.3201$$

☐ STEP 8.   Compute the F value as:

$$F = \frac{\text{Treatment adjusted } MS \text{ of } Y}{\text{Error adjusted } MS \text{ of } Y}$$

$$= \frac{2.7951}{0.3201} = 8.73$$

☐ STEP 9.   Compare the computed F value to the tabular F values, from Appendix E, with f_1 = adjusted treatment $d.f.$ and f_2 = adjusted error $d.f.$ For our example, the tabular F values with f_1 = 14 and f_2 = 27 degrees of freedom are 2.08 at the 5% level of significance and 2.83 at the 1% level. Because the computed F value is greater than the corresponding tabular F value at the 1% level of significance, it is concluded that the difference between the treatment means of Y adjusted for X are significant at the 1% level of significance.

☐ STEP 10.   Follow steps 8 to 12 of the procedure for CRD described in section 10.2.1.1.

10.2.1.3 Split-Plot Design.   We illustrate the procedure for use of covariance analysis with a split-plot design, by a field trial to evaluate eight different managements of nitrogen fertilizer on three rice varieties. The experiment was in a split-plot design with variety as main-plot and fertilizer management as subplot factors. The Y variable is the number of filled grains per panicle and

Table 10.6 Number of Filled Grains per Panicle (Y) and Score of Brown Planthopper Damage[a] (X) of Three Varieties (V_1 to V_3) tested with Eight Different Fertilizer Managements (F_1 to F_8)

Variety	Fertilizer Management	Rep. I		Rep. II		Rep. III		Rep. IV	
		X	Y	X	Y	X	Y	X	Y
V_1	F_1	3	46.9	1	37.3	1	28.9	3	58.1
	F_2	3	81.9	1	45.5	3	49.0	1	55.1
	F_3	5	50.0	1	78.1	3	71.6	3	79.1
	F_4	5	98.7	1	91.2	5	76.0	5	61.2
	F_5	5	65.3	1	87.3	3	80.4	5	61.1
	F_6	3	55.5	3	66.5	5	63.2	5	70.2
	F_7	5	51.0	5	54.1	5	62.1	5	58.3
	F_8	1	49.5	3	45.6	5	37.4	3	50.9
V_2	F_1	1	57.7	3	40.3	1	51.5	3	31.3
	F_2	5	69.6	1	81.6	3	56.1	5	73.5
	F_3	5	38.7	1	71.2	7	37.4	3	69.5
	F_4	7	53.8	3	64.5	7	52.5	3	53.5
	F_5	7	53.4	3	64.8	7	19.7	5	39.3
	F_6	5	73.2	1	86.0	7	41.7	5	61.3
	F_7	7	57.7	1	87.6	5	63.5	5	45.8
	F_8	7	61.8	1	58.7	5	22.8	7	35.8
V_3	F_1	3	37.5	3	63.8	7	36.8	4	95.5
	F_2	5	71.2	3	88.7	5	60.1	3	142.8
	F_3	3	64.8	1	139.5	3	50.7	3	110.6
	F_4	5	57.8	1	104.3	7	51.8	1	106.9
	F_5	3	63.7	1	77.8	3	125.4	1	98.5
	F_6	5	49.2	5	52.4	7	92.6	9	0.0
	F_7	1	60.5	7	30.4	7	16.0	9	0.0
	F_8	5	43.7	7	66.0	5	91.5	9	0.0

[a] Brown planthopper damage is based on scores 0 to 9 with 1 referring to the lowest, and 9 to the greatest, damage.

the X variable is the score of brown planthopper damage on the rice plants. Data are shown in Table 10.6.

The step-by-step procedures for the computation of the covariance analysis for data in Table 10.6 are:

☐ STEP 1.   For each of the two variables Y and X, compute the various sums of squares, following the standard analysis of variance procedure for a split-plot design, described in Chapter 3, Section 3.4.2. For our example, the results are shown in Table 10.7, under column YY for the Y variable and under column XX for the X variable.

Table 10.7 Analysis of Covariance of Split-Plot Data in Table 10.6[a]

Source of Variation	Degree of Freedom	Sum of Cross Products			Y Adjusted for X			
		XX	XY	YY	d.f.	SS	MS	F^b
Replication	3	82.8646	−470.1989	2,751.0086				
Main-plot factor (A)	2	22.3958	26.6667	2,194.0602				
Error(a)	6	27.8542	64.4583	2,262.3598	5	2,113.1948	422.639	
A + Error(a)	8	50.2500	91.1250	4,456.4200	7	4,291.1709		
A adjusted					2	2,177.9761	1,088.988	2.58[ns]
Subplot factor (B)	7	73.7396	−384.6781	11,567.1132				
A × B	14	77.1042	−523.0500	13,367.8915				
Error(b)	63	171.0312	−1,285.0094	31,448.5441	62	21,793.8778	351.5142	
B + Error(b)	70	244.7708	−1,669.6875	43,015.6573	69	31,625.9967		
B adjusted					7	9,832.1189	1,404.5884	4.00**
A × B + Error(b)	77	248.1354	−1,808.0594	44,816.4356	76	31,641.8592		
A × B adjusted					14	9,847.9814	703.4272	2.00*
Total	95	454.9896	−2,571.8114	63,590.9774				

[a]$cv(a) = 33.5\%$, $cv(b) = 30.5\%$, R.E. (main plot) = 64%, R.E. (subplot) = 134%, R.E. (subplot within main plot) = 136%.

[b]** = significant at 1% level, * = significant at 5% level, ns = not significant.

443

Table 10.8   The Replication × Variety Table of Totals (*RA*) Computed from Data in Table 10.6

Replication	V_1 X	V_1 Y	V_2 X	V_2 Y	V_3 X	V_3 Y	Rep. Total (*R*) X	Rep. Total (*R*) Y
I	30	498.8	44	465.9	30	448.4	104	1,413.1
II	16	505.6	14	554.7	28	622.9	58	1,683.2
III	30	468.6	42	345.2	44	524.9	116	1,338.7
IV	30	494.0	36	410.0	39	554.3	105	1,458.3
Variety total (*A*)	106	1,967.0	136	1,775.8	141	2,150.5		
Grand total (*G*)							383	5,893.3

□ STEP 2. Construct two-way tables of totals for X and Y:

- The replication × factor A two-way table of totals, with the replication totals, factor A totals, and grand total computed. For our example, the replication × variety table of totals (*RA*), with the replication totals (*R*), variety totals (*A*), and grand total (*G*) computed are shown in Table 10.8.

- The factor A × factor B two-way table of totals, with factor B totals computed. For our example, the nitrogen × fertilizer management table of totals (*AB*) with fertilizer management totals (*B*) computed are shown in Table 10.9.

□ STEP 3. Using subscripts x and y to differentiate the totals for the X variable from those for the Y variable (G_x refers to the grand total of X, and

Table 10.9   The Variety × Fertilizer Management Table of Totals (*AB*) Computed from Data in Table 10.6

Fertilizer Management	V_1 X	V_1 Y	V_2 X	V_2 Y	V_3 X	V_3 Y	Fertilizer Total (*B*) X	Fertilizer Total (*B*) Y
F_1	8	171.2	8	180.8	17	233.6	33	585.6
F_2	8	231.5	14	280.8	16	362.8	38	875.1
F_3	12	278.8	16	216.8	10	365.6	38	861.2
F_4	16	327.1	20	224.3	14	320.8	50	872.2
F_5	14	294.1	22	177.2	8	365.4	44	836.7
F_6	16	255.4	18	262.2	26	194.2	60	711.8
F_7	20	225.5	18	254.6	24	106.9	62	587.0
F_8	12	183.4	20	179.1	26	201.2	58	563.7

G_y, the grand total of Y), compute the correction factor and the various sums of cross products (SCP) of X and Y for the main-plot analysis as:

$$C.F. = \frac{G_x G_y}{rab}$$

$$= \frac{(383)(5,893.3)}{(4)(3)(8)} = 23,511.8114$$

Total $SCP = \Sigma XY - C.F. = [(3)(46.9) + (1)(37.3)$

$$+ \cdots + (9)(0.0)] - 23,511.8114$$

$$= -2,571.8114$$

Replication $SCP = \dfrac{\Sigma R_x R_y}{ab} - C.F.$

$$= \{[(104)(1,413.1) + (58)(1,683.2) + (116)(1,338.7)$$

$$+ (105)(1,458.3)] / (3)(8)\} - 23,511.8114$$

$$= -470.1989$$

A(variety) $SCP = \dfrac{\Sigma A_x A_y}{rb} - C.F.$

$$= \frac{(106)(1,967.0) + (136)(1,775.8) + (141)(2,150.5)}{(4)(8)}$$

$$- 23,511.8114$$

$$= 26.6667$$

Error(a) $SCP = \dfrac{\Sigma (RA)_x (RA)_y}{b} - C.F. -$ Replication $SCP - A\ SCP$

$$= \left[\frac{(30)(498.8) + (44)(465.9) + \cdots + (39)(554.3)}{8} \right]$$

$$- 23,511.8114 - (-470.1989) - 26.6667$$

$$= 64.4583$$

□ STEP 4.   Compute the various sums of cross products of X and Y for the subplot analysis as:

B(fertilizer management) $SCP = \dfrac{\Sigma B_x B_y}{ra} - C.F.$

$$= \left[\frac{(33)(585.6) + (38)(875.1) + \cdots + (58)(563.7)}{(4)(3)} \right]$$

$$- 23{,}511.8114$$

$$= -384.6781$$

$A \times B$ (variety $\times$ fertilizer management) $SCP = \dfrac{\Sigma (AB)_x (AB)_y}{r}$

$$- C.F. - B\ SCP - A\ SCP$$

$$= \left[\frac{(8)(171.2) + (8)(180.8) + \cdots + (26)(201.2)}{4} \right]$$

$$- 23{,}511.8114 - (-384.6781) - 26.6667$$

$$= -523.0500$$

Error(b) SCP = Total SCP − [Replication SCP + $A\ SCP$

$$+ \text{Error}(a)\ SCP + B\ SCP + A \times B\ SCP]$$

$$= -2{,}571.8114 - (-470.1989 + 26.6667 + 64.4583$$

$$- 384.6781 - 523.0500)$$

$$= -1{,}285.0094$$

□ STEP 5.   Compute the adjusted SS of Y that involves an error term, using the formula:

$$\text{Adjusted } SS \text{ of } Y = YY - \frac{(XY)^2}{XX}$$

For our example, the various adjusted SS of Y involving either error(a) or error(b) term are computed as:

$$\text{Error}(a) \text{ adjusted } SS \text{ of } Y = \text{Error}(a)\ SS \text{ of } Y - \frac{[\text{Error}(a)\ SCP]^2}{\text{Error}(a)\ SS \text{ of } X}$$

$$= 2{,}262.3598 - \frac{(64.4583)^2}{27.8542} = 2{,}113.1948$$

$[A + \text{Error}(a)]$ adjusted SS of $Y = [A + \text{Error}(a)]$ SS of Y

$$- \frac{\{[A + \text{Error}(a)]\ SCP\}^2}{[A + \text{Error}(a)]\ SS \text{ of } X}$$

$$= 4,456.4200 - \frac{(91.1250)^2}{50.2500} = 4,291.1709$$

$\text{Error}(b)$ adjusted SS of $Y = \text{Error}(b)$ SS of $Y - \dfrac{[\text{Error}(b)\ SCP]^2}{\text{Error}(b)\ SS \text{ of } X}$

$$= 31,448.5441 - \frac{(-1,285.0094)^2}{171.0312}$$

$$= 21,793.8778$$

$[B + \text{Error}(b)]$ adjusted SS of $Y = [B + \text{Error}(b)]$ SS of Y

$$- \frac{\{[B + \text{Error}(b)]\ SCP\}^2}{[B + \text{Error}(b)]\ SS \text{ of } X}$$

$$= 43,015.6573 - \frac{(-1,669.6875)^2}{244.7708}$$

$$= 31,625.9967$$

$[(A \times B) + \text{Error}(b)]$ adjusted SS of $Y = [(A \times B) + \text{Error}(b)]$ SS of Y

$$- \frac{\{[(A \times B) + \text{Error}(b)]\ SCP\}^2}{[(A \times B) + \text{Error}(b)]\ SS \text{ of } X}$$

$$= 44,816.4356 - \frac{(-1,808.0594)^2}{248.1354}$$

$$= 31,641.8592$$

☐ STEP 6. Compute the adjusted SS of Y for the main effect of each factor and for their interaction, using the formulas:

$$A \text{ adjusted } SS \text{ of } Y = [A + \text{Error}(a)] \text{ adjusted } SS \text{ of } Y$$

$$- \text{Error}(a) \text{ adjusted } SS \text{ of } Y$$

$$B \text{ adjusted } SS \text{ of } Y = [B + \text{Error}(b)] \text{ adjusted } SS \text{ of } Y$$

$$- \text{Error}(b) \text{ adjusted } SS \text{ of } Y$$

$$A \times B \text{ adjusted } SS \text{ of } Y = [A \times B + \text{Error}(b)] \text{ adjusted } SS \text{ of } Y$$

$$- \text{Error}(b) \text{ adjusted } SS \text{ of } Y$$

For our example, the three adjusted SS of Y, one for factor A, one for factor B, and one for $A \times B$ interaction, are computed as:

$$A \text{ adjusted } SS \text{ of } Y = 4{,}291.1709 - 2{,}113.1948$$

$$= 2{,}177.9761$$

$$B \text{ adjusted } SS \text{ of } Y = 31{,}625.9967 - 21{,}793.8778$$

$$= 9{,}832.1189$$

$$A \times B \text{ adjusted } SS \text{ of } Y = 31{,}641.8592 - 21{,}793.8778$$

$$= 9{,}847.9814$$

☐ STEP 7.   For each adjusted SS of Y computed in steps 5 and 6, compute the corresponding degree of freedom:

$$\text{Adjusted error}(a) \, d.f. = \text{Error}(a) \, d.f. - 1$$

$$= 6 - 1 = 5$$

$$\text{Adjusted } [A + \text{Error}(a)] \, d.f. = A \, d.f. + \text{Error}(a) \, d.f. - 1$$

$$= 2 + 6 - 1 = 7$$

$$\text{Adjusted } A \, d.f. = A \, d.f. = 2$$

$$\text{Adjusted error}(b) \, d.f. = \text{Error}(b) \, d.f. - 1$$

$$= 63 - 1 = 62$$

$$\text{Adjusted } [B + \text{Error}(b)] \, d.f. = B \, d.f. + \text{Error}(b) \, d.f. - 1$$

$$= 7 + 63 - 1 = 69$$

$$\text{Adjusted } B \, d.f. = B \, d.f. = 7$$

$$\text{Adjusted } [A \times B + \text{Error}(b)] \, d.f. = A \times B \, d.f. + \text{Error}(b) \, d.f. - 1$$

$$= 14 + 63 - 1 = 76$$

$$\text{Adjusted } A \times B \, d.f. = A \times B \, d.f. = 14$$

☐ STEP 8.   Compute the adjusted MS of Y:

$$\text{Error}(a) \text{ adjusted } MS = \frac{\text{Error}(a) \text{ adjusted } SS}{\text{Adjusted error}(a) \, d.f.}$$

$$= \frac{2{,}113.1948}{5} = 422.6390$$

$$A \text{ adjusted } MS = \frac{A \text{ adjusted } SS}{\text{Adjusted } A \text{ d.f.}}$$

$$= \frac{2,177.9761}{2} = 1,088.9880$$

$$\text{Error}(b) \text{ adjusted } MS = \frac{\text{Error}(b) \text{ adjusted } SS}{\text{Adjusted error}(b) \text{ d.f.}}$$

$$= \frac{21,793.8778}{62} = 351.5142$$

$$B \text{ adjusted } MS = \frac{B \text{ adjusted } SS}{\text{Adjusted } B \text{ d.f.}}$$

$$= \frac{9,832.1189}{7} = 1,404.5884$$

$$A \times B \text{ adjusted } MS = \frac{A \times B \text{ adjusted } SS}{\text{Adjusted } A \times B \text{ d.f.}}$$

$$= \frac{9,847.9814}{14} = 703.4272$$

□ STEP 9. Compute the F value, for each effect that needs to be tested, by dividing the mean square by its corresponding error term:

$$F(A) = \frac{A \text{ adjusted } MS}{\text{Error}(a) \text{ adjusted } MS}$$

$$= \frac{1,088.9880}{422.6390} = 2.58$$

$$F(B) = \frac{B \text{ adjusted } MS}{\text{Error}(b) \text{ adjusted } MS}$$

$$= \frac{1,404.5884}{351.5142} = 4.00$$

$$F(A \times B) = \frac{A \times B \text{ adjusted } MS}{\text{Error}(b) \text{ adjusted } MS}$$

$$= \frac{703.4272}{351.5142} = 2.00.$$

□ STEP 10. Compare each computed F value to the tabular F value from Appendix E, with $f_1 = d.f.$ of the numerator MS and $f_2 = d.f.$ of the denominator MS, at the prescribed level of significance.

For our example, the three sets of the corresponding tabular F values are:

- Tabular $F(A)$ values, with $f_1 = 2$ and $f_2 = 5$, are 5.79 at the 5% level of significance and 13.27 at the 1% level.
- Tabular $F(B)$ values, with $f_1 = 7$ and $f_2 = 62$, are 2.16 at the 5% level of significance and 2.94 at the 1% level.
- Tabular $F(A \times B)$ values, with $f_1 = 14$ and $f_2 = 62$, are 1.86 at the 5% level of significance and 2.39 at the 1% level.

Because the computed $F(A)$ value of 2.58 is smaller than the corresponding tabular F value at the 5% level of significance of 5.79, the A main effect is not significant at the 5% level. On the other hand, the computed $F(B)$ value and the computed $F(A \times B)$ value are larger than their corresponding tabular F values at the 1% level and the 5% level, respectively. Hence, both the B main effect and the $A \times B$ interaction effect are significant.

☐ STEP 11.   Compute the coefficient of variation, corresponding to the main-plot analysis and subplot analysis:

$$cv(a) = \frac{\sqrt{\text{Error}(a) \text{ adjusted } MS \text{ of } Y}}{\text{Grand mean of } Y} \times 100$$

$$= \frac{\sqrt{422.6390}}{61.389} \times 100 = 33.5\%$$

$$cv(b) = \frac{\sqrt{\text{Error}(b) \text{ adjusted } MS \text{ of } Y}}{\text{Grand mean of } Y} \times 100$$

$$= \frac{\sqrt{351.5142}}{61.389} \times 100 = 30.5\%$$

☐ STEP 12.   Compute the error regression coefficients, corresponding to the two error terms:

$$b_{y \cdot x}(a) = \frac{\text{Error}(a) \text{ } SCP}{\text{Error}(a) \text{ } SS \text{ of } X}$$

$$= \frac{64.4583}{27.8542} = 2.314$$

$$b_{y \cdot x}(b) = \frac{\text{Error}(b) \text{ } SCP}{\text{Error}(b) \text{ } SS \text{ of } X}$$

$$= \frac{-1,285.0094}{171.0312} = -7.513$$

☐ STEP 13.   Compute the adjusted treatment means:

$$\overline{Y}'_{ij} = \overline{Y}_{ij} - \left[b_{y \cdot x}(a) \right] \left[\overline{X}_i - \overline{X} \right] - \left[b_{y \cdot x}(b) \right] \left[\overline{X}_{ij} - \overline{X}_i \right]$$

where $\overline{Y}'_{ij}$ and $\overline{Y}_{ij}$ are the adjusted and unadjusted means of the Y variable for the treatment combination involving the ith level of factor A and the jth level of factor B, respectively; $\overline{X}_i$ is the mean of the X variable for the ith level of factor A (i.e., main-plot mean); $\overline{X}_{ij}$ is the mean of the X variable for the treatment combination involving the ith level of factor A and the jth level of factor B; and $\overline{X}$ is the grand mean of the X variable.

For our example, the computation of the adjusted mean of treatment V_1F_1, for example, is obtained (using data in Tables 10.8 and 10.9) as:

$$\overline{Y}'_{11} = \overline{Y}_{11} - \left[b_{y \cdot x}(a)\right]\left[\overline{X}_1 - \overline{X}\right] - \left[b_{y \cdot x}(b)\right]\left[\overline{X}_{11} - \overline{X}_1\right]$$

where

$$\overline{Y}_{11} = \frac{171.2}{4} = 42.800$$

$$\overline{X}_{11} = \frac{8}{4} = 2.000$$

$$\overline{X}_1 = \frac{106}{(4)(8)} = 3.312$$

$$\overline{X} = \frac{383}{(4)(3)(8)} = 3.990$$

$$b_{y \cdot x}(a) = 2.314$$

$$b_{y \cdot x}(b) = -7.513$$

Thus,

$$\overline{Y}'_{11} = 42.800 - 2.314\,(3.312 - 3.990)$$

$$- (-7.513)(2.000 - 3.312)$$

$$= 34.5$$

The results of the adjusted treatment means computed for all the 3×8 treatment combinations are shown in Table 10.10.

☐ STEP 14. Test the significance of the difference among the adjusted treatment means. To make pair comparisons (Chapter 5, Section 5.1), the $s_{\overline{d}}$ values are computed as:

- For a pair of A adjusted means, at same or different levels of B:

$$s_{\overline{d}} = \sqrt{\frac{2}{rb}\left[(b-1)E^*_{byy} + E^*_{ayy}\right]\left[1 + \frac{A_{xx}}{(a-1)E_{axx}}\right]}$$

Table 10.10 Adjusted Treatment Means of the Number of Filled Grains per Panicle, Using the Score of Brown Planthopper Damage as the Covariate, Computed from Data in Tables 10.8 and 10.9

Fertilizer Management	Adjusted Mean, no./panicle			
	V_1	V_2	V_3	Av.
F_1	34.5	27.7	56.3	39.5
F_2	49.6	64.0	86.7	66.8
F_3	68.9	51.7	76.1	65.6
F_4	88.5	61.1	72.4	74.0
F_5	76.5	53.1	72.4	67.3
F_6	70.5	66.9	63.4	66.9
F_7	70.7	64.9	37.7	57.8
F_8	45.0	49.8	65.1	53.3
Av.	63.0	54.9	66.3	

where E^*_{byy} and E^*_{ayy} are the adjusted error(b) mean square and the adjusted error(a) mean square of the Y variable; and A_{xx} and E_{axx} are A SS and Error(a) SS of the X variable. For our example, the $s_{\bar{d}}$ value for comparing two A (variety) means, at same or different levels of B (fertilizer management), is computed as:

$$s_{\bar{d}} = \sqrt{\frac{2}{(4)(8)}[(8 - 1)(351.5142) + 422.6390]}$$

$$\times \sqrt{\left[1 + \frac{22.3958}{(3 - 1)(27.8542)}\right]}$$

$$= 15.895$$

- For a pair of B adjusted means at the same level of A:

$$s_{\bar{d}} = \sqrt{\frac{2}{r}E^*_{byy}\left\{1 + \frac{[B_{xx} + (AB)_{xx}]}{a(b - 1)E_{bxx}}\right\}}$$

where B_{xx}, $(AB)_{xx}$, and E_{bxx} are B SS, $A \times B$ SS, and Error(b) SS of the X variable.

For our example, the $s_{\bar{d}}$ value for comparing two B (fertilizer) means under the same A (variety) is computed as:

$$s_{\bar{d}} = \sqrt{\frac{2}{4}(351.5142)\left[1 + \frac{(73.7396 + 77.1042)}{3(8-1)(171.0312)}\right]}$$

$$= 13.533$$

- For a pair of A adjusted means, averaged over all levels of B:

$$s_{\bar{d}} = \sqrt{\frac{2}{rb}E^*_{ayy}\left[1 + \frac{A_{xx}}{(a-1)E_{axx}}\right]}$$

For our example, because of the $A \times B$ interaction effect is significant at the 5% level (Table 10.7), the comparison between A (variety) means, averaged over all levels of B (fertilizer management), should not be made (see Chapter 3, Section 3.1) and, hence, no $s_{\bar{d}}$ value for such a comparison is computed.

- For a pair of B adjusted means, averaged over all levels of A:

$$s_{\bar{d}} = \sqrt{\frac{2}{ar}E^*_{byy}\left\{1 + \frac{B_{xx}}{(b-1)E_{bxx}}\right\}}$$

For our example, because of the same reason given previously, no $s_{\bar{d}}$ value for such a comparison is computed.

☐ STEP 15. Compute the efficiency of covariance analysis relative to the standard analysis of variance as:

$$R.E.\,(\text{main plot}) = \frac{\text{Error}(a)\ SS\ \text{of}\ Y}{(a-1)(r-1)E^*_{ayy}\left[1 + \dfrac{A_{xx}}{(a-1)E_{axx}}\right]} \times 100$$

$$= \frac{2{,}262.3598}{(3-1)(4-1)(422.6390)\left[1 + \dfrac{22.3958}{(3-1)(27.8542)}\right]} \times 100$$

$$= 64\%$$

$$R.E. \text{(subplot)} = \frac{\text{Error}(b) \text{ SS of } Y}{a(r-1)(b-1)E^*_{byy}\left[1 + \dfrac{B_{xx}}{(b-1)E_{bxx}}\right]} \times 100$$

$$= \frac{31,448.5441}{3(4-1)(8-1)(351.5142)\left[1 + \dfrac{73.7396}{(8-1)(171.0312)}\right]} \times 100$$

$$= 134\%$$

$R.E.$ (subplot within main plot)

$$= \frac{\text{Error}(b) \text{ SS of } Y}{a(r-1)(b-1)E^*_{byy}\left\{1 + \dfrac{[B_{xx} + (AB)_{xx}]}{a(b-1)E_{bxx}}\right\}} \times 100$$

$$= \frac{31,448.5441}{3(4-1)(8-1)(351.5142)\left\{1 + \dfrac{[73.7396 + 77.1042]}{3(8-1)(171.0312)}\right\}} \times 100$$

$$= 136\%$$

The results indicate that the use of brown planthopper damage as the covariate (X variable) increased the precision of the B main effects and the $A \times B$ interaction effects of the number of filled grains per panicle (Y variable) by 34% and 36%, respectively, over that which would have been obtained had the standard analysis of variance been used. On the other hand, no increase in precision due to the use of covariate was observed for the A main effect.

10.2.2 Estimation of Missing Data

The only difference, between the use of covariance analysis for error control and that for estimation of missing data, is the manner in which the values of the covariate are assigned. When covariance analysis is used to control error and to adjust treatment means, the covariate is measured along with the Y variable for each experimental unit. But when covariance analysis is used to estimate missing data, the covariate is not measured but is assigned, one each, to a missing observation. We confine our discussion to the case of only one missing observation. For more than one missing observation and, thus, more than one covariate, see references such as Steel and Torrie, 1980, and Snedecor and Cochran, 1980.*

*R. G. D. Steel and J. A. Torrie. *Principles and Procedures of Statistics*, 2nd ed., USA: McGraw Hill, 1980. pp. 428–434.
G. W. Snedecor and W. G. Cochran. *Statistical Methods*. USA: The Iowa State University Press, 1980. pp. 388–391.

Table 10.11 Assignment of Values for the Missing Data of Grain Yield (*Y*) and for All Values of the Covariate (*X*), when Covariance Analysis is Applied to Estimate Missing Data

Treatment, kg seed/ha	Rep. I		Rep. II		Rep. III		Rep. IV		Treatment Total	
	X	*Y*	*X*	*Y*	*X*	*Y*	*X*	*Y*	*X*	*Y*
25	0	5,113	0	5,398	0	5,307	0	4,678	0	20,496
50	0	5,346	0	5,952	0	4,719	0	4,264	0	20,281
75	0	5,272	0	5,713	0	5,483	0	4,749	0	21,217
100	0	5,164	1[a]	0	0	4,986	0	4,410	1	14,560
125	0	4,804	0	4,848	0	4,432	0	4,748	0	18,832
150	0	5,254	0	4,542	0	4,919	0	4,098	0	18,813
Rep. total	0	30,953	1	26,453	0	29,846	0	26,947		
Grand total									1	114,199

[a] Missing data.

The rules for the application of covariance analysis to a data set with one missing observation are:

1. For the missing observation, set $Y = 0$.
2. Assign the values of the covariate as $X = 1$ for the experimental unit with the missing observation, and $X = 0$ otherwise.
3. With the complete set of data for the Y variable and the X variable as assigned in rules 1 and 2, compute the analysis of covariance following the standard procedures discussed in Section 10.2.1. However, because of the nature of the covariate used, the computational procedures for the sums of squares of the covariate and for the sums of cross products can be simplified.

We illustrate the simplified procedure with data from the RCB experiment involving six rates of seeding, discussed in Chapter 2, Section 2.2.3. We assume that the observation of the fourth treatment (100 kg seed/ha) from the second replication is missing. The step-by-step procedures to obtain the analysis of covariance are:

☐ STEP 1. Assign the Y value for the missing plot and the X values for all plots, following rules 1 and 2 above. The results are shown in Table 10.11.

☐ STEP 2. Compute the various sums of squares for the Y variable following the standard analysis of variance procedure for a RCB design described in Chapter 2, Section 2.2.3. The results are shown in Table 10.12, under column YY.

Table 10.12 Analysis of Covariance to Estimate a Missing Value for Data in Table 10.11

Source of Variation	Degree of Freedom	Sum of Cross Products				Y Adjusted for X		
		XX	XY	YY	d.f.	SS	MS	F^a
Total	23	0.9583	−4,758.2917	28,409,561				
Replication	3	0.1250	−349.4584	2,403,507				
Treatment	5	0.2083	−1,118.2917	7,141,065				
Error	15	0.6250	−3,290.5416	18,864,989	14	1,540,727	110,052	
Treatment + error	20	0.8333	−4,408.8333	26,006,054	19	2,679,748		
Treatment adjusted					5	1,139,021	227,804	2.07^{ns}

[a] ns = not significant.

☐ STEP 3. Compute the various sums of squares for the X variable as:

$$\text{Total } SS = 1 - \frac{1}{rt}$$

$$= 1 - \frac{1}{24} = 0.9583$$

$$\text{Replication } SS = \frac{1}{t} - \frac{1}{rt}$$

$$= \frac{1}{6} - \frac{1}{24} = 0.1250$$

$$\text{Treatment } SS = \frac{1}{r} - \frac{1}{rt}$$

$$= \frac{1}{4} - \frac{1}{24} = 0.2083$$

$$\text{Error } SS = \text{Total } SS - \text{Replication } SS - \text{Treatment } SS$$

$$= 0.9583 - 0.1250 - 0.2083 = 0.6250$$

☐ STEP 4. Compute the various sums of cross products as:

$$C.F. = \frac{G_y}{(r)(t)}$$

$$= \frac{114,199}{(4)(6)} = 4,758.2917$$

$$\text{Total } SCP = -(C.F.) = -4,758.2917$$

$$\text{Replication } SCP = \frac{B_y}{t} - C.F.$$

$$-\frac{26,453}{6} - 4,758.2917 = -349.4584$$

$$\text{Treatment } SCP = \frac{T_y}{r} - C.F.$$

$$= \frac{14,560}{4} - 4,758.2917 = -1,118.2917$$

$$\text{Error } SCP = \text{Total } SCP - \text{Replication } SCP - \text{Treatment } SCP$$

$$= -4,758.2917 - (-349.4584) - (-1,118.2917)$$

$$= -3,290.5416$$

where B_y is the replication total for the Y variable, of the replication in which the missing data occurred (replication II in our example), and T_y is the treatment total, for the Y variable, corresponding to the treatment with the missing data (the fourth treatment in our example).

☐ STEP 5.   Follow steps 3 to 9 of Section 10.2.1.2. The final results are shown in Table 10.12.

☐ STEP 6.   Compute the estimate of the missing data as:

$$\text{Estimate of missing data} = -b_{y \cdot x} = -\frac{\text{Error } SCP}{\text{Error } SS \text{ of } X}$$

$$= \frac{-(-3,290.5416)}{0.6250} = 5,265 \text{ kg/ha}$$

Note that the estimate of the missing data computed agrees with that obtained by the missing data formula technique of Chapter 7, Section 7.1.2.1.

CHAPTER 11

Chi-Square Test

Hypotheses about treatment means are the most common in agricultural research but they are by no means the only one of concern. A type of hypothesis commonly encountered is that concerning the frequency distribution of the population or populations being studied. Examples of questions relating to this type of hypotheses are:

- Does the frequency distribution of the kernel color in maize follow a hypothetical genetic segregation ratio?
- Do individuals from several treatments in the same experiment belong to the same population distribution?
- Are the frequency distributions of two or more populations independent of each other?

The chi-square test is most commonly used to test hypotheses concerning the frequency distribution of one or more populations. We focus on three uses of the chi-square test that are most common in agricultural research: analysis of attribute data, test for homogeneity of variance, and test for goodness of fit.

11.1 ANALYSIS OF ATTRIBUTE DATA

Data from an agricultural experiment can either be measurement data or attribute data. Measurement data is specified along a continuous numerical scale, such as yield, plant height, and protein content; but attribute data is concerned with a finite number of discrete classes. The most common types of attribute data are those having two classes, which consist of the presence or absence of an attribute such as male or female, success or failure, effective or ineffective, and dead or alive. Examples of attribute data with more than two classes are varietal classification, color classification, and tenure status of farmers.

The number of discrete classes in attribute data may be specified based on one or more classification criteria. When only one criterion is used, attribute data is referred to as a one-way classification. Presence or absence of one

458

character, color classification of a plant tissue, and tenure status of farmers are illustrations of attribute data with one-way classification. When more than one classification criterion is used to specify the classes in attribute data, such data may be referred to as a two-way classification, a three-way classification, and so on, depending on the number of classification criteria used. Attribute data with two-way classification form an $r \times c$ two-way classification, or an $r \times c$ contingency table, where r and c denote the number of classes in the two classification criteria used. For example, if rice varieties in a variety trial are classified based on two criteria—color of its leaf blade (green or purple) and varietal type (indica, japonica, or hybrid)—the resulting attribute data represent a 2×3 contingency table. Note that the contingency table progresses to three-way, four-way, and so on, with successive additions of more classification criteria.

In general, attribute data are obtained when it is not possible to use measurement data. However, in some special cases experimental materials may be classified into discrete classes despite the availability of a quantitative measurement. For example, plants can be classified into three discrete height classes (tall, intermediate, or short) instead of being measured in centimeters. Or, vertical resistance to an insect pest may be scored on a scale from 0 through 9 instead of measuring the actual percentage of plant damage or of insect incidence.

There are three important applications of the chi-square test in the analysis of attribute data:

- Test for a fixed-ratio hypothesis
- Test for independence in a contingency table
- Test for homogeneity of ratio

We provide specific examples of each of these types of application and use actual data to give the corresponding computational procedures.

11.1.1 Test for a Fixed-Ratio Hypothesis

As the name implies, the chi-square test for a fixed-ratio hypothesis is a technique for deciding whether a set of attribute data conforms to a hypothesized frequency distribution that is specified on the basis of some biological phenomenon. We give three examples where this test is commonly applied in agricultural research.

Example 1. A plant breeder is studying a cross between a sweet maize inbred line with yellow kernels and a flint maize inbred line with white kernels. He would like to know whether the ratio of kernel type and color in the F_2 population follows the normal di-hybrid ratio of $9 : 3 : 3 : 1$. From the F_1 plants produced by crossing the two inbred lines, he obtains F_2 kernels and classifies them into four categories according to kernel color (yellow or white) and kernel type (flint or ~weet) as follows: yellow flint, yellow sweet, white flint, and white sweet. Suppose he examines 800 F_2 kernels and finds that 496 are yellow flint, 158 are yellow sweet, 112 are white flint, and the rest (34) are white sweet. He then asks: does the observed ratio of $496:158:112:34$ deviate significantly from the hypothesized ratio of $9:3:3:1$?

Example 2.   In rice, the green leafhopper is suspected to differ in feeding preference between an already diseased plant and a healthy plant. The researcher, therefore, encloses a prescribed number of green leafhoppers in a cage that holds an equal number of healthy and diseased rice plants. After two hours of caging, he then counts the number of insects found on diseased and on healthy plants. Of 239 insects confined, 67 were found on the healthy plants and 172 on the diseased plants. Does the observed ratio of $67:172$ deviate significantly from the hypothesized no-preference ratio of $1:1$?

Example 3.   To determine the eating quality of a newly developed rice variety relative to that of a traditional variety, the researcher may conduct a taste-panel study and ask a number of judges to rank their preference between the two varieties. Suppose that out of a total of 40 judges, 22 judges prefer the new variety and 18 judges prefer the traditional variety. Does the observed ratio of $22:18$ deviate significantly from the hypothesized ratio of $1:1$?

From the foregoing examples, it is clear that the test for a fixed-ratio hypothesis is applicable to any number of classes derived from any number of classification criteria. The maize example, for instance, involves four classes derived from a two-way classification data; the two rice examples have two classes derived from a one-way classification data.

We give the procedures for applying the chi-square test for a fixed-ratio hypothesis using the rice test to determine the feeding preference of green leafhoppers. The step-by-step procedures are:

☐ STEP 1.   Compute the χ^2 value, depending on class number:

- With more than two classes:

$$\chi^2 = \sum_{i=1}^{p} \frac{(n_i - E_i)^2}{E_i}$$

where p is the number of classes, n_i is the observed number of units falling into class i, and E_i is the number of units expected to fall into class *i* assuming that the hypothesized ratio holds. E_i is computed as:

$$E_i = \frac{(r_i)\left(\sum\limits_{i=1}^{p} n_i\right)}{\sum\limits_{i=1}^{p} r_i}$$

where $r_1 : r_2 : \cdots : r_p$ is the hypothesized ratio.
· With two classes:

$$\chi^2 = \frac{\left(|n_1 - E_1| - 0.5\right)^2}{E_1} + \frac{\left(|n_2 - E_2| - 0.5\right)^2}{E_2}$$

where | | refers to absolute value; and n_1 and n_2 are observed values and E_1 and E_2 are the expected values, as defined previously, of class 1 and class 2, respectively.

For our example, there are two classes (67 healthy and 172 diseased plants). With $r_1 = r_2 = 1$, $n_1 = 67$, and $n_2 = 172$, the values of E_1 and E_2 are computed following the formula as:

$$E_1 = E_2 = \frac{(1)(67 + 172)}{1 + 1} = \frac{239}{2} = 119.5$$

And the χ^2 value is computed as:

$$\chi^2 = \frac{\left(|67 - 119.5| - 0.5\right)^2}{119.5} + \frac{\left(|172 - 119.5| - 0.5\right)^2}{119.5} = 45.26$$

☐ STEP 2. Compare the computed χ^2 value with the tabular χ^2 values obtained from Appendix D with $(p - 1)$ degrees of freedom. Reject the hypothesis that the hypothesized ratio holds, if the computed χ^2 value exceeds the corresponding tabular χ^2 value at a prescribed level of significance.

For our example, the tabular χ^2 values, with $(p - 1) = 2 - 1 = 1$ degree of freedom, are 3.84 at the 5% level of significance and 6.63 at the 1% level. Because the computed χ^2 value exceeds the corresponding tabular χ^2 value at the 1% level of significance, the hypothesis of no preference (i.e., the ratio of 1 : 1) is rejected at the 1% level of significance. The observed data, in which 72% of the confined insects were found on diseased plants and 28% were found on healthy plants, strongly suggest that green leafhoppers preferred diseased plants over healthy plants.

11.1.2   Test for Independence in a Contingency Table

When the number of classes in attribute data is based on two classification criteria, one with r classes and another with c classes, the resulting data form an $r \times c$ contingency table. For example, an agricultural economist studying factors affecting the adoption of the newly introduced high-yielding rice varieties wishes to know if adoption is affected by the tenure status of farmers. With three distinct classes of tenure status (the first classification criterion)—owner operator, share-rent farmer, and fixed-rent farmer—and two classes of adoption status (the second classification criterion)—adopter and nonadopter—the resulting data form a 3 × 2 contingency table. For such data, each sample farmer can be classified into one of the six possible categories: owner operator and adopter, owner operator and nonadopter, share-rent farmer and adopter, share-rent farmer and nonadopter, fixed-rent farmer and adopter, and fixed-rent farmer and nonadopter.

With such a contingency table, the question usually raised is whether the ratio of the various classes in the first classification criterion remains the same over all classes of the second classification criterion, and vice versa. If the answer is yes, the two classification criteria are said to be independent.

For our example, the question is whether the ratio of adopter to nonadopter remains the same for all the three classes of tenure status or whether the farmer's adoption of the new rices is independent of tenure status. The chi-square test for independence in a contingency table is the appropriate procedure for answering such a question. We use data from this study (Table 11.1) to show the procedure for applying the chi-square test to test the hypothesis of independence between two classification criteria in a $r \times c$ contingency table.

Note the following in the succeeding discussions:

- We use a row variable with r classes and a column variable with c classes to refer to the two classification criteria in a $r \times c$ contingency table.
- We use n_{ij} to denote the observed value in the ith class of the row variable and the jth class of the column variable; or the (i, j)th cell, for short. The step-by-step procedures are:

□ STEP 1.   Compute the row totals (R), column totals (C), and grand total (G). For our example, these totals are in Table 11.1.

□ STEP 2.   Compute the *expected value* of each of the $r \times c$ cells as:

$$E_{ij} = \frac{(R_i)(C_j)}{G}$$

where E_{ij} is the expected value of the (i, j)th cell, R_i is the total of the ith

Table 11.1   Frequencies of Farmers Classified According to Two Categories: Tenure Status and Adoption of New Rice Varieties

	Farmers, no.		
Tenure Status	Adopter	Nonadopter	Row Total (R)
Owner operator	102	26	128
Share-rent farmer	42	10	52
Fixed-rent farmer	4	3	7
Column total (C)	148	39	
Grand total (G)			187

row, C_j is the total of the jth column, and G is the grand total. For our example, the expected value of the first cell is computed as:

$$E_{11} = \frac{(R_1)(C_1)}{G} = \frac{(128)(148)}{187} = 101.3$$

The results for all six cells are shown in Table 11.2.

☐ STEP 3.   Compute the χ^2 value as:

$$\chi^2 = \sum_{i=1}^{r} \sum_{j=1}^{c} \frac{\left(n_{ij} - E_{ij}\right)^2}{E_{ij}}$$

where E_{ij} is as computed in step 2.

For our example, the χ^2 value is computed as:

$$\chi^2 = \frac{(102 - 101.3)^2}{101.3} + \frac{(42 - 41.2)^2}{41.2} + \cdots + \frac{(3 - 1.5)^2}{1.5}$$

$$= 2.01$$

☐ STEP 4.   Compare the computed χ^2 value with the tabular χ^2 values, from Appendix D, with $(r - 1)(c - 1)$ degrees of freedom; and reject the hy-

Table 11.2   The Expected Values for the Data in Table 11.1 under the Hypothesis of Independence

Tenure Status	Adopter	Nonadopter
Owner operator	101.3	26.7
Share-rent farmer	41.2	10.8
Fixed-rent farmer	5.5	1.5

pothesis of independence if the computed χ^2 value is larger than the corresponding tabular χ^2 value at the prescribed level of significance.

For our example, the tabular χ^2 values, from Appendix D, with $(r-1)$ $(c-1) = (2)(1) = 2$ degrees of freedom, are 5.99 at the 5% level of significance and 9.21 at the 1% level. Because the computed χ^2 value is smaller than the corresponding tabular χ^2 value at the 5% level of significance, the hypothesis of independence between the adoption of newly introduced high-yielding rice varieties and the tenure status of the farmers cannot be rejected.

In the foregoing procedure for testing independence between two classification criteria, there is no evaluation of the nature of the frequency distribution of each criterion. That is, the test for independence does not, and cannot, specify the type of frequency distribution that exists in each classification criterion. If this is desired, the test for a fixed-ratio hypothesis, discussed in Section 11.1.1, should be separately applied, depending upon the result of the independence test. That is, if the various classification criteria are judged independent, then the test for a fixed-ratio hypothesis—either on the row variable or column variable, or both—should be done on the marginal totals (i.e., row totals for the row variable and column totals for the column variable). However, if the two classification criteria are not independent, then the test for a fixed ratio hypothesis should be applied for the row variable separately for each column, and for the column variable separately for each row.

For our example, because the two classification criteria are judged to be independent, the chi-square test for a fixed ratio hypothesis should be done on the marginal totals. Assuming that the researcher is interested in testing a $2:1$ ratio for the adoption criterion (i.e., the column variable), the chi-square test for a fixed ratio hypothesis (Section 11.1.1), applied to the column totals, would be computed as:

$$\chi^2 = \frac{(|148 - 124.7| - 0.5)^2}{124.7} + \frac{(|39 - 62.3| - 0.5)^2}{62.3}$$

$$= 12.51$$

Because the computed χ^2 value is greater than the corresponding tabular χ^2 value at the 1% level of significance, the hypothesis that the frequency of adopters is twice that of the nonadopters is rejected at the 1% level of significance.

11.1.3 Test for Homogeneity of Ratio

Experiments whose primary character of interest is based on attribute data are commonly repeated several times. For example, the preference study of the green leafhoppers in rice (Section 11.1.1) could be repeated over time or could be repeated in several cages at the same time.

When data from several trials are available, it is essential to determine how to pool the information over all trials. For the leafhopper preference study, the questions may be: is it valid to simply add the numbers of insects on diseased plants and on healthy plants over all trials, and apply the chi-square procedure to test the hypothesized ratio of $1:1$ only once based on these totals? Or, should the chi-square test be applied to the result of each trial separately? And, if so, how would all the test results be interpreted and combined? The chi-square test for homogeneity of ratio is appropriate in answering these questions.

We illustrate the procedure for applying the chi-square test for homogeneity of ratio by using the data from the green leafhopper preference study involving four trials (Table 11.3). The step-by-step procedures are:

☐ STEP 1.   Apply the χ^2 test for a fixed-ratio hypothesis, as described in Section 11.1.1, to each trial separately. For our example, note that the χ^2 value for the first trial was computed in Section 11.1.1. The χ^2 value for the second, third, and fourth trials are computed in the same manner as that for the first. The results for all four trials are shown in the first four rows of Table 11.3.

☐ STEP 2.   Compute the sum of all the χ^2 values computed in step 1 as:

$$\chi_s^2 = \sum_{i=1}^{s} \chi_i^2$$

where χ_i^2 is the computed χ^2 value for the ith trial and s is the total number of trials involved. For our example, the sum over the four computed χ^2 values is:

$$\chi_s^2 = \chi_4^2 = 45.26 + 6.32 + 22.48 + 37.33 = 111.39$$

Table 11.3   Application of the Chi-Square Test for Homogeneity of 1:1 Ratio on the Number of Green Leafhoppers Found on Diseased and Healthy Rice Plants in Four Trials

Trial Number	Total Insects	Observed Values		Expected Values		χ^2 Value
		Healthy	Diseased	Healthy	Diseased	
1	239	67	172	119.5	119.5	45.26
2	183	74	109	91.5	91.5	6.32
3	171	54	117	85.5	85.5	22.48
4	301	97	204	150.5	150.5	37.33
Total	894	292	602	447.0	447.0	$106.80 = \chi_T^2$
Sum						$111.39 = \chi_s^2$

☐ STEP 3.   Compute the totals of observed values and of expected values over all trials, and apply the same χ^2 test procedure used in step 1 to these totals. For our example, the computed totals are shown at the bottom of Table 11.3; and the χ^2 value is computed, based on these totals as:

$$\chi_T^2 = \frac{(|292 - 447| - 0.5)^2}{447} + \frac{(|602 - 447| - 0.5)^2}{447}$$

$$= 106.80$$

☐ STEP 4.   Compute the chi-square value for additivity as the difference between the χ_s^2 value computed in step 2 and the χ_T^2 value computed in step 3:

$$\chi_d^2 = \chi_s^2 - \chi_T^2$$

For our example, the χ^2 value for additivity is computed as:

$$\chi_d^2 = 111.39 - 106.80 = 4.59$$

☐ STEP 5.   Compare the computed χ_d^2 value with the tabular χ^2 value, from Appendix D, with $(s - 1)$ degrees of freedom and at the prescribed level of significance, and take the following action depending upon the outcome:

- If the computed χ_d^2 value is smaller than, or equal to, the corresponding tabular χ^2 value, data from the total of s trials can be pooled and the χ_s^2 value used as the test criterion to test the hypothesis of the hypothesized ratio. That is, only the single χ_s^2 value needs to be compared to the tabular χ^2 value with s d.f., at the prescribed level of significance.
- If the computed χ_d^2 value is larger than the corresponding tabular χ^2 value, the indication is that data from the s trials are heterogeneous (i.e., the s data sets do not share a common ratio) and, hence, data from the s trials cannot be pooled. In such a case, the individual χ^2 values should be examined and the possible causes for the differences in the ratios between trials determined.

For our example, the tabular χ^2 values with $(s - 1) = (4 - 1) = 3$ d.f. are 7.81 at the 5% level of significance and 11.34 at the 1% level. Because the computed χ_d^2 value is smaller than the corresponding tabular χ^2 value at the 5% level of significance, data from the four trials can be pooled and the χ_4^2 value compared with the tabular χ^2 value with 4 d.f. Because the χ_4^2 value of 111.39 is larger than the corresponding tabular χ^2 value at the 1% level of significance of 13.28, the hypothesis of the 1 : 1 ratio is rejected. This result indicates that there were more green leafhoppers on diseased plants than on healthy plants and, hence, the hypothesis of no preference is rejected. In fact, the data seem to indicate that the ratio is about $2:1$ in favor of the diseased plants.

11.2   TEST FOR HOMOGENEITY OF VARIANCE

Most of the statistical procedures we have discussed are concerned with the treatment effects and with equality of treatment means as the most common hypothesis being tested. In this section, we deal with a statistical procedure for testing the equality (or homogeneity) of several variances. This is the chi-square test for homogeneity of variance, commonly known as the Bartlett's test. In agricultural research, this test is usually used to:

- Verify homogeneity of variances as a requirement for a valid analysis of variance (Chapter 7, Section 7.2).
- Verify homogeneity of error variances in combining data from a series of experiments (Chapter 8).
- Verify homogeneity of variances in a genetic study where the test materials consist of genotypes belonging to different filial generations.
- Verify homogeneity of sampling variances among samples taken from two or more populations.

The chi-square test for homogeneity of variances is applied whenever more than two variances are tested. The F test should be used when there are only two variances, with the F value computed as the ratio of the two variances—the larger variance in the numerator and the smaller variance in the denominator. This is well demonstrated through the standard F test in the analysis of variance, which is used to test the homogeneity of two mean squares—generally the treatment mean square and the error mean square.

We use two cases to illustrate the procedure for applying the chi-square test for homogeneity of variances. In one case, all the variances are estimated with the same (equal) degree of freedom; in the other case the variance estimates have unequal degrees of freedom.

11.2.1   Equal Degree of Freedom

We use data from an experiment with 11 rice varieties tested in three temperature regimes in a growth chamber, to illustrate the test for homogeneity of variances with equal degree of freedom. For each temperature, the varieties were grown in a randomized complete block design with three replications. In order to combine the data from the three trials (one corresponding to each temperature) homogeneity of error variances from the three individual RCB analyses of variance must be established. The data collected are plant height in centimeters, and the three error mean squares, each with 20 degrees of freedom, are:

$$s_1^2 = 11.459848$$

$$s_2^2 = 17.696970$$

$$s_3^2 = 10.106818$$

The step-by-step procedures to apply the chi-square test to test for homogeneity of k variances with equal $d.f.$ are:

☐ STEP 1.   For each variance estimate s^2, compute $\log s^2$, where log refers to logarithm base 10. Then, compute the totals of all k values of s^2 and of $\log s^2$.

For our example, the values of s^2, and $\log s^2$, for each of the three error mean squares and their totals are shown below:

Temperature	s^2	$\log s^2$
1	11.459848	1.059179
2	17.696970	1.247899
3	10.106818	1.004614
Total	39.263636	3.311692

☐ STEP 2.   Compute the pooled estimate of variance as:

$$s_p^2 = \frac{\sum_{i=1}^{k} s_i^2}{k} = \frac{39.263636}{3} = 13.087879$$

☐ STEP 3.   Let f be the degree of freedom of each s_i^2, compute the χ^2 value as:

$$\chi^2 = \frac{(2.3026)(f)\left(k \log s_p^2 - \sum_{i=1}^{k} \log s_i^2\right)}{1 + [(k+1)/3kf]}$$

For our example, the χ^2 value is computed as:

$$\chi^2 = \frac{(2.3026)(20)[(3)(\log 13.087879) - 3.311692]}{1 + [(3+1)/(3)(3)(20)]}$$

$$= 1.75$$

☐ STEP 4.   Compare the computed χ^2 value with the tabular χ^2 value from Appendix D with $(k-1)$ $d.f.$; and reject the hypothesis of homogeneous variance if the computed χ^2 value exceeds the corresponding tabular χ' value at the prescribed level of significance.

For our example, the computed χ^2 value is smaller than the corresponding tabular χ^2 value with $(k - 1) = (3 - 1) = 2$ $d.f.$ and at the 5% level of significance of 5.99. Thus, the hypothesis that the three error variances are homogeneous cannot be rejected.

11.2.2 Unequal Degrees of Freedom

The procedure for testing homogeneity of variances with unequal degrees of freedom is illustrated, using data on lesion length collected from rice leaves inoculated with four different isolates. For each isolate, 17 to 20 randomly selected lesions were measured. The researcher wished to determine whether there were differences in the lesion length between the different isolates. Before applying a test for mean comparison, the homogeneity of sampling variances between the isolates must first be established. Data on the four sampling variances and their corresponding degrees of freedom are shown in Table 11.4.

The step-by-step procedures for the application of the chi-square test for homogeneity of the four sampling variances with unequal degrees of freedom are outlined:

☐ STEP 1.   Let s_i^2 be the ith variance estimate $(i = 1,\ldots,k)$ with f_i degrees of freedom, compute the following parameters:

$$(f_i)(s_i^2)$$
$$(f_i)(\log s_i^2)$$
$$\frac{1}{f_i}$$

For our example, these parameters are computed and shown in Table 11.4.

Table 11.4   Computation of the Chi-Square Test for Homogeneity of Variances with Unequal Degrees of Freedom

Isolate	Sampling Variance (s^2)	d.f. (f)	$(f)(s^2)$	$\log s^2$	$(f)(\log s^2)$	$\frac{1}{f}$
1	6.73920	19	128.0448	0.828608	15.743552	0.0526
2	1.93496	16	30.9594	0.286672	4.586752	0.0625
3	1.15500	17	19.6350	0.062582	1.063894	0.0588
4	10.58450	19	201.1055	1.024670	19.468730	0.0526
Total		71	379.7447		40.862928	0.2265
Pooled	5.34852 (s_p^2)			0.728234	51.704614	

☐ STEP 2.   For each parameter computed in step 1, compute its total over k values:

$$A = \sum_{i=1}^{k} f_i = 19 + 16 + 17 + 19 = 71$$

$$B = \sum_{i=1}^{k} (f_i)(s_i^2)$$

$$= 128.0448 + 30.9594 + 19.6350 + 201.1055$$

$$= 379.7447$$

$$C = \sum_{i=1}^{k} (f_i)(\log s_i^2)$$

$$= 15.743552 + 4.586752 + 1.063894 + 19.468730$$

$$= 40.862928$$

$$D = \sum_{i=1}^{k} \frac{1}{f_i} = 0.0526 + 0.0625 + 0.0588 + 0.0526$$

$$= 0.2265$$

☐ STEP 3.   Compute the estimate of the pooled variance as:

$$s_p^2 = \frac{B}{A}$$

where the A and B values are as defined and computed in step 2. For our example, the pooled variance is estimated as:

$$s_p^2 = \frac{379.7447}{71} = 5.34852$$

☐ STEP 4.   Compute the χ^2 value as:

$$\chi^2 = \frac{2.3026\left[(A)(\log s_p^2) - C\right]}{1 + \dfrac{1}{3(k-1)}\left(D - \dfrac{1}{A}\right)}$$

where A, C, and D are as defined in step 2. For our example, the χ^2 value is computed as:

$$\chi^2 = \frac{2.3026\left[(71)(0.728234) - 40.862928\right]}{1 + \dfrac{1}{3(4-1)}\left(0.2265 - \dfrac{1}{71}\right)}$$

$$= \frac{24.964}{1.024} = 24.38$$

□ STEP 5.   Compare the computed χ^2 value of step 4 with the tabular χ^2 value from Appendix D with $(k - 1)$ $d.f.$, and reject the hypothesis of homogeneous variance if the computed χ^2 value exceeds the corresponding tabular χ^2 value at the prescribed level of significance.

For our example, the tabular χ^2 values with $(k - 1) = 3$ $d.f.$ are 7.81 at the 5% level of significance and 11.34 at the 1% level. Because the computed χ^2 value exceeds the corresponding tabular χ^2 value at the 1% level of significance, the hypothesis of homogeneous variance is rejected. That is, the test showed significant differences between the four sampling variances.

11.3   TEST FOR GOODNESS OF FIT

The test for goodness of fit determines whether a set of observed data conforms to a specified probability distribution. For example, data on rice yield or protein content may be suspected to follow a normal distribution.* The spatial distribution of weeds in a rice field or of insects caught in traps over time may be suspected to have Poisson distribution.[†]

In a study to determine the mode of inheritance of protein content of rice, the percentage of brown rice protein in the grains of F_3 plants derived from a cross between two rice varieties was measured. Visual examination of the data (Figure 11.1) seems to suggest a substantial deviation from the normal distribu-

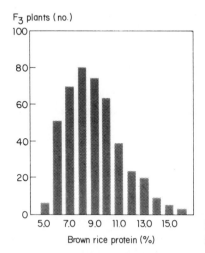

Figure 11.1 Frequency distribution of brown rice protein in 450 F_3 plants.

---

*The most important continuous distribution in agricultural research. It is bell-shaped, symmetric, and is governed by two parameters: the mean and the variance of the distribution.
†A discrete distribution which is widely used to represent occurrence of rare events over space or time. It is governed by a single parameter, which is both the mean and the variance of the distribution.

tion. The researcher wishes, therefore, to verify whether his visual observation can be substantiated by a statistical test. For this purpose, a chi-square test for goodness of fit of the observed data to the hypothesized normal distribution is most appropriate. We give the step-by-step procedures for applying the chi-square test for goodness of fit to the protein data in Figure 11.1.

☐ STEP 1.   Let *n* be the number of observations in the data set, construct the *frequency table* as follows:

A. Determine the range of the data as the difference between the largest observation and the smallest observation, and divide the range into *p* classes using the following guidelines:

- The number of classes (*p*) should be between 8 and 18 depending on the number of observations (*n*) and the range of variation. The recommended size of *p* for various ranges of *n* is:

Number of Observations (*n*)	Number of Classes (*p*)
20–100	8–12
101–500	10–15
501–1000	12–18

Note that the specific choice of *p* within each of the three specified ranges of *n* should depend upon the range of variation of the data. The wider the range, the larger *p* should be.

- There should be no class with zero entry and classes with less than three entries should be avoided, especially classes near the middle of the range. When classes with less than three entries are encountered at the middle of the range, reduce the number of classes. When they appear near either end of the range, two or three adjacent classes can be combined.
- To simplify computations, the class boundaries should be specified such that the midpoints of the class range (i.e., class values) are as close to whole numbers as possible.

For our example, with *n* = 450 and the data ranging from 4.6 to 16.3, we divide the range into *p* = 12 classes.

B. For each class, determine the class value (i.e., the midpoint of the class range) by taking the average of the lower and upper limits of the class. For our example, the class values of the 12 classes are shown in the third column of Table 11.5.

C. Determine the number of observations falling into each class and designate these as the observed frequencies (*f*). The sum of the observed frequencies over all classes should be equal to the total

Table 11.5   Frequency Distribution and Computation of Mean and Variance for Data on Brown Rice Protein of 450 F_3 Plants

Number	Limits	Value (X)	Observed Frequency (f)	(f)(X)	X²	(f)(X²)
	Class			Computation of Mean and Variance		
1	4.5–5.4	5.0	7	35	25	175
2	5.5–6.4	6.0	52	312	36	1872
3	6.5–7.4	7.0	70	490	49	3430
4	7.5–8.4	8.0	81	648	64	5184
5	8.5–9.4	9.0	74	666	81	5994
6	9.5–10.4	10.0	63	630	100	6300
7	10.5–11.4	11.0	39	429	121	4719
8	11.5–12.4	12.0	24	288	144	3456
9	12.5–13.4	13.0	21	273	169	3549
10	13.5–14.4	14.0	10	140	196	1960
11	14.5–15.4	15.0	6	90	225	1350
12	15.5–16.4	16.0	3	48	256	768
Total	—	—	450	4,049	—	38,757

number of observations. For our example, the observed frequency for each of the 12 classes is shown in the fourth column of Table 11.5. Note that their sum is equal to 450, which is the total number of F_3 plants in the study.

□ STEP 2.   Compute the mean ($\overline{X}$) and variance (s^2) from the frequency table derived in step 1:

$$\overline{X} = \frac{\sum_{i=1}^{p} (f_i)(X_i)}{\sum_{i=1}^{p} f_i}$$

$$s^2 = \frac{1}{\sum_{i=1}^{p} f_i - 1}\left\{ \sum_{i=1}^{p} (f_i)(X_i^2) - \frac{\left[\sum_{i=1}^{p} (f_i)(X_i)\right]^2}{\sum_{i=1}^{p} f_i}\right\}$$

where X_i is the class value of the ith class, f_i is the observed frequency of the ith class, and p is the total number of classes.

For our example, with the required basic computation given in the fifth through seventh column of Table 11.5, the mean and variance are computed as:

$$\overline{X} = \frac{4,049}{450} = 8.998$$

$$s^2 = \frac{1}{(450 - 1)} \left[38,757 - \frac{(4,049)^2}{450} \right] = 5.178$$

☐ STEP 3. Compute the *expected frequency* of each class based on the hypothesized probability distribution:

A. For each class, compute two standardized Z values, one for the lower limit (Z_l) and another for the higher limit (Z_h):

$$Z_l = \frac{L_l - \overline{X}}{s}$$

$$Z_h = \frac{L_h - \overline{X}}{s}$$

where L_l and L_h are the lower and upper *true class limits* of each class. The lower true class limit of a given class is defined as the average of its own lower class limit and the upper class limit of the previous class. Similarly, the upper true class limit is defined as the average of its own upper class limit and the lower class limit of the succeeding class. The lower true class limit of the first class is $-\infty$ and the upper true class limit of the last class is $+\infty$.

For our example, the lower true class limit (L_l) and the upper true class limit (L_h) of the second class in our example are:

$$L_l = \frac{5.4 + 5.5}{2} = 5.45$$

$$L_h = \frac{6.4 + 6.5}{2} = 6.45$$

and the corresponding Z_l value and Z_h value are computed as:

$$Z_l = \frac{5.45 - 8.998}{2.2755} = -1.559$$

$$Z_h = \frac{6.45 - 8.998}{2.2755} = -1.120$$

The standardized Z values for all 12 classes are shown in the third and fourth columns of Table 11.6.

Table 11.6 Computation of the Expected Frequency and the Chi-Square Test for Goodness of Fit to a Normal Distribution, from Data in Table 11.5

Class Number	Observed Frequency (f_i)	Standardized Z Values Z_l	Standardized Z Values Z_h	Probability (P_i)	Expected Frequency $F_i = (n)(P_i)$	$\dfrac{(f_i - F_i)^2}{F_i}$
1	7	$-\infty$	-1.559	$.5000 - .4405 = .0595$	26.775	14.605
2	52	-1.559	-1.120	$.4405 - .3686 = .0719$	32.355	11.928
3	70	-1.120	-0.680	$.3686 - .2517 = .1169$	52.605	5.752
4	81	-0.680	-0.241	$.2517 - .0952 = .1565$	70.425	1.588
5	74	-0.241	0.199	$.0952 + .0789 = .1741$	78.345	0.241
6	63	0.199	0.638	$.2383 - .0789 = .1594$	71.730	1.062
7	39	0.638	1.078	$.3595 - .2383 = .1212$	54.540	4.428
8	24	1.078	1.517	$.4353 - .3595 = .0758$	34.110	2.997
9	21	1.517	1.956	$.4748 - .4353 = .0395$	17.775	0.585
10	10	1.956	2.396	$.4917 - .4748 = .0169$	7.605	0.754
11	6	2.396	2.835	$.4977 - .4917 = .0060$	2.700	4.033
12	3	2.835	∞	$.5000 - .4977 = .0023$	1.035	3.731
Total	$450 = n$				450.000	$\chi^2 = 51.704$

B. Determine the probability associated with each class interval, based on the hypothesized probability distribution, as:

$$P = P(Z_l < X < Z_h)$$

where the term $P(Z_l < X < Z_h)$ refers to the probability that X lies between Z_l and Z_h.

For our example, because the normal distribution is hypothesized, the probability associated with each class is determined from Appendix B by reading the area under the standardized normal curve between Z_l and Z_h. This is done with the use of the following rules:

- The accuracy of the Z values given in Appendix B is only to two decimals. Hence, when the Z values have three or more decimals, such as in our example, linear interpolation is needed to determine the area desired.

    For example, to determine the area from 0 to $(Z = 1.559)$, we first obtain from Appendix B the area from 0 to 1.55 of .4394 and the area from 0 to 1.56 of .4406; and, through linear interpolation, compute the area from 0 to 1.559 as:

$$.4394 + \frac{(.4406 - .4394)(1.559 - 1.55)}{(1.56 - 1.55)} = .4405$$

- The area between any two Z values carrying the same sign, say, Z_l and Z_h (or $-Z_l$ and $-Z_h$), is equal to the difference between the areas from 0 to Z_l and from 0 to Z_h. For example, to compute the area between the two Z values of class 2 (Table 11.6) namely, $Z_l = -1.559$ and $Z_h = -1.120$, first determine the area from 0 to 1.559 of .4405 (computed above) and that from 0 to 1.120 of .3686 (from Appendix B) and compute their difference as:

$$P(-1.559 < X < -1.120) = P(1.559 < X < 1.120)$$

$$= .4405 - .3686$$

$$= .0719$$

- The area between any two Z values carrying different signs $(-Z_l$ and Z_h or Z_l and $-Z_h)$ is equal to the sum of the area from 0 to Z_l and the area from 0 to Z_h. For example, to compute the area between the two Z values of class 5 (Table 11.6), namely, $Z_l = -0.241$ and $Z_h = 0.199$, the areas from 0 to 0.241 and from 0 to 0.199 are first determined through linear interpolation as .0952 and .0789. Then, the area between $Z_l = -0.241$ and $Z_h = 0.199$ is computed as:

$$P(-0.241 < X < 0.199) = .0952 + .0789$$

$$= .1741$$

- The area from $-\infty$ to Z, or from Z to $+\infty$, is computed as the difference between .5 and the area from 0 to Z. For our example, to compute the area from $-\infty$ to $(Z_h = -1.559)$ of the first class, the area from 0 to 1.559 is first determined as .4405 and the area from $-\infty$ to -1.559 is then computed as $.5 - .4405 = .0595$.

The computation of probabilities for all the 12 classes is shown in the fifth column of Table 11.6.

C. Compute the expected frequency for the ith class (F_i) as the product of the probability associated with the ith class (P_i), determined in the previous step, and the total number of observations, (n):

$$F_i = (n)(P_i)$$

For our example, the expected frequency for each of the 12 classes is computed and shown in the sixth column of Table 11.6.

□ STEP 4. Compute the χ^2 value as:

$$\chi^2 = \frac{\sum\limits_{i=1}^{p} (f_i - F_i)^2}{F_i}$$

where f_i is the observed frequency and F_i is the expected frequency, for the ith class, as defined in steps 1 and 3. For our example, the computation of the χ^2 value is shown in the last column of Table 11.6. The computed χ^2 is 51.70.

□ STEP 5.   Compare the computed χ^2 value with the tabular χ^2 values, from Appendix D, with ($p - 3$) degrees of freedom, and reject the hypothesized probability distribution if the computed χ^2 value exceeds the corresponding tabular χ^2 value at the prescribed level of significance. For our example, the tabular χ^2 value with 9 *d.f.* is 21.67 at the 1% level of significance. Because the computed χ^2 value is greater than this tabular χ^2 value, the test indicates that the data set does not fit the hypothesized normal probability distribution.

CHAPTER 12

Soil Heterogeneity

Adjacent plots, planted simultaneously to the same variety and treated as alike as possible, will differ in as many characters as one would care to measure quantitatively. The causes for these differences are numerous but the most obvious, and probably the most important, is soil heterogeneity.

Experience has shown that it is almost impossible to get an experimental site that is totally homogeneous. This chapter deals with soil heterogeneity, and the techniques to cope with it.

12.1 CHOOSING A GOOD EXPERIMENTAL SITE

To choose an experimental site that has minimum soil heterogeneity, a researcher must be able to identify the features that magnify soil differences.

12.1.1 Slopes

Fertility gradients are generally most pronounced in sloping areas, with lower portions more fertile than high areas. This is because soil nutrients are soluble in water and tend to settle in lower areas. An ideal experimental site, therefore, is one that has no slope. If a level area is not available, an area with a uniform and gentle slope is preferred because such areas generally have predictable fertility gradients, which can be managed through the use of proper blocking (see Chapter 2, Section 2.2.1).

12.1.2 Areas Used for Experiments in Previous Croppings

Different treatments used in experimental planting usually increase soil heterogeneity. Thus, areas previously planted to different crops, fertilized at different levels, or subjected to varying cultural managements should be avoided, if possible. Otherwise, such areas should be planted to a uniform variety and fertilized heavily and uniformly for at least one season before conducting an experiment.

478

Another source of soil heterogeneity is the presence of nonplanted alleys, which are common in field experiments. Plants grown in previously nonplanted areas tend to perform better. Nonplanted areas should be marked so that the same areas are left as alleys in succeeding plantings.

12.1.3 Graded Areas

Grading usually removes top soil from elevated areas and dumps it in the lower areas of a site. This operation, while reducing the slope, results in an uneven depth of surface soil and at times exposes infertile subsoils. These differences persist for a long time. Thus, an area that has had any kind of soil movement should be avoided. If this is not possible, it is advisable to conduct a uniformity trial to assess the pattern of soil heterogeneity (Section 12.2.1) so that a suitable remedy can be achieved by proper blocking or by appropriate adjustment through use of the covariance technique (see Chapter 10).

12.1.4 Presence of Large Trees, Poles, and Structures

Avoid areas surrounding permanent structures. Such areas are usually undependable because the shade of the structures, and probably some soil movement during their construction, could contribute to poor crop performance.

12.1.5 Unproductive Site

A productive crop is an important prerequisite to a successful experiment. Thus, an area with poor soil should not be used, unless the experiment is set up specifically to evaluate such conditions.

12.2 MEASURING SOIL HETEROGENEITY

An adequate characterization of soil heterogeneity in an experimental site is a good guide, and at times even a prerequisite, to choosing a good experimentation technique. Based on the premise that uniform soil when cropped uniformly will produce a uniform crop, soil heterogeneity can be measured as the difference in performance of plants grown in a uniformly treated area.

12.2.1 Uniformity Trials

Uniformity trial involves planting an experimental site with a single crop variety and applying all cultural and management practices as uniformly as possible. All sources of variability, except that due to native soil differences, are kept constant. The planted area is subdivided into small units of the same size (generally referred to as *basic units*) from which separate measurements of productivity, such as grain yield, are made. Yield differences between these basic units are taken as a measure of the area's soil heterogeneity.

The size of the basic unit is governed mostly by available resources. The smaller the basic unit, the more detailed is the measurement of soil heterogeneity.

We illustrate this with a uniformity trial on rice. An area 20×38 m was planted to rice variety IR8 using a 20×20-cm spacing. No fertilizer was applied and management was kept as uniform as possible. At harvest, an area 1 meter wide on all four sides of the field was discarded as borders, leaving an effective area of 18×36 m, from which yield measurements were made separately from each basic unit of 1×1 m. Grains from each of the 648 basic units were harvested, bagged, threshed, cleaned, dried, and weighed separately. Grain yield data are in Table 12.1.

Several types of analyses are available to evaluate the pattern of soil heterogeneity based on uniformity trials. We discuss four procedures.

12.2.1.1 Soil Productivity Contour Map.

The soil productivity contour map is a simple but informative presentation of soil heterogeneity. The map describes graphically the productivity level of the experimental site based on moving averages of contiguous units.

The steps involved in constructing a fertility contour map are:

☐ STEP 1.   Decide on the number of contiguous units to go into each moving average. The object of combining several basic units into each moving average is to reduce the large random variation expected on small plots. Experience has shown that moving averages of contiguous basic units describe productivity better than the original basic units. The area involved in each moving average should be as square as possible. For this example, a 3×3 unit area is chosen as the size of the moving average.

☐ STEP 2.   Compute the 3×3 moving average, corresponding to the sth row and the tth column, as:

$$P_{s,t} = \frac{\sum\limits_{i=s-1}^{s+1} \sum\limits_{j=t-1}^{t+1} Y_{ij}}{9}$$

where $s = 2, \ldots, (r - 1)$ and $t = 2, \ldots, (c - 1)$; Y_{ij} represents the grain yield data of the (i, j)th basic unit (i.e., from the ith row and the jth column) in Table 12.1; r is the number of rows; and c is the number of

able 12.1 Grain Yield (g / m^2) of Rice Variety IR8 from a Rice Uniformity Test covering an Area 18 × 26 m

Row	1	2	3	4	5	6	7	8	9	10	11	12	13	14	15	16	17	18
							Column											
1	842	844	808	822	979	954	965	906	898	856	808	920	808	889	943	894	968	917
2	803	841	870	970	943	914	916	836	858	926	922	910	872	805	775	846	947	965
3	773	782	860	822	932	971	765	875	853	936	927	779	865	720	566	893	914	861
4	912	887	815	937	844	661	841	844	809	778	945	876	901	802	836	778	923	949
5	874	792	803	793	818	799	767	855	792	858	912	839	813	740	730	832	813	914
6	908	875	899	788	867	790	831	757	751	774	863	902	771	747	819	699	670	934
7	875	907	921	963	875	880	898	802	874	928	872	834	892	760	753	720	751	894
8	891	928	871	875	865	777	738	796	855	901	792	752	722	781	739	733	783	786
9	823	784	754	873	764	775	752	753	820	798	847	858	811	875	659	661	759	767
10	785	794	764	822	714	748	724	717	736	724	838	769	819	823	724	750	764	764
11	785	808	823	826	801	712	826	665	759	738	867	725	794	755	730	638	724	734
12	829	895	774	891	841	815	834	778	760	822	803	754	703	743	728	692	748	671
13	861	883	739	762	725	717	746	766	662	634	743	719	710	682	694	675	709	720
14	906	885	790	655	690	769	765	719	743	770	728	740	691	767	648	715	655	665
15	819	911	788	654	742	786	791	779	645	810	816	746	729	814	718	721	708	722
16	893	862	769	727	725	721	739	736	672	814	756	748	714	718	694	704	915	705
17	813	750	742	872	746	812	705	724	640	757	708	750	767	638	754	767	763	685
18	816	758	811	702	728	741	757	732	623	786	805	786	739	727	767	738	659	695
19	676	783	734	626	782	704	782	707	672	703	698	758	762	625	623	699	662	613
20	813	809	695	707	753	680	720	683	757	782	789	811	789	769	751	648	680	696
21	801	764	701	716	753	680	706	665	680	650	690	699	768	751	701	665	603	680
22	718	784	730	750	733	705	728	667	703	684	777	747	713	696	717	732	712	679
23	756	725	821	685	681	738	630	599	629	703	780	720	709	697	731	661	627	644
24	789	681	732	669	681	698	689	622	672	704	705	625	677	704	648	605	585	651
25	652	622	695	677	698	666	691	688	682	713	670	708	707	695	681	716	626	637
26	729	650	700	764	680	681	645	622	661	728	715	775	690	726	669	766	709	645
27	698	713	714	734	651	649	675	614	634	635	639	690	694	637	590	640	658	609
28	745	677	685	711	688	614	585	534	533	671	600	647	592	595	563	634	666	644
29	964	727	648	664	623	629	616	594	619	631	628	591	675	654	640	718	667	649
30	671	729	690	687	705	622	523	526	661	683	619	709	620	651	676	728	547	682
31	717	694	727	719	669	630	701	645	638	714	633	670	649	665	557	734	674	727
32	652	713	656	584	517	572	574	539	545	629	636	580	607	654	585	674	608	612
33	605	708	684	715	659	629	632	596	627	644	661	682	690	636	665	731	753	640
34	559	722	726	705	571	637	637	577	561	590	646	639	672	636	651	684	584	622
35	589	681	690	570	619	624	580	570	568	589	550	622	623	706	725	738	669	636
36	614	633	619	658	678	673	652	602	590	605	538	682	651	653	680	696	633	660

columns. For example:

$$P_{2,2} = \frac{(Y_{11} + Y_{12} + Y_{13}) + (Y_{21} + Y_{22} + Y_{23}) + (Y_{31} + Y_{32} + Y_{33})}{9}$$

$$= \frac{(842 + 844 + 808) + (803 + 841 + 870) + (773 + 782 + 860)}{9}$$

$$= 825$$

$$P_{2,3} = \frac{(Y_{12} + Y_{13} + Y_{14}) + (Y_{22} + Y_{23} + Y_{24}) + (Y_{32} + Y_{33} + Y_{34})}{9}$$

$$= \frac{(844 + 808 + 822) + (841 + 870 + 970) + (782 + 860 + 822)}{9}$$

$$= 847$$

$$\vdots$$

$$P_{2,17} = \frac{(894 + 968 + 917) + (846 + 947 + 965) + (893 + 914 + 861)}{9}$$

$$= 912$$

$$P_{3,2} = \frac{(803 + 841 + 870) + (773 + 782 + 860) + (912 + 887 + 815)}{9}$$

$$= 838$$

$$P_{3,3} = \frac{(841 + 870 + 970) + (782 + 860 + 822) + (887 + 815 + 937)}{9}$$

$$= 865$$

$$\vdots$$

$$P_{35,17} = \frac{(684 + 584 + 622) + (738 + 669 + 636) + (696 + 633 + 660)}{9}$$

$$= 658$$

The result of the computation of all moving averages is shown in Table 12.2. Note that the dimension of Table 12.2 is 34 × 16, which is two rows and two columns less than that of Table 12.1.

☐ STEP 3. Group the values of the moving averages in Table 12.2 into m classes, where m should be between five and eight. For our example, m is chosen to be six. The six classes are generated by dividing the range of the 544 moving averages (with values from 577 to 927) into six equal intervals as:

Class 1: 577–636
Class 2: 637–694
Class 3: 695–752
Class 4: 753–810
Class 5: 811–869
Class 6: 870–927

Table 12.2 Moving Averages Based on 3 × 3 Basic Units, Computed from the Uniformity Trial Data in Table 12.1

Row	2	3	4	5	6	7	8	9	10	11	12	13	14	15	16	17
								Column								
2	825	847	890	923	927	900	875	883	887	887	868	841	805	815	861	912
3	838	865	888	888	865	847	844	857	884	889	889	837	794	780	831	897
4	833	832	847	842	822	820	822	844	868	872	873	815	775	766	809	875
5	863	843	840	811	802	794	805	802	831	861	869	821	795	776	789	835
6	873	860	859	841	836	820	814	821	847	865	855	811	781	756	754	803
7	897	892	880	853	836	808	811	826	846	846	822	796	776	750	741	774
8	862	875	862	850	814	797	810	836	854	842	820	809	777	742	729	762
9	822	829	811	801	762	753	766	789	812	809	801	801	773	749	730	752
10	791	805	793	782	757	741	750	746	792	796	814	803	777	735	712	729
11	806	822	806	797	779	758	755	744	783	782	786	765	758	731	722	721
12	822	822	798	788	780	762	755	732	754	756	758	732	727	704	704	701
13	840	808	763	763	767	768	753	739	741	746	732	723	707	705	696	694
14	842	785	727	722	748	760	735	725	728	745	736	733	717	715	694	699
15	847	782	727	719	748	756	732	743	750	770	741	741	721	722	720	723
16	816	786	752	754	752	755	715	731	735	767	748	736	727	725	749	743
17	802	777	758	753	742	741	703	720	729	768	753	732	724	723	751	737
18	765	753	749	746	751	740	705	705	710	750	753	728	711	704	715	698
19	766	736	726	714	739	723	715	716	735	769	771	752	728	705	692	677
20	753	726	719	711	729	703	708	700	713	731	752	748	727	692	670	661
21	757	740	726	720	718	693	701	697	724	737	754	749	739	714	690	677
22	756	742	730	716	706	680	667	664	700	717	734	722	720	706	683	667
23	748	731	720	704	698	675	660	665	706	716	717	699	699	688	669	655
24	719	701	704	688	686	669	656	668	695	703	700	694	694	682	653	639
25	694	688	700	690	681	667	664	677	694	705	697	701	689	690	667	660
26	686	697	701	689	671	659	657	664	675	697	699	702	677	680	673	667
27	701	705	703	686	652	624	611	626	646	678	671	672	640	647	655	663
28	730	697	680	663	637	612	600	607	621	637	640	642	627	630	642	654
29	726	691	678	660	623	583	577	606	627	642	631	637	630	651	649	659
30	730	698	681	661	635	610	614	635	647	653	644	654	643	669	660	681
31	694	689	662	634	613	592	595	620	640	653	636	645	629	658	643	665
32	684	689	659	633	620	613	611	620	636	650	645	648	634	656	665	684
33	669	690	646	621	603	599	588	590	615	634	646	644	644	657	659	656
34	663	689	660	637	621	609	594	591	604	625	643	656	667	686	689	673
35	648	667	648	637	630	617	593	584	582	607	625	654	666	685	673	658

☐ STEP 4. Assign a shading pattern to each of the six classes as shown in Figure 12.1. The darkest shade should be assigned to the class with the highest yield, the second darkest shade to the class with the second highest yield, and the lightest shade is assigned to the class with the lowest yield. This allows for an easier visualization of the fertility pattern.

☐ STEP 5. Draw the productivity map as shown in Figure 12.1. The top left corner of the figure corresponds to row 2 and column 2, and the bottom left

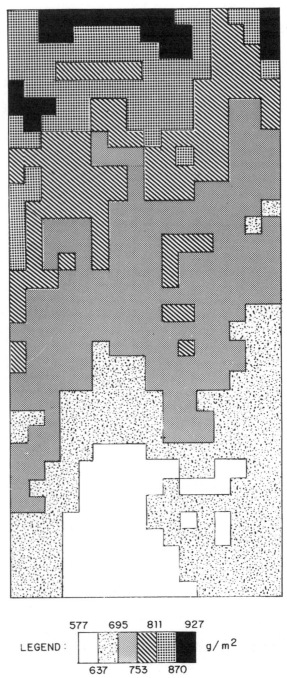

Figure 12.1 Fertility contour map of a field measuring 34 × 16 basic units, constructed from moving averages of 3 × 3 basic units in Table 12.2.

corner corresponds to row 35 and column 2, of Table 12.2. The map shows a unidirectional fertility gradient. There is a gradual change of productivity from one side of the field to another with yields in the upper portion higher than in the lower portion.

12.2.1.2 Serial Correlation. Serial correlation procedure is generally used to test the randomness of a data set. However, it is also useful in the characterization of the trend in soil fertility using uniformity trial data. The formula for computing a serial correlation coefficient of n observations $(X_1, X_2, \ldots, X_n)$ is:

$$r_s = \frac{\sum_{i=1}^{n} X_i X_{i+1} - \dfrac{\left(\sum_{i=1}^{n} X_i\right)^2}{n}}{\sum_{i=1}^{n} X_i^2 - \dfrac{\left(\sum_{i=1}^{n} X_i\right)^2}{n}}$$

where $X_{n+1} = X_1$. A serial correlation can be viewed simply as a simple correlation between two variables, one at site i and another at site $(i + 1)$. A low serial correlation coefficient indicates that fertile areas occur in spots, and a high value indicates a fertility gradient.

Two serial correlation coefficients—one for the horizontal and another for the vertical arrangement—can be computed from one set of uniformity trial data. Using the uniformity data in Table 12.1, the procedure for computing serial correlation is:

☐ STEP 1. Tabulate, separately for vertical and horizontal arrangement, the data of Table 12.1 in pairs of X_i and X_{i+1} as shown in Table 12.3. The total number of pairs for each arrangement equals the total number of observations ($r \times c = 36 \times 18 = 648$).

For example, in the vertical arrangement, the first pair is composed of the first two observations in the first column, namely, $X_i = Y_{11} = 842$ and $X_{i+1} = Y_{21} = 803$; the second pair is composed of the second and third observations in the first column, namely, $X_i = Y_{21} = 803$ and $X_{i+1} = Y_{31} = 773$; and so on. The last observation of the first column is then paired with the last observation of the second column, that is, the thirty-sixth pair is composed of $X_i = Y_{36,1} = 614$ and $X_{i+1} = Y_{36,2} = 633$. However, the first observation of the second column is paired with the first observation of the third column, that is, the seventy-second pair is composed of $X_i = Y_{12} = 844$ and $X_{i+1} = Y_{13} = 808$, and so on.

In the horizontal arrangement, the first pair is composed of the first two observations in the first row, namely, $X_i = Y_{11} = 842$ and $X_{i+1} = Y_{12} = 844$; the second pair is composed of the second and third observations in the first row, namely, $X_i = Y_{12} = 844$ and $X_{i+1} = Y_{13} = 808$; and so on. The last observation of the first row is then paired with the last observation of the second row, that is, the eighteenth pair is composed of $X_i = Y_{1,18} = 917$ and $X_{i+1} = Y_{2,18} = 965$. However, the first observation of the second row is

Table 12.3   Data Tabulation to Facilitate the Computation of Two Serial Correlation Coefficients, One for Vertical and Another for Horizontal Arrangement, Based on Uniformity Data in Table 12.1

	Vertical			Horizontal	
Pair (i)	X_i	X_{i+1}	Pair (i)	X_i	X_{i+1}
1	842	803	1	842	844
2	803	773	2	844	808
3	773	912	3	808	822
⋮	⋮	⋮	⋮	⋮	⋮
35	589	614	17	968	917
36	614	633	18	917	965
37	633	681	19	965	947
⋮	⋮	⋮	⋮	⋮	⋮
646	861	965	646	619	633
647	965	917	647	633	614
648	917	842	648	614	842

paired with the first observation of the third row, that is, the thirty-sixth pair is composed of $X_i = Y_{21} = 803$ and $X_{i+1} = Y_{31} = 773$, and so on.

☐ STEP 2.   Using the formula, compute the serial correlation coefficients for vertical and horizontal arrangement as:

$$\text{Vertical } r_s = \frac{\left[(842)(803) + (803)(773) + \cdots + (917)(842)\right] - \dfrac{G^2}{648}}{\left[(842)^2 + (803)^2 + (773)^2 + \cdots + (917)^2\right] - \dfrac{G^2}{648}}$$

$$= 0.7324$$

$$\text{Horizontal } r_s = \frac{\left[(842)(844) + (844)(808) + \cdots + (614)(842)\right] - \dfrac{G^2}{648}}{\left[(842)^2 + (844)^2 + (808)^2 + \cdots + (614)^2\right] - \dfrac{G^2}{648}}$$

$$= 0.7427$$

where G is the grand total over all observations in Table 12.1 (i.e., $G = \sum_{i+1}^{r} \sum_{j=1}^{c} Y_{ij} = 475,277$).

Both coefficients are high, indicating the presence of fertility gradients in both directions. The relative magnitude of the two serial correlations should not, however, be used to indicate the relative degree of the gradients in the two directions. For instance, the contour map (Figure 12.1) indicates that the gradient was more pronounced horizontally than vertically, even though the two serial correlation coefficients are of the same size.

12.2.1.3 Mean Square Between Strips. Mean square between strips has an objective similar to that of the serial correlation but is simpler to compute. Instead of working with the original basic units, the units are first combined into horizontal and vertical strips. Variability between the strips in each direction is then measured by the mean square between strips. The relative size of the two *MS*, one for horizontal strips and another for vertical strips, indicates the possible direction of the fertility gradient and the suitable orientation for both plots and blocks.

Using the data in Table 12.1, the steps in the computation of the *MS* between strips are:

☐ STEP 1. For each of the $c = 18$ columns (i.e., vertical strips) compute the sum of the $r = 36$ values, one corresponding to each row. For example, the sum of the first column is computed as:

$$\sum_{i=1}^{36} Y_{ij} = 842 + 803 + 773 + \cdots + 589 + 614 = 27,956$$

The totals for the 18 vertical strips are:

27,956	28,001	27,248	27,095	26,740	26,173
26,126	25,090	25,182	26,668	26,926	26,762
26,409	25,936	25,130	25,925	25,836	26,074

And, the grand total is 475,277.

☐ STEP 2. For each of the $r = 36$ rows (i.e., horizontal strips) compute the sum of the $c = 18$ values, one corresponding to each column. For example, the sum of the first row is computed as:

$$\sum_{i=1}^{18} Y_{ij} = 842 + 844 + 808 + \cdots + 968 + 917 = 16,021$$

The totals for the 36 horizontal strips are:

16,021	15,919	15,094	15,338	14,744	14,645
15,399	14,585	14,133	13,779	13,710	14,081
13,147	13,301	13,699	13,612	13,393	13,370
12,609	13,332	12,673	12,975	12,536	12,137
12,224	12,555	11,874	11,384	11,937	11,729
12,163	10,937	11,957	11,419	11,349	11,517

□ STEP 3.   Denote the total of the ith vertical strip by V_i, the total of the jth horizontal strip by H_j, and the grand total by G; compute the vertical-strip SS and horizontal-strip SS as:

$$\text{Vertical-strip } SS = \frac{\sum_{i=1}^{c} V_i^2}{r} - \frac{G^2}{rc}$$

$$= \frac{(27.956)^2 + (28.001)^2 + \cdots + (26.074)^2}{36} - \frac{(475,277)^2}{(36)(18)}$$

$$= 349,605$$

$$\text{Horizontal-strip } SS = \frac{\sum_{j=1}^{r} H_j^2}{c} - \frac{G^2}{rc}$$

$$= \frac{(16,021)^2 + (15,919)^2 + \cdots + (11,517)^2}{18} - \frac{(475,277)^2}{(36)(18)}$$

$$= 3,727,595$$

□ STEP 4.   Compute the vertical-strip MS and horizontal-strip MS as:

$$\text{Vertical-strip } MS = \frac{\text{Vertical-strip } SS}{c - 1}$$

$$= \frac{349,605}{18 - 1} = 20,565$$

$$\text{Horizontal-strip } MS = \frac{\text{Horizontal-strip } SS}{r - 1}$$

$$= \frac{3,727,595}{36 - 1} = 106,503$$

Results show that the horizontal-strip MS is more than five times higher than the vertical-strip MS, indicating that the trend of soil fertility was more pronounced along the length than along the width of the field. This trend is confirmed through the visual examination of the contour map (Figure 12.1).

12.2.1.4   Smith's Index of Soil Heterogeneity.

Smith's index of soil heterogeneity is used primarily to derive optimum plot size. The index gives a single value as a quantitative measure of soil heterogeneity in an area. The value of the index indicates the degree of correlation between adjacent experimental plots. Its value varies between unity and zero. The larger the value of the index, the lower is the correlation between adjacent plots, indicating that fertile spots are distributed randomly or in patches.

Smith's index of soil heterogeneity is obtained from the empirical relationship between plot variance and plot size:

$$V_x = \frac{V_1}{x^b}$$

where V_1 is the variance between basic units, V_x is the variance per unit area for plot size of x basic units, and b is the Smith's index of soil heterogeneity. The steps involved in the computation of the index b are:

☐ STEP 1.   Combine the $r \times c$ basic units to simulate plots of different sizes and shapes. Use only the combinations that fit exactly into the whole area, that is, the product of the number of simulated plots and the number of basic units per plot must equal the total number of basic units.

For our example, the simulated plots of different sizes and shapes are shown in Table 12.4. The simulated plot sizes range from 1 to 27 m². Note that only combinations that fit exactly into the whole area are included. For example, for plot size of 12 m² with the total number of plots of 54, its product is 12 × 54 = 648. On the other hand, a simulated plot of size 15 m² is not included because the maximum number of plots that can be derived is 42, and 15 × 42 is not equal to 648.

☐ STEP 2.   For each of the simulated plots constructed in step 1, compute the yield total T as the sum of the x basic units combined to construct that plot, and compute the between-plot variance $V_{(x)}$ as:

$$V_{(x)} = \frac{\sum\limits_{i=1}^{w} T_i^2}{x} - \frac{G^2}{rc}$$

where $w = rc/x$ is the total number of simulated plots of size x basic units.

For example, the between-plot variance for the plot of the same size as the basic unit (i.e., 1 × 1 m) is computed as:

$$V_{(1)} = V_1 = \sum_{i=1}^{r} \sum_{j=1}^{c} Y_{ij}^2 - \frac{G^2}{rc}$$

$$= (842)^2 + (844)^2 + \cdots + (660)^2 - \frac{(475,277)^2}{(36)(18)}$$

$$= 9,041$$

The between-plot variance for the plot of size 2×1 m is computed as:

$$V_{(2)} = \sum_{i=1}^{324} \frac{T_i^2}{2} - \frac{G^2}{(36)(18)}$$

$$= \frac{(1{,}645)^2 + (1{,}685)^2 + \cdots + (1{,}296)^2}{2} - \frac{(475{,}277)^2}{(36)(18)}$$

$$= 31{,}370$$

Table 12.4   Between-Plot Variance [$V_{(x)}$], Variance per Unit Area (V_x), and Coefficient of Variability (cv) for Plots of Various Sizes and Shapes, Calculated from Rice Uniformity Data in Table 12.1

Plot Size and Shape						
Size,[a] m²	Width,[b] m	Length,[c] m	Plots, no.	$V_{(x)}$	V_x	cv, %
1	1	1	648	9,041	9,041	13.0
2	2	1	324	31,370	7,842	12.1
3	3	1	216	66,396	7,377	11.7
6	6	1	108	235,112	6,531	11.0
9	9	1	72	494,497	6,105	10.7
2	1	2	324	31,309	7,827	12.1
4	2	2	162	114,515	7,157	11.5
6	3	2	108	247,140	6,865	11.3
12	6	2	54	908,174	6,307	10.8
18	9	2	36	1,928,177	5,951	10.5
3	1	3	216	66,330	7,370	11.7
6	2	3	108	247,657	6,879	11.3
9	3	3	72	537,201	6,632	11.1
18	6	3	36	1,981,408	6,115	10.7
27	9	3	24	4,231,622	5,805	10.4
4	1	4	162	113,272	7,080	11.5
8	2	4	81	427,709	6,683	11.1
12	3	4	54	943,047	6,549	11.0
24	6	4	27	3,526,179	6,121	10.7
36	9	4	18	7,586,647	5,854	10.4
6	1	6	108	238,384	6,622	11.1
12	2	6	54	913,966	6,347	10.9
18	3	6	36	2,021,308	6,239	10.8
36	6	6	18	7,757,823	5,986	10.5
9	1	9	72	514,710	6,354	10.9
18	2	9	36	2,017,537	6,227	10.8
27	3	9	24	4,513,900	6,192	10.7

[a] Number of basic units combined.
[b] Number of rows in Table 12.1.
[c] Number of columns in Table 12.1.

The results of the between-plot variances for all plot sizes and shapes constructed in step 1 are shown in the fifth column of Table 12.4.

☐ STEP 3. For each plot size and shape, compute the variance per unit area as:

$$V_x = \frac{V_{(x)}}{x^2}$$

For example, the variance per unit area for the plot of size 2×1 m is computed as:

$$V_2 = \frac{V_{(2)}}{2^2} = \frac{31,370}{4} = 7,842$$

The results for all plot sizes and shapes are shown in the sixth column of Table 12.4.

☐ STEP 4. For each plot size having more than one shape, test the homogeneity of between-plot variances $V_{(x)}$, to determine the significance of plot-orientation (plot-shape) effect, by using the F test or the chi-square test (see Chapter 11, Section 11.2.1). For each plot size whose plot-shape effect is nonsignificant, compute the average of V_x values over all plot shapes. For others, use the lowest value.

For example, there are two plot shapes (2×1 m and 1×2 m) for plot of size 2 m^2, in our example. Hence, the F test is applied as

$$F = \frac{31,370}{31,309} = 1.00^{\text{ns}}$$

For plot size of 6 m^2, there are four plot shapes (6×1 m, 3×2 m, 2×3 m, and 1×6 m) and, hence, the chi-square test is applied as

$$\chi^2 = \frac{(2.3026)(108)[(4)(\log 242073.25) - 21.535346]}{1 + \dfrac{(4 + 1)}{(3)(4)(108)}}$$

$$= .11^{\text{ns}}$$

The tests for homogeneity of variances associated with different plot shapes of the same size, applied to 7 plot sizes ranging from 2 to 27 m^2, are all nonsignificant. The average of V_x values over all plot shapes of a given size is, therefore, computed.

☐ STEP 5. Using the values of the variance per unit area V_x computed in steps 3 and 4, estimate the regression coefficient between V_x and plot size x.

Because the relationship between V_x and x, namely, $V_x = V_1/x^b$, is not linear, the first step is to linearize the function into a linear form of $Y = cX$, where $Y = \log V_x - \log V_1$, $c = -b$, and $X = \log x$; and compute the regression coefficient as:

$$c = \frac{\displaystyle\sum_{i=1}^{m} w_i x_i Y_i}{\displaystyle\sum_{i=1}^{m} w_i x_i^2}$$

where w_i is the number of plot shapes used in computing the average variance per unit area of the ith plot in step 4, and m is the total number of plots of different sizes.

For our example, using the $m = 12$ pairs of values of V_x and x, taken from Table 12.4 for plots having only one shape and from those computed

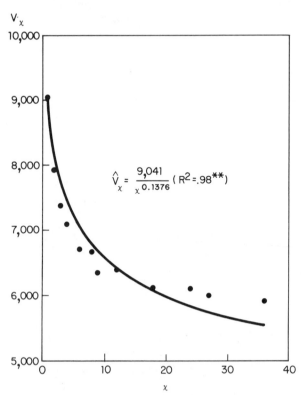

Figure 12.2 Relationship between variance per unit area (V_x) and plot size (x), computed from data in Table 12.4.

in step 4 for plots having more than one shape, we have

$$\sum_{i=1}^{12} w_i x_i Y_i = -3.8503$$

$$\sum_{i=1}^{12} w_i x_i^2 = 27.9743$$

$$c = \frac{-3.8503}{27.9743} = -0.1376.$$

Thus, the estimated regression equation is computed as:

$$\hat{V}_x = \frac{9,041}{x^{0.1376}}$$

The coefficient of determination R^2 is .98, which is highly significant. The result is represented graphically in Figure 12.2.

☐ STEP 6. Obtain the adjusted b value from Table 12.5 based on the computed b value from step 5 and the value of x_1/n, where x_1 is the size of basic unit and n is the size of the whole area. For our example, because the size of the basic unit is 1 m² and the total area is 648 m², the value of $x_1/n = 1/648$

Table 12.5 The Adjusted Values of Smith's Index of Soil Heterogeneity (b)

Computed b	Adjusted b in Range x_1/n^a	
	0.001 to 0.01	0.01 to 0.1
1.0	1.000	1.000
0.8	0.804	0.822
0.7	0.710	0.738
0.6	0.617	0.656
0.5	0.528	0.578
0.4	0.443	0.504
0.35	0.403	0.469
0.3	0.364	0.434
0.25	0.326	0.402
0.2	0.291	0.371
0.15	0.257	0.343
0.1	0.226	0.312

$^a x_1$ is the size of the basic unit and n is the size of the total area.

which, based on Table 12.5, is in the range of 0.001 to 0.01. To obtain the adjusted value of the computed b value of 0.1376 (from step 5), we interpolate between the adjusted value of 0.226 (corresponding to the computed b value of 0.10) and the adjusted value of 0.257 (corresponding to the computed b value of 0.15) from Table 12.5 as:

$$\text{Adjusted } b = 0.226 + \frac{(0.0376)(0.257 - 0.226)}{0.05}$$

$$= 0.249$$

The relatively low value of the adjusted Smith's index of soil heterogeneity of 0.249 indicates a relatively high degree of correlation among adjacent plots in the study area: the change in the level of soil fertility tends to be gradual rather than in patches. This result agrees closely with the fertility contour map (Figure 12.1) where it can be seen that the fertility level is the highest with row 1 and reduces gradually with the increase in the row number.

12.2.2.   Data from Field Experiments

Data from replicated field experiments can also be used to measure the Smith's index of soil heterogeneity. Experiments most suitable for this procedure are those involving designs with several plot sizes, such as split-plot and split-split-plot design.

The procedure involves the use of the basic analysis of variance to estimate the variances for plots of different sizes, and the use of these estimates to derive a relationship between plot variance and plot size. The number of plot variances that can be estimated through this procedure is only as many as the number of plot sizes available in the design used. For example, in a randomized complete block (RCB) design, there are two plot sizes—the unit plot and the block; in a split-plot design there are three—the subplot, the main plot, and the block; and in a split-split-plot design there are four—the sub-subplot, the subplot, the main plot, and the block. Correspondingly, the number of variance estimates for different plot sizes is two for RCB, three for split-plot, and four for split-split-plot design. Clearly the estimate of the index of soil heterogeneity b improves as the number of plot sizes increases.

Using data from field experiments to estimate the index of soil heterogeneity has two main advantages:

· This procedure is less expensive than conducting the uniformity trial because the time and money necessary for the trial are saved.

· Several experiments can be used without entailing too much cost and, thus, more areas and seasons are covered instead of the limited areas of the uniformity trial.

Using data from replicated field experiments, however, has a few disadvantages:

- Because variance estimates can usually be obtained for only a few plot sizes from each experiment, the fitting of the regression, and thus the estimate of *b*, have low accuracy.
- Because blocks in most field experiments are set up so that a large portion of the soil variability remains between blocks, the variance between plots the size of a block is generally overestimated.

Because of such disadvantages in the use of data from field experiments, a researcher should conduct at least one uniformity trial to assess and confirm the validity of the variance and plot size relationship before attempting to use data from existing field experiments in estimating the index of soil heterogeneity.

To illustrate the use of data from replicated field experiments in measuring the index of soil heterogeneity, we use an experiment on rice from a split-plot design with six nitrogen levels as main plot and thirteen rice varieties as subplot treatments in three replications. The size of the subplot is 3.7 × 5.0 m, the main plot is 3.7 × 65.0 m, and the replication is 22.2 × 65.0 m. The step-by-step procedures are:

☐ STEP 1.   The basic formats of the analysis of variance for a RCB, a split-plot, and a split-split-plot design are shown in Table 12.6. Construct an analysis of variance according to the experimental design used, following the standard procedures outlined in Chapters 2 to 4. For our example, the analysis of variance is given in Table 12.7.

☐ STEP 2.   Compute estimates of the variances associated with the different plot sizes, following the formulas given in Table 12.8. For our example, the design is a split-plot. Hence, there are three between-plot variances corresponding to the three plot sizes as follows:

$$V_1' = \text{the variance between plots the size of a block}$$

$$V_2' = \text{the variance between plots the size of a main plot}$$

$$V_3' = \text{the variance between plots the size of a subplot}$$

The computation of these variances is based on the mean square values in the analysis of variance (Table 12.7) and the formulas given in Table 12.8.

Table 12.6 Basic Format of the Analysis of Variance for Randomized Complete Block, Split-Plot, and Split-Split-Plot Design

Randomized Complete Block

Source of Variation	Degree of Freedom	Mean Square
Total	$rt - 1$	
Replication	$r - 1$	M_1
Treatment	$t - 1$	
Error	$(r - 1)(t - 1)$	M_2

Split-Plot

Source of Variation	Degree of Freedom	Mean Square
Total	$abr - 1$	
Replication	$r - 1$	M_1
A	$a - 1$	
Error(a)	$(a - 1)(r - 1)$	M_2
B	$b - 1$	
$A \times B$	$(a - 1)(b - 1)$	
Error(b)	$a(r - 1)(b - 1)$	M_3

Split-Split-Plot

Source of Variation	Degree of Freedom	Mean Square
Total	$rabc - 1$	
Replication	$r - 1$	M_1
A	$a - 1$	
Error(a)	$(a - 1)(r - 1)$	M_2
B	$b - 1$	
$A \times B$	$(a - 1)(b - 1)$	
Error(b)	$a(r - 1)(b - 1)$	M_3
C	$c - 1$	
$A \times C$	$(a - 1)(c - 1)$	
$B \times C$	$(b - 1)(c - 1)$	
$A \times B \times C$	$(a - 1)(b - 1)(c - 1)$	
Error(c)	$ab(r - 1)(c - 1)$	M_4

Thus:

$$V_1' = M_1 = 1,141,670$$

$$V_2' = \frac{r(a-1)M_2 + (r-1)M_1}{ra-1}$$

$$= \frac{3(5)(433,767) + (2)(1,141,670)}{3(6)-1}$$

$$= 517,050$$

$$V_3' = \frac{ra(b-1)M_3 + r(a-1)M_2 + (r-1)M_1}{rab-1}$$

$$= \frac{3(6)(12)(330,593) + 3(5)(433,767) + 2(1,141,670)}{[3(6)(13)]-1}$$

$$= 344,197$$

☐ STEP 3. For each variance estimate V_i' obtained in step 2, compute the corresponding *comparable variance* V_i with the size of the smallest plot in the particular experiment as the base:

$$V_i = \frac{V_i'}{x}$$

where x is the size of the ith plot in terms of the smallest plot involved.

Table 12.7 Analysis of Variance of a Split-Plot Design with Six Main-Plot and 13 Subplot Treatments

Source of Variation	Degree of Freedom	Sum of Squares	Mean Square	Computed F^a
Total	233	330,087,531		
Replication	2	2,283,340	1,141,670	
Nitrogen (N)	5	27,404,236	5,480,847	12.64**
Error(a)	10	4,337,669	433,767	
Variety (V)	12	207,534,092	17,294,508	52.31**
$V \times N$	60	40,922,735	682,046	2.06**
Error(b)	144	47,605,459	330,593	

^{a}** = significant at 1% level.

Table 12.8 Formulas for the Computation of Variances between Plots of Various Sizes, Using Data from Existing Experiments for a Randomized Complete Block, Split-Plot, and Split-Split-Plot Design[a]

Variance	Randomized Complete Block	Split-Plot	Split-Split-Plot
V_1'	$\dfrac{(r-1)M_1 + r(t-1)M_2}{rt-1}$	M_1	M_1
V_2'	—	$\dfrac{r(a-1)M_2 + (r-1)M_1}{ra-1}$	$\dfrac{r(a-1)M_2 + (r-1)M_1}{ra-1}$
V_3'	—	$\dfrac{ra(b-1)M_3 + r(a-1)M_2 + (r-1)M_1}{rab-1}$	$\dfrac{ra(b-1)M_3 + r(a-1)M_2 + (r-1)M_1}{rab-1}$
V_4'	—	—	$\dfrac{rab(c-1)M_4 + ra(b-1)M_3 + r(a-1)M_2 + (r-1)M_1}{rabc-1}$

[a]M_1, M_2, M_3, and M_4 are as defined in Table 12.6.

For our example, the smallest plot is the subplot, the size of the main plot is 13 times that of the subplot, and the size of the block is $13 \times 6 = 78$ times that of the subplot. Hence, the comparable variances are computed as:

$$V_1 = \frac{V_1'}{x_1} = \frac{1,141,670}{78} = 14,637$$

$$V_2 = \frac{V_2'}{x_2} = \frac{517,050}{13} = 39,773$$

$$V_3 = \frac{V_3'}{x_3} = \frac{344,197}{1} = 344,197$$

☐ STEP 4. Apply the appropriate regression technique (see Section 12.2.1.4, step 5) to estimate the regression coefficient b from the equation

$$\log V_i = \log V_3 - b \log x_i$$

where V_i and x_i are as defined in step 3.

For our example, there are three pairs of (V_i, x_i) values, namely, $[(14,637), 78]$, $[(39,773), 13]$ and $[(344,197), 1]$. The estimate of the b value obtained is 0.755, with a highly significant coefficient of determination R^2 of .995.

Note that when data from field experiments are used to estimate the index of soil heterogeneity, the number of plots, and, hence, the number of pair values used in the regression estimate, are small. For most purposes, therefore, the estimate of b value can be obtained through the graphical method:

- Plot the logarithm of the comparable variance ($\log V_i$) against the logarithm of the plot size ($\log x_i$).
- Draw a straight line through the plotted points.
- Compute the b value as Y_0/x_0, where Y_0 is the y intercept (the point where the line meets the Y axis) and x_0 is the x intercept (the point where the line meets the X axis).

For our example, the three pairs of ($\log V_i$, x_i) values are plotted and the line drawn through them, as shown in Figure 12.3. The line meets the Y axis at $Y_0 = 5.45$ and meets the X axis at $x_0 = 7.35$. The b value is, then, computed as:

$$b = \frac{Y_0}{x_0} = \frac{5.45}{7.35} = 0.74$$

which agrees fairly well with the b value of 0.755 obtained earlier through regression technique.

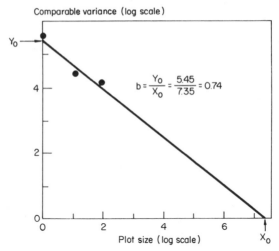

Figure 12.3   Estimation of the index of soil heterogeneity *b* by graphical method.

12.3 COPING WITH SOIL HETEROGENEITY

Once the fertility pattern of an experimental area is described, several options are available for reducing the effect of soil heterogeneity. Three options that are commonly used involved the proper choice of plot size and shape, block size and shape, and number of replications. These options can be inexpensive, involving only a change of plot or block orientation, but at times, the option may involve enlarging the experimental area or increasing the total number of plots.

12.3.1 Plot Size and Shape

The contribution of soil heterogeneity to experimental error stems from differences in soil fertility between plots within a block. The smaller this difference is, the smaller is the experimental error. The choice of suitable plot size and shape, therefore, should reduce the differences in soil productivity from plot to plot within a block and consequently reduce experimental error.

12.3.1.1 Optimum Plot Size.   Two major considerations are involved in choosing plot size, namely, practical considerations and the nature and size of variability. Practical considerations generally include ease of management in the field. The nature and size of variability is generally related to soil heterogeneity. From the empirical relationship between plot size and between-plot variance (Section 12.2.1.4), it can be seen that while variability becomes smaller as plot size becomes larger, the gain in precision decreases as plot size becomes increasingly large (Figure 12.2). Furthermore, higher costs are involved when large plots are used. Hence, the plot size that a researcher should aim for is one that balances precision and cost. This is commonly referred to as *optimum plot size*.

Given an estimate of the soil heterogeneity index b and the cost of conducting the experiment, optimum plot size can be calculated as:

$$x_{opt} = \frac{b(K_1 + K_g A)}{(1 - b)(K_2 + K_g B)}$$

where K_1 is the part of the cost associated with the number of plots only, K_2 is the cost per unit area, K_g is the cost associated with the borders, B is the ratio of the side-borders to the test area, A is the area of the plot end borders, and b is the index of soil heterogeneity. If nonbordered plots are used, K_g is zero.

Note that such optimum plot size can only be computed if a good estimate of costs is available. Dividing costs into three categories may be confusing at times because the cost items, especially K_1 and K_2, are not easy to separate. For example, it is easy to visualize why the cost of labeling the plots should depend on the number of plots alone and not at all on area; or that cost of plowing, harrowing, and so on, depends on area alone because these costs are incurred even before the area is subdivided into plots. However, for other items, such as harvesting or threshing, the line of demarcation is not as clear. Even though harvesting is done plot by plot, it is also true that the larger the plot area, the longer it takes to harvest and the higher the cost. Thus, one part of this expense is included in K_1 and another part is in K_2. The same is true for threshing and seed cleaning.

An example of the cost estimates in labor-hours for the various operations in an agronomic rice experiment is given in Table 12.9. These estimates were obtained based on the following conditions:

- Costs other than labor were ignored.
- Relative monetary costs of labor-hours for the various operations were not considered.
- The collection of data included tiller count, measurement of plant height, and determination of dry matter production at three growth stages.

With the cost estimates in Table 12.9 and the estimated b value of 0.249 (Section 12.2.1.4), optimum plot size is computed as:

- For Nonbordered Plots:

$$x_{opt} = \frac{bK_1}{(1 - b)K_2}$$

$$= \frac{(0.249)(2.9505)}{(1 - 0.249)(0.13686)}$$

$$= 7.1 \text{ m}^2$$

Table 12.9   Estimates of Costs in Labor-Hours for Completing a Rice Field Experiment

Operation	Cost K_2, h/m²	Cost K_1, h/plot
1. Land preparation	0.02583	—
2. Seedbed preparation	0.00111	—
3. Laying out of plots	0.00120[a]	0.2200
4. Transplanting	0.01700	—
5. Fertilizer application	0.00110	—
6. Insecticide application	0.00080	—
7. Hand weeding	0.02225	—
8. Labeling tags and making stakes	—	0.1033
9. Plot observation[b]	0.00394[a]	0.8925
10. Measurements of plant characteristics besides yield	—	1.2330
11. Harvesting	0.00943[a]	0.1893
12. Threshing	0.00417[a]	0.0622
13. Grain cleaning	0.05003[a]	0.0419
14. Weighing grains and determining moisture content	—	0.0333
15. Statistical analysis	—	0.1750
Total	0.13686	2.9505

[a] Not applicable to border areas.
[b] Observing panicle initiation date, flowering date, lodging, pest and disease incidence, and others.

- For Bordered Plots: With two border rows and two end hills designated as borders in each plot, and the width of plot is fixed at 3 m or 15 rows using a 20 × 20-cm spacing, the area of the plot end-borders A is 2(3)(0.4) = 2.4 m², and the ratio of the side borders to the test area B is 2(0.4)/[3 − 2(0.4)] = 0.36.

An estimated cost associated with the area of borders (K_g) can be obtained as the sum of K_2 *minus* costs of items 1, 2, 4, 5, 6, and 7 of Table 12.9, which results in the value of 0.06877. The optimum plot size is computed as:

$$x_{opt} = \frac{(0.249)[2.9505 + (0.06877)(2.4)]}{(1 - 0.249)[0.13686 + (0.06877)(0.36)]}$$

$$= 6.4 \text{ m}^2$$

The resulting optimum plot size of 6.4 m² refers to the test area only and does not include the borders. Thus, for this example, the total plot size including border areas is 11.1 m².

12.3.1.2 Plot Shape. Once the optimum plot size is determined, the choice of plot shape is governed by the following considerations:

- Long and narrow plots should be used for areas with distinct fertility gradient, with the length of the plot parallel to the fertility gradient of the field.
- Plot should be as square as possible whenever the fertility pattern of the area is spotty or not known, or when border effects (see Chapter 13, Section 13.1) are large.

12.3.2 Block Size and Shape

Block size is governed by the plot size chosen, the number of treatments tested, and the experimental design used. Once these factors are fixed, only the choice of block shape is left to the researcher.

The primary objective in choosing the shape of blocks is to reduce the differences in productivity levels among plots within a block so that most of the soil variability in the area is accounted for by variability between blocks. Information on the pattern of soil heterogeneity in the area is helpful in making this choice. When the fertility pattern of the area is known, orient the blocks so that soil differences between blocks are maximized and those within the same block are minimized. For example, in an area with a unidirectional fertility gradient, the length of the block should be oriented perpendicular to the direction of the fertility gradient. On the other hand, when the fertility pattern of the area is spotty, or is not known to the researcher, blocks should be kept as compact, or as nearly square, as possible.

Because block size, for most experimental designs, increases proportionately with the number of treatments and because it is difficult to maintain homogeneity in large blocks, a researcher must also be concerned with the number of treatments. If the number of treatments is so large that uniform area within a block cannot be attained, incomplete block designs (see Chapter 2, Section 2.4) may be used.

12.3.3 Number of Replications

The number of replications that is appropriate for any field experiment is affected by:

- The inherent variability of the experimental material
- The experimental design used
- The number of treatments to be tested
- The degree of precision desired

Because experimental variability is a major factor affecting the number of replications, soil heterogeneity clearly plays a major role in determining the number of replications in field experiments. In general, fewer replications are required with uniform soil.

Given an estimate of Smith's index of soil heterogeneity (b) and an estimated variance between plots of basic unit size (V_1), the number of replications required to satisfy a given degree of precision expressed as variance of a treatment mean (V_0) can be estimated as:

$$r = \frac{V_1}{(V_0)(x^b)}$$

where x is the plot size under consideration expressed in terms of the number of basic units.

For example, consider an experiment to be conducted in the area on which the uniformity trial data of Table 12.1 were obtained. If the estimate of each treatment mean should not deviate from the true value by more than 10%, to estimate the number of replications required, the following quantities are needed:

- The variance between plots of the size of the basic unit (V_1). This is taken from Table 12.4 to be 9,041.
- The degree of precision desired expressed in terms of variance of the treatment mean (V_0). Using the mean yield of 733.5 g/m^2 as the true value, the specification of 10% margin of error (10% of 733.5) is equivalent to specifying the variance of the mean to be $(73.35/2)^2 = 1,345.06$.
- The number of basic units in the plot (x). If the researcher plans to use 15 m^2 plot size in his experiment, x is 15 times the size of the basic unit.
- The Smith's index of soil heterogeneity b. This was computed as 0.249 (see Section 12.2.1.4).

With these quantities, the number of replications required is computed as:

$$r = \frac{9{,}041}{(1{,}345.06)(15)^{0.249}} = 3.4$$

This indicates that the experiment should be conducted with four replications to attain the level of precision specified (i.e., 10% margin of error).

CHAPTER 13

Competition Effects

A plant's growth is affected greatly by the size and proximity of adjacent plants. Those surrounded by large and vigorous plants can be expected to produce less than those surrounded by less vigorous ones. Plants a greater distance from one another generally produce more than those nearer to each other. Interdependence of adjacent plants because of their common need for limited sunshine, soil nutrients, moisture, carbon dioxide, oxygen, and so on, is commonly referred to as *competition effects*.

Because field experiments are usually set up to assess the effects on crop performance of several management factors or genetic factors, or both, experimental plots planted to different varieties and subjected to different production techniques are commonly placed side by side in a field. As a consequence border plants have an environment different from those in the plot's center: plants in the same plot are exposed to differing competitive environments.

Competition effects between plants within a plot should be kept at the same level to ensure that the measurement of plant response really represents the condition being tested and to reduce experimental error and sampling error. In this chapter we identify some competition effects important in field experiments, give techniques for measuring those effects, and describe procedures for coping with them.

13.1   TYPES OF COMPETITION EFFECT

For a given experiment, the significance of any competition effect depends primarily on the treatments being tested and the experimental layout.

13.1.1   Nonplanted Borders

Nonplanted borders are areas between plots or around the experimental areas that are left without plants and serve as markers or walkways. These areas are generally wider than the area between rows or between plants in a row, and plants adjacent to these nonplanted borders have relatively more space. They are, therefore, exposed to less competition than plants in the plot's center.

13.1.2　Varietal Competition

In trials involving different varieties of a given crop, adjacent plots are necessarily planted to different varieties. Because varieties generally differ in their ability to compete, plants in a plot will be subjected to different environments depending upon location relative to adjacent plots. This effect is called varietal competition.

Plants generally affected by varietal competition effect are the ones near the plot's perimeter. The size of the difference in plant characters between the varieties included in the trial plays an important role in determining the extent of varietal competition effects. In rice, for example, tall and high-tillering varieties compete better than the short and low-tillering ones. Thus, a short and low-tillering variety would be at a disadvantage when planted adjacent to a plot with a tall and high-tillering variety. The larger the varietal difference is, the greater is the expected disadvantage. The disadvantage of one plot is usually accompanied by a corresponding advantage to the adjacent plot.

13.1.3　Fertilizer Competition

Fertilizer competition effect is similar to the varietal competition effect except that adjacent plots receive different levels of fertilizer instead of being planted to different varieties. Here the competition effect has two sources. First, plots with higher fertilizer application will be more vigorous and can probably compete better for sunshine and carbon dioxide. Second, the fertilizer could spread to the root zone of an adjacent plot, putting the plot with higher fertilizer at a disadvantage. Because these two effects are of different direction, their difference constitutes the net competition effect. In most instances, the effect of fertilizer dispersion is larger than that due to the difference in plant vigor, and the net advantage is usually with the plot receiving lower fertilizer.

13.1.4　Missing Hills

A spot in an experimental plot where a living plant is supposed to be but is absent because of poor germination, insect or disease damage, physical mutilation, and so on, is called a missing hill or a missing plant. Because of the numerous factors that can kill a plant, even the most careful researcher cannot be assured of a complete stand for all plots in an experiment. A missing hill causes the plants surrounding its position to be exposed to less competition than the other plants. These plants, therefore, usually perform better than those surrounding a living plant.

13.2　MEASURING COMPETITION EFFECTS

Because the type and size of the competition effects can be expected to vary considerably from crop to crop and from one type of experiment to another, competition effects should be evaluated separately for different crops grown in different environments. Measurements of competition effects can be obtained from experiments planned specifically for that purpose or from those set up for other objectives.

13.2.1   Experiments to Measure Competition Effects

Experiments can be set up specifically to measure competition effects by using treatments that simulate different types of competition. At least two treatments, representing the extreme types of competitor, should be used. For example, to evaluate fertilizer competition effects, use a no-fertilizer application and a high-fertilizer rate to represent the two extremes. For varietal competition, use a short and a tall variety, a high-tillering and a low-tillering variety, or an early-maturing and a late-maturing variety. When resources are not limited, intermediate treatments can be included to assess the trend in the effects under investigation. For example, to find out whether border effects are affected by the width of an unplanted alley, test several sizes of unplanted alley. Too many treatments should be avoided, however, to cut costs and to simplify the design and the interpretation of results.

Experiments specifically set up to measure competition effects have two distinctive features:

1. Because the competition effects are usually small relative to the treatment effects in the regular experiments, the number of replications is usually large.
2. The plot size that is optimum for the regular experiments may not necessarily be optimum for experiments to measure competition effects. First, the unit of measurement is usually the individual rows, or individual plants, rather than the whole plot. Second, the plot must be large enough to ensure sufficient number of rows (mostly in the center of the plot) that are completely free of competition effects.

For example, experiments at IRRI to specifically study varietal competition effects in rice used a plot size of 10 rows 6 m long instead of the usual six rows 5 m long, and with nine replications instead of the usual four for the regular variety trials.

We give examples of rice experiments designed to study the four types of competition effect described in Section 13.1. For researchers working with crops other than rice, slight modifications are needed to adapt these techniques to their own work.

13.2.1.1 *Nonplanted Borders.*   For illustration, we consider a rice experiment to examine the border effects of several sizes of nonplanted alleys. Alleys 20, 40, 60, 100, and 140 cm wide were evaluated, with variety IR8 planted at

20 × 20-cm spacing. Thus, the 20-cm wide alley, with the same spacing as that between rows, is the control treatment.

The experiment was in a randomized complete block design with each of the five treatments replicated 10 times. The plot size was 3.2 × 6.0 m, or 16 rows, each 6 m long. The actual layout for the first three replications of the experiment is shown in Figure 13.1. In each replication, there are two plots that are exposed to each alley width, one on each side of the nonplanted alley. Hence, the actual experimental unit in this trial is not the plot as seen in the layout. Instead, it is composed of two parts, one from the right half of the plot on the left of the alley and another from the left half of the plot on the right of the alley. For each experimental unit, grain yield data were measured from each of the eight row positions with the first row position representing the outermost row, the second row position representing the second outermost row, and so on. Each row position is composed of a pair of rows, one on each side of the nonplanted alley. Eight hills at each end of the row were removed before plants were harvested for yield determination.

The raw data, therefore, consist of 400 observations representing eight row positions, five alley widths, and 10 replications. The standard analysis of variance procedure for a split-plot design (Chapter 3, Section 3.4.2) is applied with the alley widths as main-plot treatments and the row positions as subplot

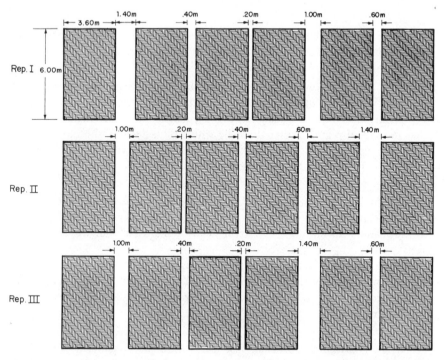

Figure 13.1 A part of the field layout in a study of the effects of nonplanted alley at five alley widths: 20, 40, 60, 100, and 140 cm (only three replications of a total of 10 replications are shown here).

Table 13.1 Analysis of Variance for Data from an Experiment to Study the Effects of Nonplanted Alley, the Layout of Which is Shown in Figure 13.1[a]

Source of Variation	Degree of Freedom	Mean Square	Computed F[b]	Tabular F 5%	Tabular F 1%
Replication	9	7,167			
Alley width (W)	4	34,008	13.82**	2.63	3.89
Error(a)	36	2,460			
Row position (R)	7	260,383	382.92**	2.04	2.71
$W \times R$	28	25,974	38.20**	1.52	1.79
Error(b)	315	680			
Total	399				

[a]$cv(a) = 14.0\%$, $cv(b) = 7.4\%$.
[b]** = significant at 1% level.

treatments. The results are shown in Table 13.1 There is a highly significant interaction between alley width and row position, indicating that the yield differences between the row positions varied with the width of the nonplanted alley. Suitable partitioning of the sums of squares is made, following procedures for group comparison and factorial comparison of Chapter 5. Results (Table 13.2) indicate that only the outermost row (row 1) gave yield significantly different from the inside rows, and this yield difference varied with alley widths. From the mean yields shown in Table 13.3, the yield differences between row 1 and the seven inside rows for the five different alley widths can be summarized as:

Alley Width, cm	Mean Yield, g/m^2 Row 1	Mean Yield, g/m^2 Inside Rows	Yield Increase in Row 1 g/m^2	Yield Increase in Row 1 %*
20	314	325	-11^{ns}	—
40	475	334	141**	42 a
60	556	324	232**	72 b
100	667	337	330**	98 c
140	654	326	328**	101 c

As expected, no significant difference between row positions is observed with the 20-cm alley width, whereas with all other alley widths the yield of the outermost row is significantly higher than the inside rows. This yield difference becomes larger as the nonplanted alley is widened up to 100 cm, after which no further increase is observed. The yield advantage exhibited in the outermost row ranges from 42% for alley width of 40 cm to about 100% for alley width of 100 cm or wider.

*Mean separation by DMRT at 5% level.

Table 13.2   Partitioned Sums of Squares of Row Position, and of Its Interaction with Alley Width, of the Analysis of Variance in Table 13.1

Source of Variation	Degree of Freedom	Mean Square	Computed F^a	Tabular F 5%	1%
Row position (R)	7	260,383			
R_1: Row 1 vs. inside rows	(1)	1,820,190	2,676.75**	3.87	6.73
R_2: Between inside rows	(6)	415	< 1	—	—
$W \times R$	28	25,974			
$W \times R_1$	(4)	179,533	264.02**	2.40	3.38
$W \times R_2$	(24)	381	< 1	—	—
Error(b)	315	680			

^{a}** = significant at 1% level.

13.2.1.2   Varietal Competition.   To illustrate varietal competition we consider an experiment to investigate the performance of border plants when different rice varieties were planted in adjacent plots. Three rice varieties that differ in tillering ability and plant height were used—IR8 (short and high-tillering), Peta (tall and medium-tillering), and IR127-80-1 (medium-short and low-tillering). All possible pairs of the three varieties were planted side by side without any nonplanted alley between adjacent plots. Plot size is 6 × 2 m, or 10 rows each 6 m long, and plant spacing is 20 × 20 cm. The field layout is shown in Figure 13.2 The layout consists of three strips of 11 plots each. The sequence of varieties planted in those plots remains the same in each strip but

Table 13.3   Mean Grain Yield^a (g / m^2) of Border Rows of IR8 Plots Adjacent to Varying Widths of Nonplanted Alley between Plots

Row Position	Alley Width, cm 20 (control)	40	60	100	140
1	314	475	556	667	654
2	325	329	329	326	317
3	322	323	326	336	317
4	330	333	315	352	326
5	322	331	318	335	334
6	327	337	318	338	328
7	318	342	322	342	329
8	330	341	343	332	330

^{a}Average of 10 replications.

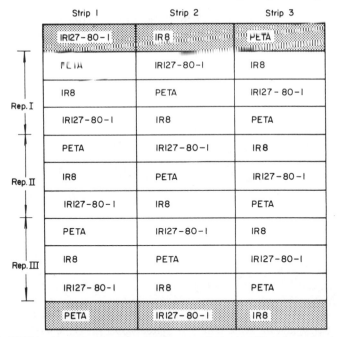

Figure 13.2 Field layout of a varietal competition experiment involving three rice varieties and nine replications (three replications per strip). Each plot consists of 10 rows 6 m long. Shaded plots are not measured.

differs from strip to strip. Excluding the first and last plots of each strip, the rest of the plots in each strip is grouped into three replications (as shown in the first strip of Figure 13.2). For each replication, a variety is adjacent to each of the other two varieties once. Harvests were made on a per-row basis and only from the 20 hills in the center of each row (i.e., 5 hills at each end of the row are considered as borders). Thus, for each variety, and in each replication, there are 10 row observations—five row observations corresponding to each of the two adjacent varieties. With nine replications, the total number of row observations for each variety is $9 \times 10 = 90$.

The analysis of variance is made, separately for each variety, following the standard procedure for a split-plot design (Chapter 3, Section 3.4.2) with the two halves of the plot (each corresponding to one of the two adjacent varieties) as main-plot treatments and the five row positions within each half as subplot treatments. The results are shown in Table 13.4. The significant difference between row positions is indicative of the presence of varietal competition effect, and the presence of interaction between row position and adjacent variety indicates that the competition effect is affected by the plant characters (namely tillering ability and plant height) of the adjacent variety.

Results of the analyses of variance show a highly significant difference between row positions in all three varieties. Significant interaction between row

Table 13.4 Analysis of Variance for Data from an Experiment to Study the Effects of Varietal Competition, the Layout of Which is Shown in Figure 13.2

Source of Variation	Degree of Freedom	Mean Square[a]		
		IR8	Peta	IR127-80-1
Total	89			
Replication	8	3,446	2,052	1,376
Adjacent variety (A)	1	37,465**	3,738^{ns}	1,914^{ns}
Error(a)	8	1,310	4,075	1,523
Row position (R)	4	5,713*	28,292**	21,804**
R_1: Row 1 vs. inside rows	(1)	11,264*	105,145**	82,778**
R_2: Between inside rows	(3)	3,863^{ns}	2,674^{ns}	1,479^{ns}
$A \times R$	4	18,314**	12,654**	2,748**
$A \times R_1$	(1)	72,287**	46,017**	4,633*
$A \times R_2$	(3)	323^{ns}	1,533^{ns}	2,120^{ns}
Error(b)	64	1,606	1,756	964
$cv(a)$, %		9.1	13.7	9.2
$cv(b)$, %		10.0	9.0	7.3

[a]** = F test significant at 1% level, * = F test significant at 5% level, ^{ns} = F test not significant.

Table 13.5 Varietal Competition Effects of Three Rice Varieties, as Affected by the Plant Characters of the Adjacent Variety

Variety	Variety in Adjacent Plot	Mean Yield, g/m^2 [a]						
		Row 1	Row 2	Row 3	Row 4	Row 5	Inside Rows[b] ($\bar{R}$)	Row 1 − $\bar{R}$[c]
IR8	IR127-80-1	499	390	387	408	417	400	99**
	Peta	345	377	380	384	412	388	−43**
Peta	IR8	482	476	434	459	442	453	29^{ns}
	IR127-80-1	585	453	437	432	452	444	141**
IR127-80-1	IR8	382	438	460	445	419	440	−58**
	Peta	345	429	432	452	441	438	−93**

[a]Average of nine replications.
[b]Mean of row 2 through row 5.
[c]** = significant at 1% level, ^{ns} = not significant.

position and adjacent variety is also indicated. Results of the partitioning of the sum of squares (following procedures of Chapter 5) indicate clearly that, in all three varieties, the competition effect reached only the outermost row (i.e., row 1), but the size of these effects differed depending on the characters of the adjacent variety.

Yield difference between row 1 and the inside rows of IR8, for instance, was positive and large when the adjacent variety was IR127-80-1 (the medium-short and low-tillering variety) but was negative when the adjacent variety was Peta (the tall and medium-tillering variety) (Table 13.5). On the other hand, the outermost row of Peta gave a significantly higher yield than the inside rows only when adjacent to IR127-80-1. And, the outermost row of IR127-80-1 gave significantly lower yields than the inside rows regardless of whether it was adjacent to IR8 or Peta.

13.2.1.3   Fertilizer Competition.   We illustrate fertilizer competition with a rice experiment to study the effects of applying two widely different nitrogen levels on adjacent plots separated by a 40-cm nonplanted alley. Fertilized (120 kg N/ha) and nonfertilized 28-row plots were arranged systematically in an alternating series. There was a total of 16 plots. Each plot was bordered on one side by plot of the same nitrogen level (control) and on the other side by plot of a different level (Figure 13.3). Grain yields were determined from the center 15 hills of each of the 28 rows per plot. The first and the last nonfertilized plots were excluded from the determination of grain yield. Of the 14 plots harvested, six are nonfertilized plots and eight are fertilized plots. Each 28-row plot is divided into two parts, each consisting of 14 row positions—with row 1 being immediately adjacent to the adjacent plot, and so on.

The analysis of variance is performed, separately for the fertilized plots and the nonfertilized plots, following the standard procedure for a split-plot design (Chapter 3, Section 3.4.2) with plots as replications, the two nitrogen levels of adjacent plot as main-plot treatments, and the 14 row positions as subplot treatments. The results, with suitable partitioning of sums of squares, following procedures of Chapter 5, are shown in Table 13.6. Note that, unlike the case of

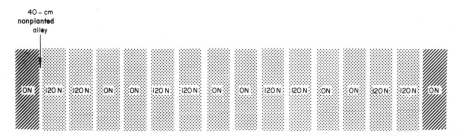

Figure 13.3 Field layout of a nitrogen competition experiment involving nonfertilized and fertilized (120 kg N/ha) plots separated by a 40-cm nonplanted alley. Two end plots are not measured.

Table 13.6 Analysis of Variance of Data from an Experiment to Study the Effects of Fertilizer Competition, the Layout of Which is Shown in Figure 13.3

Source of Variation	Nonfertilized Plots Degree of Freedom	Nonfertilized Plots Mean Square[a]	Fertilized Plots Degree of Freedom	Fertilized Plots Mean Square[a]
Total	167		223	
Between plots	5	81,268	7	301,446
Adjacent nitrogen rate (N)	1	764	1	145
Error(a)	5	20,556	7	2,851
Row position (R)	13	65,490**	13	54,832**
R_1: Row 1 vs. inside rows	(1)	823,862**	(1)	694,894**
R_2: Between inside rows	(12)	2,292^{ns}	(12)	1,494^{ns}
$N \times R$	13	6,027**	13	4,695^{ns}
$N \times R_1$	(1)	62,732**	(1)	20,200*
$N \times R_2$	(12)	1,302^{ns}	(12)	3,403^{ns}
Error(b)	130	2,324	182	3,553
$cv(a)$, %		31.8		8.3
$cv(b)$, %		10.7		9.3

[a]** = F test significant at 1% level, * = F test significant at 5% level, ^{ns} = F test not significant.

Table 13.7 Fertilizer Competition Effects in IR8 Plots, as Affected by the Rate of Fertilizer Applied in Adjacent Plots

Nitrogen Rate of Adjacent Plot, kg/ha	Mean Yield, g/m^2 [a] Row 1	Mean Yield, g/m^2 [a] Inside Rows
Nonfertilized plots		
0 (control)	631.3	434.5
120	775.0	428.1
Difference	− 143.7**	6.4^{ns}
Fertilized plots		
120 (control)	877.0	624.0
0	810.1	630.8
Difference	66.9*	− 6.8^{ns}

[a]Six nonfertilized plots and eight fertilized plots. Adjacent plots are separated by a 40-cm nonplanted alley. Inside rows are average of thirteen row positions; ** = significant at 1% level, * = significant at 5% level, ^{ns} = not significant.

514

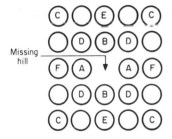

Missing hill

Figure 13.4   Positions of the measured hills (*A*, *B*, *D*, *E*, and *F*) surrounding a missing hill (20 × 25-cm spacing) in a study of the effects of missing hill. Hills marked *C* are control.

varietal competition (Section 13.2.1.2) where the difference between row positions measures the varietal competition, in this case some differences between row positions are expected because of the 40-cm nonplanted alley used to separate adjacent plots (see Section 13.2.1.1). Thus, the presence of fertilizer competition effect is indicated only by the presence of the interaction between row position and adjacent nitrogen rate.

Results (Table 13.6) show significant interaction between row position and adjacent nitrogen rate, indicating the presence of fertilizer competition effect. The competition effects, however, were confined only to the outermost row (row 1): the outermost row of the nonfertilized plot adjacent to a fertilized plot gave significantly higher grain yield than the outermost row adjacent to a similarly nonfertilized plot (Table 13.7). Likewise, the grain yield of the outermost row of a fertilized plot adjacent to a nonfertilized plot was significantly lower than that adjacent to a similarly fertilized plot.

13.2.1.4   Missing Hill.   To examine the effect of a missing hill on the performance of adjacent plants, an experiment was set up with two rice varieties, IR22 and IR127-80-1, that differ in tillering ability. Planting distance was 20 × 25 cm. Two weeks after transplanting, the center hill of each test plot (1.6 × 2.0 m) was pulled. Because grain yields have to be measured from small units (i.e., 2 to 4 hills), 36 replications were used for each variety. From each plot, the different hill positions surrounding the missing hill, as illustrated in Figure 13.4, were harvested.

The analysis of variance with *SS* partitioning is presented in Table 13.8, and the mean grain yields in Table 13.9. The results show that yields of the four hills immediately adjacent to a missing hill (hill positions *A* and *B*) are significantly higher than that of control. The effects are consistent for both varieties.

13.2.2   Experiments Set Up for Other Purposes

Any experiment undertaken for other purposes, such as fertilizer trials or variety trials, can also be used to test or measure competition effects. Because division of plots into subsections is an integral part in measuring competition effects, the experiments suited for this purpose are those having relatively large

Table 13.8 Analysis of Variance of Data from an Experiment to Study the Effects of Missing Hill[a]

Source of Variation	Degree of Freedom	Mean Square	Computed F[b]	Tabular F 5%	Tabular F 1%
Replication	35	49.164			
Variety (V)	1	9.481	< 1	—	—
Error(a)	35	26.920			
Hill position[c] (H)	5	565.793	32.02**	2.24	3.07
Control vs. other hill positions	(1)	332.987	18.85**	3.87	6.72
Between other hill positions	(4)	623.996	35.32**	2.40	3.37
(A, B) vs. (D, E, F)	1	2,373.007	134.30**	3.87	6.72
Between A and B	1	61.361	3.47[ns]	3.87	6.72
Between D, E, and F	2	30.807	1.74[ns]	3.02	4.67
$H \times V$	5	22.237	1.26[ns]	2.24	3.07
Error(b)	350	17.669			
Total	431				

[a] $cv(a) = 22.0\%$, $cv(b) = 17.8\%$.
[b] ** = significant at 1% level, [ns] = not significant.
[c] For definition, see Figure 13.4.

Table 13.9 Mean Grain Yields of Hills Surrounding a Missing Hill, for Two Varieties

Hill Position[a]	Mean Yield, g/hill[b] IR22	Mean Yield, g/hill[b] IR127-80-1	Mean Yield, g/hill[b] Av.[c]
A	27.8	27.9	27.8**
B	27.0	26.0	26.5**
D	22.2	23.1	22.6[ns]
E	20.4	22.6	21.5[ns]
F	21.9	21.3	21.6[ns]
Control	21.6	21.8	21.7

[a] For definition, see Figure 13.4.
[b] Average of 36 replications.
[c] ** = significantly different from control at 1% level, [ns] = not significantly different from control.

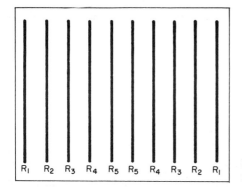

R_1 R_2 R_3 R_4 R_5 R_5 R_4 R_3 R_2 R_1

Figure 13.5 The subdivision of a 10-row plot for measuring competition effects.

plots. For example, in experiments where the plants are grown in rows, and both sides of a plot are subjected to the same type of competition effect, a plot can be subdivided into a pair of outermost rows, a pair of second outermost rows, and so on, up to the center rows. Figure 13.5 shows a plot consisting of 10 rows divided into five row positions: R_1 refers to the outermost pair of rows, R_2 the second outermost pair, and so on. For crops not planted in rows, area delineation can be used.

Measurements of plant response, such as grain yield, are then made separately for each subunit (i.e., row position or area delineation). Because subunits in each plot differ primarily in their exposure to competition, their differences measure the competition effect.

Aside from quantifying competition effects, this procedure also measures the interaction between competition effects and treatments that are tested in the experiment (i.e., varieties, fertilizers, etc.). The measurement of competition effects indicates whether certain rows or subunits are favored or are at a disadvantage because of their plot position, and the interaction shows whether such effects are consistent from one treatment to another. For example, in rice, plants in the outermost row of a plot adjacent to nonplanted area yield more than the other plants. This yield increase, however, differs from variety to variety. In general, varieties with high-tillering ability can make better use of the nonplanted area than the low-tillering ones. Such differential effects, if not corrected, can alter the varietal rankings in the experiment.

We use a study superimposed on a rice variety trial at IRRI to illustrate the procedure. The trial had 178 rice varieties replicated three times. Plot size was six rows, each 5 m long. Plant spacing was 30×15 cm. For varietal comparison, which was the major objective of the trial, yield measurement was made by harvesting the four center rows in each plot. Because of cost consideration, investigation of varietal competition effect was confined to only 12 out of 178 varieties. For each of the 12 selected varieties, the two border rows—one on each side of the plot—were harvested separately. The two end hills on each row were not harvested.

To test the presence of varietal competition and its effect on varietal comparisons, we constructed an analysis of variance based on a split-plot

Table 13.10   Analysis of Variance for Data from a Variety Trial to Study the Effects of Varietal Competition[a]

Source of Variation	Degree of Freedom	Sum of Squares	Mean Square	Computed F[b]	Tabular F 5%	Tabular F 1%
Replication	2	27,959	13,980			
Variety (V)	11	202,625	18,421	5.66**	2.26	3.18
Error(a)	22	71,568	3,253			
Method of yield determination (M)	1	204	204	< 1	—	—
$V \times M$	11	3,938	358	1.36[ns]	2.22	3.09
Error(b)	24	6,338	264			
Total	71	312,632				

[a]$cv(a) = 8.5\%$, $cv(b) = 2.4\%$.
[b]** = significant at 1% level, ns = not significant.

design (Chapter 3, Section 3.4.2) treating the 12 varieties as main-plot treatments and the two methods of yield determination (including and excluding border rows) as subplot treatments. The results (Table 13.10) indicate no significant difference between the two methods of yield determination nor significant interaction between variety and method. This means that varietal competition was not appreciable in this trial. Variety means for each of the two methods of determination are shown in Table 13.11.

Table 13.11   Grain Yields of 12 Rice Varieties, Including and Excluding Border Rows

Variety Number	Mean Yield, g/m² [a] Excluding Border Rows	Including Border Rows	Difference[b]
1	564	590	− 26
2	674	683	− 9
3	599	593	6
4	628	656	− 28
5	639	635	4
6	665	663	2
7	672	684	− 12
8	680	665	15
9	777	753	24
10	695	694	1
11	723	726	− 3
12	734	747	− 13
Av.	671	674	− 3[ns]

[a]Average of three replications.
[b]ns = not significant.

13.2.3   Uniformity Trials or Production Plots

Large production plots or uniformity trials are usually subdivided into blocks using nonplanted alleys as markers. In such plantings, the effect of nonplanted alleys can be evaluated by comparing the performance of the outermost rows with the center rows. However, nonplanted alleys for such plantings are usually wide and because the effects in certain crops such as rice become larger as the nonplanted alley becomes wider, such a study could result in a much larger effect than that actually existing in a standard field experiment. Nevertheless, such a study can be used as a guide for planning a more detailed study.

For illustration, we present grain yield data collected from the border rows of a rice uniformity trial. The nonplanted alley surrounding the experimental area was about 1 m wide. Hill spacing was 20 × 20 cm. From the perimeter of the trial, grain yield data were collected from a total of 22 sections each 30 hills long. In each section, five successive rows starting from the nonplanted side were harvested separately. The data are presented by row positions in Table

Table 13.12   Grain Yields of Rice Variety IR8 from Five Successive Rows[a] Adjacent to Nonplanted Alley

Section Number	Grain Yield, g/30 hills					
	Row 1	Row 2	Row 3	Row 4	Row 5	Av.
1	2,321	775	765	852	731	1,089
2	1,740	571	662	615	530	824
3	1,627	472	493	500	474	713
4	1,400	421	480	545	497	669
5	1,633	571	544	541	564	771
6	1,883	562	588	675	580	858
7	1,657	579	613	638	632	824
8	1,658	590	642	680	610	836
9	931	441	483	487	456	560
10	929	430	455	475	450	548
11	850	431	415	379	380	491
12	1,636	589	764	831	680	900
13	1,417	589	692	694	709	820
14	1,172	546	581	558	575	686
15	1,057	531	520	522	613	649
16	1,221	478	526	635	594	691
17	1,297	474	558	504	555	678
18	1,723	499	515	529	557	765
19	1,637	584	505	533	555	763
20	1,676	893	823	717	834	989
21	1,677	845	792	748	802	973
22	1,730	919	817	873	889	1,046
Av.	1,494	581	602	615	603	779

[a]Row 1 refers to the outermost row, row 2 the second outermost row, and so on.

Table 13.13   Analysis of Variance for Data in Table 13.12 from an Experiment to Study the Effects of Nonplanted Alley[a]

Source of Variation	Degree of Freedom	Mean Square	Computed F[b]	Tabular F 5%	1%
Replication	21	124,651			
Row position	4	3,519,409	190.05**	2.48	3.55
Row 1 vs. inside rows	(1)	14,064,766	759.52**	3.96	6.95
Between inside rows	(3)	4,289	< 1	—	—
Error	84	18,518			
Total	109				

[a]$cv = 17.5\%$.
[b]** = significant at 1% level.

13.12. The analysis of variance to test differences between the row positions is shown in Table 13.13. The computation of this analysis of variance followed the procedures for randomized complete block design (Chapter 2, Section 2.2.3) with the section treated as replication and row position as treatment. The results indicate a highly significant difference between yield of the outermost row (row 1) and that of the four inside rows, but there was no significant difference between the yields of the four inside rows. That is, the effect of the 1-m nonplanted alley was shown to reach only the outermost row.

13.3   CONTROL OF COMPETITION EFFECTS

Competition effects can be a major source of experimental error in field experiments. At times, competition effects may even alter the treatment comparisons—for example, in variety tests where the competition effect favors certain varieties more than others. The need to minimize, if not entirely eliminate, competition effects in field experiments is clear.

13.3.1   Removal of Border Plants

Because the effects of varietal competition, fertilizer competition, and non-planted borders are usually shown only on plants in the outer rows, an obvious solution is to exclude those from plot measurements. The width of borders or the number of rows (or plants) to be discarded on each side of the plot depends primarily upon the size of competition effects expected. In general, competition between plots is greater in grain crops where plant spacing is narrow. For example, in rice experiments, varietal competition has been found to reach the second outermost row and, sometimes, even the third row when the row spacing is 20 or 25 cm. Thus, in most rice variety trials, the yield data from the two outermost rows on either side of a plot are discarded. In crops, such as maize or sorghum, where plant spacing is wide (row spacing is generally wider than 50 cm), removal of only one outermost row is usually sufficient.

Naturally, border width or the number of rows to be discarded also affects plot size. The plot must be large enough so that sufficient plants are left for the desired measurements after border plants are discarded. The larger the border area, the bigger is the plot size and consequently the block size. This will bring about a corresponding increase in variability or experimental error, especially with a large number of treatments. The selection of border width should, therefore, be based on the knowledge of the size of the particular type of border effect encountered. Too large a border implies waste of experimental resources and possible increase in experimental error, whereas too small a border biases the results.

We caution researchers: to minimize the possibility of mixing the border plants with the inner ones, especially when unskilled labor is used during harvest, cut border plants *before* the experimental plants are harvested.

13.3.2   Grouping of Homogeneous Treatments

Because competition between adjacent plots in a variety trial is magnified by large morphological differences of test varieties, and in a fertilizer trial by large differences in the fertilizer rates applied, an obvious remedy is to ensure that adjacent plots are planted to varieties of fairly similar morphology or are subjected to similar fertilizer rates.

In the case of variety trials, this could be done by grouping together varieties that are fairly homogeneous in competition ability, and use a group balanced block design (Chapter 2, Section 2.5).

In the case of fertilizer trials, fertilizer is generally tested together with several varieties or several management practices, in a factorial experiment. The use of a split-plot type of designs (Chapters 3 and 4) with fertilizer as main-plot factor would allow the grouping together of plots having the same fertilizer rate and thus minimize fertilizer competition.

13.3.3   Stand Correction

When there is one or more missing hill in an experimental plot, measurement of data is affected. This is so because plants immediately adjacent to a missing hill usually perform better than the normal plants. For plant characters whose measurement is made on a per-plant basis, plot sampling (Chapter 15) should be taken such that all plants surrounding the missing hill are excluded from the sample. For yield determination, however, one of the following two procedures can be chosen:

1. Mathematical Correction.   The corrected yield (i.e., yield that should have been with perfect stand) is computed as:

$$Y_c = fY_a$$

where Y_c is the corrected yield, Y_a is the actual yield harvested from the plot with missing data, and f is the correction factor. Note that if the total number of plants in the plot is supposed to be n but there are m missing hills, then Y_a is the grain weight or yield from the $(n - m)$ plants in the plot.

The correction factor f must be derived from the estimate of yield increase over normal plants of those plants adjacent to one or more missing hills. Because yield compensation (or yield advantage) varies not only with crops but also with other factors, such as crop variety, fertilizer rate, plant spacing, and crop season; an appropriate correction factor needs to be worked out for each crop and for different conditions. For some crops, such as rice, where there is a large variation in the yield advantage between different varieties and different management practices, the required set of correction factors is too large to be practical. Hence, this procedure of mathematical correction is not effective. For these crops, the second procedure of excluding plants surrounding the missing hill from harvest is more appropriate.

2.   Exclude from Harvest Plants Surrounding the Missing Hill.   Because the increase in yields of plants surrounding one or more missing hills depends on so many factors, it is sometimes impossible to obtain an appropriate correction factor. An alternative, therefore, is to discard all plants immediately adjacent to a missing hill and harvest only those that are fully competitive (i.e., hills surrounded by living hills). This procedure is especially fitted for experiments with a large plot size and a large number of plants per plot.

CHAPTER 14
Mechanical Errors

Errors that occur in the execution of an experiment are called mechanical errors. In field experiments, these can occur from the time the plots are laid out (such as error in plot measurement), during the management and care of the experiment (such as the use of mixed or unpure seeds), and during collection of data (such as errors in the measurement of characters and in the transcription of data).

In contrast to other sources of error in field experiments, such as soil heterogeneity (Chapter 12) and competition effects (Chapter 13), mechanical errors are primarily human errors and, thus, cannot be effectively controlled by statistical techniques. Mechanical errors can make a substantial difference in experimental results, however, especially if the errors are committed during critical periods of the experiment. The most effective prevention of mechanical errors is the researcher's awareness of potential errors in relation to the various operations of the experiment.

It is not possible to enumerate all mechanical errors that could occur in all types of field experiment. We, therefore, pinpoint primarily the critical stages of an experiment when large mechanical errors are most likely to occur and suggest some practical procedures that can minimize or eliminate them. In addition, we mention areas in which uniformity of the experimental material can be enhanced.

14.1 FURROWING FOR ROW SPACING

For row crops, any errors in row spacing are reflected as error in plot measurement. For example, consider a sorghum variety test sown on rows 0.75 m apart. A plot size consisting of four 5-m-long rows would measure 3×5 m. If the spacing between rows in one plot were incorrectly set at 0.80 m, instead of the 0.75 m intended, then a plot size of 3.2×5.0 m would result. In such a case, yield determination would definitely be affected by the difference in the area size planted.

Errors in row spacing for directly seeded row crops are sometimes caused by faulty furrowing. To avoid this error, the following procedures are suggested.

- If an animal-drawn furrower is used (single furrow per pass), the area width must be measured on both ends and marker sticks set for each furrow so that every prospective furrow has a stick on each end before furrowing starts.
- If a tractor-drawn furrower is used (usually with a three-furrow attachment), the distance between the attachments should be checked before furrowing. Furthermore, although there is no need to mark the area for every furrow, the distance between adjacent furrows must be checked every time the tractor turns.

Despite these suggested precautions, if anomalies in row spacing are not discovered early enough for re-furrowing, then one of the following alternatives should be chosen:

- Harvest only from portions of the plot where row spacing is correct.
- Harvest the whole plot and mathematically correct plot yield based on the discrepancy in area between normal and anomalous plots. This alternative should be used only when discrepancy in row spacing is not large.

To illustrate, consider an experiment with plot size 5 m long and 6 rows wide. If five rows in one plot were spaced correctly at 50 cm, but the sixth was spaced at 40 cm, then the total area occupied by the anomalous plot is 5.0×2.9 m instead of the intended 5×3 m.

If a mathematical correction is to be made, the corrected plot yield is computed as:

$$y = \frac{a}{b}x$$

where y and x are the corrected and uncorrected plot yields, respectively, and a and b are the intended and actual plot areas, respectively. Thus, in this example, the corrected plot yield, when no rows are considered as borders, is

$$y = \frac{(5)(3)}{(5)(2.9)}x = 1.034x$$

and the corrected plot yield, when the first and sixth rows are considered as borders, is

$$y = \frac{(5)(2)}{(5)(1.95)}x = 1.026x$$

If the partial-harvest alternative is chosen, yields should be harvested only from the first four rows for a nonbordered plot; and from the second, third, and fourth rows for a bordered plot.

Note that the adequacy of the mathematical correction diminishes as the ratio between a and b deviates farther from unity (one). On the other hand, anomalous rows can be excluded from harvest only when the plot is large enough.

14.2   SELECTION OF SEEDLINGS

For transplanted crops, seedlings are raised in seedbeds and transplanted to the experimental area. To assure sufficient plants, more seedlings are usually grown than are actually required. This allows for some type of seedling selection. The usual practice in transplanting is to use the best looking seedlings first and the poorer seedlings are leftovers. Although this procedure may be adequate when there are few leftover seedlings or when seedboxes are small, as in many vegetable crops, a more systematic selection should be used for such crops as rice where seedbeds are generally large and the ratio of available seedlings to those actually used in transplanting is much larger. For example, rice seedlings adjacent to the seedbed's edge are usually more healthy and vigorous than those in the middle. Use of such seedlings is not recommended because their inclusion could increase the variability among plants within an experimental plot. To ensure uniform plants, all seedlings used should be taken from the center of the seedbed where no undue advantage is afforded to any seedling. Furthermore, the seedbed should be visually examined before removing the seedlings so that unusually poor or unusually healthy seedlings can be excluded. At any rate, the basis for selection should be uniformity and not maximum vigor.

14.3   THINNING

For most directly seeded crops, a high seeding rate is used to ensure that enough seeds will germinate. Several days after germination, however, the number of plants per plot is reduced to a constant number by thinning (pulling out) the excess plants. Some of the more common factors in the thinning process that could increase experimental error are:

- *Selecting Plants to be Retained.* This decision is easy for crops planted in hills (e.g., maize). All that needs to be done is to remove the poorest looking plant in a hill until only the desired number (usually one or two) is left. The problem becomes more complicated for crops drilled in the row. Aside from trying to retain the best-looking seedlings, plants have to be kept at the prescribed spacing. In sorghum, for example, the prescribed distance between plants in the row, after thinning, is 10 cm. At this spacing, it is not easy to determine visually which plants to remove and which to retain. Usually each worker is provided with a stick about the length of a plot row that is properly marked every 10 cm. The worker then sets this stick beside every row to be thinned and only plants growing closest to the 10 cm marks are retained. Because this practice still depends to a certain degree on the judgment of the particular worker, the same worker should thin all plots in a replication and should finish the whole replication within a day (see Chapter 2, Section 2.2.1).

- *Presence of Vacant Spaces in the Row.* In row-sown crops, it is common to find spots in a row with no plants. While some excess seedlings can be transplanted in these vacant spots or gaps, such practice is not recommended because the transplants are usually weaker and less productive than the normal plants. What is usually done is to manipulate the thinning process so that the areas adjacent to the vacant space are left with more plants. For example, if there is a vacant space of about 30 cm in a row of sorghum that is to be thinned to one plant per 10 cm, the researcher can retain one plant more than required in the normal thinning at each end of the gap. The resulting spacing of the plants in a 1-m row will be as in Figure 14.1. This procedure is appropriate only if deficient spaces are not too wide or too numerous. For example, it would be impractical to use this procedure if the empty space in a sorghum row were 2 to 3 m.

The decision as to how much vacant space can be corrected by this procedure has to be left to the researcher. The vacant space must be small enough to be substantially compensated for by increasing the population in the surrounding area. Thus, the smaller the vacant area, the more complete is the compensation, and as the vacant space becomes larger, much space is wasted and the compensation becomes smaller. In sorghum, with 20-cm spacing between plants, the usual practice is to use this procedure only when the vacant space in a row is no more than 40 cm long. Otherwise, long vacant spots are excluded from the plot and appropriate correction is made on plot size.

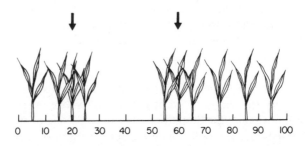

Figure 14.1 The spacing of plants in a 1-m row of sorghum, after thinning to adjust for a 30-cm gap when the intended spacing between plants is 10 cm. Filler plants are indicated by arrows.

- *Presence of Hills with Fewer Plants than Intended.* For directly seeded crops, despite all precautions to assure enough seedlings per hill, it is not uncommon to have some hills with fewer plants than needed. A researcher has two alternatives in dealing with this problem. He can ignore it at the seedling stage and then adjust to the same number of plants at harvest or he can attempt to correct the problem at the seedling stage by thinning. If the latter alternative is applied, the procedure is to retain more plants per hill in the adjacent hills so that the desired plant population is obtained. For example, two plants per hill are normally retained in maize performance tests. If a hill has only one plant, three plants could be retained in the next hill.

- *The Correct Stage for Thinning.* Thinning should be done before any tillers are formed so that counting of primary seedling stalks is not confused with tillers. Another consideration is the incidence of serious seedling insects and diseases. For experiments not primarily concerned with evaluating insect and disease reaction, allow the seedlings to pass through the most susceptible stage before completing the final thinning. In maize, for example, downy mildew is a serious problem, and during planting when the disease is a serious threat, thinning can be delayed for as long as a month after planting. Meanwhile any seedling affected by the disease is removed as soon as it is detected.

In sorghum, shoot fly is a serious seedling pest. Sorghum, however, develops resistance to shoot fly soon after the seedling stage. During plantings when the pest is serious, thinning can be done more than once. The first thinning is about a week after germination when twice as many plants as needed are left, and the second and final thinning about 2 weeks later. The first thinning reduces the variability in density of seedlings within and between plots while still allowing for flexibility in correcting for seedling damage in the susceptible period.

It should be realized, however, that removing damaged plants during the thinning process will tend to favor treatments that are more susceptible to the insect or disease. Thus, when many damaged plants are removed during thinning, the experimental results should be interpreted as applicable only to cases where the incidence of the insect or disease is negligible.

14.4 TRANSPLANTING

For transplanted crops, the number of seedlings per hill should be uniform. In most vegetable crops this is easy because seedlings are usually large and easily separated from each other. In rice, however, seedlings are crowded in the seedbed, and extra effort is needed to separate them into equal groups of seedlings. Thus, it is not uncommon in rice experiments for the number of transplanted seedlings to vary from two to five per hill. It has been alleged that these differences are corrected by tillering; that is, hills with fewer seedlings produce more tillers per plant and consequently their yields per hill remain constant despite the difference in the number of seedlings transplanted. Even if this were true, it does not necessarily follow that other plant characters will not also be affected. In fact, it has been shown that tiller number and number of productive panicles are higher for hills with more seedlings planted to a hill. Consequently, variability among hills for these characters would be larger in plots where the number of seedlings transplanted to a hill is not uniform. Therefore, an equal number of seedlings should be used per hill in rice experiments, especially when plant characters other than yield are to be measured through plot sampling (see Chapter 15).

14.5   FERTILIZER APPLICATION

Uniform application of fertilizer, even in large areas, is not much of a problem if mechanical fertilizer spreaders are available. All that is needed is a constant watch to ensure that the equipment is functioning properly and mechanical defects and imbalances are properly corrected. In most developing countries, however, fertilizer application is primarily done manually and uniformity is difficult to ensure, especially when large areas are involved. It is almost impossible to uniformly fertilize a hectare by hand, but it is a much simpler task to do so when only 10 to 50 m^2 areas are involved. Therefore, a large experimental area should be subdivided into smaller units before fertilizer is applied. This can be conveniently done by fertilizing each plot, or each row of a plot, separately. Fertilizer for each subunit can then be measured and applied separately. The usual procedure is to weigh the fertilizers in the laboratory, put them in small paper bags or in plastic sacks, transport them to the field, and spread the contents of each sack uniformly over a small unit.

This procedure, however, becomes laborious and time consuming when the subunit is small and the number of subunits is large. An alternative is to measure the fertilizers, usually by volume, right in the field. To avoid both errors and delays in the process, the measurement must be accurate and fast. The field worker carries bulked fertilizer and a volume measuring container (empty milk can, a spoon, etc.). He fills the measuring container, spreads the contents over the specified unit area, moves to the next unit, fills the same container again, and empties it over this area, and so on. This procedure saves sacks as well as weighing time.

14.6   SEED MIXTURES AND OFF-TYPE PLANTS

The seed materials used in an experiment can be contaminated mechanically or biologically from the previous harvest. This cannot be detected in the seeds but is usually easy to pinpoint a month or so after planting. Thus in a solid stand of pure lines of rice, some exceptionally tall or vigorous plants may be detected. These plants are referred to as mixtures or off-types. These plants in experimental plots, especially when detected after thinning, pose several problems:

- Off-types cannot be treated as normal plants because their performance is definitely affected not only by the treatments being evaluated but also substantially by genetic makeup different from the others.
- Off-types cannot simply be ignored, pulled, or removed, because by the time they are detected they could already have affected surrounding plants through a competition effect that is different from the normal plants. Even if the detection is early enough, their removal creates missing hills and would probably affect the surrounding plants.

Thus, the usual procedure is to allow the off-types to grow up to maturity, and count and remove them from the field just before harvest. The yield of the plot is then computed as

$$Y = \left(\frac{a + b}{a} \right) X$$

where Y is the corrected plot yield, X is the actual grain weight from harvest of a normal plants in the plot, and b is the number of off-types.

For example, consider a plot with a complete stand of 50, in which 2 plants turned out to be off-types. The 2 off-types were removed and the remaining 48 plants yielded a harvest of 500 g. Thus, the corrected plot yield is

$$500 \times \frac{50}{48} = 521 \ g$$

This correction assumes that the competition effects provided by the off-types to the surrounding plants are similar to those of the normal plants. Such a correction procedure is applicable, for instance, in field experiments with rice.

14.7 PLOT LAYOUT AND LABELING

One of the most common errors in laying out field experiments is in measuring plot dimensions. For plots with the length of 5 to 10 m, an error of 0.1 to 0.5 m is not easily visible. But because even a small error in plot dimension can greatly affect the experimental results, it is important to double-check plot dimensions as plots are laid out in the field.

Confusion sometimes occurs in making decisions regarding the border line of a plot when crops are planted in rows. For example, for a 6-row plot with a 30-cm distance between rows, the width of the plot is 1.8 m, with the starting and ending points as shown in Figure 14.2. Note that the plot width is *not*

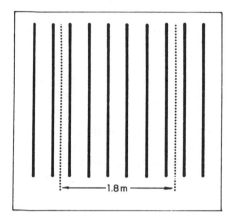

Figure 14.2 Boundaries of plot width in a 6-row plot with a 30-cm distance between rows.

measured from the first to the last row of the plot. Instead, the width of the plot should include half of the space between rows on each side of the plot.

Mistakes in plot labeling can occur as early as seed preparation or plot marking. An incorrect label can be attached either to the seed packet or to the plot. Although these errors are sometimes easily detected and corrected by an experimenter familiar with his experimental materials, it is not uncommon that the error remains undetected up to the end of the experiment. Only extreme care or counter-checking, or both, by at least two persons can reduce this source of error.

14.8 MEASUREMENT ERRORS

Reading of weight, height, and other such physical measurements in experiments can become monotonous if repeated for long periods of time. Once boredom sets in, measurements can be easily misread, misheard, or miscopied. Many of these mistakes can be corrected by rechecking the collected data at the end of each day. For example, in measuring plant height, the researcher should review the data collected for the day immediately after completing the measurement, to detect any unusually high or unusually low readings. The plants will still be in the field the following day, and doubtful figures can be rechecked by actual measurements of the specific plants.

The same procedure can be followed for any other character in which a repeat measurement is still possible. In determining grain weight, the grains from separate plots should not be mixed immediately after weighing but should be kept separate until at least the editing of data is completed and, if possible, until the analysis of data has been completed and the final report made. This practice will allow for rechecking whenever necessary during data processing.

Another common source of errors in measurement is the differential effects of the enumerators, that is, the persons making the measurements. In recording height of rice plants, for instance, plant height is measured from the ground level to the tip of the tallest leaf for young plants and to the tip of the tallest panicle for adult plants. The decision on the definite position of either the *ground level* or the *tip of the tallest leaf* (*or panicle*) leaves room for personal judgment. The possible differences in measurements of such characters among observers have been recognized. To minimize such errors, the number of persons involved in the measurement process must be kept at minimum. When several enumerators are involved, the same person should measure all plots in one replication (see Chapter 2, Section 2.2.1).

14.9 TRANSCRIPTION OF DATA

Data in the field book is primarily arranged for ease of measurement in the field. At times, this may not be the form convenient for statistical analysis. Thus, transcribing of data from the field book to the form required for data analysis may be necessary. The number of times data are transcribed depends on the complexity of the computation desired and, to a certain extent, on the whims of the researcher. The researcher should minimize the number of times data are transcribed. Aside from saving time and expense, the chance for errors is also reduced. Indeed, it is strongly advised that the field book should be designed so that statistical analysis can be done directly from it. If this is not possible, a plan should be evolved so that only one data transcription is needed for all analyses. For all transcriptions, two persons must proofread any data transcribed.

CHAPTER 15

Sampling in Experimental Plots

Plot size for field experiments is usually selected to achieve a prescribed degree of precision for measurement of the character of primary interest. Because the character of primary interest—usually economic yield such as grain yield for grain crops, stover yield for forage crops, and cane yield for sugar cane—is usually the most difficult to measure, the plot size required is often larger than that needed to measure other characters. Thus, expense and time can be saved if the measurements of additional characters of interest are made by sampling a fraction of the whole plot. For example, make measurements for plant height from only 10 of the 200 plants in the plot; for tiller number, count only 1 m² of the 15 m² plot; and for leaf area, measure from only 20 of the approximately 2,000 leaves in the plot.

There are times, however, when the choice of plot size may be greatly influenced by the management practices used or the treatments tested. In an insecticide trial, for example, relatively large plots may be required to minimize the effect of spray drift or to reduce the insect movement caused by insecticide treatments in adjacent plots. In such cases, plot size would be larger than that otherwise required by the character of primary interest. Consequently, even for the primary character such as grain yield, it may still be desirable to sample from a fraction of the whole plot.

An appropriate sample is one that provides an estimate, or a sample value, that is as close as possible to the value that would have been obtained had all plants in the plot been measured—the plot value. The difference between the sample value and the plot value constitutes the *sampling error*.

Thus, a good sampling technique is one that gives a small sampling error. In this chapter we deal with the basic features of sampling technique as applied to replicated field trials (plot sampling) and the development of an appropriate sampling technique for a given field experiment.

15.1 COMPONENTS OF A PLOT SAMPLING TECHNIQUE

For plot sampling, each experimental plot is a *population*. *Population value*, which is the same as the plot value, is estimated from a few plants selected from each plot. The procedure for selecting the plants to be measured and used for estimating the plot value is called the *plot sampling technique*. To develop a plot sampling technique for the measurement of a character in a given trial, the researcher must clearly specify the sampling unit, the sample size, and the sampling design.

15.1.1 Sampling Unit

The *sampling unit* is the unit on which actual measurement is made. Where each plot is a population, the sampling unit must necessarily be smaller than a plot. Some commonly used sampling units in replicated field trials are a leaf, a plant, a group of plants, or a unit area. The appropriate sampling unit will differ among crops, among characters to be measured, and among cultural practices. Thus, in the development of a sampling technique, the choice of an appropriate sampling unit should be made to fit the requirements and specific conditions of the individual experiments.

The important features of an appropriate sampling unit are:

- *Ease of Identifications.* A sampling unit is easy to identify if its boundary with the surrounding units can be easily recognized. For example, a single hill is easy to identify in transplanted rice because each hill is equally spaced and is clearly separated from any of the surrounding hills. In contrast, plant spacing in broadcasted rice is not uniform and a single hill is, therefore, not always easy to identify. Consequently, a single hill may be suitable as a sampling unit for transplanted rice but not for broadcasted rice.

- *Ease of Measurement.* The measurement of the character of interest should be made easy by the choice of sampling unit. For example, in transplanted rice, counting tillers from a 2 × 2-hill sampling unit can be done quite easily and can be recorded by a single number. However, the measurement of plant height for the same sampling unit requires independent measurements of the four hills, the recording of four numbers and, finally, the computation of an average of those four numbers.

- *High Precision and Low Cost.* Precision is usually measured by the reciprocal of the variance of the sample estimate; while cost is primarily based on the time spent in making measurements of the sample. The smaller the variance, the more precise the estimate is; the faster the measurement process, the lower the cost is. To maintain a high degree of precision at a reasonable cost, the variability among sampling units within a plot should be kept small. For example, in transplanted rice, variation between single-hill sampling units for tiller count is much larger than that for plant height. Hence, although a single-hill sampling unit may be appropriate for plant height, it may not be so for tiller count. Thus, for transplanted rice, the use of a 2×2-hill sampling unit for tiller count and a single-hill sampling unit for the measurement of plant height is common.

15.1.2   Sample Size

The number of sampling units taken from the population is *sample size*. In a replicated field trial where each plot is a population, sample size could be the number of plants per plot used for measuring plant height, the number of leaves per plot used for measuring leaf area, or the number of hills per plot used for counting tillers. The required sample size for a particular experiment is governed by:

- The size of the variability among sampling units within the same plot (sampling variance)
- The degree of precision desired for the character of interest

In practice, the size of the sampling variance for most plant characters is generally not known to the researcher. We describe procedures for obtaining estimates of such variances in Section 15.2. The desired level of precision can, however, be prescribed by the researcher based on experimental objective and previous experience. The usual practice is for the researcher to prescribe the desired level of precision in terms of the margin of error, either of the plot mean or of the treatment mean. For example, the researcher may prescribe that the sample estimate should not deviate from the true value by more than 5 or 10%.

With an estimate of the sampling variance, the required sample size can be determined based on the prescribed margin of error, of the plot mean, or of the treatment mean.

15.1.2.1   Margin of Error of the Plot Mean.   The sample size for a simple random sampling design (Section 15.1.3.1) that can satisfy a prescribed margin of error of the plot mean is computed as:

$$n = \frac{(Z_\alpha^2)(v_s)}{(d^2)(\overline{X}^2)}$$

where n is the required sample size, Z_α is the value of the standardized normal variate corresponding to the level of significance α (the value Z_α can be obtained from Appendix B), v_s is the sampling variance, $\overline{X}$ is the mean value, and d is the margin of error expressed as a fraction of the plot mean.

For example, a researcher may wish to measure the number of panicles per hill in transplanted rice plots with a single hill as the sampling unit. Using data from previous experiments (see Section 15.2.1) he estimates the variance in panicle number between individual hills within the same plot (v_s) to be 5.0429 —or a *cv* of 28.4% based on the average number of panicles per hill of 17.8. He prescribes that the estimate of the plot mean should be within 8% of the true value. The sample size that can satisfy the foregoing requirement, at the 5% level of significance, can be computed as:

$$n = \frac{(1.96)^2(5.0429)}{(0.08)^2(17.8)^2}$$

$$= 9.6 \approx 10 \text{ hills/plot}$$

Thus, panicle number should be counted from 10 single-hill sampling units per plot to ensure that the sample estimate of the plot mean is within 8% of the true value 95% of the time.

15.1.2.2 Margin of Error of the Treatment Mean. The information of primary interest to the researcher is usually the *treatment mean* (the average over all plots receiving the same treatment) rather than the *plot mean* (the value from a single plot). Thus, the desired degree of precision is usually specified in terms of the margin of error of the treatment mean rather than of the plot mean. In such a case, sample size is computed as:

$$n = \frac{(Z_\alpha^2)(v_s)}{r(D^2)(\bar{X}^2) - (Z_\alpha^2)(v_p)}$$

where n is the required sample size, Z_α and v_s are as defined in the equation in Section 15.1.2.1, v_p is the variance between plots of the same treatment (i.e., experimental error), and D is the prescribed margin of error expressed as a fraction of the treatment mean. Take note that, in this case, additional information on the size of the experimental error (v_p) is needed to compute sample size.

To illustrate, consider the same example we used in Section 15.1.2.1. For an experiment with four replications, the researcher wishes to determine the sample size that can achieve an estimate of the treatment mean within 5% of the true value. Using an estimate of $v_p = 0.1964$ (see Section 15.2.1), sample size that can satisfy this requirement at the 5% level of significance can be computed as:

$$n = \frac{(1.96)^2(5.0429)}{4(0.05)^2(17.8)^2 - (1.96)^2(0.1964)}$$

$$= 8.03 \approx 8 \text{ hills/plot}$$

Thus, eight single hills per plot should be measured to satisfy the requirement that the estimate of the treatment mean would be within 5% of the true value 95% of the time.

15.1.3 Sampling Design

A *sampling design* specifies the manner in which the n sampling units are to be selected from the whole plot. There are five commonly used sampling designs in replicated field trials: simple random sampling, multistage random sampling, stratified random sampling, stratified multistage random sampling, and subsampling with an auxiliary variable.

15.1.3.1 Simple Random Sampling.   In a simple random sampling design, there is only one type of sampling unit and, hence, the sample size (n) refers to the total number of sampling units to be selected from each plot consisting of N units. The selection of the n sampling units is done in such a way that each of the N units in the plot is given the same chance of being selected. In plot sampling, two of the most commonly used random procedures for selecting n sampling units per plot are the *random-number technique* and the *random-pair technique*.

15.1.3.1.1 The Random-Number Technique.   The random-number technique is most useful when the plot can be divided into N distinct sampling units, such as N single-plant sampling units or N single-hill sampling units. We illustrate the steps in applying the random-number technique with a maize variety trial where plant height in each plot consisting of 200 distinct hills is to be measured from a simple random sample of six single-hill sampling units.

☐ STEP 1.   Divide the plot into N distinctly differentiable sampling units (e.g., N hills/plot if the sampling unit is a single hill or N 1×1 cm sub-areas per plot if the sampling unit is a 1×1 cm area) and assign a number from 1 to N to each sampling unit in the plot.

For our example, because the sampling unit is a single hill, the plot is divided into $N = 200$ hills, each of which is assigned a unique number from 1 to 200.

☐ STEP 2.   Randomly select n distinctly different numbers, each within the range of 1 to N, following a randomization scheme described in Chapter 2, Section 2.1.1.

For our example, $n = 6$ random numbers (each within the range of 1 to 200) are selected from the table of random numbers, following the procedure described in Chapter 2, Section 2.1.1. The six random numbers selected may be:

Sequence	Random Number
1	78
2	17
3	3
4	173
5	133
6	98

☐ STEP 3.   Use, as the sample, all the sampling units whose assigned numbers (step 1) correspond to the random numbers selected in step 2. For our example, the six hills in the plot whose assigned numbers are 78, 17, 3, 173, 133, and 98 are used as the sample.

15.1.3.1.2   The Random-Pair Technique. The random-pair technique is applicable whether or not the plot can be divided uniquely into N sampling units. Hence, the technique is more widely used than the random-number technique. We illustrate the procedure with two cases—one where the plot can be divided into N distinct sampling units and another where clear division cannot be done.

Case I is one with clear division of N sampling units per plot. For illustration, we use the example in Section 15.1.3.1.1. Assuming that the plot consists of 10 rows and 20 hills per row ($N = 200$ hills), the steps involved in applying the random-pair technique to select a random sample of $n = 6$ single-hill sampling units are:

☐ STEP 1.   Determine the width (W) and the length (L) of the plot in terms of the sampling unit specified, such that $W \times L = N$. For our example, the sampling unit is a single hill; and $W = 10$ rows, $L = 20$ hills, and $N = (10)(20) = 200$.

☐ STEP 2.   Select n random pairs of numbers, with the first number of each pair ranging from 1 to W and the second number ranging from 1 to L; where W and L are as defined in step 1.

For our example, $n = 6$ random pairs of numbers are selected by using the table-of-random-number procedure described in Chapter 2, Section 2.1.1, with the restrictions that the first number of the pair must not exceed W (i.e., it must be within the range from 1 to 10) and the second number of the pair must not exceed L (i.e., it must be in the range from 1 to 20). The six random pairs of numbers may be as follows:

> 7, 6
> 6, 20
> 2, 3
> 3, 9
> 9, 15
> 1, 10

☐ STEP 3. Use the point of intersection of each random pair of numbers, derived in step 2, to represent each selected sampling unit. For our example, the first selected sampling unit is the sixth hill in the seventh row, the second selected sampling unit is the twentieth hill in the sixth row, and so on. The location of the six selected single-hill sampling units in the plot is shown in Figure 15.1.

Case II is one without clear division of N sampling units per plot. For illustration, consider a case where a sample of six 20×20-cm sampling units is to be selected at random from a broadcast-rice experimental plot measuring 4×5 m (after exclusion of border plants). The steps involved in applying the random-pair technique to select a random sample of $n = 6$ sampling units are:

☐ STEP 1. Specify the width (W) and length (L) of the plot using the same measurement unit as that of the sampling unit. For our example, the centimeter is used as the measurement unit because the sampling unit is defined in that scale. Thus, the plot width (W) and length (L) are specified as 400 cm and 500 cm. Note that with this specification, the division of the plot into N distinct sampling units cannot be made.

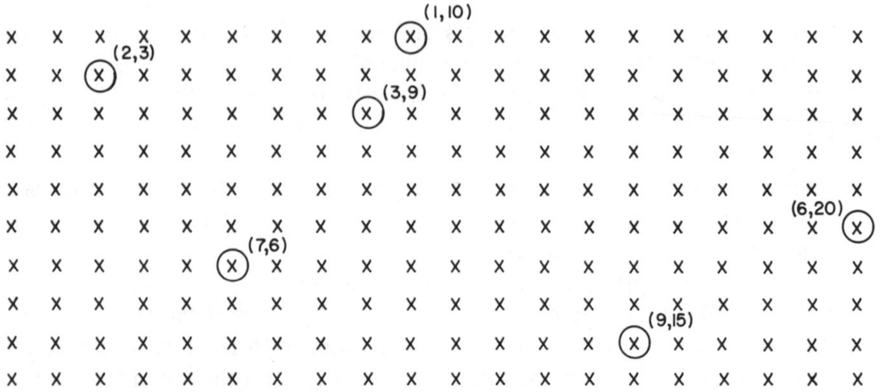

Figure 15.1 The location of six randomly selected sample hills, using the random-pair technique, for a plot consisting of 10 rows and 20 hills per row.

☐ STEP 2.   Select *n* random pairs of numbers, following the table-of-random-number procedure described in Chapter 2, Section 2.1.1, with the first number of the pair lying between 1 and *W* and the second number lying between 1 and *L*.

For our example, the six random pairs of numbers may be:

253, 74

92, 187

178, 167

397, 394

186, 371

313, 228

☐ STEP 3.   Use the point of intersection of each of the random pairs of numbers (derived in step 2) to represent the starting point of each selected sampling unit. For our example, we consider the starting point to be the uppermost left corner of each sampling unit. Thus, with the first random pair of (253, 74) the first selected sampling unit is the 20 × 20-cm area whose uppermost left corner is at the intersection of the 253 cm along the width of the plot and the 74 cm along the length of the plot (see Figure 15.2). The rest of the selected sampling units can be identified in the similar manner. The locations of the six selected 20 × 20-cm sampling units in the plot is shown in Figure 15.2.

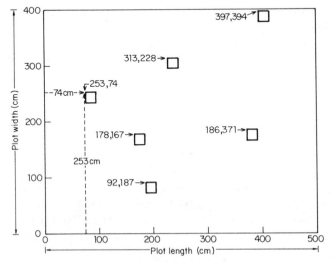

Figure 15.2   The location of six randomly selected 20 × 20-cm sampling units, using the random-pair technique for a plot measuring 4 × 5 m.

15.1.3.2   *Multistage Random Sampling.*

In contrast to the simple random sampling design, where only one type of sampling unit is involved, the multistage random sampling design is characterized by a series of sampling stages. Each stage has its own unique sampling unit. This design is suited for cases where the best sampling unit is not the same as the measurement unit. For example, in a rice field experiment, the unit of measurement for panicle length is a panicle and that for leaf area is a leaf. The use of either the panicle or the leaf as the sampling unit, however, would require the counting and listing of all panicles or all leaves in the plot—a time-consuming task that would definitely not be practical.

In such cases, a single hill may still be used as the basic sampling unit. However, to avoid difficulty of measuring all leaves or all panicles in each sample hill, a multistage random sampling design could be used to identify a few sample leaves or sample panicles that need to be measured in each sample hill. Thus, such a design provides more than one type of sampling unit and, subsequently, more than one stage of sampling.

In the measurement of panicle length, for example, a two-stage sampling design with individual hills as the primary sampling unit and individual panicles as the secondary sampling unit can be employed. This would involve the application of a simple random sampling design twice—once to the primary sampling unit (the hill) and another to the secondary sampling unit (the panicle). To get this, a simple random sample of n_1 hills would first be taken from the plot (first-stage sampling) and a simple random sample of n_2 panicles would then be taken from each of the selected n_1 hills (second-stage sampling). This would result in $n = (n_1)(n_2)$ sample panicles per plot.

The extension of the multistage random sampling design to three, four, or more stages is straightforward. For example, in the case where leaf area is to be measured, a three-stage sampling design, with individual hills as the primary sampling unit, individual tillers as the secondary sampling unit, and individual leaves as the tertiary sampling unit, would be appropriate.

The selection of the sample is done separately and independently at each stage of sampling, starting with the first-stage sampling, then the second-stage sampling, and so on, in the proper sequence. At each sampling stage, the random selection procedure follows that of the simple random sampling design described in Section 15.1.3.1. For example, in the case of the two-stage sampling design for the measurement of the panicle length, the selection process starts with the random selection of n_1 single-hill sampling units from the plot. Then, for each of the n_1 sample hills, the total number of panicles is determined and the random-number technique is applied to select a random sample of n_2 panicles from the particular hill. This random selection process is repeated n_1 times, separately and independently for each of the n_1 sample hills, resulting in the total of $n = (n_1)(n_2)$ panicles, on which panicle length is to be measured.

15.1.3.3   *Stratified Random Sampling.*

In a stratified random sampling design, sampling units within a plot are first grouped into k strata before a set of m sampling units is selected randomly from each stratum. Thus, the total number of sampling units per plot (n) is equal to (m)(k).

The stratified random sampling design is useful where there is large variation between sampling units and where important sources of variability follow a consistent pattern. In such cases, the precision of the sample estimate can be improved by first grouping the sampling units into different strata in such a way that variability between sampling units within a stratum is smaller than that between sampling units from different strata. Some examples of stratification criterion used in agricultural experiments are:

- *Soil Fertility Pattern.* In an insecticide trial where blocking was based primarily on the direction of insect migration, known patterns of soil fertility cause substantial variability among plants in the same plot. In such a case, a stratified random sampling design may be used so that each plot is first divided into several strata based on the known fertility patterns and sample plants are then randomly selected from each stratum.
- *Stress Level.* In a varietal screening trial for tolerance for soil salinity, areas within the same plot may be stratified according to the salinity level before sample plants are randomly selected from each stratum.
- *Within-Plant Variation.* In a rice hill, panicles from the taller tillers are generally larger than those from the shorter ones. Hence, in measuring such yield components as panicle length or number of grains per panicle, panicles within a hill are stratified according to the relative height of the tillers before sample panicles are randomly selected from each position (or stratum).

It should be noted at this point that the stratification technique is similar to the blocking technique, described in Chapter 2, Section 2.2.1. It is effective only if it can ensure that the sampling units from the same stratum are more similar than those from different strata. Thus, the efficiency of the stratified random sampling design, relative to the simple random sampling design, will be high only if an appropriate stratification technique is used.

15.1.3.4 Stratified Multistage Random Sampling.
When the stratification technique of Section 15.1.3.3 is combined with the multistage sampling technique of Section 15.1.3.2, the resulting design is known as stratified multistage random sampling. In it, multistage sampling is first applied and then stratification is used on one or more of the identified sampling stages.

For example, consider the case where a rice researcher wishes to measure the average number of grains per panicle through the use of a two-stage sampling design with individual hills in the plot as the primary sampling unit and individual panicles in a hill as the secondary sampling unit. He realizes that the number of grains per panicle varies greatly between the different panicles of the same hill. Hence, if the m panicles from each selected hill were selected entirely at random (i.e., a multistage random sampling design), the high variability between panicles within hill would cause the precision of the sample estimate to be low. A logical alternative is to apply the stratification technique by dividing the panicles in each selected hill (i.e., primary sampling unit) into k strata, based on their relative position in the hill, before a simple random sample of m panicles from each stratum is taken separately and independently for the k strata.

For example, if the panicles in each selected hill are divided into two strata based on the height of the respective tillers—the taller and shorter strata—and a random sample of three panicles taken from each stratum, the total number of sample panicles per plot would be $(2)(3)(a) = 6a$, where a is the total number of randomly selected hills per plot. In this case, the sampling technique is based on a two-stage sampling design with stratification applied on the secondary unit. Of course, instead of the secondary unit (panicles) the researcher could have stratified the primary unit (i.e., single-hill) based on any source of variation pertinent to his experiment (see also Section 15.1.3.3). In that case, the sampling technique would have been a two-stage sampling design with stratification of the primary unit. Or, the researcher could have applied both stratification criteria—one on the hills and another on the panicles—and the resulting sampling design would have been a two-stage sampling with stratification of both the primary and secondary units.

15.1.3.5 Subsampling With an Auxiliary Variable. The main features of a design for subsampling with an auxiliary variable are:

- In addition to the character of interest, say X, another character, say Z, which is closely associated with and is easier to measure than X, is chosen.
- Character Z is measured both on the main sampling unit and on the subunit, whereas variable X is measured only on the subunit. The subunit is smaller than the main sampling unit and is embedded in the main sampling unit.

This design is usually used when the character of interest, say X, is so variable that the large size of sampling unit or the large sample size required to achieve a reasonable degree of precision, or both, would be impractical. To improve the precision in the measurement of X, without unduly increasing either the sample size or the size of sampling unit, the subsampling with an auxiliary variable design can be used.

Improvement is achieved by measuring Z from a unit that is larger than the unit where X is measured. By using the known relationship between Z and X, it is as if X were measured from the large unit. With the proper choice of the auxiliary variable Z, a large increase in the degree of precision can be achieved with only a small increase in the cost of measuring Z from a larger unit. This means that Z must be chosen to best satisfy two conditions:

- Z must be closely associated with X and its relationship known.
- Measurement of Z must be with minimum cost.

For example, weed count is usually one of the characters of primary interest in evaluating the effect of weed infestation. Weed count is, however, highly variable and requires a relatively large sampling unit to attain a reasonable degree of accuracy. Furthermore, the task of counting weeds is tedious and time consuming, and its cost increases proportionally with the size of the sampling unit. On the other hand, weed weight, which is closely related to weed count, is simpler to measure and its measurement cost is only slightly affected by the size of the sampling unit. Thus, weed weight offers an ideal choice as the auxiliary variable for weed count.

To count weeds in a replicated field trial, the following sampling plan, based on the subsampling design with weed weight as the auxiliary variable, may be used:

- A sample of n 60 × 60-cm sampling units is randomly selected from each plot.
- From each of the n units, weed weight (Z) and weed count (X) is measured on a subsample (say 20 × 20-cm subunit) while the rest of the weeds in the main sampling unit is used only for measuring weed weight (Z).

15.1.4 Supplementary Techniques

So far, we have discussed sampling techniques for individual plots, each of which is treated independently and without reference to other plots in the same experiment. However, in a replicated field trial where the sampling technique is to be applied to each and all plots in the trial, a question usually raised is whether the same set of random sample can be repeated in all plots or whether different random processes are needed for different plots. And, when data of a plant character are measured more than once over time, the question is whether the measurements should be made on the same samples at all stages of observation or should rerandomization be applied.

The two techniques aimed at answering these questions are *block sampling* and *sampling for repeated measurements*.

15.1.4.1 Block Sampling.

Block sampling is a technique in which all plots of the same block (i.e., replication) are subjected to the same randomization scheme (i.e., using the same sample locations in the plot) and different sampling schemes are applied separately and independently for different blocks. For example, for a RCB experiment, the total number of times that the randomization process is applied is the number of replications (r). Consider a case where the researcher wishes to measure panicle number in a RCB trial with eight treatments and four replications. He decides to use a simple random sampling design with the single-hill sampling unit and a sample size of six. With block sampling, he needs only to apply the randomization scheme four times to obtain four different sample locations, as shown in Figure 15.3. The first is used for all eight plots of replication I, the second for all eight plots of replication II, the third for all eight plots of replication III, and the fourth for all eight plots of replication IV.

The block sampling technique has four desirable features. They are:

1. Randomization is minimized. With block sampling, randomization is done only r times instead of $(r)(t)$ times as it is when randomization is done separately for each and all plots.

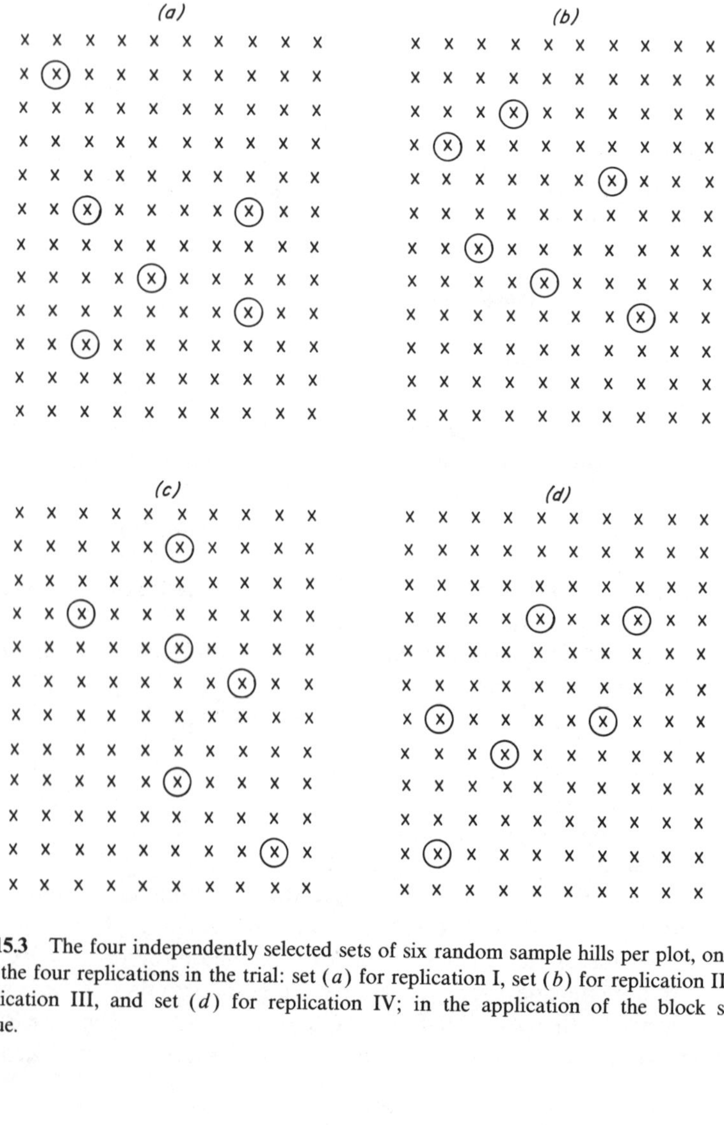

Figure 15.3 The four independently selected sets of six random sample hills per plot, one set for each of the four replications in the trial: set (a) for replication I, set (b) for replication II, set (c) for replication III, and set (d) for replication IV; in the application of the block sampling technique.

2. Data collection is facilitated. With block sampling, all plots in the same block have the same pattern of sample locations so that an observer (data collector) can easily move from plot to plot within a block without the need to reorient himself to a new pattern of sample location.

3. Uniformity between plots of the same block is enhanced because there is no added variation due to changes in sample locations from plot to plot.

4. Data collection by block is encouraged. For example, if data collection is to be done by several persons, each can be conveniently assigned to a particular block which facilitates the speed and uniformity of data collection. Even if there is only one observer for the whole experiment, he can complete the task one block at a time, taking advantage of the similar sample locations of plots in the same block and minimizing one source of variation among plots, namely, the time span in data collection.

15.1.4.2 Sampling for Repeated Measurements.

Plant characters are commonly measured at different growth stages of the crop. For example, tiller number in rice may be measured at 30, 60, 90, and 120 days after transplanting or at the tillering, flowering, and harvesting stages. If such measurements are made on the same plants at all stages of observation, the resulting data may be biased because plants that are subjected to frequent handlings may behave differently from others. In irrigated wetland rice, for example, frequent trampling around plants, or frequent handling of plants not only affect the plant characters being measured but also affect the plants' final yields. On the other hand, the use of an entirely different set of sample plants at different growth stages could introduce variation due to differences between sample plants.

The *partial replacement procedure* provides for a satisfactory compromise between the two conflicting situations noted. With partial replacement, only a portion p of the sample plants used in one growth stage is retained for measurement in the succeeding stage. The other portion of $(1 - p)$ sample plants is randomly obtained from the remaining plants in the plot. The size of p depends on the size of the estimated undesirable effect of repeated measurements of the sample plants in a particular experiment. The smaller this effect, the larger p should be. For example, in the measurement of plant height and tiller number in transplanted rice, p is usually about .75. That is, about 75% of the sample plants measured at a given growth stage is retained for measurement in the succeeding stage and the remaining 25% is obtained at random from the other plants in the plot. Figure 15.4 shows the locations of the 15 single-hill sampling units per plot for the measurement of plant height at two growth stages, using the partial replacement procedure with 25% replacement $(1 - p = .25)$. At each growth stage, measurement is made from a total of 12 sample hills, out of which nine hills are common to both growth stages.

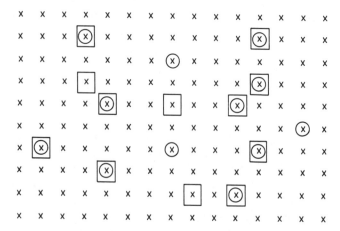

Figure 15.4 Fifteen sample hills (randomly selected with 25% replacement) for height measurement at two growth stages: 12 sample hills for the first stage are ⊗ and ⊗̄ , and 12 sample hills for the second stage are ⊠ and ⊗̄ .

15.2 DEVELOPING AN APPROPRIATE PLOT SAMPLING TECHNIQUE

A plot sampling technique that is appropriate for a particular character in a particular field experiment must satisfy the following requirements:

- The precision of the estimate obtained must be as good as, or better than, a level needed to achieve the experimental objective.
- The cost of its implementation must be within the resources available to the researcher.

Precision of a sample estimate generally increases with the size of sampling unit, the number of sampling units selected per plot, and the complexity of the sampling design used. However, an increase in either the size or the number of sampling units almost always results in an increase in cost. Hence, the choice of an appropriate sampling technique is primarily concerned with maintaining the proper balance between the size of sampling unit, sample size, and a sampling design to achieve the minimum cost.

How, then, does a researcher select the proper combination of sampling unit, sample size, and sampling design? The task requires information on the variability of the character of interest so that the precisions that will result from the various types of sampling technique can be estimated. There are three sources of data from which the required information can be obtained: data from previous experiments, additional data from on-going experiments, and data from specifically planned sampling studies.

Table 15.1 Data on Panicle Number per Hill Obtained from a Simple Random Sample of 12 Single-Hill Sampling Units per Plot in a Variety Trial Involving Eight Rice Varieties and Three Replications

Variety	Rep. I	Rep. II	Rep. III
	Panicles, no./hill		
IR22	5, 8, 12, 14, 10, 10	10, 13, 10, 13, 11, 11	7, 6, 11, 10, 7, 8
	6, 10, 8, 11, 11, 8	12, 5, 10, 7, 14, 5	8, 8, 10, 10, 6, 11
IR160-27-3	11, 11, 11, 12, 4, 12	13, 4, 4, 7, 5, 7	8, 7, 9, 10, 5, 5
	8, 14, 8, 7, 9, 9	11, 8, 7, 8, 10, 9	9, 10, 4, 9, 12, 11
BPI-76-1	4, 5, 8, 5, 8, 4	6, 8, 4, 5, 6, 10	8, 7, 6, 5, 6, 7
	5, 9, 6, 6, 7, 10	8, 3, 7, 8, 7, 11	6, 8, 6, 6, 5, 4
C4-63	8, 10, 9, 7, 9, 7	9, 7, 9, 5, 8, 9	8, 10, 7, 6, 7, 6
	9, 13, 13, 5, 7, 5	8, 10, 6, 5, 6, 5	9, 8, 6, 4, 5, 7
RD-3	7, 12, 7, 11, 12, 7	9, 7, 6, 8, 4, 8	9, 3, 4, 6, 5, 3
	7, 6, 5, 9, 8, 9	8, 9, 8, 9, 6, 7	9, 7, 9, 6, 6, 7
IR480-5-9	7, 7, 6, 11, 7, 8	8, 10, 7, 6, 8, 8	7, 6, 9, 7, 11, 8
	8, 8, 9, 6, 4, 14	10, 5, 7, 5, 8, 7	12, 7, 8, 9, 8, 9
Jaya	8, 9, 12, 7, 7, 3	8, 6, 7, 8, 9, 9	10, 4, 8, 9, 4, 6
	10, 10, 8, 7, 9, 8	14, 8, 9, 11, 6, 7	7, 4, 3, 4, 4, 6
IR20	5, 5, 10, 9, 7, 5	8, 8, 8, 3, 13, 13	5, 12, 10, 9, 7, 9
	9, 10, 9, 6, 12, 8	7, 12, 9, 9, 8, 11	8, 7, 5, 8, 10, 7

15.2.1 Data from Previous Experiments

Data from previous experiments can usually be analyzed at minimum cost, to provide valuable information for use in improving plot sampling techniques in subsequent experiments. To illustrate, we use a rice trial with eight rice varieties tested in a randomized complete block design with three replications. Data on panicle number were collected based on a simple random sampling design with 12 single-hill sampling units per plot (Table 15.1). Using these data, the variability between single-hill sampling units can be estimated and the efficiencies of various sample sizes evaluated. The step-by-step procedures

Table 15.2 Analysis of Variance (RCB Design) of Data (Table 15.1) from Plot Sampling

Source of Variation	Degree of Freedom	Sum of Squares	Mean Square
Replication	2	53.5208	26.7604
Variety	7	191.0556	27.2937
Experimental error	14	103.5903	7.3993
Sampling error	264	1,331.3333	5.0429

for doing so are:

☐ STEP 1.   Compute the analysis of variance of data from plot sampling based on a RCB design, following the procedure described in Chapter 6, Section 6.1. The result is shown in Table 15.2.

☐ STEP 2.   Compute the estimates of sampling variance (i.e., the variance between hills within a plot) and of experimental error (i.e., the variance between plots of the same treatment) as:

$$s_1^2 = MS_1$$

$$s_2^2 = \frac{MS_2 - MS_1}{n}$$

where MS_1 is the sampling error mean square and MS_2 is the experimental error mean square in the analysis of variance computed in step 1, and n is the sample size (i.e., the number of sample hills per plot). For our example, the estimates of sampling variance and experimental error are computed as:

$$s_1^2 = 5.0429$$

$$s_2^2 = \frac{7.3993 - 5.0429}{12} = 0.1964$$

☐ STEP 3.   Compute the estimates of the variance of a treatment mean and of the corresponding *cv* value (i.e., standard error of the treatment mean expressed as percent of the mean value) as:

$$v(\overline{X}) = \frac{s_1^2 + ns_2^2}{rn}$$

$$cv(\overline{X}) = \frac{100\sqrt{v(\overline{X})}}{\overline{X}}$$

where r is the number of replications, $\overline{X}$ is the grand mean, and s_1^2 and s_2^2 are as computed in step 2. For our example, the estimates of the variance of a treatment mean and its *cv* value are:

$$v(\overline{X}) = \frac{5.0429 + (12)(0.1964)}{(3)(12)}$$

$$= 0.2055$$

$$cv(\overline{X}) = \frac{(100)\sqrt{0.2055}}{8}$$

$$= 5.7\%$$

Note that the values of $v(\overline{X})$ and $cv(\overline{X})$ can be directly related to the concept of margin of error (as described in Section 15.1.2) because the margin of error (d or D) is about twice the value of $cv(\overline{X})$. Thus, for the measurement of panicle number with $cv(\overline{X}) = 5.7\%$, a variety mean can be expected to be within plus or minus 11% of the true mean. If such a level of the margin of error is satisfactory, then the researcher can go on with the present sampling procedure for panicle count (i.e., that based on a simple random sampling design with 12 sample hills per plot). If the margin of error of 11% is too high, however, then the researcher could take either or both of the following approaches:

- Increase the number of replications.
- Use a different sampling procedure with a change in sampling design, type of sampling unit, or sample size.

Because the change in either the sampling design or the type of sampling unit must be based on information that is generally not available, the two choices that are left for the researcher in practice are the increase in the number of replications and the increase in the sample size. The decision on whether to increase the number of replications or the sample size, or both, depends on the relative magnitude of the sampling variance and the experimental error and their relative costs. The higher the experimental error, the higher the number of replications that should be used; the higher the sampling variance, the greater the number of samples per plot that should be taken.

In general, adding more samples is likely to cost less than adding more replications. Thus, increasing the number of sample hills per plot is generally a preferred choice. In our example, the experimental error s_2^2 of 0.1964 is much smaller than the sampling variance s_1^2 of 5.0429 and, hence, increasing the sample size would be the best alternative to increase the precision of the sample estimate.

☐ STEP 4. As an aid in deciding whether to increase the number of replications (r) or the sample size (n) to achieve a desired degree of precision, compute the standard error of the treatment mean, expressed as percent of the mean value [i.e., $cv(\overline{X})$], for different combinations of r and n values, using the formula in step 3. For our example, the results are shown in Table 15.3.

For example, an increase in the number of sample hills per plot from 8 to 12 is shown to be able to achieve the same improvement in the degree of precision (i.e., reducing the standard error of the treatment mean from 6.6 to 5.7%) as an increase in the number of replications from 3 to 4. Thus, to improve the precision of the estimate of the treatment mean in this case, it is better to increase the sample size than to increase the number of replications; especially so if cost is considered.

Table 15.3   Estimated Standard Error of Treatment Mean for Panicle Count, Computed from Data in Table 15.2, for Different Numbers of Sample Hills per Plot and Different Numbers of Replications (r)

Sample Hills, no./plot	Estimated Standard Error, %		
	$r = 3$	$r = 4$	$r = 5$
8	6.6	5.7	5.1
10	6.0	5.2	4.7
12	5.7	4.9	4.4
14	5.4	4.7	4.2
16	5.2	4.5	4.0

Note at this point that while the foregoing data can provide information that could aid in the proper choice of sample size, it does not allow for the evaluation of the proper choice of sampling unit. For example, based on these data, we cannot tell whether the single-hill sampling unit currently used is better or worse than a 2 × 2-hill sampling unit for panicle count. To be able to make such a decision, sources of data such as those described in Sections 15.2.2 and 15.2.3 must be considered.

15.2.2   Additional Data from On-Going Experiments

To evaluate the efficiency of various types of sampling unit, additional data may be collected from on-going experiments. Even though the collection of these additional data requires more of the researcher's time and effort, there is generally sufficient flexibility for planning the collection to suit available resources. For example, even if there are many plots in an on-going experiment, not all plots need to be included in the additional data collection scheme. Or, if resources are limited, the types of sampling unit could be limited to only a few.

As an illustration, we use a two-factor experiment with four rice varieties and three levels of nitrogen in a randomized complete block design with three replications. The original plot sampling technique is based on a simple random sampling design that calls for the measurement of tiller number from eight single-hill sampling units per plot consisting of 150 hills. We wish to evaluate the efficiency of the 1 × 2-hill sampling unit relative to that of the single-hill sampling unit. The additional data needed would be the tiller count of eight more hills, each immediately adjacent to the original eight single-hill sampling units in each of the 36 plots. The positions of the eight original sample hills, and those of the eight additional hills for one of the plots, may be as shown in Figure 15.5.

Although it is usually advisable to incorporate the additional measurement in all plots, if resources are limited it is possible to measure the additional data on only a few plots. For example, if resources allow for the measurement of only $\frac{1}{6}$ of the experiment (i.e., a total of six plots), the researcher may choose to collect the additional data from only the plots with two (the two extreme levels) of the three nitrogen treatments and only one of the four varieties.

The choice of the particular set of factors or treatments to be included in the study should be based on its expected large effects on the sampling variance. In this example, the decision to include two nitrogen levels and only one variety is based on the assumption that nitrogen rather than variety is the larger contributor to the variability in tiller number. The raw data on tiller count from such a scheme of additional data from on-going experiments are shown in Table 15.4.

Note that the additional work in data collection is modest but, with the added data, two alternative types of sampling unit—single hill and two adjacent hills—can be evaluated. Computational procedures for evaluating the efficiency of alternative sampling units, using the data in Table 15.4, are:

☐ STEP 1. For each treatment, construct the analysis of variance for a three-stage nested classification, with the first stage corresponding to plots, the second stage corresponding to the larger sampling units, and the third stage corresponding to the smaller sampling units. The format of such an analysis of variance is shown below, with p = number of plots, s = number of the large sampling units per plot, and k = number of the small sampling

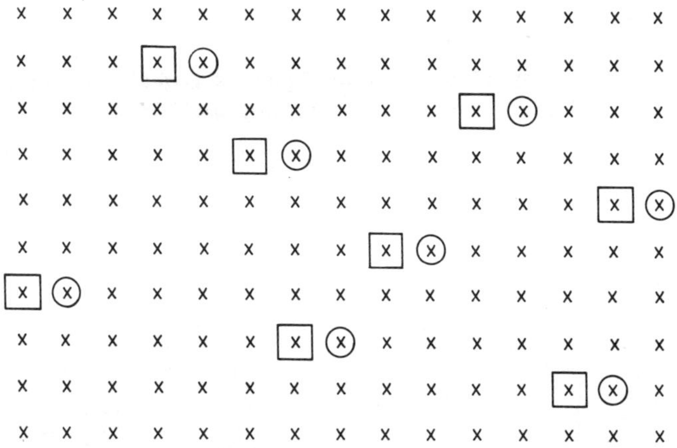

Figure 15.5   The location of the additional sample hills, ⊠ , relative to the original sample hills, ⊗, in the sampling study to compare two types of sampling unit, namely, 1×1 hill and 1×2 hill.

units within a large sampling unit:

Source of Variation	Degree of Freedom	Sum of Squares	Mean Square
Between plots	$p-1$	SS_1	MS_1
Between large sampling units within plot	$p(s-1)$	SS_2	MS_2
Between small sampling units within large sampling unit	$ps(k-1)$	SS_3	MS_3

For our example, the large sampling unit is the two-hill sampling unit, the small sampling unit is the single-hill sampling unit, $p = 3$, $s = 8$, and $k = 2$. Thus, the outline of the analysis of variance is:

Source of Variation	Degree of Freedom	Sum of Squares	Mean Square
Between plots	2		
Between 2-hill units within plot	21		
Between single-hill units within 2-hill unit	24		
Total	47		

Table 15.4 Tiller Count, Measured from Eight 1 × 2-Hill Sampling Units per Plot, from Fertilized (120 kg N / ha) and Nonfertilized Plots of IR22 Rice

	Tillers, no./hill											
	0 kg N/ha						120 kg N/ha					
Sampling	Plot I		Plot II		Plot III		Plot I		Plot II		Plot III	
Unit, no.	Hill 1	Hill 2	Hill 1	Hill 2	Hill 1	Hill 2	Hill 1	Hill 2	Hill 1	Hill 2	Hill 1	Hill 2
1	9	9	6	7	8	7	15	9	10	9	12	11
2	9	8	11	6	11	6	14	13	12	11	15	9
3	5	8	9	9	8	3	11	15	20	9	13	7
4	9	3	4	7	6	7	11	9	15	10	14	7
5	9	8	8	7	8	6	11	18	16	6	13	9
6	11	8	9	8	8	5	13	10	13	6	12	10
7	7	9	10	8	8	7	10	11	12	10	13	16
8	10	8	10	9	9	5	15	13	12	9	14	18

☐ STEP 2.   For each treatment, construct the plot × large sampling unit table of totals (AB), with the plot totals (A), the large sampling unit totals (B), and the grand total (G) computed. For our example, the results are shown in Table 15.5.

☐ STEP 3.   Compute the various SS as follows:

$$C.F. = \frac{G^2}{psk}$$

$$\text{Total } SS = \sum X^2 - C.F.$$

$$\text{Between plots } SS = \sum_{i-1}^{p} \frac{A_i^2}{sk} - C.F.$$

Between large sampling units within plot SS

$$= \sum_{i=1}^{p} \sum_{j=1}^{s} \frac{(AB)_{ij}^2}{k} - C.F. - \text{Between plots } SS$$

Between small sampling units within large sampling unit SS

$$= \text{Total } SS - (\text{sum of all other } SS)$$

For our example, we describe the computation of SS, using only data from the nonfertilized plots, as follows:

$$C.F. = \frac{(370)^2}{(3)(8)(2)} = 2{,}852.0833$$

$$\text{Total } SS = (9^2 + 9^2 + \cdots + 9^2 + 5^2) - 2{,}852.0833$$

$$= 165.9167$$

$$\text{Between plots } SS = \frac{(130)^2 + (128)^2 + (112)^2}{(8)(2)} - 2{,}852.0833$$

$$= 12.1667$$

Table 15.5 The Plot × 2-hill Sampling Unit Table of Totals from Data in Table 15.4

Tiller Number Total (AB)

Sampling Unit, no.	0 kg N/ha				120 kg N/ha			
	Plot I	Plot II	Plot III	Sampling Unit Total (B)	Plot I	Plot II	Plot III	Sampling Unit Total (B)
1	18	13	15	46	24	19	23	66
2	17	17	17	51	27	23	24	74
3	13	18	11	42	26	29	20	75
4	12	11	13	36	20	25	21	66
5	17	15	14	46	29	22	22	73
6	19	17	13	49	23	19	22	64
7	16	18	15	49	21	22	29	72
8	18	19	14	51	28	21	32	81
Plot total (A)	130	128	112	370	198	180	193	571
Grand total (G)				370				571

Between two-hill sampling units within plot SS

$$= \frac{(18)^2 + (17)^2 + \cdots + (15)^2 + (14)^2}{2}$$

$$-2,852.0833 - 12.1667$$

$$= 59.7500$$

Between single-hill sampling units within two-hill sampling unit SS

$$= 165.9167 - (12.1667 + 59.7500)$$

$$= 94.0000$$

☐ STEP 4. Compute the MS for each SS, by dividing the SS by its $d.f.$. The results of the nonfertilized plots, together with the results of the fertilized plots, are shown in Table 15.6.

☐ STEP 5. Compute the estimates of the sampling variance corresponding to the two alternative sampling units—the large and the small sampling units —as:

$$s_1^2 = \frac{[(N-1)MS_2 + N(k-1)MS_3]}{Nk-1}$$

$$s_2^2 = MS_2$$

where s_1^2 and s_2^2 are the estimates of the sampling variances based on the small and the large sampling units, N is the total number of the large sampling units in the plot, and MS_2 and MS_3 are as defined in step 1.

Table 15.6 Analysis of Variance for Data in Table 15.4

Source of Variation	Degree of Freedom	Mean Square	
		0 kg N/ha	120 kg N/ha
Between plots	2	6.0833	5.3958
Between 2-hill units within plot	21	2.8452	6.5327
Between single-hill units within 2-hill unit	24	3.9167	12.4375

For our example, each plot consists of 150 hills and the total number of the two-hill sampling units is $N = \frac{150}{2} = 75$. Thus, the values of the two variances are computed as:

- For nonfertilized plots:

$$s_1^2 = \frac{74(2.8452) + 75(3.9167)}{149} = 3.3845$$

$$s_2^2 = 2.8452$$

- For fertilized plots:

$$s_1^2 = \frac{74(6.5327) + 75(12.4375)}{149} = 9.5049$$

$$s_2^2 = 6.5327$$

☐ STEP 6.   Compute the relative efficiency of the two alternative sampling units, with and without cost consideration:

- Without cost consideration. The relative efficiency of the two alternative sampling units, without cost consideration, is computed as:

$$R.E. = \frac{100 s_1^2}{s_2^2}$$

For our example, these relative efficiency values are computed as:

$$R.E. = \frac{(100)(3.3845)}{2.8452} = 119\% \text{ for nonfertilized plots}$$

$$R.E. = \frac{(100)(9.5049)}{6.5327} = 146\% \text{ for fertilized plots}$$

Thus, for the measurement of tiller number, the two-hill sampling unit gives about 19% more precision than the single-hill sampling unit for the nonfertilized plots, and 46% more for the fertilized plots.

- With cost consideration. If the costs of gathering data for the various sizes of sampling unit are considered, the relative efficiency of two sizes of sampling unit, with the size of the larger unit being k times that of the smaller unit, is computed as:

$$R.E.(k) = \frac{100 k c_1 s_1^2}{c_k s_2^2}$$

where c_1 estimates the time required to make measurement on the smaller sampling unit and c_k estimates the time required on the large unit of size k, and s_1^2 and s_2^2 are their corresponding sampling variances. It is presumed that the cost associated with the alternative sampling units is proportional to the time needed to measure the various sampling units.

For our example, the estimate of the time spent in locating the sampling units and in counting their tiller numbers will be expressed in terms of the total number of hills that can be counted in 10 minutes. For a variety with medium tillering ability, it is estimated that 15 hills can be counted in 10 minutes with single-hill sampling units and 22 hills can be counted in 10 minutes with two-hill sampling units. More hills can be counted in a given time for larger sampling units because less time is spent in moving from one unit to another. Thus, the cost of counting tillers with a two-hill sampling unit relative to that with a single-hill sampling unit is computed as:

$$c_2 = \frac{15c_1}{22} = 0.6818c_1$$

Based on this relative cost estimate, the relative efficiency of the two alternative sampling units is computed as:

$$R.E. = \frac{(100)(2)(3.3845)}{(0.6818)(2.8452)} = 349\% \text{ for nonfertilized plots}$$

$$R.E. = \frac{(100)(2)(9.5049)}{(0.6818)(6.5327)} = 427\% \text{ for fertilized plots}$$

Note that the inclusion of cost consideration greatly increases the efficiency of the two-hill sampling unit over the single-hill sampling unit. With cost consideration, the two-hill sampling unit gave about 249% more precision than the single-hill sampling unit for the nonfertilized plots, and 327% more precision for the fertilized plots.

Based on the data we used, it is clear that the sampling plan for tiller count in such a trial should use the two-hill sampling unit rather than the single-hill sampling unit, especially in fertilized plots.

15.2.3 Specifically Planned Sampling Studies

Experiments planned specifically for sampling studies are usually set up to answer the many questions related to the development of a sampling technique. Aside from providing information to evaluate the different types of sampling unit, different sample sizes, and different sampling designs, such experiments may also be used to identify some of the important factors that should be considered in developing a sampling technique.

Consider a sampling study whose objectives are:

- To determine the optimum sampling unit, sample size, and sampling design for measuring important agronomic characters in transplanted rice field experiments.
- To determine whether such factors as varieties and fertilizer rates influence the efficiencies of the different sampling techniques.

Two rice varieties differing in growth duration and tillering capacity were tested with three nitrogen rates (0, 60, and 120 kg/ha). The 2×3 factorial experiment was laid out in a randomized complete block design with three replications. Plot size was 3.8×2.8 m. Each plot had 19×14 hills. Measurement of several plant characters was made on all the 12×8 hills in the center of each plot (i.e., excluding border plants). Because the data analysis required is similar for all the characters measured, only data of panicle number from the three nonfertilized IR22 plots (Table 15.7) are used to illustrate the computational procedures.

☐ STEP 1.   Construct the different sizes and shapes of sampling unit that the researcher wishes to evaluate, by combining the measurements of adjacent hills. The five types of sampling unit we consider are the 1×2 hill, 1×3 hill, 1×4 hill, 2×2 hill, and 1×6 hill.

☐ STEP 2.   For each of the sampling units constructed in step 1, construct an analysis of variance of a three-stage nested classification (following the procedure outlined in Section 15.2.2, steps 1 to 4) by treating each type of sampling unit as the large unit and the single-hill sampling unit as the small unit. The results of the five analyses of variance, one for each type of sampling unit, are shown in Table 15.8.

☐ STEP 3.   Using the result of any one of the five analyses of variance computed in step 2, compute the estimate of the sampling variance of the single-hill sampling unit as:

$$s_1^2 = \frac{(MS_2)(df_2) + (MS_3)(df_3)}{df_2 + df_3}$$

where MS_2 and df_2 are the mean square and degree of freedom corresponding to the variation between the large sampling units within plot, and MS_3 and df_3 are the mean square and degree of freedom corresponding to the

Table 15.7 Complete Enumeration of Panicle Number per Hill of IR22 Rice from Three Nonfertilized Plots, Each Consisting of 8 × 12 Hills

Row Number	Panicles, no./hill											
	Col. 1	Col. 2	Col. 3	Col. 4	Col. 5	Col. 6	Col. 7	Col. 8	Col. 9	Col. 10	Col. 11	Col. 12
Plot I												
1	10	9	7	11	5	13	10	2	14	5	9	3
2	4	5	9	12	5	10	14	12	10	8	12	5
3	9	10	12	4	8	10	3	5	5	6	8	11
4	6	10	13	10	11	3	5	11	5	6	7	3
5	10	8	6	10	11	5	8	9	8	5	12	4
6	13	8	7	6	10	11	9	8	11	6	9	7
7	11	13	4	8	9	9	6	10	7	7	5	6
8	3	4	11	13	9	8	9	9	4	9	7	11
Plot II												
1	9	4	8	4	8	9	6	6	4	10	7	8
2	8	7	4	8	5	4	8	9	8	10	6	8
3	7	7	6	6	8	8	9	4	8	7	7	8
4	7	3	8	4	6	6	6	9	8	10	6	5
5	8	8	8	3	8	7	6	9	7	4	8	9
6	4	9	10	5	7	7	3	8	9	4	6	10
7	8	5	6	7	6	7	5	9	9	7	6	5
8	9	9	9	9	7	7	8	8	6	7	5	8
Plot III												
1	6	9	4	9	10	10	8	5	5	10	6	5
2	4	9	11	10	5	12	11	8	9	7	7	10
3	10	6	4	5	12	5	10	5	10	7	13	3
4	9	7	5	9	10	5	10	6	6	4	4	6
5	8	3	11	9	7	7	9	9	13	10	9	10
6	13	7	3	5	7	7	7	7	6	3	5	4
7	4	9	10	10	8	8	9	9	3	4	13	12
8	6	12	8	10	7	12	12	5	13	7	7	9

Table 15.8 Analysis of Variance to Evaluate Sampling Variances of Five Sizes and Shapes of Sampling Unit

Source of Variation	Sampling Unit									
	1 × 2 hill		1 × 3 hill		1 × 4 hill		2 × 2 hill		1 × 6 hill	
	d.f.	MS	d.f.	MS	d.f.	MS	d.f.	MS	d.f.	MS
Between plots	2	32.21	2	32.21	2	32.21	2	32.21	2	32.21
Between large units within plot	141	6.36	93	5.63	69	4.66	69	4.00	45	6.26
Between single-hill units within large unit	144	6.72	192	6.99	216	7.15	216	7.36	240	6.60

variation between single-hill sampling units within the large sampling unit. For our example, applying the formula to the values from the analysis of variance of the 1 × 2 hill sampling unit in Table 15.8, we have:

$$s_1^2 = \frac{(6.36)(141) + (6.72)(144)}{(141 + 144)} = 6.54$$

□ STEP 4.   For each of the five sizes and shapes of sampling unit, compute its efficiency relative to the single-hill sampling unit as:

$$R.E. = \frac{100 s_1^2}{s_2^2}$$

where s_1^2 is as computed in step 3 and s_2^2 is the MS_2 value (i.e., MS between the large sampling units within plot) from the corresponding analysis of variance. For our example, the efficiency of the 1 × 2-hill sampling unit

Table 15.9 Efficiency of Various Sizes and Shapes of Sampling Unit Relative to a Single-Hill Unit for a Panicle Count, Computed from Data in Tables 15.7 and 15.8

Sampling Unit, hill × hill	Relative Efficiency, %
1 × 2	102.8
1 × 3	116.2
1 × 4	140.3
2 × 2	163.5
1 × 6	104.5

relative to the single-hill sampling unit is

$$R.E. = \frac{(6.54)(100)}{6.36} = 102.8\%$$

The computed relative efficiencies for the five sizes and shapes of sampling unit are shown in Table 15.9.

Results indicate that the 2 × 2-hill sampling unit gave the largest relative efficiency and, thus, seems to be optimum for measuring panicle number in experimental rice plots.

CHAPTER 16
Experiments in Farmers' Fields

Agricultural research has traditionally been in research stations where facilities for experimentation are excellent and accessibility to researchers is favorable. The assumption has often been that the best technology in research stations is also the best in farmers' fields. But the assumption of consistency or repeatability of technology performance between research stations and farmers' fields may not hold universally. The validity of this assumption is doubtful, for example, in the developing countries of the humid tropics where variability among farm conditions is high, crop yield is generally low, and response to improved crop management is less favorable than that in the research stations.

If there is inconsistency in technology performance between research stations and farmers' fields, selection of the best technologies for farmers cannot be based solely on research-station trials. Such a selection process should, in fact, be based on farm trials in which the new technology is compared to the farmer's existing practice under the growing conditions of his farm.

Experiments in farmers' fields can be classified either as technology-generation or technology-verification experiments. Technology-generation experiments are designed to develop new production technologies that can increase biological yield or reduce cost of production; whereas technology-verification experiments are designed to compare the superiority of new technologies— identified as promising by technology-generation experiments—over that of the farmer's existing practice.

The standard experimental and statistical procedures discussed in previous chapters are primarily for technology-generation experiments in research stations. Because of the distinctive differences between the research station and the farmer's field as the test site, not all of those procedures are suited for use with experiments in farmers' fields. However, because the primary objective of the technology-generation experiment is the same whether at a research station or in a farmer's field, only a slight modification of the statistical procedures discussed is needed for use in an on-farm technology-generation experiment. The process of technology verification, on the other hand, is greatly different from that for generating technology and, hence, requires an entirely different statistical approach.

562

16.1   FARMER'S FIELD AS THE TEST SITE

A farm is basically a production unit whose primary objective is to increase productivity, profit, and the well-being of the farm household. Consequently, researchers who experiment in farmers' fields must recognize and cope with the characteristics of an experimental area where production, not research, is the top priority.

Some of the distinctive features of the farmer's field as a test site, relative to research station, are:

- Lack of experimental facilities such as good water control, pest control, and equipments for such operation as land preparation and processing of harvest
- Large variation between farms and between fields in a farm
- Poor accessibility, which creates problems of supervision by researchers
- Lack of data describing the soil and climate of the experimental field
- Availability of the farmer and his practices for use in experimentation

The first three features suggest that an on-farm experiment should be small and should be conducted on several farms. For an on-farm technology-generation experiment, the size of the trial is usually controlled by maintaining the number of treatments at a reasonable size while retaining the same standard number of replications as used in research-station trials. For technology-verification trials, a large number of farms is needed to adequately sample the variation within the area in which the new technology may be recommended for adoption—the *target area*. Thus, there is a greater need to control the size of a technology-verification trial than there is for a technology-generation trial. For a technology-verification trial, both the number of treatments and the number of replications are kept at the minimum level.

Lack of information on soil and climate requires the collection of soil and weather data for the farm where the technology-generation trial is conducted.

The farmer and his practices are generally used in technology-verification trials as a basis for comparison with the test technology. There is a question, however, about the merit of considering the *farmer* as a component of the comparable farmer's practice. A farmer's management of his practice, for instance, makes the practice more realistic; a farmer's management of the test technology provides a practical assessment of its potential level of acceptability by farmers. However, in either case, the integration of farmer's management with the standard field-plot experimental techniques is not simple. In such cases, experimental error is usually increased and the chance of total experimental failure is high.

We discuss procedures for on-farm technology-generation experiments and technology-verification experiments, taking into consideration the distinctive objectives of the two types of trial and the distinctive features of the farmer's field as the test site. Most of the procedures are modified versions of the procedures given in previous chapters. There are, however, a few procedures developed specifically for technology-verification trials.

16.2   TECHNOLOGY-GENERATION EXPERIMENTS

The farmer's field provides a convenient and economical way to sample a wide array of physical and biological conditions in the generation of technology. Procedures for on-farm technology-generation trials are similar to those at research stations and most of the experimental procedures discussed in the previous chapters can be used. Consequently, we concentrate on the modification of existing procedures to allow researchers to cope with features of the farm that are distinctly different from that of the research station.

16.2.1   Selection of Test Site

The test site for a technology-generation trial is selected to provide a set of physical and biological conditions, under which the trial is to be conducted or for which the technology is to be developed. Thus, the method of site selection is deliberate, rather than at random. The selection procedures are:

☐ STEP 1.   Clearly specify the desired test environment in terms of the specific physical and biological characteristics such as soil, climate, topography, landscape, water regime, and so on.

☐ STEP 2.   Classify each of the specified environmental characteristics according to:
  - The relative size of contiguous area in which homogeneity of a given characteristic is expected. For example, areas with the same climate will be larger than those having the same landscape or water regime.
  - The availability of existing information, or the relative ease in obtaining the information, on the desired characteristics. For example, climatic data is usually more readily available than information on landscape, water regime, and cropping pattern. The latter is usually obtained through farm visits by the researchers.

☐ STEP 3.   Select a large contiguous area that satisfies those environmental features that are usually homogeneous over a wide area; and, within that area, identify sub-areas (or farms) that satisfy those environmental conditions that are more variable. For example, a large contiguous area can be first selected to satisfy the required climate and soil. These can be based on weather-station records and a soil map. Within the selected area, farm visits and interviews of selected farmers will help identify farms that have the topography, landscape, water regime, and cropping pattern that most closely approximate the required test environment.

If more than one farm is found to satisfy the specified test environment, select those most accessible, have more available resources, and are managed by cooperative farmers.

☐ STEP 4. For each selected farm, choose an area or a field that is large enough to accommodate the experiment and has the least soil heterogeneity (through visual judgment, based on information of past crops, etc.). If no single field is large enough to accommodate the whole experiment, select the smallest number of fields that can accommodate it.

16.2.2 Experimental Design and Field Layout

The design of experiments in farmers' fields must aim at keeping the size of experiment small. This is done by keeping the number of treatments and number of replications at the minimum. To reduce the number of treatments, a fractional factorial design (Chapter 4, Section 4.5) may be used. To determine the number of replications, two contrasting considerations should be examined:

- Errors can be expected to increase in on-farm trials because the fields are less accessible and more difficult for researchers to supervise. In addition, damage by rats, stray animals, vandalism, and theft are more apt to occur in on-farm trials and will increase experimental error.
- The generally low insect and disease pressure and the more uniformly managed farmers' fields that are free of residual effects from previous treatments may, on the other hand, result to less experimental error.

Thus, the choice of the number of replications to be used depends on the relative importance of these two conflicting features. For example, if the chance for increase in experimental error overshadows the chance for less error, the number of replications should be greater than that used in research-station trials. Experience in rice research has indicated that, with proper management, experimental error of an on-farm technology-generation experiment can be smaller than that in a research station; and, subsequently, the number of replications need not be larger than that used at research stations.

For plot layout in a farmer's field, the techniques used in research stations generally apply. However, the following considerations should be used in laying out plots:

- Plot shape may have to be adjusted to suit the irregular shape of the field and the manner in which the leveling of land has been done.
- If one paddy or a distinct parcel of land is not enough to accommodate the whole experiment, each paddy or parcel must accommodate at least one whole replication.

16.2.3 Data Collection

All data normally collected in research-station trials should be collected in an on-farm technology-generation trial. In addition, data such as those on weather, soil, and history of plot management that are usually available in research stations but not for farmers' fields must be collected. Data collection must be flexible enough to handle unexpected incidents such as damage by rats, stray cattle, or theft.

Because a primary objective of technology-generation trials in farmers'

Table 16.1   Grain Yield of Dry-Seeded Rice Tested in RCB Design with Five Weed-Control Methods on Four Farms and Three Replications per Farm

| Treatment[a] | Grain Yield, t/ha | | | |
	Rep. I	Rep. II	Rep. III	Av.
	Farm 1			
T_1	5.5	4.8	5.3	5.2
T_2	0.6	2.5	1.2	1.4
T_3	0.2	0.1	0.4	0.2
T_4	5.7	5.2	4.9	5.3
T_5	0.3	0.2	0.3	0.3
	Farm 2			
T_1	5.6	5.8	6.0	5.8
T_2	6.8	3.1	4.4	4.8
T_3	5.7	5.7	5.1	5.5
T_4	5.4	4.8	4.0	4.7
T_5	0.4	0.6	0.6	0.5
	Farm 3			
T_1	1.8	2.4	3.1	2.4
T_2	3.3	3.7	2.9	3.3
T_3	2.1	3.0	2.4	2.5
T_4	2.9	1.1	4.0	2.7
T_5	1.4	0.7	1.6	1.2
	Farm 4			
T_1	5.1	5.4	5.6	5.4
T_2	3.7	4.6	5.0	4.4
T_3	3.6	4.2	4.5	4.1
T_4	5.1	4.7	5.2	5.0
T_5	2.8	4.2	5.8	4.3

[a] T_1 = Butachlor + proponil, T_2 = Butralin + proponil, T_3 = Proponil, T_4 = Hand weeding, and T_5 = Untreated.

fields is to sample a wide array of environments, it is common for a technology-generation trial to be on more than one farm. To effectively assess the interaction effect between treatment and farm environment, a uniform set of data should be collected on all farms.

16.2.4. Data Analysis

The objective of technology-generation trials remains the same whether on research stations or in farmers' fields. Consequently, data analysis on a per-trial basis is the same. When the trial is on more than one farm, the procedures for combining data over locations described in Chapter 8, Section 8.2 apply. Because test farms represent different environments, emphasis should be placed on explaining the nature of the interaction effect between treatment and farm in terms of the *relevant environmental factors*—to gain a better understanding of the influence of those factors on the effectiveness of the treatments tested.

To illustrate the procedures, we use data from a weed-control trial on four farms, based on a RCB design with five treatments and three replications per farm (Table 16.1). The step-by-step procedures for examining interaction effect between treatment and farm are:

☐ STEP 1. Compute a combined analysis of variance for data from all farms, following the appropriate procedure for combining data over sites outlined in Chapter 8, Section 8.2. For our example, the results are shown in Table 16.2. The interaction effect between farm and treatment is highly significant, indicating that the relative performance of the weed-control treatments varied among farms.

☐ STEP 2. If the farm × treatment interaction in step 1 is not significant, indicating that treatment performance was consistent over all test farms, make a comparison of treatment means (averaged over all farms), following

Table 16.2 Combined Analysis of Variance over Farms, from Data in Table 16.1[a]

Source of Variation	Degree of Freedom	Sum of Squares	Mean Square	Computed F[b]
Farm (F)	3	60.8298	20.2766	25.05**
Replications within farm	8	6.4760	0.8095	
Treatment (T)	4	73.7773	18.4443	35.07**
$F \times T$	12	69.8161	5.8180	11.06**
Pooled error	32	16.8306	0.5260	
Total	59	227.7298		

[a] $cv = 21.0\%$.
[b] ** = significant at 1% level.

the standard procedures of Chapter 5. Otherwise, examine the nature of the interaction through suitable partitionings of the interaction *SS* (see Chapter 8, Section 8.2.1, step 3):

A. Plot the mean value of each treatment (*Y* axis) against the farm (*X* axis), as shown in Figure 16.1. For our example, the result shows:

   (i) Three distinct groups of treatments based on their relative performance with respect to the test farms. One group consists of T_1 and T_4, another consists of T_2 and T_3, and the last consists of T_5 (untreated). The major difference between the first two groups is the performance on farm 1: T_1 and T_4 gave high yields while T_2 and T_3 gave low yields. That is, T_1 and T_4 were highly effective in controlling weeds on farm 1 while T_2 ands T_3 were not.
   (ii) Only on farm 4 where the yield of T_5 (untreated) was not different from any of the other treatments, indicating that weeds were not the major problem on this farm.
   (iii) All weed-control treatments (T_1, T_2, T_3, and T_4) were effective on both farms 2 and 3 but the response was much higher on farm 2 than on farm 3.

B. Confirm the visual observation in A through an appropriate mean comparison, or through partitioning of the treatment × farm interaction *SS* in the combined analysis of variance of step 1. For our example, observation (ii) is confirmed through the nonsignificant result of the pair comparison among the five treatment means on farm 4 (Table 16.3). Observations (i) and (iii), on the other hand, are confirmed by the results of the partitioning of the interaction *SS* (Table 16.4).

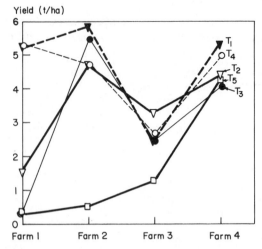

Figure 16.1 Examination of the farm × treatment interaction by plotting mean yield of each treatment against farm, from data in Table 16.1.

Table 16.3   Yield Comparison Between Five Weed-Control Treatments Tested on Four Farms

| Treatment | Mean Yield, t/ha[a] | | | | |
	Farm 1	Farm 2	Farm 3	Farm 4	Av.
T_1	5.20 a	5.80 a	2.43 ab	5.37 a	4.70
T_2	1.43 b	4.77 a	3.30 a	4.43 a	3.48
T_3	0.23 b	5.50 a	2.50 ab	4.10 a	3.08
T_4	5.27 a	4.73 a	2.67 a	5.00 a	4.42
T_5 (untreated)	0.27 b	0.53 b	1.23 b	4.27 a	1.58
Av.	2.48	4.27	2.43	4.63	3.45

[a]Average of three replications per farm. In each farm, means followed by a common letter are not significantly different at the 5% level of significance.

Table 16.4   Partitioning of the Treatment × Farm Interaction *SS* in Table 16.2, to Confirm Observations Made from Figure 16.1

Source of Variation	Degree of Freedom	Sum of Squares	Mean Square	Computed F[a]
(T_5 vs. others) × Farm	3	23.6535	7.8845	14.99**
(T_5 vs. others) × (Farm 2 vs. Farm 3)	(1)	12.0968	12.0968	23.00**
(T_5 vs. others) × others	(2)	11.5567	5.7784	10.99**
(T_1, T_4 vs. T_2, T_3) × Farm	3	41.5142	13.8381	26.31**
(T_1 vs. T_4) × Farm	3	1.5150	0.5050	< 1
(T_2 vs. T_3) × Farm	3	3.1333	1.0444	1.99[ns]

[a]** = significant at 1% level, [ns] = not significant.

Table 16.5   Mean Total Weed Weight in the Weed-Control Trial on Four Farms

| Treatment | Mean Total Weed Weight, g/m^2[a] | | | |
	Farm 1	Farm 2	Farm 3	Farm 4
T_1	28.0	44.4	176.8	7.2
T_2	290.6	93.8	197.4	12.2
T_3	528.6	19.2	351.6	45.2
T_4	16.4	39.2	32.4	5.8
T_5 (untreated)	828.2	421.8	430.4	49.6
Av.	338.4	123.7	237.7	24.0

[a]Weed data measured at 62 days after emergence and average of three replications per farm; the corresponding yield data are in Table 16.1.

569

☐ STEP 3.   Identify the environmental factors that are the most likely cause of the interaction effect between treatment and farm obtained in step 2. For our example, biological inference would lead us to suspect that weed incidence is the most likely cause. Thus, total weed weight is first examined in relation to the performance of the weed-control treatments from farm to farm (Table 16.5). The data indicate that:

- On farm 4, weeds, as represented by data in the untreated plots, were very few. This is probably the reason why the yields of untreated plots on this farm were high and no significant response to any weed-control method was observed (Table 16.3).

- On farm 1, weed weights in the untreated plots were the highest (almost double those on farms 2 and 3). This is probably the reason for the large variation among the treated plots, that is, T_2 and T_3 gave yields that were much lower than that of T_1 and T_4 (Table 16.3). Apparently, grasses and sedges were the dominant weed types on this particular farm; and while T_1 and T_4 were effective in controlling these types of weed, T_2 and T_3 were not (Table 16.6).

Table 16.6   Mean Weight of Weeds, Classified by Type (Broadleaves, Grasses, and Sedges) in a Weed Control Trial on Four Farms

Treatment	Weed Weight, g/m^2			
	Farm 1	Farm 2	Farm 3	Farm 4
Broadleaves				
T_1	7.8	16.8	0.2	0.2
T_2	9.6	26.8	4.2	1.2
T_3	4.0	5.8	9.4	0.0
T_4	0.4	0.2	3.4	0.4
T_5	0.0	13.8	32.2	2.0
Grasses				
T_1	20.2	20.8	173.0	6.4
T_2	224.0	27.2	164.0	8.0
T_3	461.2	10.8	328.0	28.4
T_4	6.2	30.0	25.6	4.8
T_5	696.0	390.6	387.0	40.8
Sedges				
T_1	0.0	6.8	3.6	0.6
T_2	57.0	39.8	29.2	0.6
T_3	63.4	2.6	14.2	8.0
T_4	9.8	8.8	3.4	0.4
T_5	132.2	17.6	11.2	6.6

• The relatively lower yields and the smaller response to weed-control treatments on farm 3, relative to farm 2, could not be satisfactorily explained by the weed data in Tables 16.5 and 16.6. First, total weed weights in untreated plots were similar on these two farms (Table 16.5) and even though farm 3 had more broadleaves in the untreated plots than farm 2, the control of this type of weed on farm 3 was even better than on farm 2 (Table 16.6). On the other hand, although total weight of grasses was the same on farms 2 and 3, the three chemical treatments T_1, T_2, and T_3 were more effective in controlling grasses on farm 2 than on farm 3. However, despite the effectiveness in reducing the amount of grasses through hand weeding (T_4) on farm 3, its yield were not significantly different from those of the chemical treatments. Thus, it seems that there were other yield limiting factors present on farm 3 besides weeds. These factors need to be examined further.

16.3 TECHNOLOGY-VERIFICATION EXPERIMENTS

The primary objective of a technology-verification trial is to compare the performance of the farmer's technology and the new technology in the farmers' fields. The primary bases of performance are biological yield and profitability. Thus, in contrast to a technology-generation trial, a technology-verification trial has the following distinctive features:

1. Farmer's Field as the Test Site. Because superiority of a new technology over that of the farmers' must be established with the actual farm conditions, technology verification must be done in farmers' fields. Because of the large variation among farms, even in a small contiguous area, the number of farms required for technology verification is generally larger than that for technology generation.

2. Farmer's Practice as the Basis of Comparison. In a technology-generation trial, it is customary to have a control treatment, which represents a zero level, such as no fertilizer application in a fertilizer trial or no insect control in an insecticide trial, as the basis for comparing the treatments tested. In the technology-verification trial, the control is the *farmer's practice*, which hardly ever represents zero level. Farmers usually use some fertilizers and some weed and insect control, but either at a lower rate or with a different method of application than that recommended.

3. Farmer's Practice as the Level of Management for Growing the Experimental Crops. In a technology-generation trial, all practices except the treatments tested are almost always prescribed at an optimum level. In a technology-verification trial, all practices except the *test factors* (the components of the new technology that differ from that of the farmer's practice) are maintained at the farmer's level.

16.3.1   Selection of Test Farms

The test farms for technology verification should adequately represent the farms in a target area—the area in which the new technology may be recommended for adoption. This makes it necessary to identify the specific boundaries of the area and choose the appropriate sampling technique for selecting test farms.

16.3.1.1   Target Area.   The target area is defined by one or more specific environmental components (physical, biological, social, and economic) that are considered critical to the superior performance and eventual adoption of the new technology. The farms in a target area may not create a single contiguous area, but a contiguous area is desirable because it facilitates the identification and selection of test farms and the diffusion of the recommended technology.

16.3.1.2   Sampling Plan.   To ensure that the test farms are representative of the target area, an adequate number of test farms (sample size) must be selected following a valid sampling technique. Sample size should increase with the variability between farms in the target area. Diversity in cropping pattern is a good indicator of the variability between farms. As a general rule, the number of test farms should be about 10 times the number of dominant cropping patterns in the target area.

The most commonly used sampling design for selecting the test farms for a technology-verification trial is a stratified random sampling (Chapter 15, Section 15.1.3.3) with farms as the primary sampling unit and fields within farm as the secondary sampling unit. Cropping pattern and farm size are the most commonly used stratification criteria.

16.3.2   Experimental Design

A technology-verification trial with k test factors can be viewed as a 2^k factorial experiment in which the two levels of each test factor represent the level prescribed in the new technology and the level of the farmer's practice. For example, consider a case where the existing farmer's practice is to grow rice in the wet season to be followed by maize in the dry season, and the new technology consists of the same cropping pattern but with a different variety, fertilizer rate, weed control, and insect control for rice and a different insect control and land preparation for maize. The new technology differs from the existing farmer's practice by six component technologies—four for rice and two for maize—hence, there are six test factors (i.e., $k = 6$). With two levels per test factor, the technology-verification trial is a 2^6 factorial experiment.

When k is large, as is the usual case for a technology-verification trial, a complete set of 2^k factorial treatments becomes too large to be practical. To reduce the size of the experiment, three different sets of treatments are generally tested. These are:

1. Set X, consisting of two treatments: the new technology (the test factors are all at the new-technology level) and the farmer's practice (the test factors are all at the farmer's level).
2. Set Y, consisting of $(k + 2)$ treatments: the two treatments of set X plus the k intermediate treatments, each of which represents a treatment combination in which all test factors but one are at the new-technology level.
3. Set Z, consisting of either the 2^k complete factorial treatment combinations or an appropriate fractional factorial set (Chapter 4, Section 4.5).

To illustrate, consider a technology-verification trial involving a single rice crop in which the new technology represents an *improvement* over the farmer's practice in four components: fertilizer (F), variety (V), weed control (W), and

Table 16.7   The Three Sets of Treatment X, Y, and Z Associated with a Technology-Verification Trial in Rice Involving Four Test Factors: Fertilizer (F), Variety (V), Weed Control (W), and Insect Control (I)

| Treatment Number | Factor Level[a] | | | | Treatment[b] | | | |
	F	V	W	I	Set X	Set Y	Set Z Complete Factorial	Set Z Fractional Factorial[c]
1	n	n	n	n	*	*	*	*
2	f	n	n	n	—	*	*	—
3	n	f	n	n	—	*	*	—
4	f	f	n	n	—	—	*	*
5	n	n	f	n	—	*	*	—
6	f	n	f	n	—	—	*	*
7	n	f	f	n	—	—	*	*
8	f	f	f	n	—	—	*	—
9	n	n	n	f	—	*	*	—
10	f	n	n	f	—	—	*	*
11	n	f	n	f	—	—	*	*
12	f	f	n	f	—	—	*	—
13	n	n	f	f	—	—	*	*
14	f	n	f	f	—	—	*	—
15	n	f	f	f	—	—	*	—
16	f	f	f	f	*	*	*	*

[a] n = level of the new technology and f = farmer's level.
[b] — = not tested and * = tested.
[c] A $\frac{1}{2}$ of 2^4 fractional factorial design (Chapter 4, Section 4.5).

insect control (I). That is, there are four test factors and the technology-verification trial is a 2^4 factorial experiment. The three sets of treatments (X, Y, and Z) associated with this technology-verification trial are shown in Table 16.7.

Because the number of treatments is smallest in set X and largest in set Z, the number of farms testing the X set of treatments (n_x) is usually the largest, followed by n_y, the number of farms testing the Y set of treatments, and finally, n_z, the number of farms testing the Z set of treatments. The size of n_z is dependent on the degree of importance of the interaction effects among test factors anticipated. If the interaction is expected to be large and to involve a large number of test factors, then n_z should also be large. The proportion of $n_x : n_y : n_z$ commonly used for technology-verification trials in rice is 3:1:1.

16.3.3   Field-Plot Technique

The field-plot technique for a technology-verification trial must cope with several potential problems:

·   Relatively large variation between and within farms
·   Diversity in the farmer's practice
·   Implementation of the farmer's practice as a treatment
·   Farmer's practice as the underlying test condition

Because none of these problems is relevant to a technology-generation trial, the standard field plot techniques used in research-station trials are not effective in dealing with these problems. Development of appropriate field plot techniques for technology-verification trials is in the early stages and is confined primarily to rice. We describe these newly developed procedures, together with their potential application to other crops and to multiple cropping.

16.3.3.1 Number of Replications.   The precision of the technology-verification trial depends primarily on the magnitude of the variation between farms and variation within farms. While the inclusion of several test farms is used to cope with variation between farms, the use of replications in each farm is used to cope with the within-farm variation. Because variation within farms is generally expected to be much smaller than that between farms, the number of replications per farm need not be as large as that normally used for a technology-generation trial. In fact, two replications per farm is most commonly used in a technology-verification trial.

16.3.3.2 Determination and Implementation of the Farmer's Practice.   Although the level associated with the new technology is known beforehand and is common in all test farms that test the same new technology, the level associated with the farmer's practice is not known in advance and has to be determined independently for each test farm. This is because, in a target area, the actual farmer's practice could vary greatly between farms; and the use of the *average* or *representative* farmer's practice, although simple, is not valid.

In each of the n_x test farms in which the set X treatments (composed of the new technology and the farmer's practice) are tested (Section 16.3.2), the farmer's practice can be obtained directly from an adjacent area cultivated by the farmer as a part of his normal farming. Thus, only the new-technology plot is actually implemented and an accurate assessment and repetition of the farmer's level is not necessary.

For farms testing the set Y and set Z treatments, however, the farmer's level must be established on each test farm because the experimental plots receiving the intermediate technologies have some test factors at the farmer's level and other factors at the new-technology level.

Ideally, the farmer's level on each test farm should be monitored throughout the cropping season or throughout the duration of the experiment. However, such a procedure is tedious and time consuming. An alternative procedure frequently used is one in which the farmer's practice is based on his recollection of what he did last year, what he plans to do this year, and what factors may alter such a plan.

There are two ways the farmer's practice can be incorporated in the test plots. First, the farmer himself can be requested to apply his practice to the experimental plot. Secondly, the researcher may implement the farmer's practice on the basis of his assessed perception of such a practice.

16.3.3.2.1 Farmer's Implementation. With farmer's implementation, the whole experimental area is managed by the farmer as a part of his farm. To establish treatments other than the farmer's practice, the researcher applies the additional inputs and practices required by the new technology for the test factors specified. This procedure is most suited to cases where the new-technology levels of all test factors require the level of management that is higher than the farmer's levels; for example, higher rate of fertilizer application, higher dose and frequency of insecticide application, and so on.

Farmer's implementation is not applicable to cases where the new-technology level and the farmer's level of a test factor share no common practice; for example, when the farmer's weed control is a manual one while the weed control of the new-technology involves only herbicide application.

With farmer's implementation, the difficult task of assessing the farmer's practice in advance is avoided. Although information on the farmer's practice is still needed for the interpretation of results, this need can be satisfied by simply keeping records of the operation after it is actually performed, which is a much simpler task.

On the other hand, with the experimental area managed by the farmer, the plot layout must be set up to facilitate the farmer's operations. For example, in wetland rice, large plots may be needed to eliminate the need for having bunds (or levees) to separate experimental plots. The presence of bunded plots obstructs such farmer's tasks as land preparation and irrigation.

16.3.3.2.2 Researcher's Implementation. With researcher's implementation, all treatments are implemented by the researcher according to the specified level of each test factor. This procedure is applicable only if the researcher has accurate information, on the farmer's level of each test factor, prior to its implementation. For example, if one of the test factors is fertilizer, the researcher must have information on the farmer's anticipated fertilizer practice before the time of the first fertilizer application.

16.3.3.3 Management of the Test Condition.

Test condition in a technology-verification trial refers to all management and cultural practices other than the test factors. For example, for the trial with four test factors of fertilizer, variety, insect control, and weed control, the test condition includes all other practices such as land preparation, method of planting, plant density, and water management.

The test condition of a technology-verification trial can be implemented by the farmer, or the researcher, or both. The choice is greatly related to that of the procedure for implementing the farmer's level of the test factors (Section 16.3.3.2). If the farmer implements the farmer's level of the test factors, then it follows that he should implement the test condition. On the other hand, even if the researcher implements the farmer's level of the test factors, the test condition can be implemented by either the researcher or the farmer. The latter is preferred because of ease in implementation (there is no need for the researcher to gather the information on all aspects of the farmer's practice) and accuracy of the test condition (see Section 16.3.3.4). However, with the researcher managing the test condition and, consequently, the whole experiment, the usual field plot technique does not need much modification, although an accurate assessment of the farmer's practice prior to the start of the experiment is still essential.

16.3.3.4 Verification and Simulation of the Farmer's Practice.

In a technology-verification experiment, the establishment of each treatment becomes more difficult than that in a technology-generation experiment because the farmer's practice differs from farm to farm and from season to season. One common difficulty occurs when the farmer's practice, obtained before the trial, is modified by the farmer in the course of experimentation. Consequently, the farmer's practice used in the trial differs from that for the rest of the farm, thus reducing the validity and applicability of the results of the trial.

An important step is to evaluate how accurately the farmer's practice has been simulated in the trial. This is done by measuring the performance of the farmer's crop from areas immediately adjacent to the experimental plots. These measurements are compared to those from the experimental plot (or plots) receiving the farmer's practice. A large difference between the two values would indicate a failure in the simulation of the farmer's practice.

16.3.4 Data Collection

Data collection for technology-verification trials is designed primarily to compare the new technology and the farmer's practice and to determine the potential for adoption of the new technology. Because of such a specific objective, the type of data to be collected generally differs from that of technology generation. Some of the important data to be collected for a technology-verification trial are:

- Productivity or yield measured in each plot.
- Physical and biological environment of the farm, such as the soil characteristics, the history of crops grown and the management applied, and the climatic and weather data recorded on a per-trial basis.
- Social and economic data, which are essential in the evaluation of the potential for adoption of the new technology, such as farm size, household income and labor, cost of agricultural inputs, and prices of the farm products, that are collected in the target area.
- Current farmer's practice and its productivity. This type of data is usually obtained through interview of the farmer, monitoring of the farmer's operations, and crop-cut sample from appropriate areas of the farm. This information is needed for a meaningful interpretation of results and to verify the validity of the simulation of the farmer's practice in the trial.
- Data that are expected to help explain the performance of each test factor. For example, if insect control is a test factor, data on insect and disease incidences, on a per-farm basis, could be helpful in explaining the variation in performance of insect control over test farms. Similarly, if weed control is a test factor, data on weed incidence should be collected.

Note that there are several types of data, normally collected for technology-generation trials, which need not be collected in a technology-verification trial. A good example is the detailed data on agronomic traits (including the yield components) which are standard data for technology-generation trials in rice. Such information is not so useful for technology verification.

16.3.5 Data Analysis

Two types of data analysis are necessary in a technology-verification trial; yield gap analysis and cost and return analysis.

16.3.5.1 Yield Gap Analysis. The yield gap analysis measures the difference in yield (or any other index of productivity) between the new technology and the farmer's practice, and partitions this difference into components representing contributions from the individual test factors. Because, in most cases, only the yields from the same crop, or crops, can be compared, the yield gap analysis is generally applicable only on a per-crop basis.

When the new technology involves more than one crop, yield gap analysis is applicable only to crops that are common to both the new technology and the farmer's practice. For example, when the technology-verification experiment involves the same cropping pattern (for example rice followed by maize) in both the new technology and the farmer's practice, yield gap analysis is applied independently to each of the two crops. On the other hand, if the new technology involves a cropping pattern of rice followed by rice and the farmer's cropping pattern is a rice followed by maize, then the yield gap analysis is applicable only to the first rice crop, because it is common to both patterns.

We illustrate the computational procedure for technology verification with yield data from a rice trial involving three test factors: fertilizer (F), insect control (I), and weed control (W). The trial involves 19 test farms, with set X treatments tested on 12 farms (Table 16.8), set Y treatments on 3 farms (Table 16.9), and set Z treatments (i.e., the 2^3 complete factorial treatments) on 4 farms (Table 16.10). On each farm, the treatments are tested in two replications. The step-by-step procedures for yield gap analysis are:

☐ STEP 1.   For each test farm, compute the yield gap as the difference between the yield of the new technology and that of the farmer's practice. Then,

Table 16.8   Grain Yield of the New Technology and the Farmer's Practice Tested on 12 Test Farms, with Two Replications per Farm, in a Technology-Verification Trial in Rice

Farm Number	Grain Yield, t/ha					
	New Technology			Farmer's Practice		
	Rep. I	Rep. II	Av.	Rep. I	Rep. II	Av.
1	6.60	6.64	6.62	3.56	4.94	4.25
2	6.50	5.96	6.23	3.88	4.62	4.25
3	7.61	6.78	7.20	6.02	5.30	5.66
4	7.20	7.02	7.11	5.26	5.15	5.20
5	3.68	3.62	3.65	3.13	2.73	2.93
6	4.18	4.47	4.32	3.00	3.05	3.02
7	6.54	5.98	6.26	5.36	5.10	5.23
8	7.26	6.20	6.73	5.76	4.69	5.22
9	7.30	6.28	6.79	5.35	5.17	5.26
10	7.60	7.60	7.60	5.96	5.74	5.85
11	6.71	6.29	6.50	5.65	5.89	5.77
12	7.14	6.16	6.65	5.02	4.56	4.79

'able 16.9   Grain Yield of Set *Y* Treatments of a Technology-Verification Trial in lice, Involving Three Test Factors *F*, *I*, and *W*; tested on Three Test Farms with Two teplications per Farm

Treatment			Grain Yield, t/ha								
Level[a]			Farm 13			Farm 14			Farm 15		
No.	*F*	*I*	*W*	Rep. I	Rep. II	Av.	Rep. I	Rep. II	Av.	Rep. I	Rep. II	Av.
1	n	n	n	6.83	7.51	7.17	6.76	7.02	6.89	6.93	6.91	6.92
2	f	n	n	6.26	5.92	6.09	6.74	6.23	6.48	6.06	6.06	6.06
2	n	f	n	6.16	6.29	6.22	5.92	6.14	6.03	6.21	6.72	6.46
4	n	n	f	6.97	7.15	7.06	7.57	6.73	7.15	6.50	7.27	6.88
5	f	f	f	5.36	4.84	5.10	5.05	4.37	4.71	4.53	4.99	4.76

Fertilizer (*F*), insect control (*I*), and weed control (*W*); n = new-technology level and = farmer's level.

compute the average of these yield gaps over all *n* test farms as:

$$G = \frac{\sum_{i=1}^{n} (\bar{P}_i - \bar{Q}_i)}{n}$$

where $\bar{P}_i$ is the mean yield of the new technology, and $\bar{Q}_i$ is the mean yield of the farmer's practice, in the *i*th farm.

For our example, the results for the 19 test farms are shown in Table 16.11. The values of yield gap range from 0.72 t/ha in farm 5 to 3.45 t/ha in farm 17, with the mean yield gap over all test farms of 1.87 t/ha.

☐ STEP 2.   Evaluate the interaction among test factors by computing a combined analysis of variance of data over all n_z farms (where set *Z* treatments are tested), following the procedure for combining data from a RCB experiment over years outlined in Chapter 8, Section 8.1.2. That is, test farms in the technology-verification experiment are considered as a random variable.

If no data from n_z farms are available, either because no farm has successfully tested set *Z* treatments or there is no n_z farm, disregard this step and proceed to Analysis I of step 3.

For our example, there are four n_z farms (Table 16.10). The results of the combined analysis of variance over four farms, based on RCB design, are shown in Table 16.12. A suitable factorial partitioning of the treatment *SS* into main effects of, and interactions between, the three test factors is performed. Results indicate no significant interaction effects between the

Table 16.10 Grain Yield of Set Z Treatments of a Technology-Verification Trial in Rice, Involving Three Test Factors F, I, and W; Tested on Four Test Farms with Two Replications per Farm

Treatment				Grain Yield, t/ha										
No.	Level[a] F	I	W	Farm 16			Farm 17			Farm 18			Farm 19		
				Rep. I	Rep. II	Av.	Rep. I	Rep. II	Av.	Rep. I	Rep. II	Av.	Rep. I	Rep. II	Av.
1	n	n	n	8.76	6.90	7.83	7.21	7.05	7.13	6.98	6.36	6.67	6.28	5.98	6.13
2	n	n	f	4.98	6.60	5.79	6.20	6.62	6.41	6.43	6.23	6.33	6.14	6.00	6.07
3	n	f	n	6.50	6.50	6.50	5.74	5.76	5.75	6.00	6.16	6.08	4.96	4.44	4.70
4	n	f	f	7.29	6.31	6.80	4.74	5.54	5.14	5.87	6.03	5.95	4.96	4.21	4.58
5	f	n	n	7.63	6.73	7.18	5.14	5.52	5.33	5.80	5.81	5.80	4.86	4.39	4.62
6	f	n	f	7.02	6.27	6.64	4.93	5.12	5.02	5.47	5.57	5.52	4.88	4.30	4.59
7	f	f	n	5.58	5.12	5.35	4.40	5.06	4.73	4.69	4.67	4.68	3.80	3.16	3.48
8	f	f	f	5.44	5.01	5.22	3.49	3.88	3.68	4.64	4.70	4.67	3.49	3.04	3.26

[a]Fertilizer (F), insect control (I), and weed control (W); n = new-technology level and f = farmer's level.

Table 16.11 Computation of Yield Gap as the Difference between the New Technology and the Farmer's Practice, Computed from Data in Tables 16.8 to 16.10

Farm Number	Mean Yield, t/ha		
	New Technology	Farmer's Practice	Yield Gap
1	6.62	4.25	2.37
2	6.23	4.25	1.98
3	7.20	5.66	1.54
4	7.11	5.20	1.91
5	3.65	2.93	0.72
6	4.32	3.02	1.30
7	6.26	5.23	1.03
8	6.73	5.22	1.51
9	6.79	5.26	1.53
10	7.60	5.85	1.75
11	6.50	5.77	0.73
12	6.65	4.79	1.86
13	7.17	5.10	2.07
14	6.89	4.71	2.18
15	6.92	4.76	2.16
16	7.83	5.22	2.61
17	7.13	3.68	3.45
18	6.67	4.67	2.00
19	6.13	3.26	2.87
Av.	6.55	4.68	1.87

three test factors and that the main effects of all three test factors are significant.

□ STEP 3.   Calculate the contribution of each test factor to the yield gap by performing either one of two analyses: Analysis I when interaction effect between test factors is not significant, and Analysis II, otherwise.

For our example, no interaction between the three test factors is significant (step 2) and, hence, Analysis I is appropriate. However, computational steps of both analyses will be illustrated using data of the present example.

• *Analysis I. Average Contribution.* Compute the average contribution for each test factor as:

(i)   For each of the n_y farms (each testing set Y treatments), compute the contribution of each test factor, say factor A, as:

$$C_A = \bar{P} - \bar{T}_A$$

Table 16.12   Combined Analysis of Variance of Data in Table 16.10 over Four Test Farms; with Each Trial Involving 2^3 Complete Factorial Treatment Combinations

Source of Variation	Degree of Freedom	Sum of Squares	Mean Square	Computed F^a
Farm	$n_z - 1 = 3$	24.85006	8.28335	
Reps. within farm	$n_z(r - 1) = 4$	2.27329	0.56832	
Treatment	$2^k - 1 = 7$	41.25957	5.89422	15.35**
F	(1)	20.38522	20.38522	53.10**
I	(1)	16.99501	16.99501	44.27**
W	(1)	2.45706	2.45706	6.40*
$F \times I$	(1)	0.48302	0.48302	1.26^{ns}
$F \times W$	(1)	0.08410	0.08410	< 1
$I \times W$	(1)	0.35106	0.35106	< 1
$F \times I \times W$	(1)	0.50410	0.50410	1.31^{ns}
Treatment $\times$ Farm	$(2^k - 1)(n_z - 1) = 21$	8.06227	0.38392	
Pooled error	$n_z(2^k - 1)(r - 1) = 28$	4.23951	0.15141	
Total	$(n_z)(r)(2^k) - 1 = 63$	80.68470		

^{a}** = significant at 1% level, * = significant at 5% level, ^{ns} = not significant.

where $\bar{P}$ is the mean yield of the new technology and $\bar{T}_A$ is the mean yield of the treatment in which only the level of factor A is at the farmer's level (i.e., all other test factors are at the new-technology level).

For our example, using data in Table 16.9, the contribution of, say factor F, of farm 13 is computed as:

$$C_F = \bar{P} - \bar{T}_F$$

$$= 7.17 - 6.09 = 1.08 \text{ t/ha}$$

And, the contribution of factor W of farm 14 is computed as:

$$C_W = \bar{P} - \bar{T}_W$$

$$= 6.89 - 7.15 = -0.26 \text{ t/ha}$$

The results for all three test factors and for all three test farms (farms 13, 14, and 15) are shown in the first three rows of Table 16.13.

(ii)   For each of the n_z farms (each testing set Z treatments) compute the contribution for each test factor, say factor A, as the difference

Table 16.13 Average Contribution of Each of the Three Test Factors,[a] Tested in a Technology-Verification Trial; Computed from Data in Tables 16.9 and 16.10

Farm Number	Av. Contribution, t/ha		
	F	I	W
13	1.08	0.95	0.11
14	0.41	0.86	−0.26
15	0.86	0.46	0.04
16	0.63	0.89	0.60
17	1.42	1.15	0.68
18	1.09	0.74	0.19
19	1.38	1.35	0.11
Av.	0.98	0.91	0.21

[a] Fertilizer (F), insect control (I), and weed control (W).

between the mean yield of all treatments with factor A at the new-technology level ($\bar{A}_n$) and the mean yield of all treatments with factor A at the farmer's level ($\bar{A}_f$).

For our example, using data in Table 16.10, the contribution of, say factor F, of farm 16, is computed as:

$$C_F = \bar{F}_n - \bar{F}_f$$

$$= \frac{7.83 + 5.79 + 6.50 + 6.80}{4} - \frac{7.18 + 6.64 + 5.35 + 5.22}{4}$$

$$= 6.73 - 6.10 = 0.63 \text{ t/ha}$$

And, the contribution of factor W of farm 17 is computed as:

$$C_W = \bar{W}_n - \bar{W}_f$$

$$= \frac{7.13 + 5.75 + 5.33 + 4.73}{4} - \frac{6.41 + 5.14 + 5.02 + 3.68}{4}$$

$$= 5.74 - 5.06 = 0.68 \text{ t/ha}$$

The results for all three test factors and for all four test farms (farms 16, 17, 18, and 19) are shown in Table 16.13.

(iii) For each test factor, compute the mean, over all ($n_y + n_z$) farms, of all contributions computed in (i) and (ii). For our example, the mean contribution for each test factor is computed over $n_y + n_z = 3 + 4 = 7$ farms. The results are shown in the bottom row of Table 16.13. The mean contributions are 0.98 t/ha for fertilizer, 0.91 t/ha for insect control, and 0.21 t/ha for weed control.

(iv) Compute the adjusted mean contribution of each test factor as the product of the adjustment factor (g) and the mean contribution computed in (iii):

$$\text{Adjusted mean contribution} = (g)(\text{mean contribution})$$

$$g = 1 - \frac{(S - G)}{S}$$

where S is the sum of the mean contributions over all test factors, and G is the mean yield gap of step 1.

Note that if $S = G$, no adjustment is needed. The difference between S and G should be small when there is no interaction between test factors, and large, otherwise. In cases where there is no data from n_z farms and, hence, no assessment of interaction (step 2) can be made, the difference $D = S - G$ that is greater than 20% of the mean yield gap can be taken as an indicator of the presence of interaction. Thus, if D is greater than $(0.2)(G)$, Analysis I should be aborted and Analysis II applied instead.

For our example, the difference D is computed as:

$$D = S - G$$

$$= (0.98 + 0.91 + 0.21) - 1.87$$

$$= 0.23 \text{ t/ha}$$

which is less than 20% of the mean yield gap. This collaborates with the result of no significant interaction between test factors obtained in step 2. Hence, the adjusted mean contribution of each test factor is computed as:

$$\text{Adjusted mean contribution of fertilizer} = (0.89048)(0.98) = 0.87$$

$$\text{Adjusted mean contribution of insect control} = (0.89048)(0.91) = 0.81$$

$$\text{Adjusted mean contribution of weed control} = (0.89048)(0.21) = 0.19$$

Note that the sum of the adjusted mean contributions equals the mean yield gap.

- *Analysis II. Individual and joint contributions.* When one or more of the interaction effects between test factors is significant (step 2), compute the *individual contribution* of each test factor based on data from n_z farms which tested the complete factorial treatments, and the *joint contribution* of each combination of two or more test factors based on data from both the n_y farms and the n_z farms.

(i) *Individual contribution* of a test factor, say factor A, measures the added yield over that of the farmer's practice, obtained from applying factor A at the new-technology level while maintaining all other test factors at the farmer's level. Thus, it is computed as the difference between the yield of the farmer's practice and the yield of the treatment that has only factor A at the new-technology level (i.e., all other test factors are at the farmer's level). For example, consider set Z with complete factorial treatments in Table 16.7. Here, the individual contribution of factor F is computed as:

$$H_F = \overline{Y}_{15} - \overline{Y}_{16}$$

where $\overline{Y}_{15}$ and $\overline{Y}_{16}$ are the mean yields of treatments 15 and 16.

For our example, using data in Table 16.10, the individual contribution of factor F for farm 16 is computed as:

$$H_F = 6.80 - 5.22 = 1.58 \text{ t/ha}$$

(ii) *Joint contribution* measures the added yield, over that of the farmer's practice, obtained from applying a combination of two or more test factors at the new-technology level. For example, the joint contribution of a set of two test factors, say, factors F and W, based on the list of treatments in Table 16.7, is computed as:

$$J_{FW} = \overline{Y}_{11} - \overline{Y}_{16}$$

where $\overline{Y}_{11}$ and $\overline{Y}_{16}$ are the mean yields of treatments 11 and 16. The joint contribution of a set of three test factors F, W, and I of Table 16.7 is computed as:

$$J_{FWI} = \overline{Y}_3 - \overline{Y}_{16}$$

where $\overline{Y}_3$ and $\overline{Y}_{16}$ are the mean yields of treatments 3 and 16.

For our example, using data in Table 16.10, the joint contribution of the test factors F and W, for farm 16, is computed as:

$$J_{FW} = 6.50 - 5.22 = 1.28 \text{ t/ha}$$

16.3.5.2 Cost and Return Analysis.

The cost and return analysis measures the profitability of the new technology and its components represented by the test factors. Unlike the yield gap analysis (Section 16.3.5.1) which is applied on a per-crop basis, the cost and return analysis is applied at the cropping-pattern level. Thus, the cost and return analysis is applicable even when there is no crop in common between the new technology and the farmer's practice. This is possible because the cost and return analysis is expressed in monetary values, which are comparable even among different crops.

We illustrate the computational procedure for a cost and return analysis with the example we used in Section 16.3.5.1. The step-by-step procedures are:

☐ STEP 1.   For each farm, determine the added cost for each test factor by computing the added cost of applying the new-technology level over and above the farmer's level. Then, sum the added costs over all test factors to represent the added cost for the new technology. That is:

$$C = \sum_{i=1}^{k} c_i$$

where c_i is the added cost of the ith factor and k is the number of test factors. For our example, the added costs, for each of the three test factors and for each of the 19 test farms, are shown in Table 16.14.

☐ STEP 2.   Determine the price of each product by determining the price per unit of yield for each crop. In most cases, a single value per crop, usually the mean value if moderate price variation exists in the target area, is adequate. If large price variation exists, stratification of the target area may be necessary and one price per stratum is established. For our example, a single price of rice of $0.28/kg is used.

☐ STEP 3.   Determine the added returns for the new technology, and for each of its component test factor, as:
  • For each test farm, multiply the yield gap (determined in step 1 of the yield gap analysis, Section 16.3.5.1) by the price of the product for each crop and sum over all crops. That is, the added return of the new technology over the farmer's practice is computed as:

$$R = \Sigma(\text{price})(\text{yield gap})$$

where the summation is over all crops in the cropping pattern.

For our example, the values of added return for each of the 19 farms are computed from the yield gap values in Table 16.11 and the price value determined in step 2. The results are shown in the last column of Table 16.14. The added returns from the new technology over the farmer's practice range from $202 to $966/ha, with an average over all test farms of $524/ha.

Table 16.14 Added Costs and Added Returns of the New-Technology Level over the Farmer's Level, for Each Test Factor, on 19 Test Farms

| Farm Number | Added Cost, $/ha | | | | Added Return, $/ha[a] |
	Fertilizer	Insect Control	Weed Control	Total	
1	0	72	23	95	664
2	35	87	11	133	554
3	−22	93	11	82	431
4	−2	114	23	135	535
5	−19	56	39	76	202
6	−2	118	23	139	364
7	1	175	23	199	288
8	12	136	11	159	423
9	15	120	11	146	428
10	4	130	11	145	490
11	6	168	7	181	204
12	10	112	23	145	521
13	6	138	15	159	580
14	1	172	15	188	610
15	15	127	11	153	605
16	−4	206	15	217	731
17	4	141	27	172	966
18	9	78	23	110	560
19	17	145	39	201	804
Av.	5	126	19	149	524

[a] Computed from yield gap values in Table 16.11 and the price of rice of $0.28/kg.

- For each of the n_y and n_z farms, compute the added returns for each test factor, or for each combination of test factors, as the product of price and average contribution or individual or joint contribution, as the case may be (see Section 16.3.5.1, step 3). That is, the added return for each test factor, say factor A, is computed as:

$$R_A = pM_A$$

where p is the price (step 2) and M_A is the contribution of factor A computed in step 3 of Section 16.3.5.1.

For our example, the added returns are computed for each test factor from the average contributions in Table 16.13. The results for each of the 7 farms are shown in Table 16.15. The added returns averaged $275/ha from fertilizer, $256/ha from insect control, and $59/ha from weed

Table 16.15   Added Return for Each Test Factor, Computed from Data in Table 16.13

Farm Number	Added Return, $/ha[a]		
	F	I	W
13	302	266	31
14	115	241	-73
15	241	129	11
16	176	249	168
17	398	322	190
18	305	207	53
19	386	378	31
Av.	275	256	59

[a] Fertilizer (F), insect control (I), and weed control (W).

control. That makes the proportion of contribution by the three test factors to the total added returns from the new technology 47% for fertilizer, 43% for insecticide, and 10% for weed control.

Because there is a difference of $66 between the average total added return in Table 16.14 ($524/ha) and the sum of the added returns of the three test factors in Table 16.15 ($590/ha), the added returns per hectare of the three test factors are adjusted to be $244 for fertilizer, $228 for insect control and $52 for weed control.

☐ STEP 4.   In each farm, compute the added profit as follows:

- Added profit from the new technology over the farmer's practice is computed as:

$$P = R - C$$

where R is the added return of step 3 and C is the added cost of step 1.

- Added profit of each test factor, say factor A, is computed as:

$$P_A = R_A - c_A$$

where R_A and c_A are the added returns and added cost corresponding to factor A, computed in step 3 and step 1, respectively.

For our example, the results are shown in Tables 16.16 and 16.17. The added profit from the new technology ranges from $23/ha for farm 11 to $794/ha for farm 17, with an average of $375/ha over all test farms. When classified by individual test factors, the highest added profit comes from fertilizer (with a mean of $268/ha), followed by insect control (with a mean of $112/ha), and weed control (with a mean of $38/ha).

Table 16.16 Added Profit of the New Technology over the Farmer's Practice, on 19 Test Farms, Computed from Data in Table 16.14

Farm Number	Added Profit, $/ha	Farm Number	Added Profit, $/ha	Farm Number	Added Profit, $/ha
1	569	8	264	14	422
2	421	9	282	15	452
3	349	10	345	16	514
4	400	11	23	17	794
5	126	12	376	18	450
6	225	13	421	19	603
7	89				

Table 16.17 Added Profit for Each Test Factor, Computed from Data in Tables 16.14 and 16.15

Farm Number	Added Profit, $/ha[a]		
	F	I	W
13	296	128	16
14	114	69	-88
15	226	2	0
16	180	43	153
17	394	181	163
18	296	129	30
19	369	233	-8
Av.	268	112	38

[a]Fertilizer (F), insect control (I), and weed control (W).

□ STEP 5.   Determine the marginal benefit-cost ratio for each farm as:

- Marginal benefit-cost ratio of the new technology over the farmer's practice is computed as:

$$B = \frac{R}{C}$$

- Marginal benefit-cost ratio of the new-technology level over the farmer's level of a test factor, say, factor A, is computed as:

$$B_A = \frac{R_A}{c_A}$$

Table 16.18   Marginal Benefit-Cost Ratio (*B*) of the New Technology over the Farmer's Practice, for Each of the 19 Test Farms, Computed from Data in Table 16.14

Farm Number	B	Farm Number	B	Farm Number	B
1	6.99	8	2.66	14	3.24
2	4.17	9	2.93	15	3.95
3	5.26	10	3.38	16	3.37
4	3.96	11	1.13	17	5.62
5	2.66	12	3.59	18	5.09
6	2.62	13	3.65	19	4.00
7	1.45				

Table 16.19   Marginal Benefit-Cost Ratio for each Test Factor, for Each of the Seven Test Farms, Computed from Data in Tables 16.14 and 16.15

Farm Number	Marginal Benefit-Cost Ratio[a]		
	F	I	W
13	50.33	1.93	2.07
14	115.00	1.40	−4.87
15	16.07	1.02	1.00
16	−44.00	1.21	11.20
17	99.50	2.28	7.04
18	33.89	2.65	2.30
19	22.71	2.61	0.79
Av.	41.93	1.87	2.79

[a] Fertilizer (F), insect control (I), and weed control (W).

For our example, the results are shown in Tables 16.18 and 16.19. The marginal benefit-cost ratio of the new technology ranges from 1.13 in farm 11 to 6.99 in farm 1, with an average of 3.67 over all test farms. Based on individual test factors, the new-technology level of fertilizer gave the highest mean marginal benefit-cost ratio of 41.93 followed by weed control (mean of 2.79); the lowest is that for insect control with the mean of 1.87.

CHAPTER 17

Presentation of Research Results

The last and most important task for any researcher is the summarization and presentation of research results. The primary objective in this task is to put research findings in a form that can be easily understood by all interested parties. Raw data left unsummarized will generally remain in obscurity and valueless in the researchers' files; data well documented have the best chance of reaching intended users.

Proper presentation of research results is a task well accepted by serious researchers; however, which techniques are appropriate for such a presentation is not well established. Although most scientific societies prescribe the style and format of articles they publish, those guidelines deal primarily with standardization and uniformity. They assume that each author will choose the best technique for presenting data.

There are two types of information that are generally presented in agricultural research:

- The first type of information is the description of experimental materials and environmental factors for the experimental site. Correct presentation involves primarily the choice of the characters to present, the type of statistics to use, and the appropriate measure of precision. For discrete characters, the arithmetic mean is the most commonly used statistic, with the standard error of the mean as the indicator of precision (Table 17.1). For characters measured over time, such as rainfall and solar radiation, the line graph or bar chart, or both, are commonly used (Figure 17.1).

- The second type of information presented concerns the results of data analysis. Its presentation varies greatly with the treatments tested (e.g., discrete or quantitative and single-factor or factorial treatments), the characters measured (e.g., measurements over time or multicharacter data), and the statistical procedures used (e.g., mean comparison, regression, or chi-square test).

591

Table 17.1 Chemical and Physical Properties of the Surface Soil at the Site of a Long-Term Fertility Experiment

Soil Character	Unit	Mean[a]
pH	—	5.8 ± 0.1
Organic matter	%	4.15 ± 0.01
Total nitrogen	%	0.31 ± 0.01
Extractable phosphorus	ppm	7.3 ± 1.0
Exchangeable potassium	meq/100 g	1.46 ± 0.07
Exchangeable calcium	meq/100 g	9.18 ± 0.27
Cation exchange capacity	meq/100 g	73.3 ± 0.6

[a]Average of eight samples ± standard error of the mean.

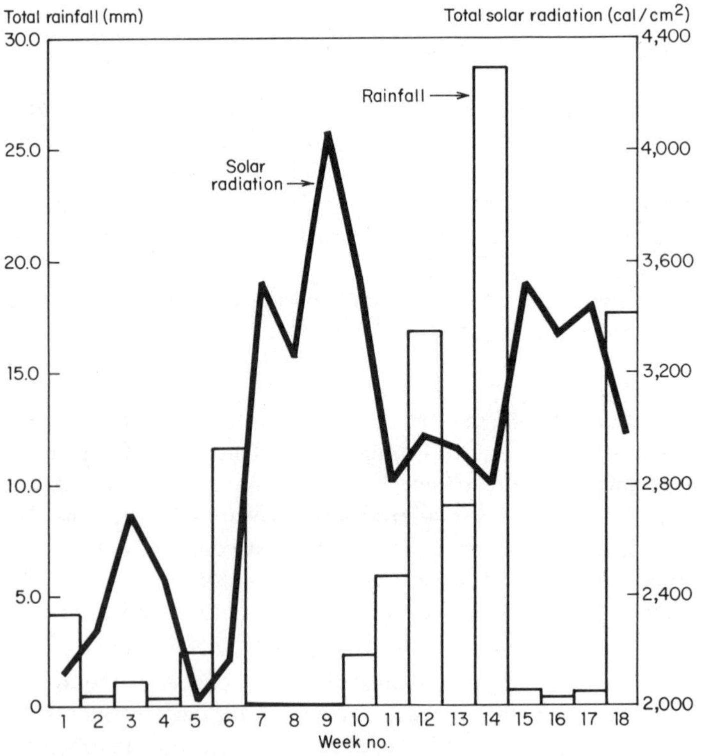

Figure 17.1 Rainfall and solar radiation during the period of an experiment, by a combination of line graph and bar chart.

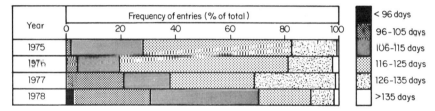

Figure 17.2   Bar chart showing the frequency distribution of growth duration of rice varieties tested in IRRI trials from 1975 to 1978. The chart emphasizes the increasing frequency of shorter-growth-duration varieties over years.

The tabular form is the most commonly used form for presenting research results. A table is flexible and can be used to present a wide variety of results. In practice, any type of data that cannot be suitably summarized graphically can be presented in tabular form. Mean comparison of discrete treatments (see Chapter 5, Tables 5.1, 5.7, 5.9, and 5.11 for LSD test; and Chapter 6, Tables 6.17 and 6.22 for DMRT) and the description of some key features of the environment (Table 17.1) are two common uses of the tabular form.

The most commonly used graphic presentations for agricultural research, arranged according to the level of accuracy and frequency of use, are the line graph, the bar chart, and the pie chart. A line graph is suited for presenting the relationship between two quantitative (continuous) variables such as that between crop response and a quantitative treatment variable (see Chapter 6, Figure 6.1; Chapter 8, Figure 8.1; Chapter 9, Figure 9.9). A bar chart is generally used for discrete (discontinuous) data such as frequency distribution and percentage data (Figure 17.2). A pie chart is generally used to present striking differences in the relative magnitudes of a few components of a whole unit (Figure 17.3).

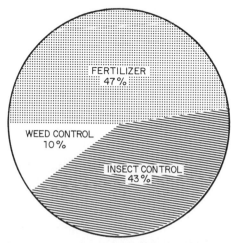

Figure 17.3   Pie chart showing the relative contribution of insect control, fertilizer, and weed control to improved rice yields in farmers' fields (data from Chapter 16, Table 16.13).

The primary focus of this chapter is to present guidelines for appropriate summarization and presentation of experimental results, with emphasis on mean comparison—the most commonly used data analysis in agricultural research.

17.1   SINGLE-FACTOR EXPERIMENT

In a single-factor experiment, there are two types of comparison between treatment means: pair comparison for discrete treatments and trend comparison for quantitative treatments.

17.1.1   Discrete Treatments

The least significant difference (LSD) test and the Duncan's multiple range test (DMRT) are the two most commonly used procedures for comparing means of discrete treatments (Chapter 5, Section 5.1). The results of such comparisons can be presented either in tables or in bar charts.

17.1.1.1   Tabular Form With LSD Test.   The presentation of mean comparison based on the LSD test is simple and should be used whenever use of the LSD test is valid (Chapter 5, Section 5.1.1). Some rules for the proper presentation of mean comparison using the LSD test are:

RULE 1.   Use the LSD test only when the F test in the analysis of variance is significant.

RULE 2.   Use the LSD test only when there are no more than five treatments tested (Table 17.2) or whenever the comparison is between the control treatment and every other treatment (see Chapter 5, Table 5.1).

Table 17.2   Tabular Presentation to Illustrate the Use of the LSD Test, in Comparing the Yields of Three Promising Maize Hybrids *A*, *B*, and *D*, and a Check Variety *C*[a]

Maize Hybrid/Variety	Mean Yield, t/ha[b]
A	1.46
B	1.47
C (check)	1.07
D	1.34
$LSD_{.05}$	0.25

[a]Source: Chapter 2, Tables 2.7 and 2.8.
[b]Average of four replications.

RULE 3.   Use only one test. Do not give both LSD test and DMRT for the same set of treatment means.

RULE 4.   The LSD value must have the same unit of measurement and the same number of significant digits as those of the mean values.

RULE 5.   When the analysis of variance is made with data transformation (Chapter 7, Section 7.2.2.1), the LSD test can be presented only if the mean values are presented in the transformed scale.

RULE 6.   For comparison of all possible pairs of means, present the LSD value either as the last (bottom) row in the table (Table 17.2) or as a footnote. For comparison between the control treatment and every other treatment, mark each treatment with **, *, or ns (or no mark) depending on the level of the significance of the LSD test (see Chapter 5, Table 5.1).

17.1.1.2   Tabular Form with DMRT.   Some rules for the proper use and presentation of the DMRT are:

RULE 1.   For comparing all possible pairs of treatments, use DMRT when the number of treatments exceeds five. For less than six treatments, the LSD test can be used (see Section 17.1.1.1). When data transformation is used in the analysis of variance and treatment means are presented in the original scale, use DMRT regardless of the number of treatments.

RULE 2.   Use *line notation* (see Chapter 5, Table 5.12) when the treatments are arranged according to their rank, and use *alphabet notation*, otherwise (see Chapter 6, Tables 6.17 and 6.22).

RULE 3.   With the *alphabet notation*:

- Start the first letter for each mean at the same position (see Chapter 6, Tables 6.17 and 6.22).
- Whenever four or more letters are required by more than two treatments, use the dash notation to shorten the string of letters. For example, the string *abcd* is written as *a–d*, the string *bcdefg* is written as *b–g*, and so on (see Chapter 7, Table 7.22).

RULE 4.   When only a fraction of total treatments are to be presented in the table, adjust the DMRT letters so that letters are consecutive and as few letters as possible are used. For illustration, see Table 17.3.

RULE 5.   Means with the same numerical value should not be presented with different DMRT letters. Such cases are usually caused by rounding errors resulting from the reduction of significant digits. When that happens, the remedy is to increase the significant digit of the mean values (Table 17.4). Note that it is possible for the mean values to be presented with a larger significant digit than the original data. In Table 17.4, for example, the mean values are presented with one decimal more than the raw data.

RULE 6.   Reduce the number of significant digits if the DMRT so indicates. For example, it is clear in Table 17.5 that the decimal in the mean values is not

Table 17.3 Adjustment of DMRT Letters when Means of Only Six of the 10 Varieties Tested are Presented

Variety[b]	Grain Yield				Plant Height			
	Mean,[a] t/ha	DMRT			Mean,[a] cm	DMRT		
		Original	Unadjusted	Adjusted		Original	Unadjusted	Adjusted
CPM-13-32-41	1.16	a	a	a	56.5	f	f	d
CR214-JS-52-102	0.44	c	c	b	53.2	f	f	d
D6-2-2	0.98	ab			120.2	a		
Bala	1.16	a			72.6	d		
Annapuma	0.49	c	c	b	64.7	e	e	c
CR245-1	1.14	a	a	a	105.9	b	b	a
OR165-28-14-B	0.60	bc	bc	b	66.3	de	de	c
IAC25	0.62	bc			104.6	b		
KS109	0.51	c			67.4	de		
IR12787-3	0.81	abc	abc	ab	84.3	c	c	b

[a]Average of four replications. Mean separation by DMRT at 5% level.
[b]Varieties in boldface are to be excluded from presentation.

Table 17.4 Adjustment in the Number of Significant Digits to Avoid the Occurrence of Different DMRT Letters for Means with the Same Value

Treatment Number	Brown Planthopper, no./hill[a]	
	Before Adjustment	After Adjustment
1	1 ab	1.1 ab
2	1 ab	1.2 ab
3	1 ab	1.3 ab
4	1 a	1.4 a
5	1 ab	1.2 ab
6	1 b	0.6 b
7	2 a	1.5 a

[a]Average of four replications. Mean separation by DMRT at 5% level.

necessary because significant differences among means that are less than 10% cannot be detected.

RULE 7. Do not present the DMRT when all treatments have the same letters. A footnote indicating no significant differences among treatments should suffice.

17.1.1.3 Bar Chart. The bar chart is appropriate when the treatments are discrete and relatively few. A bar chart is generally used to emphasize striking

Table 17.5 Adjustment in the Number of Significant Digits to Reflect the Degree of Precision Indicated by the DMRT

Treatment Number	Germination, %[a]	
	Before Adjustment	After Adjustment
1	17.4 b	17 b
2	23.5 b	24 b
3	21.3 b	21 b
4	51.6 a	52 a
5	54.7 a	55 a
6	55.7 a	56 a
7	57.5 a	58 a
8	58.2 a	58 a
9	61.4 a	61 a

[a]Average of four replications. Mean separation by DMRT at 5% level.

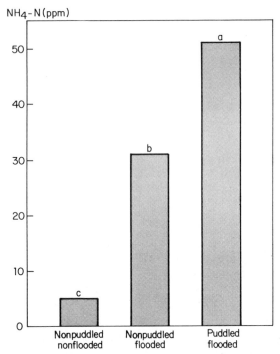

Figure 17.4 Bar chart showing the use of DMRT in comparing available soil ammonium nitrogen (NH$_4$-N) in three types of rice culture; bars with no common letter are significantly different at the 5% level.

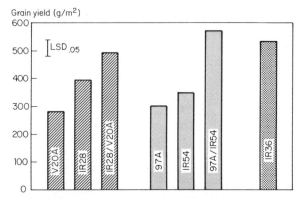

Figure 17.5 Bar chart showing the relative yields of rice hybrids, their parents, and a commercial variety IR36: the bars are grouped and the LSD value is presented.

598

differences between treatments (Figure 17.4), or to show some distinctive patterns of change among groups of treatments (Figure 17.5).

Some rules for the use of bar chart to present the results of mean comparison involving discrete treatments are:

RULE 1.   Use a bar chart when a striking difference or relative pattern of change is to be emphasized, and when it is not important to maintain a high degree of precision of the individual mean values.

RULE 2.   When the DMRT is applied, use alphabet notation by placing DMRT letters on top of each treatment bar (see Figure 17.4).

RULE 3.   When the LSD test is used, show the size of the LSD as illustrated in Figure 17.5.

RULE 4.   Always begin the Y axis at the zero level so that both the absolute and the relative bar heights reflect accurately the magnitude of the treatment means and the treatment differences (Figure 17.6).

RULE 5.   Avoid use of a cut-off bar for the purpose of shortening the height of the bar when there are extremely large differences in the bar height (Figure 17.7).

RULE 6.   Avoid writing the actual data on top of each bar (see Figure 17.7). In a bar chart, the mean value should be read from the Y axis based on the

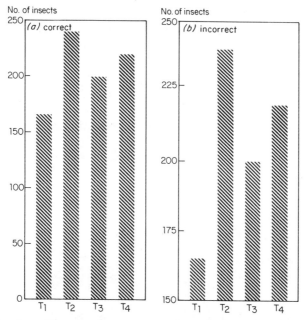

Figure 17.6 Bar charts illustrating the correct procedure (*a*) of starting the Y axis at the zero level. The incorrect procedure (*b*) truncates the Y axis and exaggerates the differences among treatments.

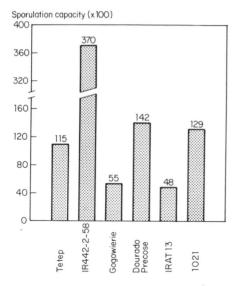

Figure 17.7   Bar chart showing the use of a cut-off bar to shorten the height and the placement of data on top of each bar. Both practices should be avoided.

height of the corresponding bar. If greater precision than what is obtainable from the bar chart is required, the bar chart is not the suitable form of presentation (see Rule 1).

RULE 7.   Avoid joining the tips of adjacent bars with a line, as illustrated in Figure 17.8. Such lines across the bars to show trend is inappropriate for discontinuous (discrete) variable. If the variable on the X axis is a continuous variable, use the line graph and not the bar chart.

RULE 8.   Determine the sequence of the bars based on the type of treatments and the specific point that needs to be emphasized. When there is a natural grouping of treatments, bars may be presented in treatment groups (Figures 17.5 and 17.9). Otherwise, bars can be arranged according to their rank—from highest to lowest or lowest to highest (Figure 17.10).

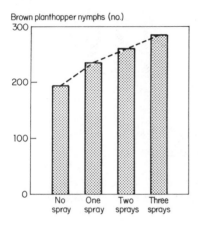

Figure 17.8   Illustration of the combination of bar chart and line graph for a single character. This practice should be avoided.

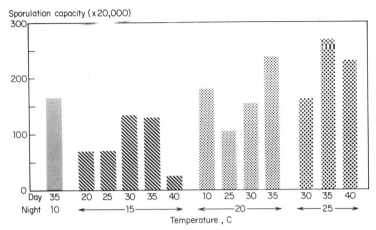

Figure 17.9 Bar chart showing the effect of fluctuating day and night temperatures on sporulation of Gibberella. This illustrates a group arrangement of bars based on night temperature.

17.1.2 Quantitative Treatments: Line Graph

Trend comparison showing the functional relationship between the treatment level and the corresponding biological response is the most appropriate method of mean comparison among quantitative treatments. For such comparisons, the line graph is most appropriate because the values of the response that are of

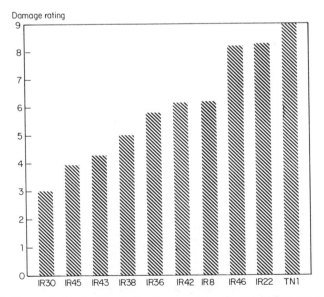

Figure 17.10 Bar graph showing the relative damage ratings of green leafhoppers on selected rice varieties. This illustrates the arrangement of bars according to rank.

interest are not limited to the treatment levels tested but are also on all points within the range of the treatments tested (see Chapter 5, Section 5.2.3). Note that with the tables or bar charts (Section 17.1.1), the response is specified only for each level of treatments actually tested. But with a line graph the response is specified for all points between the highest and the lowest treatment levels.

Some rules for the use of line graph in presenting mean comparison of quantitative treatments are:

RULE 1.   Use the Y axis to represent response and the X axis to represent treatment levels. Choose the scales for the X axis and the Y axis to highlight important points, but avoid distortion of the results. Some guidelines for doing this are:

- Choose the scale on the Y axis that allows the reader to *see* the differences that are shown to be significant, and *not to see* nonsignificant differences as real. Difference is exaggerated with wide scale and is reduced with narrow scale.
- There is no need to start the Y axis of a line graph at the zero level (Figures 17.11 and 17.12). Similarly, the X axis should start at the lowest treatment level (Figure 17.13).

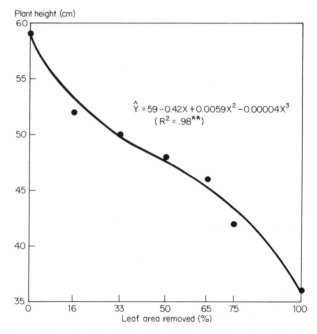

Figure 17.11   Line graph showing plant height of rice variety IR36 as affected by the extent of leaf removal: an illustration of an X axis with marks that are not equidistant and a Y axis that does not start at zero.

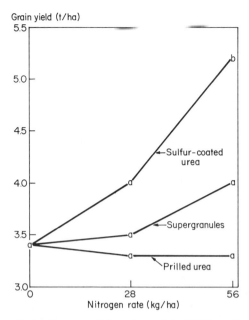

Figure 17.12 Line graph showing mean yield of rice variety RD19 as affected by rates and forms of urea: an illustration of a case where the regression equation cannot be derived because of inadequate treatment levels, and DMRT is used instead. Means with common letter are not significantly different at the 5% level.

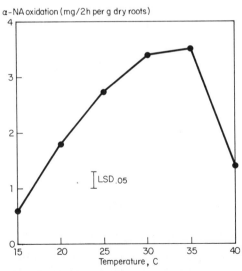

Figure 17.13 Line graph showing the effect of temperature on α-naphthylamine oxidation by rice plants: an illustration of a case where the regression equation cannot be derived because of the unspecified nature of the relationship.

603

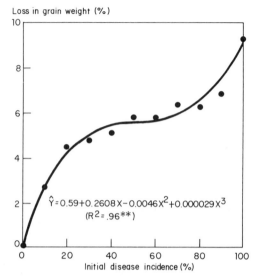

Loss in grain weight (%)

$$\hat{Y}=0.59+0.2608\,X-0.0046\,X^2+0.000029\,X^3$$
$$(R^2=.96^{**})$$

Initial disease incidence (%)

Figure 17.14 Line graph showing the estimated yield loss due to disease: an illustration of an X axis with fewer marks than the levels of disease tested.

- The X axis should cover only the range of treatment levels tested. As much as possible, the *mark* on the X axis (i.e., the spot where the treatment level is indicated) should correspond to the treatment level (Figure 17.11). However, if the number of treatments is large, not all levels need to be marked (Figure 17.14), or if the treatments are of unequal intervals, the marks need not be equidistant (Figure 17.11). The marks on the Y axis should be equidistant. Minimize the number of marks to those necessary for clear readability.

RULE 2.   Use a line graph when there are at least three treatments.

RULE 3.   Whenever possible, estimate an appropriate regression equation, using the regression technique of Chapter 9, as shown in Figures 17.11 and 17.14. Make sure that the line graph includes:

- The observed points (i.e., data used in fitting the regression)
- The estimated regression line, or curve, drawn over the range of the treatment levels tested
- The estimated regression function, its significance, and its correlation coefficient

RULE 4.   When a proper regression equation cannot be derived, either because the number of treatments is not adequate (Figure 17.12) or because no meaningful functional relationship can be prescribed (Figure 17.13), simply draw a line to connect observed points and incorporate the appropriate test of significance.

17.2   FACTORIAL EXPERIMENT

The presentation of mean comparison in a factorial experiment should be based on the following considerations:

- *Interaction between Test Factors.* When interactions between test factors are significant and large, the presentation should emphasize the nature and magnitude of the interactions. This can be done by constructing a multidimensional graph or a multidimensional table involving the factors that interact. For example, if two discrete factors A and B interact, then a two-way table involving factors A and B should be presented.
- *Type of Test Factors.* As in the case of a single-factor experiment (Section 17.1), the appropriate mode of data presentation is a table or a bar chart for discrete factors and a line graph when at least one of the factors is quantitative.
- *Number of Test Factors.* Data presentation increases in complexity as the number of test factors increases. When the number of test factors (k) is not more than three, a k-dimensional table or graph is generally used. Otherwise, only those factors that interact with each other should be included in the same table or graph.

We limit our discussion to multidimensional tables and multidimensional graphs with no more than three dimensions. This is usually adequate in crop research because interactions of an order higher than three are infrequent.

17.2.1   Tabular Form

Rules for constructing a multidimensional table for presenting data from a factorial experiment are:

RULE 1.   Use a tabular form when all factors to be presented are discrete. Otherwise, consider using a line graph.

RULE 2.   Avoid presenting data from a factorial experiment in a one-dimensional table (Table 17.6). Such a format does not facilitate evaluation of the interaction among test factors.

RULE 3.   The number of factors to be included in a multidimensional table should be determined as follows:

- Whenever possible, the dimension of the table should equal the number of test factors. This is usually done when the number of factors is not more than three and the levels in each factor are not too large (Table 17.7).
- If the number of test factors exceeds three, include in the same table only those factors that interacted significantly with each other. For example, in a four-factor experiment, if $A \times B \times C$ interaction is significant and no higher-order interaction is significant, then only a three-dimensional table involving three factors A, B, and C is needed.

Table 17.6   An Improper Presentation of Data from a Factorial Experiment in a One-Dimensional Table (see Table 17.7)

Treatment			Mean Yield, t/ha[a]
Lime	Variety	Manganese Dioxide	
Without	IR26	Without	3.6 d
		With	3.9 d
	IR43	Without	4.0 cd
		With	6.2 a
With	IR26	Without	4.3 cd
		With	4.8 bcd
	IR43	Without	5.3 b
		With	6.2 a

[a]Average of four replications. Mean separation by DMRT at 5% level.

- When there is more than one significant interaction of the same order, construct one table for each interaction or one table involving all interacting factors. For example, if both $A \times B$ and $A \times C$ interactions are significant, either construct two two-way tables of means, $A \times B$ and $A \times C$, or present one $A \times B \times C$ three-way table of means.

RULE 4.   Whenever one or more factors included in the same table have only two levels each, consider presenting the difference between the two levels, in addition to, or in place of, the means (Table 17.8). This facilitates the assessment of the magnitude, as well as the significance, of the effect of each factor. For example, from Table 17.8, the following conclusions can be easily arrived at:

- In IR26, the effect of either lime or manganese dioxide was not significant.
- In IR43, the effect of manganese dioxide was enhanced when lime was not applied, and the effect of lime was observed only when manganese dioxide was not applied.

Table 17.7   A Three-way Table of Means for Presenting Data from a Three-Factor Experiment (This Format is Preferred to That in Table 17.6)

	Mean Yield, t/ha[a]			
	IR26		IR43	
Manganese Dioxide	With Lime	Without Lime	With Lime	Without Lime
With	4.8 bcd	3.9 d	6.2 a	6.2 a
Without	4.3 cd	3.6 d	5.3 b	4.0 cd

[a]Average of four replications. Mean separation by DMRT at 5% level.

Table 17.8 A Three-way Table of Means for a 2^3 Factorial Experiment, Showing Differences Between the Two Levels in Each of Two Factors

	Mean Yield, t/ha[a]					
	IR26			IR43		
Manganese Dioxide	With Lime	Without Lime	Difference	With Lime	Without Lime	Difference
With	4.8	3.9	0.9 ns	6.2	6.2	0.0
Without	4.3	3.6	0.7 ns	5.3	4.0	1.3*
Difference	0.5 ns	0.3 ns		0.9*	2.2**	

[a]Average of 4 replications. ** = significant at 1% level, * = significant at 5% level, ns = not significant.

RULE 5. With a complete block design (CRD, RCB, or Latin square), use the letter notation to present the DMRT results such that mean comparison can be made, either vertically or horizontally, for all treatments (Table 17.7).

RULE 6. For a two-factor experiment in either a split-plot or a strip-plot design, the test criterion for the row factor differs from that of the column factor. The guidelines for presenting the results of such types of data are:

- If the $A \times B$ interaction is significant and the level of factor A is less than six while that of factor B is not, assign factor A as the column factor and factor B as the row factor (Table 17.9). Place appropriate DMRT letters for

Table 17.9 Effects of Weed Control and Land Preparation on Yield of Mungbean: an Illustration of the Use of DMRT for the Column Factor and of LSD Test for Row Factor

	Mean Yield, kg/ha[a]		
Weed Control		Stale Seedbed with	
	Conventional	Glyphosate	One Rototillage
Trifluralin	114 abc	274 ab	104 b
Butralin	101 bcd	265 ab	84 b
Butachlor	26 d	232 ab	37 b
Alachlor	48 cd	201 bc	48 b
Pendimenthalin	46 cd	200 bc	58 b
Thiobencarb	94 bcd	137 c	44 b
Hand weeding (2 ×)	182 a	289 a	230 a
Hand weeding (1 ×)	160 ab	263 ab	224 a
No weeding	75 cd	148 c	54 b
Av.	94	223	98

[a]Average of four replications. Mean separation in a column by DMRT at 5% level. $LSD_{.05}$ value for comparing land-preparation means in a row is 73 kg/ha.

Table 17.10 Effect of Three Sources of Urea and Five Tillage Methods on Panicle Length of Rice: an Illustration of the Use of DMRT Instead of LSD Test for the Factor Needing Emphasis

Tillage frequency			Mean Panicle Length, cm[a]		
Plowing	Harrowing	Rototillage	Prilled	Sulfur-coated	Super-granule
1	1	0	20.8 a	21.6 a	22.1 a
1	3	0	19.6 b	21.2 a	21.9 a
1	1	1	20.5 b	20.4 b	22.4 a
1	2	2	20.9 a	20.1 a	21.3 a
2	2	0	21.6 a	21.7 a	21.2 a

[a]Average of two varieties and four replications. Mean separation in a row by DMRT at 5% level. $LSD_{.05}$ for comparing means in a column is 1.5 cm.

comparing means of factor B at each level of factor A. To compare means of factor A at each level of factor B, use the LSD test and present the appropriate LSD value as a footnote (Table 17.9).

- If the $A \times B$ interaction is significant and the levels of factors A and B are both less than six but one factor, say A, is more important or requires more emphasis than the other factor, use DMRT on factor A and the LSD test on factor B (Table 17.10). If both factors need equal emphasis, use the LSD test on both factors and prescribe the appropriate LSD values, one for each factor, as a footnote (Table 17.11).

- If the $A \times B$ interaction is significant and both factors A and B have more than five levels each, use two sets of DMRT letters, one for the column factor and another for the row factor (Table 17.12).

Table 17.11 Straw Weight of Different Rice Varieties with Various Methods of Straw Incorporation: an Illustration of the Use of LSD Test for Both Factors

Straw Treatment	Straw Weight, t/ha[a]					
	IR38	IR40	IR42	IR44	IR46	Av.
Straw removed	3.44	5.00	3.56	3.72	3.80	3.90
Straw spread	3.30	4.28	4.08	3.60	3.24	3.70
Straw burned	3.14	3.68	3.98	3.94	3.32	3.61
Straw composted	2.88	3.90	3.60	4.12	3.84	3.67
Av.	3.19	4.22	3.80	3.84	3.55	3.72

[a]Average of five replications. To compare means in a column, $LSD_{.05} = 0.71$ t/ha; and in a row, $LSD_{.05} = 0.65$ t/ha.

Table 17.12 Effects of Different Isolates of Bacterial Leaf Blight on Lesion Length of Six Rice Varieties: an Illustration of the Use of Two Sets of DMRT Letters — *a* to *e* for Comparing Means in a Column and *w* to *z* for Comparing Means in a Row

Isolate	\multicolumn{7}{c}{Lesion Length, cm[a]}
	IR8	IR20	IR1565	DV85	RD7	RD9	Av.
TB7803	23.5 b w	18.4 a x	4.9 ab y	5.3 ab y	18.9 b x	23.0 bc w	15.7
TB7805	16.0 c xy	15.0 b y	6.5 a z	6.2 a z	18.8 b wx	20.5 c w	13.8
TB7807	1.4 d w	1.4 f w	2.2 bc w	2.2 bc w	1.1 d w	1.0 d w	1.6
TB7808	30.2 a w	4.4 de y	2.7 bc y	3.2 abc y	5.5 c y	25.6 b x	11.9
TB7810	24.3 b x	5.3 de y	1.5 c z	3.2 abc yz	5.7 c y	30.0 a w	11.7
TB7814	23.4 b w	8.6 c y	3.6 abc z	4.8 abc z	18.3 b x	25.2 b w	14.0
TB7831	24.2 b x	2.7 ef y	1.9 bc y	3.0 abc y	4.8 c y	28.7 a w	10.9
TB7833	1.4 d w	1.2 f w	2.0 bc w	1.8 c w	2.6 cd w	2.2 d w	1.9
TB7841	29.0 a w	6.1 cd y	3.6 abc y	4.9 abc y	23.3 a x	30.4 a w	16.2
Av.	19.3	7.0	3.2	3.8	11.0	20.7	10.8

[a]Average of three replications. Mean separation by DMRT at 5% level.

- If the $A \times B$ interaction is significant and one of the factors, say A, has two levels and the other, say B, has six or more levels, assign factor A as the column factor and present the difference between the two levels of A following rule 4. For factor B, use DMRT (Table 17.13).

  If the $A \times B$ interaction is not significant and the $A \times B$ table of means is to be presented, simply compare the A means averaged over all levels of factor B and the B means averaged over all levels of factor A, using DMRT (Table 17.14). Note that, in this case, the same set of DMRT letters (a, b, c, ...) can be used for both factors.

RULE 7. For a three-factor experiment whose design is either a split-plot, a strip-plot, a split-split-plot, or a strip-split-plot design, two or more test criteria are involved. The appropriate method of presentation depends on the factor interactions that are significant, the relative importance of the factors, and the number of levels tested in each factor.

When the three-factor interaction is not significant, consider presenting one or more two-way table of means, one for each of the significant two-factor interactions. For example, consider a $5 \times 2 \times 3$ factorial experiment, the analysis of variance of which is shown in Table 17.15. Because only the $A \times C$ (tillage $\times$ nitrogen source) interaction is significant, only the $A \times C$ two-way table of means as shown in Table 17.10 needs to be presented.

When all three factors are presented in one table, allocate the factors as either row or column factor as follows:

- Combine factors that are tested with the same degree of precision, and allocate the combination of these factors as the row or the column factor.

Table 17.13 Plant Height of the Weed _Cyperus rotundus_ at Harvest as Affected by Elevation and Water Regime: an Illustration of the Use of Difference Column for the Factor with Two Levels

Water Regime[a]	Plant Height, cm[b]		
	Upland	Lowland	Difference
Well drained	17.50 a	34.58 c	-17.08^*
Saturated	19.82 a	33.46 c	-13.64^{ns}
Flooded 7 DAE	17.92 a	84.60 a	-66.68^{**}
Flooded 14 DAE	24.85 a	85.07 a	-60.22^{**}
Flooded 21 DAE	23.82 a	71.60 ab	-47.78^{**}
Flooded 28 DAE	26.83 a	67.04 b	-40.21^{**}

[a] DAE = days after emergence.

[b] Average of six replications. Mean separation in a column by DMRT at 5% level. ** = significant at 1% level, * = significant at 5% level, [ns] = not significant.

Table 17.14 Effects of Different Herbicide Combinations (W_1 to W_6) and Different Methods of Land Preparation (M_1 to M_{10}) on Yield of Maize: an Illustration of the Proper Comparison Between Means when the Two-Factor Interaction Is Not Significant

Land Preparation	Mean Yield, t/ha[a]						
	W_1	W_2	W_3	W_4	W_5	W_6	Av.[b]
M_1	3.45	4.42	4.01	3.84	4.04	4.15	3.98 a
M_2	3.18	3.99	4.03	3.78	4.17	3.84	3.83 a
M_3	3.16	4.36	4.11	4.07	4.28	3.75	3.96 a
M_4	3.33	4.51	4.48	3.52	3.66	4.36	3.98 a
M_5	3.46	4.13	4.06	4.25	4.05	4.14	4.02 a
M_6	3.77	4.07	4.17	4.33	4.72	4.46	4.25 a
M_7	2.88	3.89	3.52	3.69	3.62	4.06	3.61 a
M_8	2.94	4.41	3.68	3.68	3.61	4.20	3.75 a
M_9	3.15	4.26	4.51	4.27	4.26	3.98	4.07 a
M_{10}	3.32	4.91	4.33	4.41	4.60	4.15	4.29 a
Av.[b]	3.26 c	4.30 a	4.09 ab	3.98 b	4.10 ab	4.11 ab	

[a] Average of three replications.

[b] In a row (or column), means followed by a common letter are not significantly different at 5% level.

610

Table 17.15 Analysis of Variance^a of Data on Panicle Length from a 5 × 2 × 3 Factorial Experiment in a Split-Split-Plot Design with Four Replications

Source of Variation	Degree of Freedom	Sum of Squares	Mean Square	Computed F^b
Replication	3	22.94025		
Tillage (*A*)	4	10.38033	2.59508	< 1
Error(*a*)	12	38.35433	3.19619	
Variety (*B*)	1	304.96408	304.96408	154.53**
A × *B*	4	5.56467	1.39117	< 1
Error(*b*)	15	29.60292	1.97353	
Nitrogen source (*C*)	2	26.72017	13.36008	8.33**
A × *C*	8	31.94317	3.99290	2.49*
B × *C*	2	10.01217	5.00608	3.12^{ns}
A × *B* × *C*	8	17.31783	2.16473	1.35^{ns}
Error(*c*)	60	96.18000	1.60300	
Total	119	593.97992		

$^a cv(a) = 8.5\%$, $cv(b) = 6.6\%$, and $cv(c) = 6.0\%$.
^{b}** = significant at 1% level, * = significant at 5% level, ^{ns} = not significant.

For example, when the split-plot design is used in an experiment with more than two factors, a combination of two or more factors may be used as the main-plot or the subplot treatments (see Chapter 4, Section 4.2.2). This is the case in Table 17.16 where the six treatment combinations of water regime and straw treatment were used as the main-plot treatments and the four varieties as subplot treatments. Thus, in presenting the three-dimensional table of means, water regime and straw treatments are combined together and treated as the row factor while the four varieties are assigned as the column factor.

- When all three factors have three different test criteria, identify the least important factor and combine it with one other factor for assignment as either row factor or column factor.

17.2.2 Bar Chart

For data from a factorial experiment where none of the factors is continuous, a bar chart provides an alternative to use of a table. In addition to the rules discussed in Section 17.1.1.3 for a single-factor experiment, the primary concern for factorial experiments is the proper sequencing and grouping of the factors and the levels to be presented. For example, in a 2 × 2 factorial experiment involving variety as factor *A* and manganese dioxide as factor *B*, there would be four bars—one corresponding to each of the four treatment combinations. The bars are usually arranged such that the levels of the factor of major interest are adjacent to each other; the levels of the less important

Table 17.16   Effects of Straw Treatment on Grain Yields of Four Rice Varieties Grown in Different Water Regimes: an Illustration of the Combinations of Two Factors into a Single Dimension

| Water Regime | | Straw Treatment | Mean Yield, t/ha[a] | | | |
Fallow	Midseason Drainage		IR40	IR42	IR44	IR46
Dry	Without	Yes	3.5 b	3.5 b	3.0 b	2.7 b
		No	3.4 b	3.8 ab	2.5 b	2.8 b
Dry	With	Yes	4.0 ab	4.4 a	3.4 a	3.3 b
		No	3.7 ab	3.5 b	3.4 a	2.8 b
Flood	With	Yes	4.3 a	3.8 ab	4.0 a	3.7 a
		No	3.9 ab	4.4 a	3.7 a	2.8 b

[a]Average of four replications. Mean separation in a column by DMRT at 5% level. To compare means in a row, $LSD_{.05} = 0.7$ t/ha.

factor are not. Thus, if the researcher's primary interest is to examine the effect of manganese dioxide, Figure 17.15a is the more appropriate presentation. On the other hand, if the researcher wishes to emphasize varietal difference, Figure 17.15b is more appropriate.

When one of the factors has more than two levels, the proper sequencing of the levels within the factor becomes an issue. Some examples of the sequencing

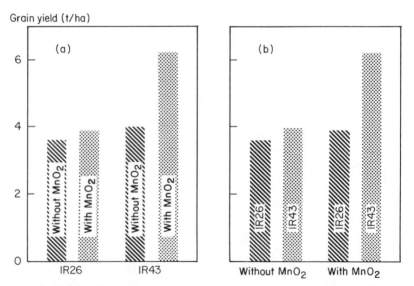

Figure 17.15   Bar charts showing two alternatives for the grouping of factors: alternative (a) to emphasize the effect of manganese dioxide application and alternative (b) to emphasize varietal differences.

criteria commonly used are:

- The natural sequence of the treatments (Figure 17 16)
- The ranking of the data according to one of the levels of the other factor—usually the first level (Figures 17.17 and 17.18a). In Figure 17.17, the sequence of the six varieties is based on their rankings with the first

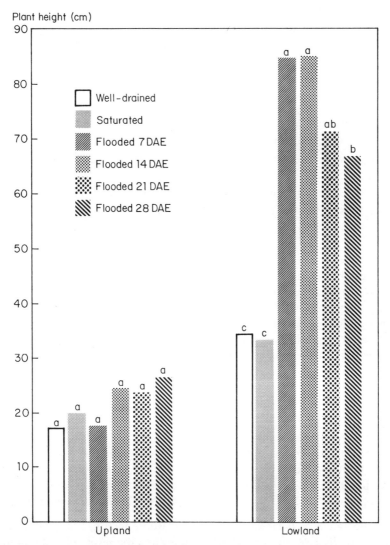

Figure 17.16 Bar chart showing the sequential arrangement of bars according to the nature of treatments (levels within the factor). Means with the same rice culture (upland or lowland) with a common letter are not significantly different at the 5% level.

Sporulation capacity (x100)

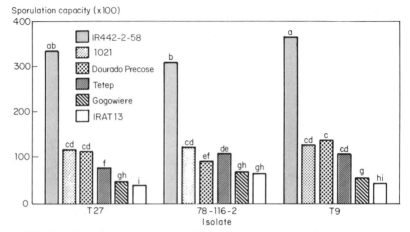

Figure 17.17 Bar chart showing the sporulation capacity of three isolates of *Pyricularia oryzae Cav.* on six rice varieties: an illustration of the arrangement of bars (varieties) according to their rankings based on the first isolate (T27).

isolate while the sequence of the isolates is not determined by any set criterion. In Figure 17.18a, on the other hand, the sequence of the seven varieties is based on their yield ranking when no zinc is applied.

- The ranking of the *treatment difference* if the other factor has two levels (Figure 17.18b).

A proper choice of sequencing criterion is generally dependent upon the nature of the interaction between the factors. For example, Figure 17.18a is preferred over Figure 17.18b if the researcher wishes to emphasize that the significant response to zinc application is obtained only with varieties with intermediate yield levels.

However, regardless of the sequencing criterion used, the same sequence must be used throughout (i.e., over all levels of the other factor). For example, in Figure 17.17, the same sequence for the six varieties at the first isolate is followed at the other two isolates.

17.2.3 Line Graph

When at least one of the factors in a factorial experiment is quantitative, use of a line graph should be considered. The basic considerations and rules for the use of line graph, described for single-factor experiments in Section 17.1.2, are also applicable for factorial experiments. Some additional guidelines are:

1. For an $A \times B$ factorial experiment where factor A is quantitative and factor B is discrete, the line graph would have the levels of the quantitative factor A on the X axis, the response on the Y axis, and one *line* is drawn for each level of the discrete factor B (Figures 17.19 and 17.20).

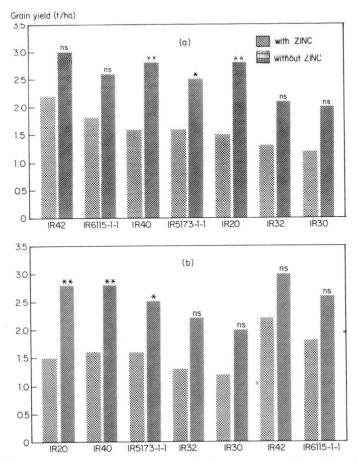

Figure 17.18 Bar charts showing two alternatives for the arrangement of bars: alternative (*a*) is based on the rank of yields without zinc and alternative (*b*) is based on the rank of zinc effect (difference between yields with and without zinc).

In Figure 17.19, there are two factors: variety with four levels and nitrogen rate with six levels. Because the nitrogen rate is quantitative, a regression line to represent the response to nitrogen is fitted for each of the four varieties. If, on the other hand, no regression equation can be estimated, straight lines may be drawn to join adjacent points (Figure 17.20).

2. For an $A \times B$ factorial experiment where both factors are quantitative, follow procedure 1, by treating one of the factors as discrete. The factor that is treated as discrete should correspond to that with fewer levels and whose importance is lower, or to that whose relationship to crop response is not well defined (see Figure 17.21).

3. For an $A \times B \times C$ factorial experiment where factor A is the only factor that is quantitative, follow procedure 1, treating the $m = b \times c$ factorial

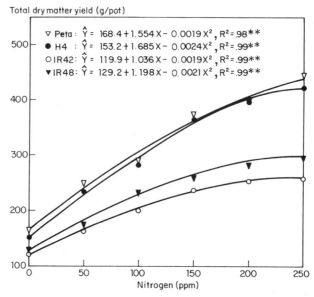

Figure 17.19 Response to nitrogen of four rice varieties: an illustration of the use of line graph for factorial experiments in which one factor is quantitative and regression equation is fitted.

treatment combinations as levels of the discrete factor, where b and c are the levels of factors B and C. In addition, use appropriate line-identification to clearly distinguish between the factors involved. For example, the results of a $4 \times 2 \times 2$ factorial experiment involving four levels of inoculum in the system, two types of soil (soil of wetland and dryland origin), and two water regimes (dry and submerged) are shown in Figure 17.22; with solid lines used for the

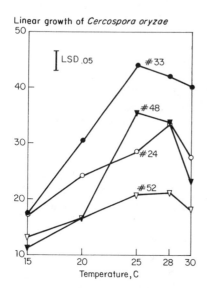

Figure 17.20 Linear growth of four different isolates of *Cercospora oryzae* as affected by temperature: an illustration of the use of line graph for factorial experiments in which one factor is quantitative but no regression equation is fitted.

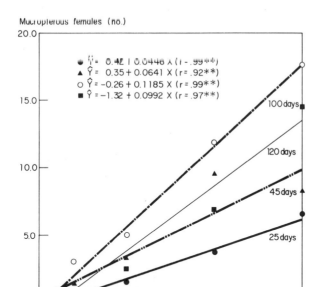

Figure 17.21 The effect of nymphal density on the number of macropterous females, as affected by four plant ages: an illustration of the use of line graph for factorial experiments where both factors are continuous.

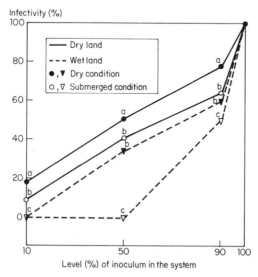

Figure 17.22 Infectivity of rice straw buried in soil of wetland and dryland origin: an illustration of the use of appropriate line identification to differentiate factors; mean separation at each inoculum level by DMRT at 5% level.

617

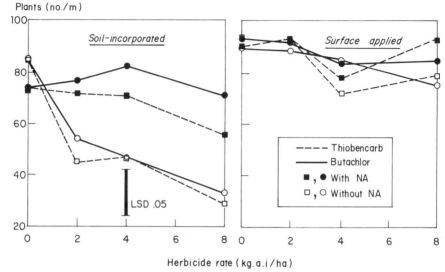

Figure 17.23 Effect of rate, type, and method of application of herbicide and naphthalic anhydride (NA) on the germination of dry-seeded rice: an illustration of the use of more than one graph for data from a factorial experiment.

dryland origin and broken lines for wetland origin, while solid points (● and ▼) refer to a dry condition and empty points (○ and ▽) refer to a submerged condition.

4. For an experiment with three factors or more, use more than one graph if needed. The factor, in which each level is represented by a graph, is usually one in which the main effect is not of primary interest, or whose effect is large (Figure 17.23).

17.3   MORE-THAN-ONE SET OF DATA

The effect of treatments is not confined to a single plant character at a single growth stage but rather to a series of characters (multicharacter data) at various growth stages (measurement over time, Chapter 6, Section 6.2), both of the crop and its immediate environment. Thus, for most experiments, the data that are gathered consist of the measurement of as many characters and in as many stages of observation as are expected to be affected by the treatments.

Although the basic guidelines for the presentation of one character and one set of data, as discussed in Sections 17.1 and 17.2, are still applicable, the added volume and diversity of data to be presented can limit the number of feasible alternatives. For example, when different characters do not have the same unit of measurement, the use of a single bar chart (or line graph) is not appropriate. Even with the tabular form, which is more flexible and can easily

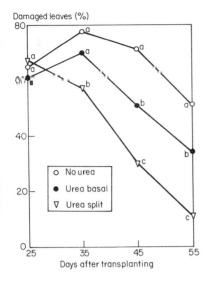

Figure 17.24 Rate of recovery from whorl maggot damage, as affected by the method of urea application: an illustration of the use of line graph for measurements over time from a single-factor experiment with a discrete factor. Mean separation at each day by DMRT at 5% level.

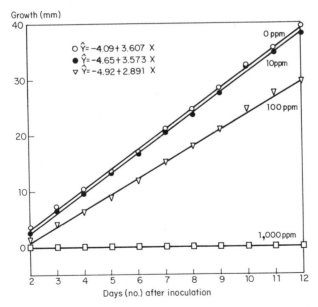

Figure 17.25 Growth rate of the blast pathogen (isolate T27) in nutrient agar plates containing the extract of an indigenous plant at different concentrations: an illustration of the use of line graph for presenting measurements over time from a single-factor experiment with a continuous factor.

619

accommodate additional characters, clarity and readability decreases rapidly as the volume of data included in the same table is increased.

17.3.1   Measurement Over Time

Data for characters that are measured repeatedly over time may be classified into three groups:

- Development-rate data in which measurements are made at regular time intervals in order to assess, over time, the rate of change of the character of interest.
- Growth-stage data in which the period of measurement is associated to a stage of growth instead of a specified time period.
- Occurrence-date data, which measure the time of occurrence of biological phenomenon whose rate of completion is fast. An example is leaf rust infestation, the occurrence of which comes in a flash and the information of interest is the date the flash occurred.

In Chapter 6 (Section 6.2), we discussed the need for a combined analysis over time to evaluate the interaction between treatment and time of observa-

Table 17.17   Weight of Dry Matter of Rice, Measured at Three Growth Stages, with Different Fertilizer Treatments: a Tabular Form of Presentation for Measurements over Time

Fertilizer Treatment	Dry Matter Weight, kg/m^{2a}		
	At 40 Days after Transplanting	At Panicle Initiation	At Harvest
T_1	0.7 f	1.4 f	3.9 d
T_2	0.9 ef	2.1 def	6.3 cd
T_3	1.0 ef	1.7 f	4.8 d
T_4	2.3 ab	3.5 bc	8.4 bc
T_5	1.4 de	2.4 def	6.0 cd
T_6	2.0 bc	3.0 bcd	7.5 c
T_7	1.1 def	2.0 ef	6.5 cd
T_8	2.6 ab	4.8 a	10.8 ab
T_9	1.7 cd	2.8 cde	7.8 c
T_{10}	2.7 a	3.9 ab	8.3 bc
T_{11}	1.0 ef	2.0 ef	4.8 d
T_{12}	2.4 ab	4.7 a	12.2 a
Av.	1.6	2.9	7.3

[a]Average of four replications. Mean separation in a column by DMRT at 5% level.

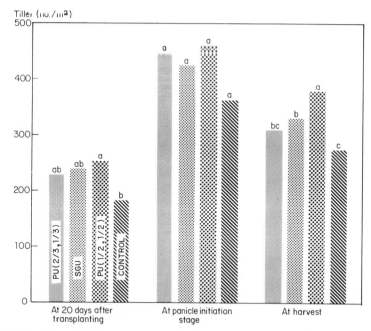

Figure 17.26 Effect of different sources (prilled urea = PU, super granule urea = SGU) and methods of application of urea on tiller number of rice plants measured at different growth stages. Mean separation at each growth stage by DMRT at 5% level.

Table 17.18 Effect of Weeding on Grain Yield and Other Characters of Mungbean: an Illustration of the Flexibility of the Tabular Form in Presenting Multicharacter Data

Character	Weeded	Not Weeded	Difference[a]
Yield, kg/ha	969	516	453**
Plant height, cm[b]	79	87	−8**
Leaf-area index[b]	4.1	3.4	0.7**
Pod length, cm	10.0	9.5	0.5[ns]
Pods, no./plant	13.2	10.5	2.7**
Seeds, no./pod	12.2	11.7	0.5[ns]
100-seed weight, g	5.3	5.3	0.0

[a]** = significant at 1% level, [ns] = not significant.
[b]Measured at 5 weeks after emergence.

621

Table 17.19 Nitrogen, Phosphorus, and Potassium Content of Maize Ear Leaves at Silking, as Influenced by Tillage and Mulch Treatments: an Illustration of the Flexibility of the Tabular Form in Presenting Multicharacter Data[a]

Tillage	Nitrogen Content, %			Phosphorus Content, %			Potassium Content, %		
	Straw Mulch	No Mulch	Difference	Straw Mulch	No Mulch	Difference	Straw Mulch	No Mulch	Difference
No tillage	2.57 a	2.27 a	0.30	0.31 a	0.24 b	0.07**	2.12 a	2.12 ab	0.00
Moldboard plowing	2.53 a	2.16 a	0.37*	0.28 a	0.27 ab	0.01	2.12 a	2.06 b	0.06
Chisel plowing, shallow	2.71 a	2.15 a	0.56**	0.30 a	0.28 ab	0.02	2.10 a	2.27 a	−0.17*
Chisel plowing, deep	2.29 a	2.16 a	0.13	0.28 a	0.30 a	−0.02	2.10 a	2.03 b	0.07
Rototilling	2.47 a	2.17 a	0.30	0.27 a	0.26 ab	0.01	2.13 a	2.15 ab	−0.02
Av.	2.51	2.18		0.29	0.27		2.11	2.13	

[a]Average of four replications. In each column, means followed by a common letter are not significantly different at the 5% level. ** = significant at 1% level, * = significant at 5% level.

tion. Data presentation should, therefore, reflect this important information.

As a general rule, the *time of observations* is considered as an additional factor. Hence, for a single factor experiment, presentation of measurements over time follows that of a two-factor experiment; for a two-factor experiment, presentation follows that of a three-factor experiment; and so on. Thus, the guidelines discussed for factorial experiments in Section 17.2 are directly applicable. In addition, the following guidelines should be considered:

· If the time of observation is quantitative, line graphs such as shown in Figure 17.24 for discrete treatments, and in Figure 17.25 for quantitative treatments, are appropriate. Note that the time of observation is usually placed on the *X* axis, even though the treatments are themselves quantitative, to reflect the emphasis placed on the response curves (or trends over time) and how they are affected by the treatments tested.

· If the time of observation is discrete, a table (Table 17.17) or a bar chart (Figure 17.26) can be used.

17.3.2   Multicharacter Data

In addition to crop yield, which is usually the character of primary interest, other characters such as plant height, crop maturity, pest incidence, soil

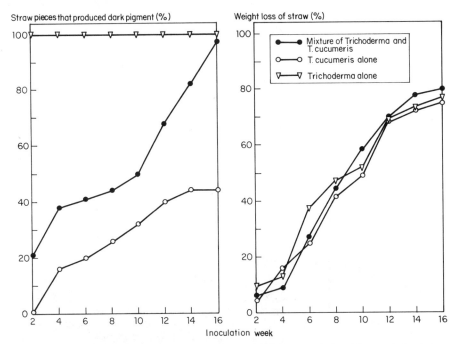

Figure 17.27   Two line graphs showing the effect of *Trichoderma sp.* (isolated from a dryland rice field) on rice straw decomposition and weight loss.

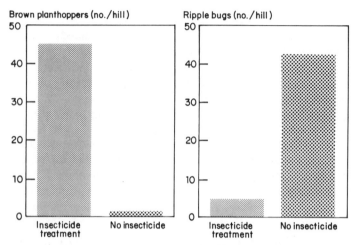

Figure 17.28 Effect of diazinon treatment followed by decamethrin sprays on the populations of brown planthopper and ripple bug, a predator, on a susceptible rice variety.

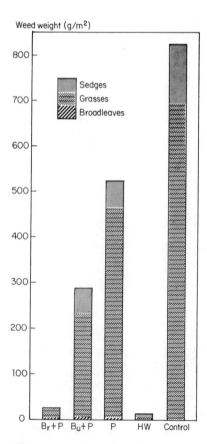

Figure 17.29 Bar chart showing the effect of different weed control methods on weed incidence. (Data are from farm 1 of Table 16.6, Chapter 16).

properties, and weather are usually measured to study their responses to the treatments or to explain their effect on yield. Some examples of multicharacter data in crop research are.

- In a variety trial, such characters as grain yield, plant height, growth duration, pest incidence, and grain quality are measured simultaneously to identify the desirable traits associated with each variety.
- In an insect-control trial, several insect pests such as brown planthopper, whorl maggot, and green leafhopper are measured simultaneously, in as many stages of crop growth as are deemed necessary, to identify the manner with which pest species are affected by the pest control treatment.
- In a nitrogen-response trial, grain yield and its components such as panicle number, number of grains per panicle, and weight per grain may be examined, to quantify the contribution of the yield components, individually or jointly, to the variation in grain yield.

Some rules for presenting mean-comparison results of multicharacter data, either from a single-factor or a factorial experiment, are:

RULE 1.   Because of its flexibility with regards to the number of characters involved, and its ease in handling different units of measurement, the table is

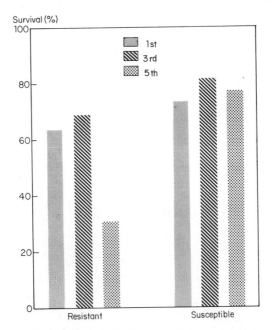

Figure 17.30   Bar chart showing the survival of brown planthopper nymph instars on resistant and susceptible rice varieties.

Table 17.20 Tabular Presentation of Multicharacter Data Using DMRT[a] to Compare Insect-Control Treatments when Some Characters Require Data Transformation

Preplanting Carbofuran Treatment	Plant Height, cm	Caterpillar Damage, %	Deadhearts, %	Green Leafhopper	Insects, no./10 sweeps Brown Planthopper	Cyrtorhinus	White-backed Planthopper	Whorl Maggot Adults
Seedling soak	45 b	13 b	7.5 c	32 c	5 a	28 bc	7 b	87 bc
Seedling soak + capsule 20 DT	53 a	7 a	0.4 a	2 a	3 a	0 a	2 a	34 a
Soil incorporation	54 a	9 a	4.1 b	8 b	5 a	19 b	7 b	60 b
No control	35 c	14 b	4.4 b	81 d	12 b	50 c	25 c	114 c

[a]Average of four replications. Mean separation by DMRT at 5% level.

the most common method for presenting multicharacter data. Once the best tabular form for one character is determined, either from a single-factor experiment (Sections 17.1.1.1 and Section 17.1.1.2) or from a factorial experiment (Section 17.2.1), other characters can be added either along the rows (Table 17.18) or along the columns (Table 17.19).

RULE 2. Graphical presentation of multicharacter data is usually done with separate bar charts or line graphs—one for each character (Figures 17.27 and 17.28). Exceptions to this guideline are:

- Two characters with different units of measurement can be accommodated in one graph (see Figure 17.1).
- Multicharacter data with additive features, such as dry weight for different plant parts or classification of weed by types, can be presented in a bar chart with *subdivided* bars (Figure 17.29).
- Multicharacter data with the same unit of measurements can be placed in the same graph (Figure 17.30).

RULE 3. Whatever form of presentation is used, the same test criterion should be used for mean comparisons of all characters presented. For example, if the number of treatments is five or less but one or more of the characters were analyzed using data transformation (Chapter 7, Section 7.2.2.1), use DMRT for mean comparisons of all characters (see Table 17.20).

RULE 4. Do not present more characters than needed:

- If only yield response to treatments is discussed, present only data on yield. There is no need to present data on yield components or other agronomic characters that are not discussed in the text.
- Avoid presenting data of characters whose treatment effect is not significant.

Appendixes

Appendix A   Table of Random Numbers

14620	95430	12951	81953	17629	83603	09137	26453	02148	30742
09724	85125	48477	42783	70473	52491	66875	93650	91487	37190
56919	17803	95781	85069	61594	85437	92086	53045	31847	36207
97310	78209	51263	52396	82681	82611	70858	78195	47615	23721
07585	28040	26939	64531	70570	98412	74070	83468	18295	32585
25950	85189	69374	37904	06759	70799	59249	63461	75108	45703
82973	16405	81497	20863	94072	83615	09701	47920	46857	31924
60819	27364	59081	72635	49180	72537	46950	81736	53290	81736
59041	38475	03615	84093	49731	62748	39206	47315	84697	30853
74208	69516	79530	47649	53046	95420	41857	69420	79762	01935
39412	03642	87497	29735	14308	46309	28493	75091	82753	15040
48480	50075	11804	24956	72182	59649	16284	83538	53920	47192
95318	28749	49512	35408	21814	07564	70949	50969	15395	26081
72094	16385	90185	72635	86259	38352	94710	36853	94969	38405
63158	49753	84279	56496	30618	23973	25354	25237	48544	20405
19082	73645	09182	73649	56823	95208	49635	01420	46768	45362
15232	84146	87729	65584	83641	19468	34739	57052	43056	29950
94252	77489	62434	20965	20247	03994	25989	19609	74372	74151
72020	18895	84948	53072	74573	19520	92764	85397	52095	18079
48392	06359	47040	05695	79799	05342	54212	21539	48207	95920
37950	77387	35495	48192	84518	30210	23805	27837	24953	42610
09394	59842	39573	51630	78548	06461	06566	21752	78967	45692
34800	28055	91570	99154	39603	76846	77183	50369	16501	68867
36435	75946	85712	06293	85621	97764	53126	37396	57039	06096
28187	31824	52265	80494	66428	15703	05792	53376	54205	91590
13838	79940	97007	67511	87939	68417	21786	09822	67510	23817
72201	08423	41489	15498	94911	79392	65362	19672	93682	84190
63435	45192	62020	47358	32286	41659	31842	47269	70904	62972
59038	96983	49218	57179	08062	25074	06374	96484	59159	23749
62367	45627	58317	76928	50274	28705	45060	50903	66578	41465
71254	81686	85861	63973	96086	89681	50212	92829	27698	62284
07896	62924	35682	42820	43646	37385	37236	16496	51396	77975
71433	54331	58437	03542	76797	50437	13576	72876	02323	95237
54614	19092	83860	11351	32533	56032	42009	49745	14651	80128
30176	71248	37983	06073	89096	43498	95782	70452	90804	12042
79072	87795	23294	61602	62921	38385	69546	47104	72917	66273
75014	96754	67151	82741	24283	64276	78438	70757	40749	85183
37390	75846	74579	94606	54959	35310	31249	15101	95390	73432
24524	32751	28350	43090	79672	94672	07091	42920	46046	38083
26316	20378	16474	62438	42496	35191	49368	30074	93436	29425
61085	96937	02520	86801	30980	58479	34924	25101	87373	61560
45836	41086	41283	97460	51798	29852	47271	42480	94156	49341
92103	19679	16921	65924	12521	31724	60336	01968	15971	07963
10317	82592	65205	12528	24367	15817	12479	52021	02350	76394
39764	21251	41749	43789	70565	35496	87172	76830	41843	83489
83594	95692	52910	23202	93736	10817	53164	10724	27035	67562
08087	01753	01787	51631	74978	79608	01242	07525	72656	80854
57819	39689	32509	87540	38150	47872	14614	18427	06725	69326
96957	81060	28587	60905	67404	80450	21082	16074	61437	24961
48426	43513	82950	79838	45149	07143	73967	23723	06909	75375

630

57856	87037	57196	47916	15960	13036	84639	30186	48347	40780
61684	96598	28043	25325	81767	20792	39823	48749	79489	39329
06847	83825	12858	18689	41319	15959	38030	80057	67617	18501
40810	85323	18076	02821	94728	96808	11072	39823	63756	04478
06461	45073	88350	35246	15851	16129	57460	34512	10243	47635
82197	35028	96295	95795	76553	50223	37215	07692	76527	80764
47430	50260	03643	72259	71294	69176	21753	58341	07468	19219
25043	52002	84476	69512	95036	69095	96340	89713	06381	61522
34718	11667	96345	60791	06387	54221	40422	93251	43456	89176
23965	59598	09746	48646	47409	32406	80874	74010	91548	79394
67207	47166	44917	94177	31846	73872	92835	12596	64807	23978
08261	71627	96865	75380	42735	19446	78478	35681	07769	18230
10289	93145	14456	32978	82587	64377	54270	47869	66444	68728
75622	83203	14951	46603	84176	17564	53965	80771	10453	87972
62557	05584	27879	08081	01467	19691	39814	66538	65243	76009
51695	70743	68481	57937	62634	86727	69563	29308	51729	10453
54839	69596	25201	56536	54517	86909	92927	07827	28271	52075
75284	36241	59749	81958	44318	28067	67638	72196	54648	36886
64082	68375	30361	32627	38970	82481	94725	56930	34939	27641
94649	33784	84691	48334	74667	48289	29629	61248	47276	76162
25261	28316	37178	82874	37083	73818	78758	97096	48508	26484
21967	90859	05692	34023	09397	55027	39897	51482	81867	81783
63749	41490	72232	71710	36489	15291	68579	83195	60186	78142
63487	42869	24783	80895	78641	50359	20497	91381	72319	83280
91729	08960	70364	14262	76861	06406	85253	57490	80497	54272
38532	52316	41320	29806	57594	59360	50929	18752	12856	09587
27650	57930	25216	67180	42352	41671	78178	09058	42479	60463
68318	14891	96592	44278	80631	82547	39787	97394	98513	29634
91423	83067	14837	03817	21850	39732	18603	27174	71319	82016
54574	54648	29265	63051	07586	78418	48489	05425	27931	84965
93987	91493	61816	09628	31397	17607	97095	47154	40798	06217
59854	13847	37190	47369	39657	45179	06178	58918	37965	32031
12636	51498	34352	52548	57125	24634	95394	71846	98148	12839
04856	80651	35242	60595	61636	97294	56276	30294	62698	47548
92417	96727	90734	84549	04236	02520	29057	22102	18358	95938
95723	05695	64543	12870	17646	25542	91526	91395	46359	52952
14398	47916	56272	10835	76054	67823	07381	96863	72547	29368
97643	48258	46058	34375	29890	71563	82459	37210	65765	82546
14020	16902	47286	27208	09898	04837	13967	24974	55274	79587
38715	36409	52324	96537	99811	60503	44262	70562	82081	64785
70051	31424	26201	88098	31019	36195	23032	92648	74724	68292
56602	58040	48323	37857	99639	10700	98176	34642	43428	39068
69874	15653	70998	02969	42103	01069	68736	52765	23824	31235
35242	79841	46481	17365	84609	26357	60470	35212	51863	00401
20364	89248	58280	41596	87712	97928	45494	78356	72100	32949
16572	14877	42927	46635	09564	45334	63012	47305	27136	19428
74256	15507	02159	21981	00649	40382	43087	34506	53229	08383
04653	48391	78424	67282	46854	61980	10745	73924	12717	25524
32077	87214	14924	45190	51808	30474	29771	51573	82713	69487
46545	23074	80308	52685	95334	12428	50970	47019	21993	43350

Appendix B Cumulative Normal Frequency Distribution.
(Area under the standard normal curve from 0 to Z)

Z	0.00	0.01	0.02	0.03	0.04	0.05	0.06	0.07	0.08	0.09
0.0	0.0000	0.0040	0.0080	0.0120	0.0160	0.0199	0.0239	0.0279	0.0319	0.0359
0.1	.0398	.0438	.0478	.0517	.0557	.0596	.0636	.0675	.0714	.0753
0.2	.0793	.0832	.0871	.0910	.0948	.0987	.1026	.1064	.1103	.1141
0.3	.1179	.1217	.1255	.1293	.1331	.1368	.1406	.1443	.1480	.1517
0.4	.1554	.1591	.1628	.1664	.1700	.1736	.1772	.1808	.1844	.1879
0.5	.1915	.1950	.1985	.2019	.2054	.2088	.2123	.2157	.2190	.2224
0.6	.2257	.2291	.2324	.2357	.2389	.2422	.2454	.2486	.2517	.2549
0.7	.2580	.2611	.2642	.2673	.2704	.2734	.2764	.2794	.2823	.2852
0.8	.2881	.2910	.2939	.2967	.2995	.3023	.3051	.3087	.3106	.3133
0.9	.3159	.3186	.3212	.3238	.3264	.3289	.3315	.3340	.3365	.3389
1.0	.3413	.3438	.3461	.3485	.3508	.3531	.3554	.3577	.3599	.3621
1.1	.3643	.3665	.3686	.3708	.3729	.3749	.3770	.3790	.3810	.3830
1.2	.3849	.3869	.3888	.3907	.3925	.3944	.3962	.3980	.3997	.4015
1.3	.4032	.4049	.4066	.4082	.4099	.4115	.4131	.4147	.4162	.4177
1.4	.4192	.4207	.4222	.4236	.4251	.4265	.4279	.4292	.4306	.4319
1.5	.4332	.4345	.4357	.4370	.4382	.4394	.4406	.4418	.4429	.4441
1.6	.4452	.4463	.4474	.4484	.4495	.4505	.4515	.4525	.4535	.4545
1.7	.4554	.4564	.4573	.4582	.4591	.4599	.4608	.4616	.4625	.4633
1.8	.4641	.4649	.4656	.4664	.4671	.4678	.4686	.4693	.4699	.4706
1.9	.4713	.4719	.4726	.4732	.4738	.4744	.4750	.4756	.4761	.4767
2.0	.4772	.4778	.4783	.4788	.4793	.4798	.4803	.4808	.4812	.4817
2.1	.4821	.4826	.4830	.4834	.4838	.4842	.4846	.4850	.4854	.4857
2.2	.4861	.4864	.4868	.4871	.4875	.4878	.4881	.4884	.4887	.4890
2.3	.4893	.4896	.4898	.4901	.4904	.4906	.4909	.4911	.4913	.4916
2.4	.4918	.4920	.4922	.4925	.4927	.4929	.4931	.4932	.4934	.4936
2.5	.4938	.4940	.4941	.4943	.4945	.4946	.4948	.4949	.4951	.4952
2.6	.4953	.4955	.4956	.4957	.4959	.4960	.4961	.4962	.4963	.4964
2.7	.4965	.4966	.4967	.4968	.4969	.4970	.4971	.4972	.4973	.4974
2.8	.4974	.4975	.4976	.4977	.4977	.4978	.4979	.4979	.4980	.4981
2.9	.4981	.4982	.4982	.4983	.4984	.4984	.4985	.4985	.4986	.4986
3.0	.4987	.4987	.4987	.4988	.4988	.4989	.4989	.4989	.4990	.4990
3.1	.4990	.4991	.4991	.4991	.4992	.4992	.4992	.4992	.4993	.4993
3.2	.4993	.4993	.4994	.4994	.4994	.4994	.4994	.4995	.4995	.4995
3.3	.4995	.4995	.4995	.4996	.4996	.4996	.4996	.4996	.4996	.4997
3.4	.4997	.4997	.4997	.4997	.4997	.4997	.4997	.4997	.4997	.4998
3.6	.4998	.4998	.4999	.4999	.4999	.4999	.4999	.4999	.4999	.4999
3.9	.5000									

Reprinted by permission from STATISTICAL METHODS by George W. Snedecor and William G. Cochran, sixth edition © 1967 by Iowa State University Press, Ames, Iowa.

Appendix C Distribution of *t* Probability

n	.9	.8	.7	.6	.5	.4	.3	.2	.1	.05	.02	.01	.001
1	.158	.325	.510	.727	1.000	1.376	1.963	3.078	6.314	12.706	31.821	63.657	636.619
2	.142	.289	.445	.617	.816	1.061	1.386	1.886	2.920	4.303	6.965	9.925	31.598
3	.137	.277	.424	.584	.765	.978	1.250	1.638	2.353	3.182	4.541	5.841	12.924
4	.134	.271	.414	.569	.741	.941	1.190	1.533	2.132	2.776	3.747	4.604	8.610
5	.132	.267	.408	.559	.727	.920	1.156	1.476	2.015	2.571	3.365	4.032	6.869
6	.131	.265	.404	.553	.718	.906	1.134	1.440	1.943	2.447	3.143	3.707	5.959
7	.130	.263	.402	.549	.711	.896	1.119	1.415	1.895	2.365	2.998	3.499	5.408
8	.130	.262	.399	.546	.706	.889	1.108	1.397	1.860	2.306	2.896	3.355	5.041
9	.129	.261	.398	.543	.703	.883	1.100	1.383	1.833	2.262	2.821	3.250	4.781
10	.129	.260	.397	.542	.700	.879	1.093	1.372	1.812	2.228	2.764	3.169	4.587
11	.129	.260	.396	.540	.697	.876	1.088	1.363	1.796	2.201	2.718	3.106	4.437
12	.128	.259	.395	.539	.695	.873	1.083	1.356	1.782	2.179	2.681	3.055	4.318
13	.128	.259	.394	.538	.694	.870	1.079	1.350	1.771	2.160	2.650	3.012	4.221
14	.128	.258	.393	.537	.692	.868	1.076	1.345	1.761	2.145	2.624	2.977	4.140
15	.128	.258	.393	.536	.691	.866	1.074	1.341	1.753	2.131	2.602	2.947	4.073
16	.128	.258	.392	.535	.690	.865	1.071	1.337	1.746	2.120	2.583	2.921	4.015
17	.128	.257	.392	.534	.689	.863	1.069	1.333	1.740	2.110	2.567	2.898	3.965
18	.127	.257	.392	.534	.688	.862	1.067	1.330	1.734	2.101	2.552	2.878	3.922
19	.127	.257	.391	.533	.688	.861	1.066	1.328	1.729	2.093	2.539	2.861	3.883
20	.127	.257	.391	.533	.687	.860	1.064	1.325	1.725	2.086	2.528	2.845	3.850
21	.127	.257	.391	.532	.686	.859	1.063	1.323	1.721	2.080	2.518	2.831	3.819
22	.127	.256	.390	.532	.686	.858	1.061	1.321	1.717	2.074	2.508	2.819	3.792
23	.127	.256	.390	.532	.685	.858	1.060	1.319	1.714	2.069	2.500	2.807	3.767
24	.127	.256	.390	.531	.685	.857	1.059	1.318	1.711	2.064	2.492	2.797	3.745
25	.127	.256	.390	.531	.684	.856	1.058	1.316	1.708	2.060	2.485	2.787	3.725
26	.127	.256	.390	.531	.684	.856	1.058	1.315	1.706	2.056	2.479	2.779	3.707
27	.127	.256	.389	.531	.684	.855	1.057	1.314	1.703	2.052	2.473	2.771	3.690
28	.127	.256	.389	.530	.683	.855	1.056	1.313	1.701	2.048	2.467	2.763	3.674
29	.127	.256	.389	.530	.683	.854	1.055	1.311	1.699	2.045	2.462	2.756	3.659
30	.127	.256	.389	.530	.683	.854	1.055	1.310	1.697	2.042	2.457	2.750	3.646
40	.126	.255	.388	.529	.681	.851	1.050	1.303	1.684	2.021	2.423	2.704	3.551
60	.126	.254	.387	.527	.679	.848	1.046	1.296	1.671	2.000	2.390	2.660	3.460
120	.126	.254	.386	.526	.677	.845	1.041	1.289	1.658	1.980	2.358	2.617	3.373
∞	.126	.253	.385	.524	.674	.842	1.036	1.282	1.645	1.960	2.326	2.576	3.291

Reproduced from STATISTICAL TABLES FOR BIOLOGICAL, AGRICULTURAL, AND MEDICAL RESEARCH by R.A. Fisher and F. Yates, sixth edition 1963. Reprinted by permission of Longman Group Limited, Essex, England.

Appendix D Percentage Points of the Chi-Square Distribution

Degrees of freedom	Probability of a larger value of χ^2								
	0.99	0.95	0.90	0.75	0.50	0.25	0.10	0.05	0.01
1	0.000	0.000	0.016	0.102	0.455	1.32	2.71	3.84	6.63
2	0.020	0.103	0.211	0.575	1.386	2.77	4.60	5.99	9.21
3	0.115	0.352	0.584	1.213	2.366	4.11	6.25	7.81	11.34
4	0.297	0.711	1.064	1.923	3.357	5.38	7.78	9.49	13.28
5	0.554	1.145	1.610	2.675	4.351	6.63	9.24	11.07	15.09
6	0.872	1.635	2.204	3.455	5.348	7.84	10.64	12.59	16.81
7	1.239	2.167	2.833	4.255	6.346	9.04	12.02	14.07	18.47
8	1.646	2.733	3.490	5.017	7.344	10.22	13.36	15.51	20.09
9	2.088	3.325	4.168	5.899	8.343	11.39	14.68	16.92	21.67
10	2.568	3.940	4.865	6.737	9.342	12.55	15.99	18.31	23.21
11	3.053	4.575	5.578	7.584	10.341	13.70	17.27	19.67	24.72
12	3.571	5.226	6.304	8.438	11.340	14.84	18.55	21.03	26.22
13	4.107	5.892	7.042	9.299	12.340	15.98	19.81	22.36	27.69
14	4.660	6.571	7.790	10.165	13.339	17.12	21.06	23.68	29.14
15	5.229	7.261	8.547	11.036	14.339	18.25	22.31	25.00	30.58
16	5.812	7.962	9.312	11.912	15.338	19.37	23.54	26.30	32.00
17	6.408	8.672	10.085	12.792	16.338	20.49	24.77	27.59	33.41
18	7.015	9.390	10.865	13.675	17.338	21.60	25.99	28.87	34.80
19	7.633	10.117	11.651	14.562	18.338	22.72	27.20	30.14	36.19
20	8.260	10.851	12.443	15.452	19.337	23.83	28.41	31.41	37.57
22	9.542	12.338	14.041	17.240	21.337	26.04	30.81	33.92	40.29
24	10.856	13.848	15.659	19.037	23.337	28.24	33.20	36.41	42.98
26	12.198	15.379	17.292	20.843	25.336	30.43	35.56	38.88	45.64
28	13.565	16.928	18.939	22.657	27.336	32.62	37.92	41.34	48.28
30	14.953	18.493	20.599	24.478	29.336	34.80	40.26	43.77	50.89
40	22.164	26.509	29.051	33.660	39.335	45.62	51.80	55.76	63.69
50	27.707	34.764	37.689	42.942	49.335	56.33	63.17	67.50	76.15
60	37.485	43.188	46.459	52.294	59.335	66.98	74.40	79.08	88.38

f_1, Degrees of freedom (for greater mean square)

f_2	1	2	3	4	5	6	7	8	9	10	11	12	14	16	20	24	30	40	50	75	100	200	500	∞
1	161	200	216	225	230	234	237	239	241	242	243	244	245	246	248	249	250	251	252	253	253	254	254	254
	4,052	**4,999**	**5,403**	**5,625**	**5,764**	**5,859**	**5,928**	**5,981**	**6,022**	**6,056**	**6,082**	**6,106**	**6,142**	**6,169**	**6,208**	**6,234**	**6,261**	**6,286**	**6,302**	**6,323**	**6,334**	**6,352**	**6,361**	**6,366**
2	18.51	19.00	19.16	19.25	19.30	19.33	19.36	19.37	19.38	19.39	19.40	19.41	19.42	19.43	19.44	19.45	19.46	19.47	19.47	19.48	19.49	19.49	19.50	19.50
	98.49	**99.00**	**99.17**	**99.25**	**99.30**	**99.33**	**99.36**	**99.37**	**99.39**	**99.40**	**99.41**	**99.42**	**99.43**	**99.44**	**99.45**	**99.46**	**99.47**	**99.48**	**99.48**	**99.49**	**99.49**	**99.49**	**99.50**	**99.50**
3	10.13	9.55	9.28	9.12	9.01	8.94	8.88	8.84	8.81	8.78	8.76	8.74	8.71	8.69	8.66	8.64	8.62	8.60	8.58	8.57	8.56	8.54	8.54	8.53
	34.12	**30.82**	**29.46**	**28.71**	**28.24**	**27.91**	**27.67**	**27.49**	**27.34**	**27.23**	**27.13**	**27.05**	**26.92**	**26.83**	**26.69**	**26.60**	**26.50**	**26.41**	**26.35**	**26.27**	**26.23**	**26.18**	**26.14**	**26.12**
4	7.71	6.94	6.59	6.39	6.26	6.16	6.09	6.04	6.00	5.96	5.93	5.91	5.87	5.84	5.80	5.77	5.74	5.71	5.70	5.68	5.66	5.65	5.64	5.63
	21.20	**18.00**	**16.69**	**15.98**	**15.52**	**15.21**	**14.98**	**14.80**	**14.66**	**14.54**	**14.45**	**14.37**	**14.24**	**14.15**	**14.02**	**13.93**	**13.83**	**13.74**	**13.69**	**13.61**	**13.57**	**13.52**	**13.48**	**13.46**
5	6.61	5.79	5.41	5.19	5.05	4.95	4.88	4.82	4.78	4.74	4.70	4.68	4.64	4.60	4.56	4.53	4.50	4.46	4.44	4.42	4.40	4.38	4.37	4.36
	16.26	**13.27**	**12.06**	**11.39**	**10.97**	**10.67**	**10.45**	**10.29**	**10.15**	**10.05**	**9.96**	**9.89**	**9.77**	**9.68**	**9.55**	**9.47**	**9.38**	**9.29**	**9.24**	**9.17**	**9.13**	**9.07**	**9.04**	**9.02**
6	5.99	5.14	4.76	4.53	4.39	4.28	4.21	4.15	4.10	4.06	4.03	4.00	3.96	3.92	3.87	3.84	3.81	3.77	3.75	3.72	3.71	3.69	3.68	3.67
	13.74	**10.92**	**9.78**	**9.15**	**8.75**	**8.47**	**8.26**	**8.10**	**7.98**	**7.87**	**7.79**	**7.72**	**7.60**	**7.52**	**7.39**	**7.31**	**7.23**	**7.14**	**7.09**	**7.02**	**6.99**	**6.94**	**6.90**	**6.88**
7	5.59	4.74	4.35	4.12	3.97	3.87	3.79	3.73	3.68	3.63	3.60	3.57	3.52	3.49	3.44	3.41	3.38	3.34	3.32	3.29	3.28	3.25	3.24	3.23
	12.25	**9.55**	**8.45**	**7.85**	**7.46**	**7.19**	**7.00**	**6.84**	**6.71**	**6.62**	**6.54**	**6.47**	**6.35**	**6.27**	**6.15**	**6.07**	**5.98**	**5.90**	**5.85**	**5.78**	**5.75**	**5.70**	**5.67**	**5.65**
8	5.32	4.46	4.07	3.84	3.69	3.58	3.50	3.44	3.39	3.34	3.31	3.28	3.23	3.20	3.15	3.12	3.08	3.05	3.03	3.00	2.98	2.96	2.94	2.93
	11.26	**8.65**	**7.59**	**7.01**	**6.63**	**6.37**	**6.19**	**6.03**	**5.91**	**5.82**	**5.74**	**5.67**	**5.56**	**5.48**	**5.36**	**5.28**	**5.20**	**5.11**	**5.06**	**5.00**	**4.96**	**4.91**	**4.88**	**4.86**
9	5.12	4.26	3.86	3.63	3.48	3.37	3.29	3.23	3.18	3.13	3.10	3.07	3.02	2.98	2.93	2.90	2.86	2.82	2.80	2.77	2.76	2.73	2.72	2.71
	10.56	**8.02**	**6.99**	**6.42**	**6.06**	**5.80**	**5.62**	**5.47**	**5.35**	**5.26**	**5.18**	**5.11**	**5.00**	**4.92**	**4.80**	**4.73**	**4.64**	**4.56**	**4.51**	**4.45**	**4.41**	**4.36**	**4.33**	**4.31**
10	4.96	4.10	3.71	3.48	3.33	3.22	3.14	3.07	3.02	2.97	2.94	2.91	2.86	2.82	2.77	2.74	2.70	2.67	2.64	2.61	2.59	2.56	2.55	2.54
	10.04	**7.56**	**6.55**	**5.99**	**5.64**	**5.39**	**5.21**	**5.06**	**4.95**	**4.85**	**4.78**	**4.71**	**4.60**	**4.52**	**4.41**	**4.33**	**4.25**	**4.17**	**4.12**	**4.05**	**4.01**	**3.96**	**3.93**	**3.91**
11	4.84	3.98	3.59	3.36	3.20	3.09	3.01	2.95	2.90	2.86	2.82	2.79	2.74	2.70	2.65	2.61	2.57	2.53	2.50	2.47	2.45	2.42	2.41	2.40
	9.65	**7.20**	**6.22**	**5.67**	**5.32**	**5.07**	**4.88**	**4.74**	**4.63**	**4.54**	**4.46**	**4.40**	**4.29**	**4.21**	**4.10**	**4.02**	**3.94**	**3.86**	**3.80**	**3.74**	**3.70**	**3.66**	**3.62**	**3.60**
12	4.75	3.88	3.49	3.26	3.11	3.00	2.92	2.85	2.80	2.76	2.72	2.69	2.64	2.60	2.54	2.50	2.46	2.42	2.40	2.36	2.35	2.32	2.31	2.30
	9.33	**6.93**	**5.95**	**5.41**	**5.06**	**4.82**	**4.65**	**4.50**	**4.39**	**4.30**	**4.22**	**4.16**	**4.05**	**3.98**	**3.86**	**3.78**	**3.70**	**3.61**	**3.56**	**3.49**	**3.46**	**3.41**	**3.38**	**3.36**
13	4.67	3.80	3.41	3.18	3.02	2.92	2.84	2.77	2.72	2.67	2.63	2.60	2.55	2.51	2.46	2.42	2.38	2.34	2.32	2.28	2.26	2.24	2.22	2.21
	9.07	**6.70**	**5.74**	**5.20**	**4.86**	**4.62**	**4.44**	**4.30**	**4.19**	**4.10**	**4.02**	**3.96**	**3.85**	**3.78**	**3.67**	**3.59**	**3.51**	**3.42**	**3.37**	**3.30**	**3.27**	**3.21**	**3.18**	**3.16**

continued next page

Appendix E (Continued)

f_2	1	2	3	4	5	6	7	8	9	10	11	12	14	16	20	24	30	40	50	75	100	200	500	∞	f_2
14	4.60 **8.86**	3.74 **6.51**	3.34 **5.56**	3.11 **5.03**	2.96 **4.69**	2.85 **4.46**	2.77 **4.28**	2.70 **4.14**	2.65 **4.03**	2.60 **3.94**	2.56 **3.86**	2.53 **3.80**	2.48 **3.70**	2.44 **3.62**	2.39 **3.51**	2.35 **3.43**	2.31 **3.34**	2.27 **3.26**	2.24 **3.21**	2.21 **3.14**	2.19 **3.11**	2.16 **3.06**	2.14 **3.02**	2.13 **3.00**	14
15	4.54 **8.68**	3.68 **6.36**	3.29 **5.42**	3.06 **4.89**	2.90 **4.56**	2.79 **4.32**	2.70 **4.14**	2.64 **4.00**	2.59 **3.89**	2.55 **3.80**	2.51 **3.73**	2.48 **3.67**	2.43 **3.56**	2.39 **3.48**	2.33 **3.36**	2.29 **3.29**	2.25 **3.20**	2.21 **3.12**	2.18 **3.07**	2.15 **3.00**	2.12 **2.97**	2.10 **2.92**	2.08 **2.89**	2.07 **2.87**	15
16	4.49 **8.53**	3.63 **6.23**	3.24 **5.29**	3.01 **4.77**	2.85 **4.44**	2.74 **4.20**	2.66 **4.03**	2.59 **3.89**	2.54 **3.78**	2.49 **3.69**	2.45 **3.61**	2.42 **3.55**	2.37 **3.45**	2.33 **3.37**	2.28 **3.25**	2.24 **3.18**	2.20 **3.10**	2.16 **3.01**	2.13 **2.96**	2.09 **2.98**	2.07 **2.86**	2.04 **2.80**	2.02 **2.77**	2.01 **2.75**	16
17	4.45 **8.40**	3.59 **6.11**	3.20 **5.18**	2.96 **4.67**	2.81 **4.34**	2.70 **4.10**	2.62 **3.93**	2.55 **3.79**	2.50 **3.68**	2.45 **3.59**	2.41 **3.52**	2.38 **3.45**	2.33 **3.35**	2.29 **3.27**	2.23 **3.16**	2.19 **3.08**	2.15 **3.00**	2.11 **2.92**	2.08 **2.86**	2.04 **2.79**	2.02 **2.76**	1.99 **2.70**	1.97 **2.67**	1.96 **2.65**	17
18	4.41 **8.28**	3.55 **6.01**	3.16 **5.09**	2.93 **4.58**	2.77 **4.25**	2.66 **4.01**	2.58 **3.85**	2.51 **3.71**	2.46 **3.60**	2.41 **3.51**	2.37 **3.44**	2.34 **3.37**	2.29 **3.27**	2.25 **3.19**	2.19 **3.07**	2.15 **3.00**	2.11 **2.91**	2.07 **2.83**	2.04 **2.78**	2.00 **2.71**	1.98 **2.68**	1.95 **2.62**	1.93 **2.59**	1.92 **2.57**	18
19	4.38 **8.18**	3.52 **5.93**	3.13 **5.01**	2.90 **4.50**	2.74 **4.17**	2.63 **3.94**	2.55 **3.77**	2.48 **3.63**	2.43 **3.52**	2.38 **3.43**	2.34 **3.36**	2.31 **3.30**	2.26 **3.19**	2.21 **3.12**	2.15 **3.00**	2.11 **2.92**	2.07 **2.84**	2.02 **2.76**	2.00 **2.70**	1.96 **2.63**	1.94 **2.60**	1.91 **2.54**	1.90 **2.51**	1.88 **2.49**	19
20	4.35 **8.10**	3.49 **5.85**	3.10 **4.94**	2.87 **4.43**	2.71 **4.10**	2.60 **3.87**	2.52 **3.71**	2.45 **3.56**	2.40 **3.45**	2.35 **3.37**	2.31 **3.30**	2.28 **3.23**	2.23 **3.13**	2.18 **3.05**	2.12 **2.94**	2.08 **2.86**	2.04 **2.77**	1.99 **2.69**	1.96 **2.63**	1.92 **2.56**	1.90 **2.53**	1.87 **2.47**	1.85 **2.44**	1.84 **2.42**	20
21	4.32 **8.02**	3.47 **5.78**	3.07 **4.87**	2.84 **4.37**	2.68 **4.04**	2.57 **3.81**	2.49 **3.65**	2.42 **3.51**	2.37 **3.40**	2.32 **3.31**	2.28 **3.24**	2.25 **3.17**	2.20 **3.07**	2.15 **2.99**	2.09 **2.88**	2.05 **2.80**	2.00 **2.72**	1.96 **2.63**	1.93 **2.58**	1.89 **2.51**	1.87 **2.47**	1.84 **2.42**	1.82 **2.38**	1.81 **2.36**	21
22	4.30 **7.94**	3.44 **5.72**	3.05 **4.82**	2.82 **4.31**	2.66 **3.99**	2.55 **3.76**	2.47 **3.59**	2.40 **3.45**	2.35 **3.35**	2.30 **3.26**	2.26 **3.18**	2.23 **3.12**	2.18 **3.02**	2.13 **2.94**	2.07 **2.83**	2.03 **2.75**	1.98 **2.67**	1.93 **2.58**	1.91 **2.53**	1.87 **2.46**	1.84 **2.42**	1.81 **2.37**	1.80 **2.33**	1.78 **2.31**	22
23	4.28 **7.88**	3.42 **5.66**	3.03 **4.76**	2.80 **4.26**	2.64 **3.94**	2.53 **3.71**	2.45 **3.54**	2.38 **3.41**	2.32 **3.30**	2.28 **3.21**	2.24 **3.14**	2.20 **3.07**	2.14 **2.97**	2.10 **2.89**	2.04 **2.78**	2.00 **2.70**	1.96 **2.62**	1.91 **2.53**	1.88 **2.48**	1.84 **2.41**	1.82 **2.37**	1.79 **2.32**	1.77 **2.28**	1.76 **2.26**	23
24	4.26 **7.82**	3.40 **5.61**	3.01 **4.72**	2.78 **4.22**	2.62 **3.90**	2.51 **3.67**	2.43 **3.50**	2.36 **3.36**	2.30 **3.25**	2.26 **3.17**	2.22 **3.09**	2.18 **3.03**	2.13 **2.93**	2.09 **2.85**	2.02 **2.74**	1.98 **2.66**	1.94 **2.58**	1.89 **2.49**	1.86 **2.44**	1.82 **2.36**	1.80 **2.33**	1.76 **2.27**	1.74 **2.23**	1.73 **2.21**	24
25	4.24 **7.77**	3.38 **5.57**	2.99 **4.68**	2.76 **4.18**	2.60 **3.86**	2.49 **3.63**	2.41 **3.46**	2.34 **3.32**	2.28 **3.21**	2.24 **3.13**	2.20 **3.05**	2.16 **2.99**	2.11 **2.89**	2.06 **2.81**	2.00 **2.70**	1.96 **2.62**	1.92 **2.54**	1.87 **2.45**	1.84 **2.40**	1.80 **2.32**	1.77 **2.29**	1.74 **2.23**	1.72 **2.19**	1.71 **2.17**	25
26	4.22 **7.72**	3.37 **5.53**	2.98 **4.64**	2.74 **4.14**	2.59 **3.82**	2.47 **3.59**	2.39 **3.42**	2.32 **3.29**	2.27 **3.17**	2.22 **3.09**	2.18 **3.02**	2.15 **2.96**	2.10 **2.86**	2.05 **2.77**	1.99 **2.66**	1.95 **2.58**	1.90 **2.50**	1.85 **2.41**	1.82 **2.36**	1.78 **2.28**	1.76 **2.25**	1.72 **2.19**	1.70 **2.15**	1.69 **2.13**	26

f_1, Degrees of freedom (for greater mean square)

Appendix E (Continued)

f_1, Degrees of freedom (for greater mean square)

f_2	1	2	3	4	5	6	7	8	9	10	11	12	14	16	20	24	30	40	50	75	100	200	500	∞	f_2
27	4.21 **7.68**	3.35 **5.49**	2.96 **4.60**	2.73 **4.11**	2.57 **3.79**	2.46 **3.56**	2.37 **3.39**	2.30 **3.26**	2.25 **3.14**	2.20 **3.06**	2.16 **2.98**	2.13 **2.93**	2.08 **2.83**	2.03 **2.74**	1.97 **2.63**	1.93 **2.55**	1.88 **2.47**	1.84 **2.38**	1.80 **2.33**	1.76 **2.25**	1.74 **2.21**	1.71 **2.16**	1.68 **2.12**	1.67 **2.10**	27
28	4.20 **7.64**	3.34 **5.45**	2.95 **4.57**	2.71 **4.07**	2.56 **3.76**	2.44 **3.53**	2.36 **3.36**	2.29 **3.23**	2.24 **3.11**	2.19 **3.03**	2.15 **2.95**	2.12 **2.90**	2.06 **2.80**	2.02 **2.71**	1.96 **2.60**	1.91 **2.52**	1.87 **2.44**	1.81 **2.35**	1.78 **2.30**	1.75 **2.22**	1.72 **2.18**	1.69 **2.13**	1.67 **2.09**	1.65 **2.06**	28
29	4.18 **7.60**	3.33 **5.42**	2.93 **4.54**	2.70 **4.04**	2.54 **3.73**	2.43 **3.50**	2.35 **3.33**	2.28 **3.20**	2.22 **3.08**	2.18 **3.00**	2.14 **2.92**	2.10 **2.87**	2.05 **2.77**	2.00 **2.68**	1.94 **2.57**	1.90 **2.49**	1.85 **2.41**	1.80 **2.32**	1.77 **2.27**	1.73 **2.19**	1.71 **2.15**	1.68 **2.10**	1.65 **2.06**	1.64 **2.03**	29
30	4.17 **7.56**	3.32 **5.39**	2.92 **4.51**	2.69 **4.02**	2.53 **3.70**	2.42 **3.47**	2.34 **3.30**	2.27 **3.17**	2.21 **3.06**	2.16 **2.98**	2.12 **2.90**	2.09 **2.84**	2.04 **2.74**	1.99 **2.66**	1.93 **2.55**	1.89 **2.47**	1.84 **2.38**	1.79 **2.29**	1.76 **2.24**	1.72 **2.16**	1.69 **2.13**	1.66 **2.07**	1.64 **2.03**	1.62 **2.01**	30
32	4.15 **7.50**	3.30 **5.34**	2.90 **4.46**	2.67 **3.97**	2.51 **3.66**	2.40 **3.42**	2.32 **3.25**	2.25 **3.12**	2.19 **3.01**	2.14 **2.94**	2.10 **2.86**	2.07 **2.80**	2.02 **2.70**	1.97 **2.62**	1.91 **2.51**	1.86 **2.42**	1.82 **2.34**	1.76 **2.25**	1.74 **2.20**	1.69 **2.12**	1.67 **2.08**	1.64 **2.02**	1.61 **1.98**	1.59 **1.96**	32
34	4.13 **7.44**	3.28 **5.29**	2.88 **4.42**	2.65 **3.93**	2.49 **3.61**	2.38 **3.38**	2.30 **3.21**	2.23 **3.08**	2.17 **2.97**	2.12 **2.89**	2.08 **2.82**	2.05 **2.76**	2.00 **2.66**	1.95 **2.58**	1.89 **2.47**	1.84 **2.38**	1.80 **2.30**	1.74 **2.21**	1.71 **2.15**	1.67 **2.08**	1.64 **2.04**	1.61 **1.98**	1.59 **1.94**	1.57 **1.91**	34
36	4.11 **7.39**	3.26 **5.25**	2.86 **4.38**	2.63 **3.89**	2.48 **3.58**	2.36 **3.35**	2.28 **3.18**	2.21 **3.04**	2.15 **2.94**	2.10 **2.86**	2.06 **2.78**	2.03 **2.72**	1.98 **2.62**	1.93 **2.54**	1.87 **2.43**	1.82 **2.35**	1.78 **2.26**	1.72 **2.17**	1.69 **2.12**	1.65 **2.04**	1.62 **2.00**	1.59 **1.94**	1.56 **1.90**	1.55 **1.87**	36
38	4.10 **7.35**	3.25 **5.21**	2.85 **4.34**	2.62 **3.86**	2.46 **3.54**	2.35 **3.32**	2.26 **3.15**	2.19 **3.02**	2.14 **2.91**	2.09 **2.82**	2.05 **2.75**	2.02 **2.69**	1.96 **2.59**	1.92 **2.51**	1.85 **2.40**	1.80 **2.32**	1.76 **2.22**	1.71 **2.14**	1.67 **2.08**	1.63 **2.00**	1.60 **1.97**	1.57 **1.90**	1.54 **1.86**	1.53 **1.84**	38
40	4.08 **7.31**	3.23 **5.18**	2.84 **4.31**	2.61 **3.83**	2.45 **3.51**	2.34 **3.29**	2.25 **3.12**	2.18 **2.99**	2.12 **2.88**	2.07 **2.80**	2.04 **2.73**	2.00 **2.66**	1.95 **2.56**	1.90 **2.49**	1.84 **2.37**	1.79 **2.29**	1.74 **2.20**	1.69 **2.11**	1.66 **2.05**	1.61 **1.97**	1.59 **1.94**	1.55 **1.88**	1.53 **1.84**	1.51 **1.81**	40
42	4.07 **7.27**	3.22 **5.15**	2.83 **4.29**	2.59 **3.80**	2.44 **3.49**	2.32 **3.26**	2.24 **3.10**	2.17 **2.96**	2.11 **2.86**	2.06 **2.77**	2.02 **2.70**	1.99 **2.64**	1.94 **2.54**	1.89 **2.46**	1.82 **2.35**	1.78 **2.26**	1.73 **2.17**	1.68 **2.08**	1.64 **2.02**	1.60 **1.94**	1.57 **1.91**	1.54 **1.85**	1.51 **1.80**	1.49 **1.78**	42
44	4.06 **7.24**	3.21 **5.12**	2.82 **4.26**	2.58 **3.78**	2.43 **3.46**	2.31 **3.24**	2.23 **3.07**	2.16 **2.94**	2.10 **2.84**	2.05 **2.75**	2.01 **2.68**	1.98 **2.62**	1.92 **2.52**	1.88 **2.44**	1.81 **2.32**	1.76 **2.24**	1.72 **2.15**	1.66 **2.06**	1.63 **2.00**	1.58 **1.92**	1.56 **1.88**	1.52 **1.82**	1.50 **1.78**	1.48 **1.75**	44
46	4.05 **7.21**	3.20 **5.10**	2.81 **4.24**	2.57 **3.76**	2.42 **3.44**	2.30 **3.22**	2.22 **3.05**	2.14 **2.92**	2.09 **2.82**	2.04 **2.73**	2.00 **2.66**	1.97 **2.60**	1.91 **2.50**	1.87 **2.42**	1.80 **2.30**	1.75 **2.22**	1.71 **2.13**	1.65 **2.04**	1.62 **1.98**	1.57 **1.90**	1.54 **1.86**	1.51 **1.80**	1.48 **1.76**	1.46 **1.72**	46
48	4.04 **7.19**	3.19 **5.08**	2.80 **4.22**	2.56 **3.74**	2.41 **3.42**	2.30 **3.20**	2.21 **3.04**	2.14 **2.90**	2.08 **2.80**	2.03 **2.71**	1.99 **2.64**	1.96 **2.58**	1.90 **2.48**	1.86 **2.40**	1.79 **2.28**	1.74 **2.20**	1.70 **2.11**	1.64 **2.02**	1.61 **1.96**	1.56 **1.88**	1.53 **1.84**	1.50 **1.78**	1.47 **1.73**	1.45 **1.70**	48

continued next page

637

Appendix E (Continued)

f_1, Degrees of freedom (for greater mean square)

f_2	1	2	3	4	5	6	7	8	9	10	11	12	14	16	20	24	30	40	50	75	100	200	500	∞	f_2
50	4.03	3.18	2.79	2.56	2.40	2.29	2.20	2.13	2.07	2.02	1.98	1.95	1.90	1.85	1.78	1.74	1.69	1.63	1.60	1.55	1.52	1.48	1.46	1.44	50
	7.17	**5.06**	**4.20**	**3.72**	**3.41**	**3.18**	**3.02**	**2.88**	**2.78**	**2.70**	**2.62**	**2.56**	**2.46**	**2.39**	**2.26**	**2.18**	**2.10**	**2.00**	**1.94**	**1.86**	**1.82**	**1.76**	**1.71**	**1.68**	
55	4.02	3.17	2.78	2.54	2.38	2.27	2.18	2.11	2.05	2.00	1.97	1.93	1.88	1.83	1.76	1.72	1.67	1.61	1.58	1.52	1.50	1.46	1.43	1.41	55
	7.12	**5.01**	**4.16**	**3.68**	**3.37**	**3.15**	**2.98**	**2.85**	**2.75**	**2.66**	**2.59**	**2.53**	**2.43**	**2.35**	**2.23**	**2.15**	**2.06**	**1.96**	**1.90**	**1.82**	**1.78**	**1.71**	**1.66**	**1.64**	
60	4.00	3.15	2.76	2.52	2.37	2.25	2.17	2.10	2.04	1.99	1.95	1.92	1.86	1.81	1.75	1.70	1.65	1.59	1.56	1.50	1.48	1.44	1.41	1.39	60
	7.08	**4.98**	**4.13**	**3.65**	**3.34**	**3.12**	**2.95**	**2.82**	**2.72**	**2.63**	**2.56**	**2.50**	**2.40**	**2.32**	**2.20**	**2.12**	**2.03**	**1.93**	**1.87**	**1.79**	**1.74**	**1.68**	**1.63**	**1.60**	
65	3.99	3.14	2.75	2.51	2.36	2.24	2.15	2.08	2.02	1.98	1.94	1.90	1.85	1.80	1.73	1.68	1.63	1.57	1.54	1.49	1.46	1.42	1.39	1.37	65
	7.04	**4.95**	**4.10**	**3.62**	**3.31**	**3.09**	**2.93**	**2.79**	**2.70**	**2.61**	**2.54**	**2.47**	**2.37**	**2.30**	**2.18**	**2.09**	**2.00**	**1.90**	**1.84**	**1.76**	**1.71**	**1.64**	**1.60**	**1.56**	
70	3.98	3.13	2.74	2.50	2.35	2.23	2.14	2.07	2.01	1.97	1.93	1.89	1.84	1.79	1.72	1.67	1.62	1.56	1.53	1.47	1.45	1.40	1.37	1.35	70
	7.01	**4.92**	**4.08**	**3.60**	**3.29**	**3.07**	**2.91**	**2.77**	**2.67**	**2.59**	**2.51**	**2.45**	**2.35**	**2.28**	**2.15**	**2.07**	**1.98**	**1.88**	**1.82**	**1.74**	**1.69**	**1.62**	**1.56**	**1.53**	
80	3.96	3.11	2.72	2.48	2.33	2.21	2.12	2.05	1.99	1.95	1.91	1.88	1.82	1.77	1.70	1.65	1.60	1.54	1.51	1.45	1.42	1.38	1.35	1.32	80
	6.96	**4.88**	**4.04**	**3.56**	**3.25**	**3.04**	**2.87**	**2.74**	**2.64**	**2.55**	**2.48**	**2.41**	**2.32**	**2.24**	**2.11**	**2.03**	**1.94**	**1.84**	**1.78**	**1.70**	**1.65**	**1.57**	**1.52**	**1.49**	
100	3.94	3.09	2.70	2.46	2.30	2.19	2.10	2.03	1.97	1.92	1.88	1.85	1.79	1.75	1.68	1.63	1.57	1.51	1.48	1.42	1.39	1.34	1.30	1.28	100
	6.90	**4.82**	**3.98**	**3.51**	**3.20**	**2.99**	**2.82**	**2.69**	**2.59**	**2.51**	**2.43**	**2.36**	**2.26**	**2.19**	**2.06**	**1.98**	**1.89**	**1.79**	**1.73**	**1.64**	**1.59**	**1.51**	**1.46**	**1.43**	
125	3.92	3.07	2.68	2.44	2.29	2.17	2.08	2.01	1.95	1.90	1.86	1.83	1.77	1.72	1.65	1.60	1.55	1.49	1.45	1.39	1.36	1.31	1.27	1.25	125
	6.84	**4.78**	**3.94**	**3.47**	**3.17**	**2.95**	**2.79**	**2.65**	**2.56**	**2.47**	**2.40**	**2.33**	**2.23**	**2.15**	**2.03**	**1.94**	**1.85**	**1.75**	**1.68**	**1.59**	**1.54**	**1.46**	**1.40**	**1.37**	
150	3.91	3.06	2.67	2.43	2.27	2.16	2.07	2.00	1.94	1.89	1.85	1.82	1.76	1.71	1.64	1.59	1.54	1.47	1.44	1.37	1.34	1.29	1.25	1.22	150
	6.81	**4.75**	**3.91**	**3.44**	**3.14**	**2.92**	**2.76**	**2.62**	**2.53**	**2.44**	**2.37**	**2.30**	**2.20**	**2.12**	**2.00**	**1.91**	**1.83**	**1.72**	**1.66**	**1.56**	**1.51**	**1.43**	**1.37**	**1.33**	
200	3.89	3.04	2.65	2.41	2.26	2.14	2.05	1.98	1.92	1.87	1.83	1.80	1.74	1.69	1.62	1.57	1.52	1.45	1.42	1.35	1.32	1.26	1.22	1.19	200
	6.76	**4.71**	**3.88**	**3.41**	**3.11**	**2.90**	**2.73**	**2.60**	**2.50**	**2.41**	**2.34**	**2.28**	**2.17**	**2.09**	**1.97**	**1.88**	**1.79**	**1.69**	**1.62**	**1.53**	**1.48**	**1.39**	**1.33**	**1.28**	
400	3.86	3.02	2.62	2.39	2.23	2.12	2.03	1.96	1.90	1.85	1.81	1.78	1.72	1.67	1.60	1.54	1.49	1.42	1.38	1.32	1.28	1.22	1.16	1.13	400
	6.70	**4.66**	**3.83**	**3.36**	**3.06**	**2.85**	**2.69**	**2.55**	**2.46**	**2.37**	**2.29**	**2.23**	**2.12**	**2.04**	**1.92**	**1.84**	**1.74**	**1.64**	**1.57**	**1.47**	**1.42**	**1.32**	**1.24**	**1.19**	
1000	3.85	3.00	2.61	2.38	2.22	2.10	2.02	1.95	1.89	1.84	1.80	1.76	1.70	1.65	1.58	1.53	1.47	1.41	1.36	1.30	1.26	1.19	1.13	1.08	1000
	6.66	**4.62**	**3.80**	**3.34**	**3.04**	**2.82**	**2.66**	**2.53**	**2.43**	**2.34**	**2.26**	**2.20**	**2.09**	**2.01**	**1.89**	**1.81**	**1.71**	**1.61**	**1.54**	**1.44**	**1.38**	**1.28**	**1.19**	**1.11**	
∞	3.84	2.99	2.60	2.37	2.21	2.09	2.01	1.94	1.88	1.83	1.79	1.75	1.69	1.64	1.57	1.52	1.46	1.40	1.35	1.28	1.24	1.17	1.11	1.00	∞
	6.64	**4.60**	**3.78**	**3.32**	**3.02**	**2.80**	**2.64**	**2.51**	**2.41**	**2.32**	**2.24**	**2.18**	**2.07**	**1.99**	**1.87**	**1.79**	**1.69**	**1.59**	**1.52**	**1.41**	**1.36**	**1.25**	**1.15**	**1.00**	

Reprinted by permission from STATISTICAL METHODS by George W. Snedecor and William G. Cochran, sixth edition (c) 1967 by Iowa State University Press, Ames, Iowa.

Appendix F Significant Studentized Ranges for 5% and 1% Level New Multiple Range Test

p = number of means for range being tested

Error df	Protection level	2	3	4	5	6	7	8	9	10	12	14	16	18	20
1	.05	18.0	18.0	18.0	18.0	18.0	18.0	18.0	18.0	18.0	18.0	18.0	18.0	18.0	18.0
	.01	90.0	90.0	90.0	90.0	90.0	90.0	90.0	90.0	90.0	90.0	90.0	90.0	90.0	90.0
2	.05	6.09	6.09	6.09	'6.09	6.09	6.09	6.09	6.09	6.09	6.09	6.09	6.09	6.09	6.09
	.01	14.0	14.0	14.0	14.0	14.0	14.0	14.0	14.0	14.0	14.0	14.0	14.0	14.0	14.0
3	.05	4.50	4.50	4.50	4.50	4.50	4.50	4.50	4.50	4.50	4.50	4.50	4.50	4.50	4.50
	.01	8.26	8.5	8.6	8.7	8.8	8.9	8.9	9.0	9.0	9.0	9.1	9.2	9.3	9.3
4	.05	3.93	4.01	4.02	4.02	4.02	4.02	4.02	4.02	4.02	4.02	4.02	4.02	4.02	4.02
	.01	6.51	6.8	6.9	7.0	7.1	7.1	7.2	7.2	7.3	7.3	7.4	7.4	7.5	7.5
5	.05	3.64	3.74	3.79	3.83	3.83	3.83	3.83	3.83	3.83	3.83	3.83	3.83	3.83	3.83
	.01	5.70	5.96	6.11	6.18	6.26	6.33	6.40	6.44	6.5	6.6	6.6	6.7	6.7	6.8
6	.05	3.46	3.58	3.64	3.68	3.68	3.68	3.68	3.68	3.68	3.68	3.68	3.68	3.68	3.68
	.01	5.24	5.51	5.65	5.73	5.81	5.88	5.95	6.00	6.0	6.1	6.2	6.2	6.3	6.3
7	.05	3.35	3.47	3.54	3.58	3.60	3.61	3.61	3.61	3.61	3.61	3.61	3.61	3.61	3.61
	.01	4.95	5.22	5.37	5.45	5.53	5.61	5.69	5.73	5.8	5.8	5.9	5.9	6.0	6.0
8	.05	3.26	3.39	3.47	3.52	3.55	3.56	3.56	3.56	3.56	3.56	3.56	3.56	3.56	3.56
	.01	4.74	5.00	5.14	5.23	5.32	5.40	5.47	5.51	5.5	5.6	5.7	5.7	5.8	5.8
9	.05	3.20	3.34	3.41	3.47	3.50	3.52	3.52	3.52	3.52	3.52	3.52	3.52	3.52	3.52
	.01	4.60	4.86	4.99	5.08	5.17	5.25	5.32	5.36	5.4	5.5	5.5	5.6	5.7	5.7
10	.05	3.15	3.30	3.37	3.43	3.46	3.47	3.47	3.47	3.47	3.47	3.47	3.47	3.47	3.48
	.01	4.48	4.73	4.88	4.96	5.06	5.13	5.20	5.24	5.28	5.36	5.42	5.48	5.54	5.55
11	.05	3.11	3.27	3.35	3.39	3.43	3.44	3.45	3.46	3.46	3.46	3.46	3.46	3.47	3.48
	.01	4.39	4.63	4.77	4.86	4.94	5.01	5.06	5.12	5.15	5.24	5.28	5.34	5.38	5.39
12	.05	3.08	3.23	3.33	3.36	3.40	3.42	3.44	3.44	3.46	3.46	3.46	3.46	3.47	3.48
	.01	4.32	4.55	4.68	4.76	4.81	4.92	4.96	5.02	5.07	5.13	5.17	5.22	5.24	5.26
13	.05	3.06	3.21	3.30	3.35	3.38	3.41	3.42	3.44	3.45	3.45	3.46	3.46	3.47	3.47
	.01	4.26	4.48	4.62	4.69	4.74	4.84	4.88	4.94	4.98	5.04	5.08	5.13	5.14	5.15
14	.05	3.03	3.18	3.27	3.33	3.37	3.39	3.41	3.42	3.44	3.45	3.46	3.46	3.47	3.47
	.01	4.21	4.42	4.55	4.63	4.70	4.78	4.83	3.87	4.91	4.96	5.00	5.04	5.06	5.07
15	.05	3.01	3.16	3.25	3.31	3.36	3.38	3.40	3.42	3.43	3.44	3.45	3.46	3.47	3.47
	.01	4.17	4.37	4.50	4.58	4.64	4.72	4.77	4.81	4.84	4.90	4.94	4.97	4.99	5.00

continued next page

Appendix F (Continued)

Error df	Protection level	2	3	4	5	6	7	8	9	10	12	14	16	18	20
								p = number of means for range being tested							
16	.05	3.00	3.15	3.23	3.30	3.34	3.37	3.39	3.41	3.43	3.44	3.45	3.46	3.47	3.47
	.01	4.13	4.34	4.45	4.54	4.60	4.67	4.72	4.76	4.79	4.84	4.88	4.91	4.93	4.94
17	.05	2.98	3.13	3.22	3.28	3.33	3.36	3.38	3.40	3.42	3.44	3.45	3.46	3.47	3.47
	.01	4.10	4.30	4.41	4.50	4.56	4.63	4.68	4.72	4.75	4.80	4.83	4.86	4.88	4.89
18	.05	2.97	3.12	3.21	3.27	3.32	3.35	3.37	3.39	3.41	3.43	3.45	3.46	3.47	3.47
	.01	4.07	4.27	4.38	4.46	4.53	4.59	4.64	4.68	4.71	4.76	4.79	4.82	4.84	4.85
19	.05	2.96	3.11	3.19	3.26	3.31	3.35	3.37	3.39	3.41	3.43	3.44	3.46	3.47	3.47
	.01	4.05	4.24	4.35	4.43	4.50	4.56	4.61	4.64	4.67	4.72	4.76	4.79	4.81	4.82
20	.05	2.95	3.10	3.18	3.25	3.30	3.34	3.36	3.38	3.40	3.43	3.44	3.46	3.46	3.47
	.01	4.02	4.22	4.33	4.40	4.47	4.53	4.58	4.61	4.65	4.69	4.73	4.76	4.78	4.79
22	.05	2.93	3.08	3.17	3.24	3.29	3.32	3.35	3.37	3.39	3.42	3.44	3.45	3.46	3.47
	.01	3.99	4.17	4.28	4.36	4.42	4.48	4.53	4.57	4.60	4.65	4.68	4.71	4.74	4.75
24	.05	2.92	3.07	3.15	3.22	3.28	3.31	3.34	3.37	3.38	3.41	3.44	3.45	3.46	3.47
	.01	3.96	4.14	4.24	4.33	4.39	4.44	4.49	4.53	4.57	4.62	4.64	4.67	4.70	4.72
26	.05	2.91	3.06	3.14	3.21	3.27	3.30	3.34	3.36	3.38	3.41	3.43	3.45	3.46	3.47
	.01	3.93	4.11	4.21	4.30	4.36	4.41	4.46	4.50	4.53	4.58	4.62	4.65	4.67	4.69
28	.05	2.90	3.04	3.13	3.20	3.26	3.30	3.33	3.35	3.38	3.40	3.43	3.45	3.46	3.47
	.01	3.91	4.08	4.18	4.28	4.34	4.39	4.43	4.47	4.51	4.56	4.60	4.62	4.65	4.67
30	.05	2.89	3.04	3.12	3.20	3.25	3.29	3.32	3.35	3.37	3.40	3.43	3.44	3.46	3.47
	.01	3.89	4.06	4.16	4.22	4.32	4.36	4.41	4.45	4.48	4.54	4.58	4.61	4.63	4.65
40	.05	2.86	3.01	3.10	3.17	3.22	3.27	3.30	3.33	3.35	3.39	3.42	3.44	3.46	3.47
	.01	3.82	3.99	4.10	4.17	4.21	4.30	4.34	4.37	4.41	4.46	4.51	4.54	4.57	4.59
60	.05	2.83	2.98	3.08	3.14	3.20	3.24	3.28	3.31	3.33	3.37	3.40	3.43	3.45	3.47
	.01	3.76	3.92	4.03	4.12	4.17	4.23	4.27	4.31	4.34	4.39	4.44	4.47	4.50	4.53
100	.05	2.80	2.95	3.05	3.12	3.18	3.22	3.26	3.29	3.32	3.36	3.40	3.42	3.45	3.47
	.01	3.71	3.86	3.98	4.06	4.11	4.17	4.21	4.25	4.29	4.35	4.38	4.42	4.45	4.48
∞	.05	2.77	2.92	3.02	3.09	3.15	3.19	3.23	3.26	3.29	3.34	3.38	3.41	3.44	3.47
	.01	3.64	3.80	3.90	3.98	4.04	4.09	4.14	4.17	4.20	4.26	4.31	4.34	4.38	4.41

Source: Reproduced from *Principles and Procedures of Statistics* by R. G. D. Steel and J. H. Torrie, 1960, McGraw Hill Book Co. Inc., New York. Printed with the permission of the publisher, Biometric Society, North Carolina.

Appendix G Orthogonal Polynomial Coefficients for Comparison between Three to Six Equally Spaced Treatments

Treatments (no.)	Degree of polynomials	T_1	T_2	T_3	T_4	T_5	T_6	Sum of squares of the coefficients
3	Linear	−1	0	+1				2
	Quadratic	+1	−2	+1				6
4	Linear	−3	−1	+1	+3			20
	Quadratic	+1	−1	−1	+1			4
	Cubic	−1	+3	−3	+1			20
5	Linear	−2	−1	0	+1	+2		10
	Quadratic	+2	−1	−2	−1	+2		14
	Cubic	−1	+2	0	−2	+1		10
	Quartic	+1	−4	+6	−4	+1		70
6	Linear	−5	−3	−1	+1	+3	+5	70
	Quadratic	+5	−1	−4	−4	−1	+5	84
	Cubic	−5	+7	+4	−4	−7	+5	180
	Quartic	+1	−3	+2	+2	−3	+1	28
	Quintic	−1	+5	−10	+10	−5	+1	252

Appendix H Simple Linear Correlation Coefficients, *r*, at the 5% and 1% Levels of Significance

d.f. [1]	5%	1%	d.f.	5%	1%
1	.997	1.000	26	.374	.478
2	.950	.990	27	.367	.470
3	.878	.959	28	.361	.463
4	.811	.917	29	.355	.456
5	.754	.874	30	.349	.449
6	.707	.834	32	.339	.437
7	.666	.798	34	.329	.424
8	.632	.765	36	.321	.413
9	.602	.735	38	.312	.403
10	.576	.708	40	.304	.393
11	.553	.684	45	.288	.372
12	.532	.661	50	.273	.354
13	.514	.641	55	.262	.340
14	.497	.623	60	.250	.325
15	.482	.606	70	.232	.302
16	.468	.590	80	.217	.283
17	.456	.575	90	.205	.267
18	.444	.561	100	.195	.254
19	.433	.549	125	.174	.228
20	.423	.537	150	.159	.208
21	.413	.526	175	.148	.194
22	.404	.515	200	.138	.181
23	.396	.505	300	.113	.148
24	.388	.496	400	.098	.128
25	.381	.487	500	.088	.115

[1] d.f. = n − 2, where n is the sample size

Reprinted by permission from STATISTICAL METHODS by George W. Snedecor and William G. Cochran, sixth edition (c) 1967 by Iowa State University Press, Ames, Iowa.

641

Appendix I Table of Corresponding Values of r^1 and z^1

z	r	z	r	z	r	z	r
0.00	0.000	0.70	0.607	1.40	0.885	2.10	0.970
0.20	0.020	0.72	0.617	1.42	0.890	2.12	0.972
0.04	0.040	0.74	0.629	1.44	0.894	2.14	0.973
0.06	0.060	0.76	0.641	1.46	0.898	2.16	0.974
0.08	0.080	0.78	0.653	1.48	0.902	2.18	0.975
0.10	0.100	0.80	0.664	1.50	0.905	2.20	0.976
0.12	0.119	0.82	0.675	1.52	0.909	2.22	0.977
0.14	0.139	0.84	0.686	1.54	0.912	2.24	0.978
0.16	0.159	0.86	0.696	1.56	0.915	2.26	0.978
0.18	0.178	0.88	0.706	1.58	0.919	2.28	0.979
0.20	0.197	0.90	0.716	1.60	0.922	2.30	0.980
0.22	0.216	0.92	0.726	1.62	0.925	2.32	0.981
0.24	0.236	0.94	0.735	1.64	0.928	2.34	0.982
0.26	0.254	0.96	0.744	1.66	0.930	2.36	0.982
0.28	0.273	0.98	0.753	1.68	0.933	2.38	0.983
0.30	0.291	1.00	0.762	1.70	0.935	2.40	0.984
0.32	0.310	1.02	0.770	1.72	0.938	2.42	0.984
0.34	0.327	1.04	0.778	1.74	0.940	2.44	0.985
0.36	0.345	1.06	0.786	1.76	0.942	2.46	0.986
0.38	0.363	1.08	0.793	1.78	0.945	2.48	0.986
0.40	0.380	1.10	0.800	1.80	0.947	2.50	0.987
0.42	0.397	1.12	0.808	1.82	0.949	2.52	0.987
0.44	0.414	1.14	0.814	1.84	0.951	2.54	0.988
0.46	0.430	1.16	0.821	1.86	0.953	2.56	0.988
0.48	0.446	1.18	0.828	1.88	0.954	2.58	0.989
0.50	0.462	1.20	0.834	1.90	0.956	2.62	0.989
0.52	0.478	1.22	0.840	1.92	0.958	2.66	0.990
0.54	0.493	1.24	0.846	1.94	0.960	2.70	0.991
0.56	0.508	1.26	0.851	1.96	0.961	2.74	0.992
0.58	0.523	1.28	0.856	1.98	0.963	2.78	0.992
0.60	0.537	1.30	0.862	2.00	0.964	2.82	0.993
0.62	0.551	1.32	0.867	2.02	0.965	2.86	0.993
0.64	0.565	1.34	0.872	2.04	0.967	2.90	0.994
0.66	0.578	1.36	0.876	2.06	0.968	2.94	0.994
0.68	0.592	1.38	0.881	2.08	0.969	2.98	0.995

$^1 r = (e^{2z} - 1)/(e^{2z} + 1)$ or $z = \frac{1}{2} \ln (1 + r) / (1 - r)$

Reprinted by permission from STATISTICAL METHODS by William G. Cochran, fourth edition 1946 by Iowa State University Press, Ames, Iowa.

Appendix J The Arc Sine $\sqrt{\text{Percentage}}$ Transformation.

(Transformation of binomial percentages, in the margins, to angles of equal information in degrees. The + or − signs following angles ending in 5 are for guidance in rounding to one decimal.)

%	0	1	2	3	4	5	6	7	8	9
0.0	0	0.57	0.81	0.99	1.15−	1.28	1.40	1.52	1.62	1.72
0.1	1.81	1.90	1.99	2.07	2.14	2.22	2.29	2.36	2.43	2.50
0.2	2.56	2.63	2.69	2.75−	2.81	2.87	2.92	2.98	3.03	3.09
0.3	3.14	3.19	3.24	3.29	3.34	3.39	3.44	3.49	3.53	3.58
0.4	3.63	3.67	3.72	3.76	3.80	3.85−	3.89	3.93	3.97	4.01
0.5	4.05+	4.09	4.13	4.17	4.21	4.25+	4.29	4.33	4.37	4.40
0.6	4.44	4.48	4.52	4.55+	4.59	4.62	4.66	4.69	4.73	4.76
0.7	4.80	4.83	4.87	4.90	4.93	4.97	5.00	5.03	5.07	5.10
0.8	5.13	5.16	5.20	5.23	5.26	5.29	5.32	5.35+	5.38	5.41
0.9	5.44	5.47	5.50	5.53	5.56	5.59	5.62	5.65+	5.68	5.71
1	5.74	6.02	6.29	6.55−	6.80	7.04	7.27	7.49	7.71	7.92
2	8.13	8.33	8.53	8.72	8.91	9.10	9.28	9.46	9.63	9.81
3	9.98	10.14	10.31	10.47	10.63	10.78	10.94	11.09	11.24	11.39
4	11.54	11.68	11.83	11.97	12.11	12.25−	12.39	12.52	12.66	12.79
5	12.92	13.05+	13.18	13.31	13.44	13.56	13.69	13.81	13.94	14.06
6	14.18	14.30	14.42	14.54	14.65+	14.77	14.89	15.00	15.12	15.23
7	15.34	15.45+	15.56	15.68	15.79	15.89	16.00	16.11	16.22	16.32
8	16.43	16.54	16.64	16.74	16.85−	16.95+	17.05+	17.16	17.26	17.36
9	17.46	17.56	17.66	17.76	17.85+	17.95+	18.05−	18.15−	18.24	18.34
10	18.44	18.53	18.63	18.72	18.81	18.91	19.00	19.09	19.19	19.28
11	19.37	19.46	19.55+	19.64	19.73	19.82	19.91	20.00	20.09	20.18
12	20.27	20.36	20.44	20.53	20.62	20.70	20.79	20.88	20.96	21.05−
13	21.13	21.22	21.30	21.39	21.47	21.56	21.64	21.72	21.81	21.89
14	21.97	22.06	22.14	22.22	22.30	22.38	22.46	22.55−	22.63	22.71
15	22.79	22.87	22.95−	23.03	23.11	23.19	23.26	23.34	23.42	23.50
16	23.58	23.66	23.73	23.81	23.89	23.97	24.04	24.12	24.20	24.27
17	24.35+	24.43	24.50	24.58	24.65+	24.73	24.80	24.88	24.95+	25.03
18	25.10	25.18	25.25+	25.33	25.40	25.48	25.55−	25.62	25.70	25.77
19	25.84	25.92	25.99	26.06	26.13	26.21	26.28	26.35−	26.42	26.49
20	26.56	26.64	26.71	26.78	26.85+	26.92	26.99	27.06	27.13	27.20
21	27.28	27.35−	27.42	27.49	27.56	27.63	27.69	27.76	27.83	27.90
22	27.97	28.04	28.11	28.18	28.25−	28.32	28.38	28.45+	28.52	28.59
23	28.66	28.73	28.79	28.86	28.93	29.00	29.06	29.13	29.20	29.27
24	29.33	29.40	29.47	29.53	29.60	29.67	29.73	29.80	29.87	29.93
25	30.00	30.07	30.13	30.20	30.26	30.33	30.40	30.46	30.53	30.59
26	30.66	30.72	30.79	30.85+	30.92	30.98	31.05−	31.11	31.18	31.24
27	31.31	31.37	31.44	31.50	31.56	31.63	31.69	31.76	31.82	31.88
28	31.95−	32.01	32.08	32.14	32.20	32.27	32.33	32.39	32.46	32.52
29	32.58	32.65−	32.71	32.77	32.83	32.90	32.96	33.02	33.09	33.15−
30	33.21	33.27	33.34	33.40	33.46	33.52	33.58	33.65−	33.71	33.77
31	33.83	33.89	33.96	34.02	34.08	34.14	34.20	34.27	34.33	34.39
32	34.45−	34.51	34.57	34.63	34.70	34.76	34.82	34.88	34.94	35.00
33	35.06	35.12	35.18	35.24	35.30	35.37	35.43	35.49	35.55−	35.61
34	35.67	35.73	35.79	35.85−	35.91	35.97	36.03	36.09	36.15+	36.21
35	36.27	36.33	36.39	36.45+	36.51	36.57	36.63	36.69	36.75+	36.81
36	36.87	36.93	36.99	37.05+	37.11	37.17	37.23	37.29	37.35−	37.41
37	37.47	37.52	37.58	37.64	37.70	37.76	37.82	37.88	37.94	38.00
38	38.06	38.12	38.17	38.23	38.29	38.35+	38.41	38.47	38.53	38.59
39	38.65−	38.70	38.76	38.82	38.88	38.94	39.00	39.06	39.11	39.17
40	39.23	39.29	39.35−	39.41	39.47	39.52	39.58	39.64	39.70	39.76
41	39.82	39.87	39.93	39.99	40.05−	40.11	40.16	40.22	40.28	40.34
42	40.40	40.46	40.51	40.57	40.63	40.69	40.74	40.80	40.86	40.92
43	40.98	41.03	41.09	41.15−	41.21	41.27	41.32	41.38	41.44	41.50
44	41.55+	41.61	41.67	41.73	41.78	41.84	41.90	41.96	42.02	42.07

continued next page

Appendix J (*Continued*)

%	0	1	2	3	4	5	6	7	8	9
45	42.13	42.19	42.25−	42.30	42.36	42.42	42.48	42.53	42.59	42.65−
46	42.71	42.76	42.82	42.88	42.94	42.99	43.05−	43.11	43.17	43.22
47	43.28	43.34	43.39	43.45+	43.51	43.57	43.62	43.68	43.74	43.80
48	43.85+	43.91	43.97	44.03	44.08	44.14	44.20	44.25+	44.51	44.37
49	44.43	44.46	44.54	44.60	44.66	44.71	44.77	44.83	44.89	44.94
50	45.00	45.06	45.11	45.17	45.23	45.29	45.34	45.40	45.46	45.52
51	45.57	45.63	45.69	45.75−	45.80	45.86	45.92	45.97	46.03	46.09
52	46.15−	46.20	46.26	46.32	46.38	46.43	46.49	46.55−	46.61	46.66
53	46.72	46.78	46.83	46.89	46.95+	47.01	47.06	47.12	47.18	47.24
54	47.29	47.35+	47.41	47.47	47.52	47.58	47.64	47.70	47.75+	47.81
55	47.87	47.93	47.98	48.04	48.10	48.16	48.22	48.27	48.33	48.39
56	48.45−	48.50	48.56	48.62	48.68	48.73	48.79	48.85+	48.91	48.97
57	49.02	49.08	49.14	49.20	49.26	49.31	49.37	49.43	49.49	49.54
58	49.60	49.66	49.72	49.78	49.84	49.89	49.95+	50.01	50.07	50.13
59	50.18	50.24	50.30	50.36	50.42	50.48	50.53	50.59	50.65+	50.71
60	50.77	50.83	50.89	50.94	51.00	51.06	51.12	51.18	51.24	51.30
61	51.35+	51.41	51.47	51.53	51.59	51.65−	51.71	51.77	51.83	51.88
62	51.94	52.00	52.06	52.12	52.18	52.24	52.30	52.36	52.42	52.48
63	52.53	52.59	52.65+	52.71	52.77	52.83	52.89	52.95+	53.01	53.07
64	53.13	53.19	53.25−	53.31	53.37	53.43	53.49	53.55−	53.61	53.67
65	53.73	53.79	53.85−	53.91	53.97	54.03	54.09	54.15+	54.21	54.27
66	54.33	54.39	54.45+	54.51	54.57	54.63	54.70	54.76	54.82	54.88
67	54.94	55.00	55.06	55.12	55.18	55.24	55.30	55.37	55.43	55.49
68	55.55+	55.61	55.67	55.73	55.80	55.86	55.92	55.98	56.04	56.11
69	56.17	56.23	56.29	56.35+	56.42	56.48	56.54	56.60	56.66	56.73
70	56.79	56.85+	56.91	56.98	57.04	57.10	57.17	57.23	57.29	57.35+
71	57.42	57.48	57.54	57.61	57.67	57.73	57.80	57.86	57.92	57.99
72	58.05+	58.12	58.18	58.24	58.31	58.37	58.44	58.50	58.56	58.63
73	58.69	58.76	58.82	58.89	58.95+	59.02	59.08	59.15−	59.21	59.28
74	59.34	59.41	59.47	59.54	59.60	59.67	59.74	59.80	59.87	59.93
75	60.00	60.07	60.13	60.20	60.27	60.33	60.40	60.47	60.53	60.60
76	60.67	60.73	60.80	60.87	60.94	61.00	61.07	61.14	61.21	61.27
77	61.34	61.41	61.48	61.55−	61.62	61.68	61.75+	61.82	61.89	61.96
78	62.03	62.10	62.17	62.24	62.31	62.37	62.44	62.51	62.58	62.65+
79	62.72	62.80	62.87	62.94	63.01	63.08	63.15−	63.22	63.29	63.36
80	63.44	63.51	63.58	63.65+	63.72	63.79	63.87	63.94	64.01	64.08
81	64.16	64.23	64.30	64.38	64.45+	64.52	64.60	64.67	64.75−	64.82
82	64.90	64.97	65.05−	65.12	65.20	65.27	65.35−	65.42	65.50	65.57
83	65.65−	65.73	65.80	65.88	65.96	66.03	66.11	66.19	66.27	66.34
84	66.42	66.50	66.58	66.66	66.74	66.81	66.89	66.97	67.05+	67.13
85	67.21	67.29	67.37	67.45+	67.54	67.62	67.70	67.78	67.86	67.94
86	68.03	68.11	68.19	68.28	68.36	68.44	68.53	68.61	68.70	68.78
87	68.87	68.95+	69.04	69.12	69.21	69.30	69.38	69.47	69.56	69.64
88	69.73	69.82	69.91	70.00	70.09	70.18	70.27	70.36	70.45−	70.54
89	70.63	70.72	70.81	70.91	71.00	71.09	71.19	71.23	71.37	71.47
90	71.56	71.66	71.76	71.85+	71.95+	72.05−	72.15−	72.24	72.34	72.44
91	72.54	72.64	72.74	72.84	72.95−	73.05−	73.15+	73.26	73.36	73.46
92	73.57	73.68	73.78	73.89	74.00	74.11	74.21	74.32	74.44	74.55−
93	74.66	74.77	74.88	75.00	75.11	75.23	75.35−	75.46	75.58	75.70
94	75.82	75.94	76.06	76.19	76.31	76.44	76.56	76.69	76.82	76.95−
95	77.08	77.21	77.34	77.48	77.61	77.75+	77.89	78.03	78.17	78.32
96	78.46	78.61	78.76	78.91	79.06	79.22	79.37	79.53	79.69	79.86
97	80.02	80.19	80.37	80.54	80.72	80.90	81.09	81.28	81.47	81.67
98	81.87	82.08	82.29	82.51	82.73	82.96	83.20	83.45+	83.71	83.98

Appendix J (*Continued*)

%	0	1	2	3	4	5	6	7	8	9
00.0	84.26	84.29	84.33	84.35	84.38	84.41	84.44	84.47	84.50	84.53
99.1	84.56	84.59	84.62	84.65−	84.68	84.71	84.74	84.77	84.80	84.84
99.2	84.87	84.90	84.93	84.97	85.00	85.03	85.07	85.10	85.13	85.17
99.3	85.20	85.24	85.27	85.31	85.34	85.38	85.41	85.45−	85.48	85.52
99.4	85.56	85.60	85.63	85.67	85.71	85.75−	85.79	85.83	85.87	85.91
99.5	85.95−	85.99	86.03	86.07	86.11	86.15−	86.20	86.24	86.28	86.33
99.6	86.37	86.42	86.47	86.51	86.56	86.61	86.66	86.71	86.76	86.81
99.7	86.86	86.91	86.97	87.02	87.08	87.13	87.19	87.25+	87.31	87.37
99.8	87.44	87.50	87.57	87.64	87.71	87.78	87.86	87.93	88.01	88.10
99.9	88.19	88.28	88.38	88.48	88.60	88.72	88.85+	89.01	89.19	89.43
100.0	90.00									

Reproduced from *Principles and Procedures of Statistics* by R. G. D. Steel and J. H. Torrie, 1960. Printed with the permission of C. I. Bliss, pp. 448−449.

Appendix K Selected Latin Squares

3 × 3

```
A  B  C
B  C  A
C  A  B
```

4 × 4

1
```
A  B  C  D
B  A  D  C
C  D  B  A
D  C  A  B
```

2
```
A  B  C  D
B  C  D  A
C  D  A  B
D  A  B  C
```

3
```
A  B  C  D
B  D  A  C
C  A  D  B
D  C  B  A
```

4
```
A  B  C  D
B  A  D  C
C  D  A  B
D  C  B  A
```

5 × 5

```
A  B  C  D  E
B  A  E  C  D
C  D  A  E  B
D  E  B  A  C
E  C  D  B  A
```

6 × 6

```
A  B  C  D  E  F
B  F  D  C  A  E
C  D  E  F  B  A
D  A  F  E  C  B
E  C  A  B  F  D
F  E  B  A  D  C
```

7 × 7

```
A  B  C  D  E  F  G
B  C  D  E  F  G  A
C  D  E  F  G  A  B
D  E  F  G  A  B  C
E  F  G  A  B  C  D
F  G  A  B  C  D  E
G  A  B  C  D  E  F
```

8 × 8

```
A  B  C  D  E  F  G  H
B  C  D  E  F  G  H  A
C  D  E  F  G  H  A  B
D  E  F  G  H  A  B  C
E  F  G  H  A  B  C  D
F  G  H  A  B  C  D  E
G  H  A  B  C  D  E  F
H  A  B  C  D  E  F  G
```

9 × 9

```
A  B  C  D  E  F  G  H  I
B  C  D  E  F  G  H  I  A
C  D  E  F  G  H  I  A  B
D  E  F  G  H  I  A  B  C
E  F  G  H  I  A  B  C  D
F  G  H  I  A  B  C  D  E
G  H  I  A  B  C  D  E  F
H  I  A  B  C  D  E  F  G
I  A  B  C  D  E  F  G  H
```

10 × 10

```
A  B  C  D  E  F  G  H  I  J
B  C  D  E  F  G  H  I  J  A
C  D  E  F  G  H  I  J  A  B
D  E  F  G  H  I  J  A  B  C
E  F  G  H  I  J  A  B  C  D
F  G  H  I  J  A  B  C  D  E
G  H  I  J  A  B  C  D  E  F
H  I  J  A  B  C  D  E  F  G
I  J  A  B  C  D  E  F  G  H
J  A  B  C  D  E  F  G  H  I
```

11 × 11

```
A  B  C  D  E  F  G  H  I  J  K
B  C  D  E  F  G  H  I  J  K  A
C  D  E  F  G  H  I  J  K  A  B
D  E  F  G  H  I  J  K  A  B  C
E  F  G  H  I  J  K  A  B  C  D
F  G  H  I  J  K  A  B  C  D  E
G  H  I  J  K  A  B  C  D  E  F
H  I  J  K  A  B  C  D  E  F  G
I  J  K  A  B  C  D  E  F  G  H
J  K  A  B  C  D  E  F  G  H  I
K  A  B  C  D  E  F  G  H  I  J
```

Reproduced from EXPERIMENTAL DESIGNS by William G. Cochran and Gertrude M. Cox, second edition 1957. Copyright 1950, © 1957 by John Wiley and Sons, Inc. Reprinted by permission of John Wiley and Sons, Inc.

Appendix L   Basic Plans for Balanced and Partially Balanced Lattice Designs

4 × 4 BALANCED LATTICE

Block	Rep. I					Rep. II					Rep. III			
(1)	1	2	3	4	(5)	1	5	9	13	(9)	1	6	11	16
(2)	5	6	7	8	(6)	2	6	10	14	(10)	5	2	15	12
(3)	9	10	11	12	(7)	3	7	11	15	(11)	9	14	3	8
(4)	13	14	15	16	(8)	4	8	12	16	(12)	13	10	7	4

Block	Rep. IV					Rep. V			
(13)	1	14	7	12	(17)	1	10	15	8
(14)	13	2	11	8	(18)	9	2	7	16
(15)	5	10	3	16	(19)	13	6	3	12
(16)	9	6	15	4	(20)	5	14	11	4

5 × 5 BALANCED LATTICE

Block	Rep. I						Rep. II						Rep. III				
(1)	1	2	3	4	5	(6)	1	6	11	16	21	(11)	1	7	13	19	25
(2)	6	7	8	9	10	(7)	2	7	12	17	22	(12)	21	2	8	14	20
(3)	11	12	13	14	15	(8)	3	8	13	18	23	(13)	16	22	3	9	15
(4)	16	17	18	19	20	(9)	4	9	14	19	24	(14)	11	17	23	4	10
(5)	21	22	23	24	25	(10)	5	10	15	20	25	(15)	6	12	18	24	5

Block	Rep. IV						Rep. V						Rep. VI				
(16)	1	12	23	9	20	(21)	1	17	8	24	15	(26)	1	22	18	14	10
(17)	16	2	13	24	10	(22)	11	2	18	9	25	(27)	6	2	23	19	15
(18)	6	17	3	14	25	(23)	21	12	3	19	10	(28)	11	7	3	24	20
(19)	21	7	18	4	15	(24)	6	22	13	4	20	(29)	16	12	8	4	25
(20)	11	22	8	19	5	(25)	16	7	23	14	5	(30)	21	17	13	9	5

continued next page

6 × 6 TRIPLE LATTICE

Block	Rep. I							Rep. II							Rep. III					
(1)	1	2	3	4	5	6	(7)	1	7	13	19	25	31	(13)	1	8	15	22	29	36
(2)	7	8	9	10	11	12	(8)	2	8	14	20	26	32	(14)	31	2	9	16	23	30
(3)	13	14	15	16	17	18	(9)	3	9	15	21	27	33	(15)	25	32	3	10	17	24
(4)	19	20	21	22	23	24	(10)	4	10	16	22	28	34	(16)	19	26	33	4	11	18
(5)	25	26	27	28	29	30	(11)	5	11	17	23	29	35	(17)	13	20	27	34	5	12
(6)	31	32	33	34	35	36	(12)	6	12	18	24	30	36	(18)	7	14	21	28	35	6

7 × 7 BALANCED LATTICE

Block	Rep. I								Rep. II						
(1)	1	2	3	4	5	6	7	(8)	1	8	15	22	29	36	43
(2)	8	9	10	11	12	13	14	(9)	2	9	16	23	30	37	44
(3)	15	16	17	18	19	20	21	(10)	3	10	17	24	31	38	45
(4)	22	23	24	25	26	27	28	(11)	4	11	18	25	32	39	46
(5)	29	30	31	32	33	34	35	(12)	5	12	19	26	33	40	47
(6)	36	37	38	39	40	41	42	(13)	6	13	20	27	34	41	48
(7)	43	44	45	46	47	48	49	(14)	7	14	21	28	35	42	49

Block	Rep. III								Rep. IV						
(15)	1	9	17	25	33	41	49	(22)	1	37	24	11	47	34	21
(16)	43	2	10	18	26	34	42	(23)	15	2	38	25	12	48	35
(17)	36	44	3	11	19	27	35	(24)	29	16	3	39	26	13	49
(18)	29	37	45	4	12	20	28	(25)	43	30	17	4	40	27	14
(19)	22	30	38	46	5	13	21	(26)	8	44	31	18	5	41	28
(20)	15	23	31	39	47	6	14	(27)	22	9	45	32	19	6	42
(21)	8	16	24	32	40	48	7	(28)	36	23	10	46	33	20	7

Block	Rep. V								Rep. VI						
(29)	1	30	10	39	19	48	28	(36)	1	23	45	18	40	13	35
(30)	22	2	31	11	40	20	49	(37)	29	2	24	46	19	41	14
(31)	43	23	3	32	12	41	21	(38)	8	30	3	25	47	20	42
(32)	15	44	24	4	33	13	42	(39)	36	9	31	4	26	48	21
(33)	36	16	45	25	5	34	14	(40)	15	37	10	32	5	27	49
(34)	8	37	17	46	26	6	35	(41)	43	16	38	11	33	6	28
(35)	29	9	38	18	47	27	7	(42)	22	44	17	39	12	34	7

Block	Rep. VII								Rep. VIII						
(43)	1	16	31	46	12	27	42	(50)	1	44	38	32	26	20	14
(44)	36	2	17	32	47	13	28	(51)	8	2	45	39	33	27	21
(45)	22	37	3	18	33	48	14	(52)	15	9	3	46	40	34	28
(46)	8	23	38	4	19	34	49	(53)	22	16	10	4	47	41	35
(47)	43	9	24	39	5	20	35	(54)	29	23	17	11	5	48	42
(48)	29	44	10	25	40	6	21	(55)	36	30	24	18	12	6	49
(49)	15	30	45	11	26	41	7	(56)	43	37	31	25	19	13	7

Appendix L *(Continued)*

8 × 8 SEXTUPLE LATTICE

Block	Rep. I							
(1)	1	2	3	4	5	6	7	8
(2)	9	10	11	12	13	14	15	16
(3)	17	18	19	20	21	22	23	24
(4)	25	26	27	28	29	30	31	32
(5)	33	34	35	36	37	38	39	40
(6)	41	42	43	44	45	46	47	48
(7)	49	50	51	52	53	54	55	56
(8)	57	58	59	60	61	62	63	64

Block	Rep. II							
(9)	1	9	17	25	33	41	49	57
(10)	2	10	18	26	34	42	50	58
(11)	3	11	19	27	35	43	51	59
(12)	4	12	20	28	36	44	52	60
(13)	5	13	21	29	37	45	53	61
(14)	6	14	22	30	38	46	54	62
(15)	7	15	23	31	39	47	55	63
(16)	8	16	24	32	40	48	56	64

Block	Rep. III							
(17)	1	10	19	28	37	46	55	64
(18)	9	2	51	44	61	30	23	40
(19)	17	50	3	36	29	62	15	48
(20)	25	42	35	4	21	14	63	56
(21)	33	58	27	20	5	54	47	16
(22)	41	26	59	12	53	6	39	24
(23)	49	18	11	60	45	38	7	32
(24)	57	34	43	52	13	22	31	8

Block	Rep. IV							
(25)	1	18	27	44	13	62	39	56
(26)	17	2	35	60	53	46	31	16
(27)	25	34	3	12	45	54	23	64
(28)	41	58	11	4	29	22	55	40
(29)	9	50	43	28	5	38	63	24
(30)	57	42	51	20	37	6	15	32
(31)	33	26	19	52	61	14	7	48
(32)	49	10	59	36	21	30	47	8

Block	Rep. V							
(33)	1	26	43	60	21	54	15	40
(34)	25	2	11	52	37	62	47	24
(35)	41	10	3	20	61	38	31	56
(36)	57	50	19	4	45	30	39	16
(37)	17	34	59	44	5	14	55	32
(38)	49	58	35	28	13	6	23	48
(39)	9	42	27	36	53	22	7	64
(40)	33	18	51	12	29	46	63	8

Block	Rep. VI							
(41)	1	34	11	20	53	30	63	48
(42)	33	2	59	28	45	22	15	56
(43)	9	58	3	52	21	46	39	32
(44)	17	26	51	4	13	38	47	64
(45)	49	42	19	12	5	62	31	40
(46)	25	18	43	36	61	6	55	16
(47)	57	10	35	44	29	54	7	24
(48)	41	50	27	60	37	14	23	8

continued next page

9 × 9 SEXTUPLE LATTICE

Block	Rep. I								
(1)	1	2	3	4	5	6	7	8	9
(2)	10	11	12	13	14	15	16	17	18
(3)	19	20	21	22	23	24	25	26	27
(4)	28	29	30	31	32	33	34	35	36
(5)	37	38	39	40	41	42	43	44	45
(6)	46	47	48	49	50	51	52	53	54
(7)	55	56	57	58	59	60	61	62	63
(8)	64	65	66	67	68	69	70	71	72
(9)	73	74	75	76	77	78	79	80	81

Block	Rep. II								
(10)	1	10	19	28	37	46	55	64	73
(11)	2	11	20	29	38	47	56	65	74
(12)	3	12	21	30	39	48	57	66	75
(13)	4	13	22	31	40	49	58	67	76
(14)	5	14	23	32	41	50	59	68	77
(15)	6	15	24	33	42	51	60	69	78
(16)	7	16	25	34	43	52	61	70	79
(17)	8	17	26	35	44	53	62	71	80
(18)	9	18	27	36	45	54	63	72	81

Block	Rep. III								
(19)	1	20	12	58	77	69	34	53	45
(20)	10	2	21	67	59	78	43	35	54
(21)	19	11	3	76	68	60	52	44	36
(22)	28	47	39	4	23	15	61	80	72
(23)	37	29	48	13	5	24	70	62	81
(24)	46	38	30	22	14	6	79	71	63
(25)	55	74	66	31	50	42	7	26	18
(26)	64	56	75	40	32	51	16	8	27
(27)	73	65	57	49	41	33	25	17	9

Block	Rep. IV								
(28)	1	11	21	31	41	51	61	71	81
(29)	19	2	12	49	32	42	79	62	72
(30)	10	20	3	40	50	33	70	80	63
(31)	55	65	75	4	14	24	34	44	54
(32)	73	56	66	22	5	15	52	35	45
(33)	64	74	57	13	23	6	43	53	36
(34)	28	38	48	58	68	78	7	17	27
(35)	46	29	39	76	59	69	25	8	18
(36)	37	47	30	67	77	60	16	26	9

Block	Rep V								
(37)	1	29	57	22	50	78	16	44	72
(38)	55	2	30	76	23	51	70	17	45
(39)	28	56	3	49	77	24	43	71	18
(40)	10	38	66	4	32	60	25	53	81
(41)	64	11	39	58	5	33	79	26	54
(42)	37	65	12	31	59	6	52	80	27
(43)	19	47	75	13	41	69	7	35	63
(44)	73	20	48	67	14	42	61	8	36
(45)	46	74	21	40	68	15	34	62	9

Block	Rep. VI								
(46)	1	56	30	13	68	42	25	80	54
(47)	28	2	57	40	14	69	52	26	81
(48)	55	29	3	67	41	15	79	53	27
(49)	19	74	48	4	59	33	16	71	45
(50)	46	20	75	31	5	60	43	17	72
(51)	73	47	21	58	32	6	70	44	18
(52)	10	65	39	22	77	51	7	62	36
(53)	37	11	66	49	23	78	34	8	63
(54)	64	38	12	76	50	24	61	35	9

10 × 10 TRIPLE LATTICE

Block	Rep. I									
(1)	1	2	3	4	5	6	7	8	9	10
(2)	11	12	13	14	15	16	17	18	19	20
(3)	21	22	23	24	25	26	27	28	29	30
(4)	31	32	33	34	35	36	37	38	39	40
(5)	41	42	43	44	45	46	47	48	49	50
(6)	51	52	53	54	55	56	57	58	59	60
(7)	61	62	63	64	65	66	67	68	69	70
(8)	71	72	73	74	75	76	77	78	79	80
(9)	81	82	83	84	85	86	87	88	89	90
(10)	91	92	93	94	95	96	97	98	99	100

Block	Rep. II									
(11)	1	11	21	31	41	51	61	71	81	91
(12)	2	12	22	32	42	52	62	72	82	92
(13)	3	13	23	33	43	53	63	73	83	93
(14)	4	14	24	34	44	54	64	74	84	94
(15)	5	15	25	35	45	55	65	75	85	95
(10)	6	16	26	36	46	56	66	76	86	96
(17)	7	17	27	37	47	57	67	77	87	97
(18)	8	18	28	38	48	58	68	78	88	98
(19)	9	19	29	39	49	59	69	79	89	99
(20)	10	20	30	40	50	60	70	80	90	100

Block	Rep. III									
(21)	1	12	23	34	45	56	67	78	89	100
(22)	91	2	13	24	35	46	57	68	79	90
(23)	81	92	3	14	25	36	47	58	69	80
(24)	71	82	93	4	15	26	37	48	59	70
(25)	61	72	83	94	5	16	27	38	49	60
(26)	51	62	73	84	95	6	17	28	39	50
(27)	41	52	63	74	85	96	7	18	29	40
(28)	31	42	53	64	75	86	97	8	19	30
(29)	21	32	43	54	65	76	87	98	9	20
(30)	11	22	33	44	55	66	77	88	99	10

12 × 12 TRIPLE LATTICE

Block   Rep. I

	1	2	3	4	5	6	7	8	9	10	11	12
(1)	1	2	3	4	5	6	7	8	9	10	11	12
(2)	13	14	15	16	17	18	19	20	21	22	23	24
(3)	25	26	27	28	29	30	31	32	33	34	35	36
(4)	37	38	39	40	41	42	43	44	45	46	47	48
(5)	49	50	51	52	53	54	55	56	57	58	59	60
(6)	61	62	63	64	65	66	67	68	69	70	71	72
(7)	73	74	75	76	77	78	79	80	81	82	83	84
(8)	85	86	87	88	89	90	91	92	93	94	95	96
(9)	97	98	99	100	101	102	103	104	105	106	107	108
(10)	109	110	111	112	113	114	115	116	117	118	119	120
(11)	121	122	123	124	125	126	127	128	129	130	131	132
(12)	133	134	135	136	137	138	139	140	141	142	143	144

Rep. II

(13)	1	13	25	37	49	61	73	85	97	109	121	133
(14)	2	14	26	38	50	62	74	86	98	110	122	134
(15)	3	15	27	39	51	63	75	87	99	111	123	135
(16)	4	16	28	40	52	64	76	88	100	112	124	136
(17)	5	17	29	41	53	65	77	89	101	113	125	137
(18)	6	18	30	42	54	66	78	90	102	114	126	138
(19)	7	19	31	43	55	67	79	91	103	115	127	139
(20)	8	20	32	44	56	68	80	92	104	116	128	140
(21)	9	21	33	45	57	69	81	93	105	117	129	141
(22)	10	22	34	46	58	70	82	94	106	118	130	142
(23)	11	23	35	47	59	71	83	95	107	119	131	143
(24)	12	24	36	48	60	72	84	96	108	120	132	144

Rep. III

(25)	1	14	27	40	57	70	83	96	101	114	127	140
(26)	2	13	28	39	58	69	84	95	102	113	128	139
(27)	3	16	25	38	59	72	81	94	103	116	125	138
(28)	4	15	26	37	60	71	82	93	104	115	126	137
(29)	5	18	31	44	49	62	75	88	105	118	131	144
(30)	6	17	32	43	50	61	76	87	106	117	132	143
(31)	7	20	29	42	51	64	73	86	107	120	129	142
(32)	8	19	30	41	52	63	74	85	108	119	130	141
(33)	9	22	35	48	53	66	79	92	97	110	123	136
(34)	10	21	36	47	54	65	80	91	98	109	124	135
(35)	11	24	33	46	55	68	77	90	99	112	121	134
(36)	12	23	34	45	56	67	78	89	100	111	122	133

Appendix M Selected Plans of $\frac{1}{2}$ Fractional Factorial Design for 2^5, 2^6, and 2^7 Factorial Experiments

PLAN 1. 1/2 of 2^5 factorial in a single block of 16 units

Treatments:

Treatment no.	1	2	3	4	5	6	7	8
	9	10	11	12	13	14	15	16
	(1)	ab	ac	ad	ae	bc	bd	be
	cd	ce	de	abcd	abce	abde	acde	bcde

Conditions:

- All five main effects are estimable.
- All 10 two-factor interactions are estimable.
- No higher-order interactions are estimable.
- Must have at least two replications.

Defining contrast. ABCDE

ANOV with r replications:

SV	d.f.
Replication	$r - 1$
Main effect	5
Two-factor interaction	10
Error	$15(r - 1)$
Total	$16r - 1$

PLAN 2. 1/2 of 2^6 factorial in four blocks of eight units each

Treatments:

Treatment no.	1	2	3	4	5	6	7	8
Block I:	(1)	ab	ef	abef	acde	acdf	bcde	bcdf
Block II:	ac	bc	de	df	abde	abdf	acef	bcef
Block III:	ae	af	be	bf	cd	abcd	cdef	abcdef
Block IV:	ad	bd	ce	cf	abce	abcf	adef	bdef

Conditions:

- All six main effects are estimable.
- All two-factor interactions except CD are estimable.
- Eight higher-order interactions are estimable.
- CD, ABC, and ABD are confounded with blocks.

Defining contrast. ABCDEF

ANOV without replication and with the higher-order interactions used as error:

SV	d.f.
Block	3
Main effect	6
Two-factor interaction	14
Error	8
Total	31

ANOV with *r* replications:

SV	d.f.
Replication	$r - 1$
Block	3
Block × replication	$3(r - 1)$
Main effect	6
Two-factor interaction	14
Higher-order interaction	8
Error	$28(r - 1)$
Total	$32r - 1$

PLAN 3. 1/2 of 2^6 factorial in two blocks of 16 units each

Treatments:

Treatment no.	1	2	3	4	5	6	7	8
	9	10	11	12	13	14	15	16
Block I	(1)	ab	ef	abef	acde	acdf	bcde	bcdf
	ac	bc	de	df	abde	abdf	acef	bcef
Block II	ae	af	be	bf	cd	abcd	cdef	abcdef
	ad	bd	ce	cf	abce	abcf	adef	bdef

continued next page

Conditions:
- All six main effects are estimable.
- All 15 two-factor interactions are estimable.
- ABC is confounded with blocks.
- Of the total of 20 three-factor interactions, two are not estimable (ABC and DEF) and the rest are estimable in alias pairs as follows:

ABD = CEF	ABE = CDF	ABF = CDE
ACD = BEF	ACE = BDF	ACF = BDE
ADE = BCF	ADF = BCE	AEF = BCD

Defining contrast. ABCDEF

ANOV without replication and with the three-factor interactions used as error:

SV	d.f.
Block	1
Main effect	6
Two-factor interaction	15
Error	9
Total	31

ANOV with r replications:

SV	d.f.
Replication	$r - 1$
Block	1
Block × replication	$r - 1$
Main effect	6
Two-factor interaction	15
Three-factor interaction	9
Error	$30(r - 1)$
Total	$32r - 1$

PLAN 4. 1/2 of 2^7 factorial in four blocks of 16 units each

Treatments:

Block I:	(1)	bc	de	fg	abdf	abdg	abef	abeg
	acdf	acdg	acef	aceg	bcde	bcfg	defg	bcdefg
Block II:	ab	ac	df	dg	ef	eg	abde	abfg
	acde	acfg	bcdf	bcdg	bcef	bceg	abdefg	acdefg
Block III:	af	ag	bd	be	cd	ce	abcf	abcg
	adef	adeg	bdfg	befg	cdfg	cefg	abcdef	abcdeg
Block IV:	ad	ae	bf	bg	cf	cg	abcd	abce
	adfg	aefg	bdef	bdeg	cdef	cdeg	abcdfg	abcefg

Conditions:

- All seven main effects are estimable.
- All 21 two-factor interactions are estimable.
- ABC, ADE, and AFG are confounded with blocks.
- Only 32 higher-order interactions are estimable.

Defining contrast: ABCDEFG

ANOV without replication and with the higher-order interactions used as error:

SV	d.f.
Block	3
Main effect	7
Two-factor interaction	21
Error	32
Total	63

ANOV with r replications:

SV	d.f.
Replication	$r - 1$
Block	3
Block × replication	$3(r - 1)$
Main effect	7
Two-factor interaction	21
Higher-order interaction	32
Error	$60(r - 1)$
Total	$64r - 1$

Index